ISNM 60:
International Series of Numerical Mathematics
Internationale Schriftenreihe zur Numerischen Mathematik
Série internationale d'Analyse numérique
Vol. 60

Birkhäuser Verlag
Basel · Boston · Stuttgart

t:

Functional Analysis and Approximation

Proceedings of the Conference held at the Mathematical Research Institute at Oberwolfach, Black Forest, August 9–16, 1980

Edited by
P.L. Butzer, Aachen
B. Sz.-Nagy, Szeged
E. Görlich, Aachen

1981

Birkhäuser Verlag
Basel · Boston · Stuttgart

Editors

Prof. Dr. P. L. Butzer
Rhein.-Westf. Techn. Hochschule Aachen
Lehrstuhl A für Mathematik
Templergraben 55
D-51 Aachen (FRG)

Prof. Dr. B. Sz.-Nagy
József Attila Tudományegyetem
Aradi vértanúk tere 1
H-6720 Szeged (Hungary)

Prof. Dr. E. Görlich
Rhein.-Westf. Techn. Hochschule Aachen
Lehrstuhl A für Mathematik
Templergraben 55
D-51 Aachen (FRG)

CIP-Kurztitelaufnahme der Deutschen Bibliothek

Functional analysis and approximation : proceedings of the conference held at the Math. Research Inst. at Oberwolfach, Black Forest, August 9–16, 1980 / ed. by P. L. Butzer ... – Basel ; Boston; Stuttgart : Birkhäuser, 1981.
(International series of numerical mathematics ; Vol. 60)
ISBN 3-7643-1212-2
NE: Butzer, Paul L. [Hrsg.]; Mathematisches Forschungsinstitut ‹Oberwolfach›; GT

Printed in Switzerland by Birkhäuser AG, Grafisches Unternehmen, Basel
ISBN 3-7643-1212-2

In Memory of

Jacob Lionel Bakst Cooper

Born on December 27, 1915 in Beaufort-West, South Africa
Died on August 8, 1979 in London

Preface

These Proceedings form a record of the lectures presented at the international Conference on *Functional Analysis and Approximation* held at the Oberwolfach Mathematical Research Institute, August 9–16, 1980. They include 33 of the 38 invited conference papers, as well as three papers subsequently submitted in writing. Further, there is a report devoted to new and unsolved problems, based on two special sessions of the conference. The present volume is the sixth Oberwolfach Conference in Birkhäuser's ISNM series to be edited at Aachen*. It is once again devoted to more significant results obtained in the wide areas of approximation theory, harmonic analysis, functional analysis, and operator theory during the past three years. Many of the papers solicited not only outline fundamental advances in their fields but also focus on interconnections between the various research areas.

The papers in the present volume have been grouped into nine chapters. Chapter I, on operator theory, deals with maps on positive semidefinite operators, spectral bounds of semigroup operators, evolution equations of diffusion type, the spectral theory of propagators, and generalized inverses. Chapter II, on functional analysis, contains papers on modular approximation, interpolation spaces, and unconditional bases. In Chapter III, on abstract harmonic analysis, one may find results on approximation on compact abelian groups, minimal projections in L^1, Wiener type distributions, and analysis on local fields, whereas Chapter IV, on Fourier analysis and integral transforms, comprises papers on polynomial inequalities, classical orthogonal expansions, multiple series, and the Hilbert transform. Chapter V deals with best approximation, in general Hilbert spaces, in the complex domain, as well as in the multipoint sense. Chapter VI, on approximation by linear operators, includes an estimate for the Lebesgue function of Lagrange interpolation, a uniform boundedness theorem with rates, slow and asymptotically optimal approximations. Strong and Müntz approximation then follow in Chapter VII, whereas problems of asymptotic distribution of lattice points as well as two papers concerned with limit theorems of probabilty theory in Banach spaces appear in Chapter VIII. Chapter IX contains papers on spline functions and piecewise polynomial approximation as well as a paper on dominant integrability. The volume closes with a bibliography on Bernstein polynomials, as well as the section on 22 new and unsolved problems.

One mathematician was sorely missed at the conference. Lionel Cooper, who had actively taken part in all but one of our conferences since 1963, was again on the list of distinguished speakers who were invited. But in August of 1979 he passed away after heart operation. The loss caused by his death will surely be long felt by the scientific world, in particular by the community of mathematicians and physicists. The participants and organizing committee of the Conference wish to dedicate these Proceedings to the memory of this distinguished and independent scientist. Lionel Cooper also was a sincere friend to many of us. Two brief appreciations of his life and work appear in these Proceedings.

The editors' warm thanks are due to all of the participants and contributors: they made the conference the success it was; to Wolfgang Splettstösser for his competent handling of the greater part of the general editorial work; to Rolf J. Nessel for valuable advice during the preparations of the conference; to the coworkers and research assistants from Aachen for their help in organizing the conference, and to the secretaries of Lehrstuhl A für Mathematik for retyping many of the papers and for their aid in preparing this volume. To Carl Einsele of Birkhäuser Verlag, Basel, we extend our thanks for his cooperation over the years.

April 1981

P. L. Butzer
Aachen

E. Görlich
Aachen

B. Sz.-Nagy
Szeged

* The earlier volumes are:

1. On Approximation Theory. Oberwolfach 1963. Eds.: P.L. Butzer and J. Korevaar. ISNM, vol. 5, Basel 1964 (second edition 1972), XVI + 261 pages.
2. Abstract Spaces and Approximation. Oberwolfach 1969. Eds.: P.L. Butzer and B. Sz.-Nagy. ISNM, vol. 10, Basel 1969, 423 pages.
3. Linear Operators and Approximation I. Oberwolfach 1971. Eds.: P.L. Butzer, J.P. Kahane and B. Sz.-Nagy. ISNM, vol. 20, Basel 1972, 506 pages.
4. Linear Operators and Approximation II. Oberwolfach 1974. Eds.: P.L. Butzer and B. Sz.-Nagy. ISNM, vol. 25, Basel 1974, 585 pages.
5. Linear Spaces and Approximation. Oberwolfach 1977. Eds.: P.L. Butzer and B. Sz.-Nagy. ISNM, vol. 40, Basel 1978, 685 pages.

Contents

Zur Tagung

Vom 9. bis 16. August 1980 fand im Mathematischen Forschungsinstitut Oberwolfach eine Tagung über «Funktionalanalysis und Approximation» statt. Sie setzte die 1963 begonnene und inzwischen zur Tradition gewordene Reihe internationaler Tagungen über Approximationstheorie und angrenzende Gebiete fort. Diesmal stand sie unter der Leitung von Prof. P. L. Butzer (Aachen), Prof. E. Görlich (Aachen) und Prof. B. Szökefalvi-Nagy (Szeged, Ungarn). Es nahmen 54 Mathematiker aus 14 Nationen an der Tagung teil, darunter auch viele Kollegen, die zum ersten Mal eine Konferenz dieser Reihe besuchten, insbesondere mehrere jüngere Mathematiker. Zum Bedauern aller Teilnehmer mußten eine Reihe von Kollegen aus der UdSSR ihre Zusage in letzter Minute zurückziehen.

Das Vortragsprogramm bestand aus 38 Übersichts- und Spezialvorträgen, in denen ein breites Spektrum von Themen aus den verschiedensten Gebieten der Approximationstheorie, der harmonischen Analysis, der Funktionalanalysis und der Operatortheorie behandelt wurden. Zwei weitere Sitzungen waren aktuellen Problemstellungen gewidmet; hier wurden von den Teilnehmern 18 neue und ungelöste Probleme vorgestellt. (Der Programmablauf ist auf den Seiten 15–17 ausführlich wiedergegeben.) Der vorliegende Band enthält den größten Teil dieser Vorträge und Problemstellungen.

Neben dem Vortragsprogramm fanden zwei gesellige Abende statt, und am Mittwochnachmittag das traditionelle Ausflugsprogramm nach Baden-Baden, Freiburg, Freudenstadt und in die nähere Umgebung.

Die Tagung war gekennzeichnet durch eine kollegiale und freundschaftliche Atmosphäre, wozu die Teilnehmer durch ihr spontanes und sympatisches Mitwirken in vielfältiger Weise beigetragen haben. Allen Vortragenden, den Sitzungsleitern und besonders den Vorsitzenden der beiden «problem sessions» sei für ihr Engagement herzlich gedankt.

An dieser Stelle ist besonders die Gastfreundschaft und Hilfsbereitschaft der Mitarbeiter des Oberwolfacher Instituts zu erwähnen, ohne die solch eine Tagung kaum denkbar wäre, und für die sich die Tagungsleiter bei den Damen und Herren des Oberwolfacher Hauses und insbesondere bei dem Direktor des Instituts, Herrn Professor Dr. M. Barner, herzlich bedanken möchten.

Tagungsleiter: P. L. Butzer E. Görlich B. Sz.-Nagy

List of Participants

T. Ando, Research Institute for Applied Electricity, Hokkaido University,
Oyodenki Kenkyusho, Sapporo 060, Japan

M. Becker, Lehrstuhl A für Mathematik, Rheinisch-Westfälische Technische Hochschule Aachen,
Templergraben 55, D-5100 Aachen, Fed. Rep. Germany

C. Bennett, Dept. of Mathematics, University of South Carolina,
Columbia, SC 29208, USA

H. Berens, Mathematisches Institut, Universität Erlangen-Nürnberg,
Bismarckstrasse 1 1/2, D-8520 Erlangen, Fed. Rep. Germany

W. Bloom, Murdoch University,
Murdoch, Western Australia 6153, Australia

H. Brass, Lehrstuhl E für Mathematik, TU Braunschweig,
Pockelstr. 14, D-3300 Braunschweig, Fed. Rep. Germany

P. L. Butzer, Lehrstuhl A für Mathematik, Rheinisch-Westfälische Technische Hochschule Aachen,
Templergraben 55, D-5100 Aachen, Fed. Rep. Germany

C. K. Chui, Dept. of Mathematics, Texas A & M University,
College Station, TX, 77843, USA

Z. Ciesielski, Instytut Matematyczny, Polskiej Akademii Nauk,
Oddzial w Gdańsku, ul. Abrahama 18, 81-825 Sopot, Poland

W. Dickmeis, Lehrstuhl A für Mathematik, Rheinisch-Westfälische Technische Hochschule Aachen,
Templergraben 55, D-5100 Aachen, Fed. Rep. Germany

W. Engels, Lehrstuhl A für Mathematik, Rheinisch-Westfälische Technische Hochschule Aachen,
Templergraben 55, D-5100 Aachen, Fed. Rep. Germany

F. Fehér, Lehrstuhl A für Mathematik, Rheinisch-Westfälische Technische Hochschule Aachen,
Templergraben 55, D-5100 Aachen, Fed. Rep. Germany

H. Feichtinger, Institut für Mathematik, Universität Wien,
Studlhofgasse 4, A-1090 Wien, Austria

T. H. Ganelius, Swedish Academy of Sciences,
Box 50005, S-104 05 Stockholm, Sweden

E. Görlich, Lehrstuhl A für Mathematik, Rheinisch-Westfälische Technische Hochschule Aachen,
Templergraben 55, D-5100 Aachen, Fed. Rep. Germany

P. R. Halmos, Dept. of Mathematics, Indiana University,
Bloomington, IN, 47401, USA

L. Iliev, Institue of Mathematics, Bulgarian Akademy of Sciences, Bul. Akad. G. Bontschev, BI viii,
1113 Sofia, Bulgaria
K. Ishiguro, Dept. of Mathematics, Hokkaido University,
Oydenki Kenkyusho, Sapporo 060, Japan
J. B. Kioustelidis, Dept. of Applied Mathematics, Nat. Technical University of Athens,
Patision 42, Athens 147, Greece
J. Korevaar, Mathem. Institute, University of Amsterdam,
Roetersstraat 15, Amsterdam 1004, Netherlands
P. Lambert, Dept. of Mathematics, Limburgs Universitair Centrum, B-3610 Diepenbeek, Belgium
L. Leindler, József Attila Tudományegyetem, Aradi vértanúk tere 1, 6720 Szeged, Hungary
D. Leviatan, Dept. of Mathematics, University of Tel Aviv, Tel Aviv, Israel
G. G. Lorentz, Dept. of Mathematics, University of Texas, Austin, TX 78712, USA
R. Lorentz, Institut für Mathematik, GMD,
Postfach 1240, Schloss Birlinghoven, D-5205 St. Augustin 1, Fed. Rep. Germany
G. Lumer, Faculté des Sciences, Université de l'Etat,
Avenue Maistriau 15, B-7000 Mons, Belgium
C. Markett, Lehrstuhl A für Mathematik, Rheinisch-Westfälische Technische Hochschule Aachen,
Templergraben 55, D-5100 Aachen, Fed. Rep. Germany
P. Masani, Dept. of Mathematics, University of Pittsburgh, Pittsburgh, PA 15260, USA
W. Meyer-König, Mathematisches Institut B, Universität Stuttgart,
Postfach 506, D-7000 Stuttgart, Fed. Rep. Germany
D. Milman, Dept. of Mathematics, University of Tel Aviv, Tel Aviv, Israel
F. Móricz, József Attila Tudományegyetem, Aradi vértanúk tere 1, 6720 Szeged, Hungary
B. Muckenhoupt, Dept. of Mathematics, Rutgers University, New Brunswick, NJ 08903, USA
J. Musielak, Institute of Mathematics, A. Mickiewicz University,
Matejki 48/49, 60-830 Poznań, Poland
M. Z. Nashed, Dept. of Mathematics, University of Delaware, Newark, DE 19711, USA
R. J. Nessel, Lehrstuhl A für Mathematik, Rheinisch-Westfälische Technische Hochschule Aachen,
Templergraben 55, D-5100 Aachen, Fed. Rep. Germany
C. W. Onneweer, Dept. of Mathematics, University of New Mexiko, Albuquerque, NM 87131, USA
R. S. Phillips, Dept. of Mathematics, Stanford University, Stanford, CA 94305, USA

M. Th. Roeckerath, Lehrstuhl A für Mathematik, Rheinisch-Westfälische Technische Hochschule Aachen,
Templergraben 55, D-5100 Aachen, Fed. Rep. Germany
W. Schempp, Lehrstuhl für Mathematik, Gesamthochschule Siegen,
Hölderlinstrasse 3, D-5900 Siegen 21, Fed. Rep. Germany
R. C. Sharpley, Dept. of Mathematics, University of South Carolina, Columbia, SC 29208, USA
O. Shisha, Dept. of Mathematics, University of Rhode Island, Kingston, RI 02881, USA
P. C. Sikkema, Mathem. Institute, Technische Hogeschool Delft,
Julianalaan 132, Delft 8, Netherlands
W. Splettstösser, Lehrstuhl A für Mathematik, Rheinisch-Westfälische Technische Hochschule Aachen,
Templergraben 55, D-5100 Aachen, Fed. Rep. Germany
E. Stark, Lehrstuhl A für Mathematik, Rheinisch-Westfälische Technische Hochschule Aachen,
Templergraben 55, D-5100 Aachen, Fed. Rep. Germany
R. L. Stens, Lehrstuhl A für Mathematik, Rheinisch-Westfälische Technische Hochschule Aachen,
Templergraben 55, D-5100 Aachen, Fed. Rep. Germany
J. Szabados, Mathem. Institute, Hungarian Academy of Sciences,
Reáltanoda u. 13-15, Budapest V, Hungary
B. Sz.-Nagy, József Attila Tudományegyetem, Aradi vértanúk tere 1,
6720 Szeged, Hungary
V. Totik, József Attila Tudományegyetem, Aradi vértanúk tere 1, 6720 Szeged, Hungary
P. Vértesi, Mathem. Institute, Hungarian Academy of Sciences,
Reáltanoda u. 13-15, 1053 Budapest V, Hungary
M. Wehrens, Lehrstuhl A für Mathematik, Rheinisch-Westfälische Technische Hochschule Aachen,
Templergraben 55, D-5100 Aachen, Fed. Rep. Germany
U. Westphal-Schmidt, Institut für Mathematik, Universität Hannover,
Welfengarten 1, D-3000 Hannover 1, Fed. Rep. Germany
G. Wilmes, Lehrstuhl A für Mathematik, Rheinisch-Westfälische Technische Hochschule Aachen,
Templergraben 55, D-5100 Aachen. Fed. Rep. Germany
M. Wolff, Mathematisches Institut, Universität Tübingen,
Auf der Morgenstelle 10, D-7400 Tübingen, Fed. Rep. Germany
A. C. Zaanen, Mathem. Institute, University of Leiden, Wassenaarseweg 80, Leiden, Netherlands

Program of the Sessions

Sunday, August 10

10.00 B. Sz.-Nagy, P. L. Butzer, E. Görlich: Words of welcome

Morning session. Chairman: B. Sz.-Nagy

10.15 P. R. Halmos: Ten years in Hilbert space

11.05 P. Masani: Spectral theory of propagators

First afternoon session. Chairman: J. Musielak

4.00 M. Wolff: On the spectral bound of the generator of semigroups of positive operators

Second afternoon session. Chairman: A. C. Zaanen

4.55 M. Th. Roeckerath: On the closeness of the distributions of two weighted sums of random variables in Banach spaces with applications

5.35 F. Móricz: Convergence problems of multiple function series

Monday, August 11

First morning session. Chairman: P. L. Butzer

9.00 P. C. Sikkema: Slow approximation with convolution operators

9.45 R. S. Phillips: Scattering theory for automorphic functions

Second morning session. Chairman: W. Meyer-König

10.50 T. Ando: Fixed points of certain maps on positive semidefinite operators

11.35 H. Berens: Über ein Problem über die beste Approximation in Hilberträumen

First afternoon session. Chairman: H. Brass

4.00 L. Leindler: Strong approximation and enlarged Lipschitz classes

Second afternoon session. Chairman: P. V. Lambert

4.40 P. Vértesi: On the Lebesgue function of the Lagrange interpolation

5.25 O. Shisha: The order of magnitude of functions and their degree of approximation by piecewise interpolating polynomials

Evening session. Chairman: D. Milman

7.45 G. G. Lorentz: Probability and interpolation

Tuesday, August 12

First morning session. Chairman: L. Iliev

9.00 C. Markett: Norm estimates for partial sums of ultraspherical and Laguerre expansions with shifted parameter, consequences for projection norms

9.40 B. Muckenhoupt: Norm inequalities relating the Hilbert transform to the Hardy-Littlewood maximal function

Second morning session. Chairman: W. Schempp
10.40 M. Wehrens: Best approximation on the unit sphere in R^3
11.20 D. Leviatan: On the rate of approximation by Müntz polynomials satisfying constraints
First afternoon session. Chairman: R. S. Phillips
3.50 C. Bennett: Weak - L^∞ and BMO
4.45 R. Sharpley: Maximal operators on weak - L^∞ and BMO
Second afternoon session. Chairman: T. Ganelius
5.40 J. B. Kioustelidis: Uniqueness of optimal piecewise polynomial L_1 approximation for generalized convex functions
Evening session. Chairman: J. Korevaar
7.45 First problem session

Wednesday, August 13
First morning session. Chairman: P. Masani
9.00 J. Musielak: Modular approximation by a filtered family of linear operators
9.55 P. V. Lambert: On the minimum norm property of the Fourier projection in L^1-spaces
Second morning session. Chairman: Z. Ciesielski
11.00 J. Szabados: Bernstein and Markov type estimates for the derivative of a polynomial with real zeros
11.40 V. Totik: Strong approximation and behaviour of Fourier series

Thursday, August 14
First morning session. Chairman: B. Muckenhoupt
9.00 W. Schempp: Splines and harmonic analysis
9.55 Z. Ciesielski: Exponential estimates for periodic splines and unconditional bases in H^1
Second morning session. Chairman: M. Wolff
10.55 M. Z. Nashed: Best approximation problem arising from generalized inverse operator theory
11.45 R. A. Lorentz: Convergence of numerical differentiation formulas
First afternoon session. Chairman: H. Berens
3.45 C. W. Onneweer: Bessel potentials and generalized Lipschitz spaces on local fields
Second afternoon session. Chairman: G. G. Lorentz
4.40 W. R. Bloom: Approximation theory on the compact solenoid
5.35 H. G. Feichtinger: Banach spaces of distributions of Wiener's type
Evening session. Chairman: T. Ganelius
7.45 Second problem session

Friday, August 15
First morning session. Chairman: J. Szabados
9.45 T. Ganelius: Degree of rational approximation to entire functions

Second morning session. Chairman: T. Ando

10.45 J. Korevaar: The inverse approximation theorems of Lebedev and Tamrazov

11.40 E. Görlich: Asymptotically optimal approximation by means of Faber series

First afternoon session. Chairman: E. Görlich

3.50 W. Dickmeis: On Banach-Steinhaus theorems and uniform boundedness principles with rates

Second afternoon session. Chairman: R.J. Nessel

4.30 G. Lumer: Feller semigroups and evolution equations of diffusion type

5.35 Ch.K. Chui: Best multipoint local approximation

Evening session. Chairman: P.R. Halmos

7.45 B.Sz. -Nagy: The functional model of a contraction and the space L^1

JACOB LIONEL BAKST COOPER - IN MEMORIAM

P.L. Butzer
Lehrstuhl A für Mathematik
Rheinisch-Westfälische Technische Hochschule
Aachen

We are here together to pay tribute [1)] to Professor Lionel Cooper. He was born in Beaufort-West in the Republic of South Africa on 27. Dec. 1915. After receiving his B.Sc. degree at the University of Cape Town in 1935, he came to England as a Rhodes scholar to study at Oxford University. He wrote his doctoral dissertation under the direction of Professor E.C. Titchmarsh, and received his D.Phil. in 194o. In 1939/44 he published three papers on Fourier integrals, and shortly thereafter he wrote three further papers on operators in Hilbert space, including one on semigroup operators (Oxford Quart. J., Ann. of Math., PLMS; 1945-8). The latter three papers are cited in most books on functional analysis and established his early reputation.

It was at Oxford University that he had the great luck to meet Kathleen Cooper, also studying at Oxford. They were married in June, 194o.

During the early years of the last war he worked in the aircraft industry at Bristol before joining Birkbeck College, University of London, in 1944 as Lecturer in Mathematics, later becoming Reader.

---

1) Professor J.L.B. Cooper died in London on 8. August 1979; he had been unconscious since a heart operation on 23. July. This is an address given on the occasion of the funeral service of Professor Cooper on 14. August, 1979. The author would like to thank Kathleen Cooper, Tom Williams of London, as well as Wilhelmine Butzer, Rolf Nessel and Eberhard Stark, all of Aachen, for their help in its preparation. For an obituary emphasizing Cooper's contributions to mathematics the reader is referred to D.E. Edmunds: Jacob Lionel Bakst Cooper, 1915 - 1979, Bull. London Math. Soc. 13 (1981) (in print).

In 195o Lionel Cooper was appointed Professor of Pure Mathematics and Chairman of the Dept. of Mathematics at University College, Cardiff, Wales. There he built up a department which came to have the best reputation of any Welsh university college.

It was during that time, in 1959, that I first wrote to Professor Cooper. I was stuck on a basic problem in Fourier transform theory which I needed to solve problems in trigonometric approximation theory. Within a few weeks he replied with the complete solution. Our contacts began then and have continued ever since. In 1963 I organized my first conference on Approximation Theory at the Oberwolfach Mathematical Research Institute which is located in the Black Forest of Southern Germany. Of course the first person I thought of inviting was Professor Cooper. He accepted my invitation; what a high honour for me considering I was pretty young at the time! It was also the first time I met him in person.

He brought with him Kathleen and his family of four children; they came in a Commer caravan. Deborah was four years old at the time, David seven. What a pleasant time we had together! All of the participants lived for a week in the old, stately hunting lodge which has since been demolished. Since that time I am most fortunate to say we have been good friends, not only on a professional but also on a personal and family basis.

It was Lionel I turned to whenever I was stuck. This was not only in mathematical problems, but also in solving problems arising in contacts with other mathematicians, finding journals to publish articles, personal problems, etc. Lionel inspired me and my many students; a number of them are now Professors at various German universities. I can speak to you only of the great help I myself and my students received from Lionel, but I am sure this was the case with everyone who knew him.

When we planned some of our most difficult scientific adventures, such as writing the book on Semigroup Operators of 1967 with my former student Professor H. Berens, or on Fourier Analysis and Approximation of 1971 with my former student Professor R.J. Nessel, we owed a good deal of our confidence to Lionel. We knew Lionel was there, we knew we could always turns to him for advice, not only because these projects lay in his central fields of interest.

Lionel himself wrote three further basic papers on Fourier analysis in 196o/64. These as well as many of the basic ideas we learned from him we incorporated into these books.

Lionel came to all but the first of our subsequent triennial Oberwolfach conferences from 1965 to 1977. Each time he gave inspiring lectures and was section chairman. He was a participant who made sure that the conference were a success. He was a guiding and unifying spirit!

Lionel spent the years 1964/65 as visiting Professor at California Institute of Technology, Pasadena, and 1965 - 67 as Full Professor at the University of Toronto in Canada. He returned to England in 1967 to become Head of the Mathematics Dept. at Chelsea College of Science and Technology of the University of London.

In 1973 Lionel invited me to spend a month in Britain: he arranged a grand lecture tour which took me to ten universities in England and Scotland. My parents also came along - we were often at his home and had a wonderful time together.

Allow me to say just a few words about his mathematical publications. He wrote at least 45 papers to my knowledge in various journals and conference proceedings throughout the world. These papers are mainly concerned with two broad fields in the wide area of mathematical analysis, namely Fourier analysis and integral transform theory on the real line and on groups, and with functional analysis, essentially operators in Hilbert space. Apart from these he wrote many papers in a variety of individual topics, including measure and set theory, differential equations, quantum theory, foundations of thermodynamics.

All in all he was a mathematical analyst in the very broad sense of the word, with an international reputation.

He was an editor ot the Proceedings of the London Math.Soc. and of the Russian Mathematical Surveys-Uspehi, and gave generously of his time on numerous committees.

One can also characterize a scholar by the students he produced. Let me just mention two of them whom I know. Dr. Finbarr Holland of Cork University is one of the very active young Southern Irish mathematicians; just recently he founded the Irish Mathematical Society. Then there is Professor David Edmunds of the University of Sussex in Brighton. He is an international authority in differential equations. He studied under

Professor Cooper in Cardiff and became a university professor in Britain without ever having attended either Oxford or Cambridge. He seems to be one of the very few exceptions to the general rule. What an honour for Edmunds and Lionel!

Cooper had a sharp intellect, always interested in the basic assumptions of the problems studied. He was a scholar in the old sense of the word, widely read, having brilliant ideas, an inspiration to those who knew him.

He did not seek the limelight, and was somewhat reserved in public. He worked in a quiet way but still with great influence. He radiated authority in every situation of life, an authority based on deep respect and justice. He had a healthy self - confidence which allowed him to be composed; there was no rushing about him.

Lionel was of noble character, obliging and courteous; also in every day life, a true and reliable friend in every situation. He was encouraging and had a deep sense of humanity; he was a true gentleman. His greatness was accompanied by his real modesty.

Apart from English he read or spoke many languages; German, French, Italian, Africaans (enabling him to converse with Dutch people); he could also speak and read Russian. He was a lover of music; he was fond of poetry, even read poems in German (Rainer Maria Rilke!). It is the German mathematician Karl Weierstraß who said that a mathematician who is not a poet can never be a perfect mathematician.

While in good company, for example at the traditional wine evenings at the Oberwolfach conferences, he was a most charming entertainer. Since he was somewhat shy, the fact that he could tell stories so effectively often came as a surprise. In addition he had a dry sense of humour!

While at Oberwolfach he was a great hiker, an enthusiastic swimmer - at one meeting he was the only participant to go swimming in early spring in a lake with a temperature of about 1o degrees centigrade. He was also a determined tennis player. He had great staying power.

Lionel was a true family man. Whenever he could he would always take Kathleen along on his many trips, and when they were young, his four children. I always felt he had a very deep affection for all of them. The Cooper family always radiated harmony, which was a pleasure to observe. The family has now lost a dear husband and loving father.

A testimony of the positive image that he projected to his family is that all four of his children followed him in his study of mathematics.

In 1947 Godfrey H. Hardy of Cambridge died, in 1963 Edward C. Titchmarsh of Oxford, a short while ago John E. Littlewood of Cambridge, all three mathematical analysts belonging to an incomparable school of analysis, probably the best that Britain has ever produced. Today it is Lionel Cooper - he was brought up in this tradition of British analysis and he belongs to that category of mathematicians.

The world has lost a great mathematician, and if I may add another personal word, I have lost a great, my best friend.

A TESTIMONY FROM A FRIEND

A.J.W. Hill, Esq., M.A. (Cantab.),
Heinemann Publishers Ltd.,
22 Bedford Square
London, WC1B 3HH, England

I first met Lionel[1)] in Oxford before the war, more than forty years ago. I was not a member of that university myself, but I used to visit my future wife there, and she and Lionel belonged to the same group of friends. He was a Rhodes Scholar over from South Africa, and he struck me at once as a man of outstanding and quite unusual qualities.

Firstly, he was interested in everything. Every field of intellectual and cultural activity - from his own specialism, mathematics, right across to poetry, music, drama, languages, history, physical activities, and human beings - engaged his critical and discerning attention.

In particular, he seemed to be very interested in politics. But I felt that this interest was really a m o r a l concern; he was less interested in the politics of power than in seeing that people were treated with decency and justice. And his high view of how mankind should be treated was exemplified in his own life - as his many colleagues and friends who received his unfailing kindness and consideration will testify.

Lionel was a man of great intellectual power and integrity, and when required he could be forcible - even fierce - in his attitude. But his friends knew that beneath this exterior breathed one of the warmest-hearted of men. In fact, the longer one knew Lionel, the more one realised that his true gentleness was one of his most outstanding and endearing qualities. I

---

1) An address held at the funeral service of Professor J.L.B. Cooper on 14. August 1979 in London.

used to play tennis with him regularly and I cannot remember any occasion on which I ever won a game. But always as we walked off the court he would soften the bitterness of defeat with some kind words about how well I had played.

Lionel was a truly happy man - happy in his friends, and above all in his family. Together with Kathleen, whom he met at Oxford, he built an exceptionally united family, and a visit to the Coopers was always one of life's rewarding experiences. He enriched the lives of those who were privileged to know him. We shall always remember him with admiration and deep gratitude.

Let me close by reading to you one of the poems that Lionel cherished so much; it is one of the Holy Sonnets from John Donne's Divine Poems:

Death be not proud, though some have called thee
Mighty and dreadfull, for, thou art not soe,
For, those, whom thou think'st, thou dost overthrow,
Die not, poore death, not yet canst thou kill mee.
From rest and sleepe, which but thy pictures bee,
Much pleasure, then from thee, much more must flow,
And soonest our best men with thee doe goe,
Rest of their bones, and soules deliverie.
Thou are slave to Fate, Chance, kings and desperate men,
And dost with poyson, warre, and sicknesse dwell,
And poppie, or charmes can make us sleepe as well,
And better than thy stroake; why swell'st thou then?
One short sleepe past, wee wake eternally,
And death shall be no more; Death, thou shalt die.

I Operator Theory

FIXED POINTS OF CERTAIN MAPS ON POSITIVE SEMIDEFINITE OPERATORS

T. Ando

Research Institute of Applied Electricity
Hokkaido University
Sapporo/Japan

The usual addition $A + B$ and the parallel addition $A:B$ for pairs of positive semidefinite operators are most basic operations; $(A+B)/2$ and $2(A:B)$ are considered as the operator versions of arithmetic and harmonic means respectively. An operator version of the geometric mean is characterized as a unique solution of the equation $(A+X):(B+X) = X$.

1. Introduction and Theorems

Motivated by parallel connection of electrical networks, Anderson and Duffin [1] introduced the notion of parallel sum of positive semidefinite matrices. Subsequently Anderson and Trapp [2] extended it to the case of bounded positive semidefinite (*positive*, for short) operators on a Hilbert space. Given positive operators A, B their *parallel sum*, $A:B$ in symbol, is defined by $\lim_{\varepsilon\to 0+} \{(A+\varepsilon I)^{-1} + (B+\varepsilon I)^{-1}\}^{-1}$, where I is the identity operator; in particular, if A and B are invertible then $A:B = (A^{-1} + B^{-1})^{-1}$. In electrical network theory the resistance of a multiport network is considered to be represented by a positive operator (see [1], [2]). Given two networks, one with resistance A and the other with B, the parallel sum $A:B$ is considered to represent the joint resistance of parallel connection. On the other hand, the usual sum $A + B$ represents the joint resistance of series connection.

The operators $(A+B)/2$ and $2(A:B)$ are considered the *arithmetic* and the *harmonic means*, respectively, of positive operators A and B. Pusz and Woronowicz [9] introduced the notion of geometric mean. It is shown in [4] that the geometric mean, A#B in symbol, can be defined by the formula:

$$A\#B = \lim_{\varepsilon \downarrow 0} (A+\varepsilon I)^{\frac{1}{2}}\{(A+\varepsilon I)^{-\frac{1}{2}}B(A+\varepsilon I)^{-\frac{1}{2}}\}^{\frac{1}{2}}(A+\varepsilon I)^{\frac{1}{2}}.$$

If A commutes with B, then A#B coincides with $(AB)^{\frac{1}{2}}$ as expected. From the view point of network theory, it is natural to seek realization of A#B by using only series and parallel connections, and there are already several approaches (see [3], [7]).

Here we take up a cascade-type synthesis. Given two positive operators A and B, let us consider the map Φ defined by

(1) $$\Phi(X) = (A+X):(B+X)$$

in the set of positive operators. Starting with X_o, define successively $X_{n+1} = \Phi(X_n)$. We ask whether X_n converges to the geometric mean A#B. This is true if, for instance, $X_o = A + B$, but not clear if $X_o = 0$. In this paper we confine ourselves to determine the fixed points of Φ.

THEOREM 1. _The geometric mean_ A#B _is a unique fixed point of the map_ Φ.

Let us consider another map Ψ defined by

(2) $$\Psi(X) = A:X + B:X.$$

Though Ψ has many fixed points, for instance 0, we can prove

THEOREM 2. _The geometric mean_ A#B _is a unique fixed point of the map_ Ψ _in the set_ $\{X \mid A:B \leqslant \alpha X$ _for some_ $\alpha = \alpha_X > 0\}$.

The proofs of these theorems will be given in the final section.

2. Some Lemmas

In this section A, B, C denote positive operators on a Hilbert space

H, and order relation $A \geqslant B$ means that $A - B$ is positive. We use ran(A) and ker(A) to denote the range and the kernel of A, respectively. The generalized inverse A^{-1} is, by definition, the (unbounded) operator defined on ran(A) by $A^{-1}(Ax) = Px$ where P is the orthoprojection to the orthocomplement of ker(A). In many cases, it is useful to extend the functional $x \longmapsto \|A^{-1}x\|$ over whole H by setting $\|A^{-1}x\| = \infty$ for x outside ran(A). Thus a vector x is in ran(A) if and only if $\|A^{-1}x\| < \infty$. The following formulas hold, with convention $0/0 = 0$,

$$\|A^{-1}x\| = \sup_y \frac{|(x,y)|}{\|Ay\|}, \tag{3}$$

$$\|Ax\| = \sup_y \frac{|(x,y)|}{\|A^{-1}y\|}. \tag{4}$$

The positive square-root of A is denoted by $A^{\frac{1}{2}}$. For notational convenience, we use $A^{-\frac{1}{2}}$ instead of $(A^{\frac{1}{2}})^{-1}$. If x is in ran(A), then obviously $A^{-\frac{1}{2}}(A^{-\frac{1}{2}}x) = A^{-1}x$. The following well-known lemma (see [6]) is a bridge between the order relation of positive operators and the inclusion relations of their ranges.

LEMMA 1. <u>There is</u> $\alpha > 0$ <u>such that</u> $A \leqslant \alpha B$ <u>if and only if</u> $\mathrm{ran}(A^{\frac{1}{2}}) \subseteq \mathrm{ran}(B^{\frac{1}{2}})$.

It is known (see [2]) that parallel sum admits a variational description:

$$\|(A:B)^{\frac{1}{2}}x\|^2 = \inf_u\{\|A^{\frac{1}{2}}u\|^2 + \|B^{\frac{1}{2}}(x-u)\|^2\}. \tag{5}$$

In this connection, the following two lemmas show that usual and parallel additions are dual notions.

LEMMA 2.

$$\|(A:B)^{-\frac{1}{2}}x\|^2 = \|A^{-\frac{1}{2}}x\|^2 + \|B^{-\frac{1}{2}}x\|^2, \tag{6}$$

<u>and consequently</u>

$$\mathrm{ran}((A:B)^{\frac{1}{2}}) = \mathrm{ran}(A^{\frac{1}{2}}) \cap \mathrm{ran}(B^{\frac{1}{2}}). \tag{7}$$

PROOF. Introduce a new pre-Hilbert norm $|||\cdot|||$ in the algebraic direct sum $K = H \oplus H$ by

$$|||x \oplus y|||^2 = ||A^{\frac{1}{2}}x||^2 + ||B^{\frac{1}{2}}y||^2.$$

For each $z \in H$, consider a linear functional ϕ_z on K defined by $\phi_z(x \oplus y) = (x+y,z)$. It follows from (3) that the functional norm of ϕ_z, even in the unbounded case, is given by

$$|||\phi_z|||^2 = ||A^{-\frac{1}{2}}z||^2 + ||B^{-\frac{1}{2}}z||^2.$$

We claim that $|||\phi_z|||$ coincides with $||(A:B)^{-\frac{1}{2}}z||$. Since the linear manifold $\{u \oplus (-u) \mid u \in H\}$ is annihilated by ϕ_z, we have

$$|||\phi_z|||^2 = \sup_{x,y} \frac{|(x+y,z)|^2}{\inf_u |||x \oplus y - u \oplus (-u)|||^2} .$$

On the other hand, it follows from (5) that

$$\inf_u |||x \oplus y - u \oplus (-u)|||^2 =$$

$$= \inf_u \{||A^{\frac{1}{2}}(x+u)||^2 + ||B^{\frac{1}{2}}(y-u)||^2\}$$

$$= ||(A+B)^{\frac{1}{2}}(x+y)||^2.$$

Now the claim results from (3), which completes the proof.

LEMMA 3.

(8) $$||(A+B)^{-\frac{1}{2}}x||^2 = \inf_u \{||A^{-\frac{1}{2}}u||^2 + ||B^{-\frac{1}{2}}(x-u)||^2\},$$

and consequently

(9) $$\operatorname{ran}((A+B)^{\frac{1}{2}}) = \operatorname{ran}(A^{\frac{1}{2}}) + \operatorname{ran}(B^{\frac{1}{2}}).$$

PROOF. Introduce a new pre-Hilbert norm $|||\cdot|||$ in the algebraic direct sum

$K = \mathrm{ran}(A^{\frac{1}{2}}) \oplus \mathrm{ran}(B^{\frac{1}{2}})$ by

$$|||x \oplus y|||^2 = \|A^{-\frac{1}{2}}x\|^2 + \|B^{-\frac{1}{2}}y\|^2.$$

Since both A and B are bounded, K is complete with respect to the new norm, that is, $(K, |||\cdot|||)$ is a Hilbert space. As in the proof of Lemma 2, consider for each $z \in H$ a linear functional ϕ_z on K; $\phi_z(x + y) = (x+y,z)$. It follows from (4) that the functional norm of ϕ_z is given by

$$|||\phi_z|||^2 = \|A^{\frac{1}{2}}z\|^2 + \|B^{\frac{1}{2}}z\|^2 = \|(A+B)^{\frac{1}{2}}z\|^2.$$

Then by (3) we have

$$\|(A+B)^{-\frac{1}{2}}x\|^2 = \sup_z \frac{|(x,z)|^2}{\|(A+B)^{\frac{1}{2}}z\|^2}$$

$$= \sup_z \frac{|\phi_z(x \oplus x)/2|^2}{|||\phi_z|||^2} .$$

Since $(K, |||\cdot|||)$ is a Hilbert space, the last term in the above identity coincides with the distance from $(x \oplus x)/2$ to the subspace N consisting of all vectors that are annihilated by all ϕ_z $(z \in H)$. Obviously this subspace consists of all vectors of the form $v \oplus (-v)$ where v runs over $\mathrm{ran}(A^{\frac{1}{2}}) \cap \mathrm{ran}(B^{\frac{1}{2}})$. Therefore

$$\|(A+B)^{-\frac{1}{2}}x\|^2 = \inf_v \{\|A^{-\frac{1}{2}}(x/2+v)\|^2 + \|B^{-\frac{1}{2}}(x/2-v)\|^2\}$$

$$= \inf_u \{\|A^{-\frac{1}{2}}u\|^2 + \|B^{-\frac{1}{2}}(x-u)\|^2\},$$

which completes the proof.

Remark that (6) and (8) give quantitative improvement of (7) and (9) that were proved in [2] and [6].

LEMMA 4. <u>If</u> $A:C + B:C \leqslant C$, <u>then for all</u> $x,y \in \mathrm{ran}(C^{\frac{1}{2}})$

$$|(C^{-\frac{1}{2}}x, C^{-\frac{1}{2}}y)| \leqslant \|A^{-\frac{1}{2}}x\|\,\|B^{-\frac{1}{2}}y\|.$$

PROOF. Since assumption means that for all $z \in \mathcal{H}$

$$\|(A:C + B:C)^{\frac{1}{2}}z\| \leqslant \|C^{\frac{1}{2}}z\|,$$

we have, for $x,y \in \mathrm{ran}(C^{\frac{1}{2}})$ and $|\zeta| = 1$,

$$\begin{aligned}
&\|C^{-\frac{1}{2}}(x + \zeta y)\|^2 \\
&\quad \leqslant \|(A:C + B:C)^{-\frac{1}{2}}(x + \zeta y)\|^2 && \text{by (3)} \\
&\quad \leqslant \|(A:C)^{-\frac{1}{2}}x\|^2 + \|(B:C)^{-\frac{1}{2}}y\|^2 && \text{by (8)} \\
&\quad = \|A^{-\frac{1}{2}}x\|^2 + \|C^{-\frac{1}{2}}x\|^2 + \|B^{-\frac{1}{2}}y\|^2 + \|C^{-\frac{1}{2}}y\|^2 && \text{by (6).}
\end{aligned}$$

On the other hand, with suitable choice of ζ, we have

$$\|C^{-\frac{1}{2}}(x + \zeta y)\|^2 = \|C^{-\frac{1}{2}}x\|^2 + 2|(C^{-\frac{1}{2}}x, C^{-\frac{1}{2}}y)| + \|C^{-\frac{1}{2}}y\|^2,$$

which together with the above yields

$$2|(C^{-\frac{1}{2}}x, C^{-\frac{1}{2}}y)| \leqslant \|A^{-\frac{1}{2}}x\|^2 + \|B^{-\frac{1}{2}}y\|^2.$$

Replacing x and y by λx and $\lambda^{-1}y$ respectively in the above inequality and computing the minimum of the right hand side with respect to λ, we arrive at the assertion of the theorem.

LEMMA 5. *Suppose that the following conditions are fulfilled*;

(a) $A:C + B:C \leqslant C$

(b) $(A+C):(B+C) \leqslant \alpha C$ *for some* $\alpha > 0$.

Then for all x,y *in* $\mathrm{ran}(C^{\frac{1}{2}})$

$$|(C^{-\frac{1}{2}}x, C^{-\frac{1}{2}}y)| \leqslant \|(A:B)^{-\frac{1}{2}}x\| \, \|(A+B)^{-\frac{1}{2}}y\|.$$

PROOF. Take x,y in $\mathrm{ran}(C^{\frac{1}{2}})$. By Lemma 4 it follows from (a) that for each $u \in \mathrm{ran}(C^{\frac{1}{2}})$

$$|(C^{-\frac{1}{2}}x, C^{-\frac{1}{2}}y)| \leqslant |(C^{-\frac{1}{2}}x, C^{-\frac{1}{2}}(y-u))| + |(C^{-\frac{1}{2}}x, C^{-\frac{1}{2}}u)|$$

$$\leqslant \|A^{-\frac{1}{2}}x\| \, \|B^{-\frac{1}{2}}(y-u)\| + \|B^{-\frac{1}{2}}x\| \, \|A^{-\frac{1}{2}}u\|$$

$$\leqslant \{ \|A^{-\frac{1}{2}}x\|^2 + \|B^{-\frac{1}{2}}x\|^2 \}^{\frac{1}{2}} \{ \|A^{-\frac{1}{2}}u\|^2 + \|B^{-\frac{1}{2}}(y-u)\|^2 \}^{\frac{1}{2}} .$$

Since the first factor of the extreme right hand side is equal to $\|(A:B)^{-\frac{1}{2}}x\|$ by (6), the proof will be completed if

$$\|(A+B)^{-\frac{1}{2}}y\|^2 = \inf \{ \|A^{-\frac{1}{2}}u\|^2 + \|B^{-\frac{1}{2}}(y-u)\|^2 \mid u \in \mathrm{ran}(C^{\frac{1}{2}}) \} .$$

To prove this identity, it suffices, by (8), to show that u is in $\mathrm{ran}(C^{\frac{1}{2}})$ whenever u is in $\mathrm{ran}(A^{\frac{1}{2}})$ and $y - u$ is in $\mathrm{ran}(B^{\frac{1}{2}})$, or even more

(10) $$\{\mathrm{ran}(A^{\frac{1}{2}}) + \mathrm{ran}(C^{\frac{1}{2}})\} \cap \{\mathrm{ran}(B^{\frac{1}{2}}) + \mathrm{ran}(C^{\frac{1}{2}})\} \subseteq \mathrm{ran}(C^{\frac{1}{2}}).$$

But (10) is equivalent to (b) on the basis of Lemma 1, and (7) and (9). This completes the proof.

3. Proof of Theorems

Recall that Φ and Ψ are the maps induced by given A and B according to (1) and (2), respectively. Suppose that C is a fixed point of Φ or that it is a fixed point of Ψ and satisfies $A:B \leqslant \alpha C$ for some $\alpha > 0$. We claim that the conditions (a) and (b) of Lemma 5 are fulfilled in each case. Since (5) implies

(11) $$A:C + B:C \leqslant (A+C):(B+C)$$

(see [2]), this is immediate for the case of Φ, i.e.

(12) $$(A+C):(B+C) = C.$$

In the case of Ψ, i.e.

(13) $$A:C + B:C = C$$

the condition (a) is immediately fulfilled. It remains to show (b) or its equivalent form (10). Remark that (13) implies, on the basis of Lemma 1, (7) and (9),

$$\mathrm{ran}(A^{\frac12}) \cap \mathrm{ran}(C^{\frac12}) + \mathrm{ran}(B^{\frac12}) \cap \mathrm{ran}(C^{\frac12}) = \mathrm{ran}(C^{\frac12})$$

while the additional assumption $A:B \leqslant \alpha C$ does

$$\mathrm{ran}(A^{\frac12}) \cap \mathrm{ran}(B^{\frac12}) \subseteq \mathrm{ran}(C^{\frac12}).$$

These two inclusion relations yield immediately (10).

With the claim established, in view of Lemma 5 we are in position to assume that for all x, y in $\mathrm{ran}(C^{\frac12})$

$$|(C^{-\frac12}x, C^{-\frac12}y)| \leqslant \|(A:B)^{-\frac12}x\| \, \|(A+B)^{-\frac12}y\| , \tag{14}$$

and further that

$$\alpha(A:B) \leqslant C \leqslant (A+C):(B+C) \quad \text{for some } \alpha > 0. \tag{15}$$

Since (5) implies

$$(A+C):(B+C) \leqslant (A+C+B+C)/4$$

(see [2]), the right hand inequality of (15) implies $C \leqslant A + B$. A consequence is that the operators $(A:B)^{\frac12}(A+B)^{-\frac12}$ and $C(A+B)^{-\frac12}$ are uniquely extended to bounded operators, say K and L respectively, with the restriction that they vanish on the orthocomplement of $\mathrm{ran}((A+B)^{\frac12})$ (see [6]).

Take $w \in \mathrm{ran}((A+B)^{\frac12})$ and let $z = (A+B)^{-\frac12}w$. Since the left hand inequality of (15) implies $\mathrm{ran}((A:B)^{\frac12}) \subseteq \mathrm{ran}(C^{\frac12})$ by Lemma 1, it follows from (14) that

$$|(x,z)| = |(C^{-\frac12}x, C^{-\frac12}Cz)| \leqslant \|(A:B)^{-\frac12}x\| \, \|(A+B)^{-\frac12}Cz\|.$$

Therefore we have by (4)

$$\|(A:B)^{\frac{1}{2}}(A+B)^{-\frac{1}{2}}w\| = \|(A:B)^{\frac{1}{2}}z\|$$

$$= \sup_x \frac{|(x,z)|}{\|(A:B)^{-\frac{1}{2}}x\|} \leq \|(A+B)^{-\frac{1}{2}}Cz\|$$

$$= \|(A+B)^{-\frac{1}{2}}C(A+B)^{-\frac{1}{2}}w\|.$$

In terms of K and L the above inequalities are written in the form $K^*K \leq (L^*L)^2$. Since the square-root function preserves order relation between positive operators (see [4]), we have $(K^*K)^{\frac{1}{2}} \leq L^*L$, hence

$$(A+B)^{\frac{1}{2}}(K^*K)^{\frac{1}{2}}(A+B)^{\frac{1}{2}} \leq (A+B)^{\frac{1}{2}}L^*L(A+B)^{\frac{1}{2}}. \tag{16}$$

The left hand side of (16) is just the geometric mean of $A + B$ and $A:B$ that is known to coincide with $A\#B$ (see [4, 5, 9]) while the right hand side is equal to C by definition of L. Thus we have proved $A\#B \leq C$.

To prove the reversed inequality, remark that the right hand inequality of (15) is equivalent to an inequality between operator matrices

$$\begin{bmatrix} C & -C \\ -C & C \end{bmatrix} \leq \begin{bmatrix} A + C & 0 \\ 0 & B + C \end{bmatrix}$$

(see [2, 5]), hence the operator matrix $\begin{bmatrix} A & C \\ C & B \end{bmatrix}$ is positive. Since the geometric mean $A\#B$ is the maximum of all positive X for which the operator matrix $\begin{bmatrix} A & X \\ X & B \end{bmatrix}$ is positive (see [4,5]), we have $C \leq A\#B$. This completes the proof of the theorems.

That the geometric mean $A\#B$ is a fixed point of Φ was already pointed out by Nishio [8].

RERERENCES

[1] Anderson, W.N.Jr. - Duffin, R.J., Series and parallel addition of matrices. J. Math. Anal. Appl. 26 (1969), 576-594.

[2] Anderson, W.N.Jr. - Trapp, G.E., Shorted operators II. SIAM J. Appl. Math. 28 (1975), 61-71.

[3] Anderson, W.N.Jr. - Morley, T.D. - Trapp, G.E., Characterization of parallel subtraction. Proc. Natl. Acad. Sci. USA 76 (1979), 3599-3601.

[4] Ando, T., Topics on Operator Inequalities. Lecture Note. Hokkaido Univ., Sapporo 1978.

[5] Ando, T., Concavity of certain maps on positive definite matrices and applications to Hadamard products. Linear Alg. Appl. 26 (1979), 203-241.

[6] Fillmore, P.A. - Williams, J.P., On operator ranges. Adv. in Math. 7 (1971), 254-281.

[7] Kubo, F. - Ando, T., Means of positive linear operators. Math. Ann. 246 (1980), 205-224.

[8] Nishio, K., Characterization of Lebesgue-type decomposition of positive operators. Acta Sci. Math. 42 (1980), 143-152.

[9] Pusz, W. - Woronowicz, S.L., Functional calculus for sesquilinear forms and the purification map. Rep. Math. Phys. 8 (1975), 159-170.

A REMARK ON THE SPECTRAL BOUND OF THE GENERATOR OF SEMIGROUPS OF POSITIVE OPERATORS WITH APPLICATIONS TO STABILITY THEORY

Manfred Wolff
Mathematisches Institut
Eberhard-Karls-Universität
Tübingen

In [3] we proved that the spectral bound of the generator A of a strongly continuous semigroup of positive operators is always contained in the spectrum of A. Here we apply this result to some problems in stability theory. Moreover we give an example of an irreducible group of positive operators on a Banach lattice of continuous functions such that its type differs from the spectral bound of its generator. This solves an open problem of [3] and serves as a counter example to some conjectures in stability theory.

1. Introduction

In the last few years the theory of strongly continuous semigroups of positive linear operators on ordered Banach spaces became more and more important in its own right as well as in applications (see e. g. [1, 2, 3, 5, 6, 7, 8, 11, 12, 15, 16, 17]).

One of the most interesting questions in this field is that one about the limit behaviour of the semigroup $\mathcal{T} = (T_t)_{t \geq 0}$ (for t tending to infinity) which in turn is closely related to the size of the spectrum $\sigma(A)$ of the infinitesimal generator A of $\mathcal{T}$ (see e. g. [7, 8, 13, 14]).

As a major step towards the answer of this question we proved in [3], that for a strongly continuous semigroup $\mathcal{T} = (T_t)$ of positive operators on a (non-pathologically) ordered Banach space the well known formula for the resolvent of the generator A

$$R_z(x) = \int_0^\infty e^{-zt}\, T_t x dt$$

does not only hold for all z with Re z > ω_0 where

$$\omega_0 := \lim_{t\to\infty} t^{-1} \ln \| T_t \|$$

but for all z with Re z > s(A) where the spectral bound s(A) is given by

$$s(A) = \sup \{ \mathrm{Re}\, z : z \in \sigma(A) \}$$

Here $\sigma(A)$ denotes the spectrum of A.

In the present paper we will apply this theorem to stability theory. From the foregoing we get a feeling for the important question whether or not s(A) equals ω_0. In fact in [5] this seems to be tacitly assumed. This, however is not true in general for semigroups of positive operators as was shown by an example in [3]. But there the problem remained open whether $s(A) = \omega_0$ holds at least for all *groups* of positive operators.

This, however, is not true, too, as we shall show by an example. Surprisingly this example is quite easy and is furnished by the group of translations on a suitable Banach lattice of continuous functions on $\mathbb{R}$. (Note that in the nonpositive case examples of similar kinds are already well-established, see [4, 18], but these examples are quite more complicated than our one. On the other hand the underlying space in these cases is the Hilbert space, and here our problem remains open.)

The paper is organized in the following manner: In Section 2 we recall some notions and the most important results of [3]. Section 3 is devoted to stability theory whereas in Section 4 we give our counter-example. For notions not explained here we refer to [4] in the case of strongly continuous semigroups and to [9] ([10], resp.) for ordered vector spaces (Banach lattices, resp.).

2. A Formula for the Resolvent of the Generator

2.1 Notations. In the following let E be a real Banach space ordered by a closed, normal cone E_+ satisfying $E_+ - E_+ = E$. Denote by $E_{\mathbb{C}}$ the complexification of E, i. e. $E_{\mathbb{C}} = E \oplus iE$, equipped with an appropriate norm inducing the product topology and such that $E_{\mathbb{C}}$ becomes a complex Banach space (e. g. $\|x + iy\| = \sup \{ \|x \cos t + y \sin t\| : 0 \le t \le 2\pi \}$). Then E is called an *ordered Banach space over* $\mathbb{C}$.

A linear operator T from one ordered Banach space $E_{\mathbb{C}}$ to another one $F_{\mathbb{C}}$ is called p o s i t i v e ($T \geq 0$) if $T(E_+) \subset F_+$. Such an operator is necessarily bounded (apply [9], V.5.6 together with 5.5). We set $S \geq T$ whenever $S - T \geq 0$. As usual $\mathbb{C}$ is ordered by $\mathbb{R}_+$.

2.2 Examples. a) A complex Banach lattice is defined as the complexification of a real Banach lattice; in particular the classical Banach lattices of functions (or of measures) fit into our frame (see [10], II.11).

b) Every complexification of a real order unit space (see [9], V).

c) Every C^*-algebra A. The real space A_o consists of the selfadjoint elements, A_+ consists of the nonnegative selfadjoint elements.

The most important result of [3] now is the following one: Let $\mathcal{T} = (T_t)_{t\geq o}$ denote a strongly continuous semigroup of positive linear operators on the ordered Banach space $E_{\mathbb{C}}$ over $\mathbb{C}$. Let A be the infinitesimal generator of $\mathcal{T}$ and denote by $\sigma(A)$ its spectrum and by $s(A)$ the spectral bound.

THEOREM 2.1. a) If $\sigma(A)$ is nonempty then $s(a) \in \sigma(A)$.

b) For $u > s(A)$ the resolvent $(u - A)^{-1} =: R_u(A)$ is positive. Moreover for $\mathrm{Re}\, z > s(A)$ the net $(\int_0^t e^{-zs} T_s\, ds)_{t\geq o}$ converges to $R_z(A)$ with respect to the operator norm (for $t \to \infty$).

c) Let $s(A)$ be a pole of order m of the resolvent of A. If $z = s(A) + iv$ ($v \in \mathbb{R}$) is another pole then its order is $\leq m$.

An easy corollary is the following one:

COROLLARY 2.2. Let $\mathcal{T} = (T_t)_{t\in\mathbb{R}}$ be a strongly continuous group of positive linear operators on $E_{\mathbb{C}}$. Then $\sigma(A) \neq \emptyset$. More precisely: $\sigma(A) \cap \mathbb{R} \neq \emptyset$.

Note, that $\sigma(A) \neq \emptyset$ for uniformly bounded strongly continuous groups on an arbitrary Banach space. Thus the interesting case here is that the group may be unbounded.

3. Applications to Stability Theory

3.1 Basic Notions. Let $\mathcal{T} = (T_t)_{t\geq o}$ denote a strongly continuous semigroup on the Banach space E. Let X denote a (not necessarily closed) linear subspace of E.

DEFINITION 3.1. a) $\mathcal{T}$ is called weakly (strongly, or uniformly, resp.) asymptotically stable on X if $(T_{t/X})$[1] converges to 0 with respect to the weak (strong, uniform) topology for $t \to \infty$.

b) $\mathcal{T}$ is called exponentially asymptotically stable on X if there is $0 < u \in \mathbb{R}$ such that for every $x \in X$ there exists $M(x) \geq 0$ satisfying $\|T_t x\| \leq e^{-ut} M(x)$ for all $t > 0$. If $\sup\{M(x): x \in X, \|x\| = 1\} = M < \infty$ holds then $\mathcal{T}$ is called uniformly exponentially asymptotically stable on X.

DEFINITION 3.2. $\mathcal{T}$ is called weakly (strongly, uniformly) integrable on X if $(\int_0^t T_s \, ds_{/X})$ converges with respect to the weak (strong, uniform) topology for $t \to \infty$.

3.2 Preliminary Results. Let $\mathcal{T}$, E, X be as in 3.1. The uniform boundedness principle implies that $\mathcal{T}$ is uniformly bounded if $\mathcal{T}$ is weakly asymptotically stable on the whole space E. Thus from now on we make the assumption that $\mathcal{T}$ is uniformly bounded.

We need the following

LEMMA 3.1. Let $\mathcal{T} = (T_t)_{t \geq 0}$ be a strongly continuous semigroup on the Banach space E with infinitesimal generator A. If for an $x \in E$ and $z \in \mathbb{C}$

$$\lim_{t \to \infty} \int_0^t e^{-zs} \, T_s x \, ds = y \tag{3.1}$$

exists (in the weak topology) then y is in the domain D(A) of A and $(z - A)y = x$.

The easy proof is omitted.

The next proposition should be known, we have taken it from [8].

PROPOSITION 3.2. Let $\mathcal{T} = (T_t)_{t \geq 0}$ be a uniformly bounded strongly continuous semigroup on the Banach space E with infinitesimal generator A. The following assertions are equivalent:

a) $\mathcal{T}$ is weakly (strongly) asymptotically stable on E.

b) The image Im(A) of A is dense and $\mathcal{T}$ is weakly (strongly) integrable

1) $T_{/X}$: restriction of T to X

<u>on</u> Im(A).

c) <u>There exists a dense subspace</u> X <u>on which</u> $\mathcal{T}$ <u>is weakly (strongly) integrable</u>.

PROOF. a)$\Rightarrow$ b): Since $\mathcal{T}$ is weakly asymptotically stable, 0 is not an eigengenvalue of the adjoint A^* of A, hence Im(A) is dense. For $x \in \mathrm{Im}A$ there exists $y \in D(A)$ with $Ay = x$ hence

$$\int_0^t T_s x\,ds = T_t y - y$$

which converges to $-y$ by assymption.

c) $\Rightarrow$b): follows from Lemma 3.1.

b) $\Rightarrow$a): By Lemma 3.1 for $x \in D(A)$ $\int_0^\infty (-T_s\ Ax\ ds) = x$ holds, hence $T_t\ x = -\int_t^\infty T_x\ Ax\ ds$ converges to 0. Since D(A) is dense and $\mathcal{T}$ is uniformly bounded the assertion follows.

In general weak stability does not imply strong stability. Thus the following corollary is of interest.

COROLLARY 3.3. <u>Let</u> $\mathcal{T} = (T_t)_{t\geq 0}$ <u>be a strongly continuous semigroup of positive linear operators on the ordered Banach space</u> $E_{\mathbb{C}}$ <u>over</u> $\mathbb{C}$. <u>Let</u> A <u>denote the infinitesimal generator</u>.

<u>If</u> $\mathcal{T}$ <u>is weakly asymptotically stable and if</u> $\mathrm{Im}(A)_+ := \{y \in \mathrm{Im}A : y \geq 0\}$ <u>separates the points on the dual space</u> E' <u>then</u> $\mathcal{T}$ <u>is strongly asymptotically stable</u>.

PROOF. Since $\mathrm{Im}(A)_+$ separates the points of E', the linear hull X of $\mathrm{Im}(A)_+$ is dense in E. If $x \in \mathrm{Im}(A)_+$ then the weak limit $z := \int_0^\infty T_s x\,ds$ exists by Prop. 3.2 because of our assumption. But by the theorem of Dini-Schaefer ([9], V.4.3) this implies that $(\int_0^t T_s x\,ds)_{t>0}$ converges strongly to z. The assertion now follows from Prop. 3.2.

<u>3.3 The Main Result</u>. If a strongly continuous semigroup $\mathcal{T} = (T_t)_{t\geq 0}$ is exponentially asymptotically stable on the whole space then by the uniform boundedness principle $\mathcal{T}$ is uniformly exponentially asymptotically stable hence its type ω_0 (see Sect. 1) is strictly less than zero. So we turn to the following problem: under which conditions does there exist a dense subspace on which $\mathcal{T}$ is exponentially asymptotically stable?

First of all there may exist such a subspace even if $s(A) = \omega_o = 0$ happens. Consider the space $E = C_o(\mathbb{R}_+)$ of all complex-valued continuous functions vanishing at infinity. Define T_t by $(T_t f)(x) = f(x + t)$. Consider $X = \{f \in E : |f(x)| \leq ne^{-x}$ for all x and a suitable $n \in \mathbb{N}$ not depending on $x\}$. Clearly X is dense in E with respect to the sup-norm, and for $f \in X$ $\|T_t f\| \leq M(f)\, e^{-t}$ holds. Since every function e^{ut} $(u < 0)$ is an eigenfunction of the generator A, $s(A) = \omega_o = 0$.

F r o m n o w o n w e r e s t r i c t o u r c o n s i d e r a - t i o n s t o (c o m p l e x) B a n a c h l a t t i c e s E. Let us recall the notion of an ideal and related subject.

DEFINITION 3.4. a) <u>A linear subspace</u> J <u>of</u> E <u>is called an ideal if</u> $y \in E$ <u>and</u> $|y| \leq x$ <u>for some</u> $x \in J$ <u>always implies</u> $y \in J$.

b) <u>A linear subspace</u> X <u>of</u> E <u>is called positively generated if</u> X <u>is the linear hull of the set</u> $X_+ := X \cap E_+$ <u>of its positive elements</u>.

c) <u>Let</u> X <u>be a positively generated linear subspace</u>. <u>Then</u> $J(X) = \{y:$ <u>there exists</u> $x \in X$ <u>satisfying</u> $|y| \leq x\}$ <u>is called the ideal generated by</u> X.

<u>Note, that</u> J(X) <u>is the minimal ideal containing</u> X.

The following lemma is nearly obvious but important.

LEMMA 3.5. <u>Let</u> A <u>denote the infinitesimal generator of a strongly continuous semigroup</u> $\mathcal{T}$ <u>of positive operators on the Banach lattice</u> E. <u>Then the domain</u> D(A) <u>of</u> A <u>is positively generated</u>.

PROOF. For $u > s(A)$ we know by Theorem 2.1 that the resolvent $R_u(A) > 0$. Since E is the linear hull of E_+ and $R_u(A)(E) = D(A)$ the assertion follows.

We need one further notion. In fact it looks a little bit strange at first glance but the examples and the theorem succeeding it may justify it.

DEFINITION 3.6. <u>Let</u> E <u>be a Banach lattice</u>. <u>A linear operator</u> A <u>from</u> $D(A) \subset E$ <u>into</u> E <u>is called inverse monotonously continuous (imc for short) if every increasing sequence</u> (z_n) <u>in</u> D(A) <u>for which</u> (Az_n) <u>is decreasing and convergent itself is convergent</u>.

EXAMPLES. a) Let A be the generator of a strongly continuous semigroup of positive operators. If $s(A) < 0$ then A is imc, since then $(-A)^{-1}$ exists on E and is positive.

b) Let $\mathcal{T} = (T_t)$ be the group of shifts on $E = C_o(\mathbb{R}) = \{f \in \mathbb{C}^{\mathbb{R}} : f$ is continuous and $\lim_{|t|\to\infty} f(t) = 0\}$. Then the generator A: $f \to Af = f'$ is imc though $0 \in \sigma(A)$. For if (f_n) is increasing and (Af_n) is decreasing then $A(f_n - f_1) \leq 0$ hence $f_n = f_1$ for all n.

c) Let $E = C_o(\mathbb{R}_+)$ and $(T_t f)(x) = f(x + t)$. Then $Af = f'$, and A is not imc. For consider $f_n(x) = n(1 + x)^{-1/n}$. Then (Af_n) is decreasing and convergent, but (f_n) increases and fails to converge.

These examples show the following: $s(A) < 0$ implies A to be imc, but not conversely. Example c) shows that $\mathcal{T}$ may be exponentially asymptotically stable on a dense ideal (see the paragraph at the beginning of 3.3), but neither A is imc nor $s(A) < 0$. Nevertheless in this example $\mathcal{T}$ is strongly asymptotically stable (use Prop. 3.2).

Thus in view of these remarks the following theorem is best possible.

THEOREM 3.7. *Let $\mathcal{T} = (T_t)_{t\geq o}$ be a uniformly bounded strongly continuous semigroup of positive operators on the Banach lattice E, and denote by A its infinitesimal generator. The following assertions are equivalent:*

a) The spectral bound $s(A)$ is strictly less than 0, in particular $\mathcal{T}$ is strongly asymptotically stable.

b) $\mathcal{T}$ is exponentially asymptotically stable on the domain $D(A)$ of A and A is imc.

c) A is imc and $\mathcal{T}$ is exponentially asymptotically stable on the ideal $J(D(A))$ generated by $D(A)$.

d) A is imc and there exists a positively generated dense subspace X on which $\mathcal{T}$ is exponentially asymptotically stable.

e) A is imc and there exists a dense ideal on which $\mathcal{T}$ is exponentially asymptotically stable.

PROOF. a)$\Rightarrow$ c): Let $s(A) < u < 0$. To $y \in J(D(A))$ there exists $x \in D(A)_+$ with $|y| \leq x$; now $x = (u - A)^{-1}z$ for $z = (u - A)x$. But $(u - A)^{-1} = R_u(A) \geq 0$ by thm. 2.1, hence

$$|y| \leq x \leq R_u(A)(|z|) = \int_o^\infty e^{-us}\, T_s |z|\, ds \ .$$

This implies $|T_s y| \leq T_s|y| \leq T_s\, R_u(A)(|z|) \leq e^{us}\, R_u(|z|)$, and thus

$\|T_s y\| \leq e^{us}\|R_u(A)(|z|)\|$. Finally A is imc by example a) above.

The only remaining nontrivial implication is e) $\Rightarrow$ a): Let J denote the ideal in question. There exists $0 < u$ such that $\|T_t x\| \leq e^{-ut}M(x)$ for all $x \in J$. Hence for $v < u$ the integral

$$\int_0^\infty e^{vt}\, T_t x dt =: S_v x$$

exists (even with respect to the norm), thus there is defined a positive linear operator S_v from J into D(A) satisfying $(v - A)S_v = I$ (on J) (use Lemma 3.1).

Hence $\mathcal{T}$ is strongly asymptotically stable by Prop. 3.2. Now $Ax = 0$ implies $T_t x = x$ for all t, hence $x = 0$, thus A is injective.

We now prove that A is onto. Then A^{-1} exists on E hence $0 \notin \sigma(A)$ and Theorem 2.1 yields $s(A) < 0$, since $\omega_0 \not\leq 0$.

Now let $0 < x \in E$ be arbitrary. Since J is dense in E, there exists a sequence (y_n) in J converging to x.

Let $w_n = \inf(x, |y_n|)$. Then $0 \leq w_n \leq |y_n|$, hence $w_n \in J$ and $\lim w_n = x$, since the lattice operations are continuous. Then (v_n) defined by $v_n = \sup(w_1, \ldots, w_n)$, is in J, it is increasing and converges again to x. For $z_n = S_0 v_n$ (z_n) is increasing in D(A), and $Az_n = -v_n$ by Lemma 3.1. Now A is imc, hence $y = \lim_{n\to\infty} z_n$ exists, and since A is closed, $(-A)y = x$.

COROLLARY 3.8. _Let E be equal to the space_ C(X) _of all complex-valued continuous functions on a compact space_ X, _and let_ $\mathcal{T}$, A _be as before. Then the following assertions are equivalent:_

a) $\mathcal{T}$ _is uniformly bounded_, A _is imc, and there exists a positively generated dense subspace on which_ $\mathcal{T}$ _is exponentially asymptotically stable_.

b) $\mathcal{T}$ _is uniformly exponentially asymptotically stable on_ E.

The same equivalence is true in case E _is of type_ $L^1(X,\Sigma,\mu)$.

PROOF. In both cases $s(A) = \omega_0$ holds ([2,3]).

4. A Group of Positive Operators with $s(A) < \omega_0 = 0$

First of all we point out that for a group $\mathcal{T} = (T_t)_{t\in\mathbb{R}}$ of positive operators $s(A) \neq -\infty$ by Corollary 2.2. The idea behind our example is the following: the group will consist of all translations on the intersection of $C(\mathbb{R})$ with three weighted function spaces.

The weight functions are chosen in such a way, that (i) $\|T_t\| = 1$ for $t > 0$, (ii) the space is translation-invariant, (iii) $s(A) < 0$, i. e. we eliminate the functions e^{ut} for $-1 \leq u \leq 0$.

We give the construction in a series of particular steps.

4.1 Construction of E. Let E consist of all complex-valued continuous functions f on $\mathbb{R}$ satisfying

$$\lim_{x\to\infty} f(x) = \lim_{x\to-\infty} e^{3x} f(x) = 0 \text{ and } \int_{-\infty}^{\infty} e^{2x}|f(x)|dx =: p_1(f) < \infty$$

Set $p_2(f) = \sup\{|f(x)| : x \geq 0\}$ and $p_3(f) = \sup\{e^{3x}|f(x)| : x \leq 0\}$. Equipped with the norm $\|f\| = p_1(f) + p_2(f) + p_3(f)$ E is easily seen to be a Banach lattice.

4.2 Construction of the Group. For $f \in E$ set $(T_t f)(x) = f(x + t)$. Since

$p_1(T_t f) \leq e^{-2t} p_1(f)$, and moreover $\lim_{x\to\infty} f(x + t) = \lim_{x\to-\infty} e^{3x} f(x + t) = 0$ we get

$T_t(E) \in E$ and all T_t are positive, hence continuous (see Sect. 2.1). In fact $\|T_t\| \leq 1$ holds for $t > 0$.

To show that $\mathcal{T} = (T_t)_{t\in\mathbb{R}}$ is strongly continuous we choose w. l. o. g. $0 < f \in E$. If $\varepsilon > 0$ is given then there exists $a > 0$ such that

$$\int_{|x|>a} (f(x + t) + f(x))e^{2x} dx < \varepsilon/2 \quad \text{for } |t| \leq 1$$

Since f is uniformly continuous on $[-(a + 1), a + 1]$ there exists $0 < d < 1$ such that

$$|f(x + t) - f(x)| < \varepsilon \cdot (2 \cdot \int_{|x|\leq a} e^{2x} dx)^{-1} \quad \text{for } |t| < d,\ |x| \leq a.$$

But then $p_1(T_t f - f) < \varepsilon$.

Similarly we prove that $\mathcal{T}$ is strongly continuous with respect to p_2 and p_3. Obviously the domain $D(A)$ of the generator A equals $\{f \in E : f' \in E\}$, and $Af = f'$ (derivative of f).

4.3 $\omega_0 = 0$. More precisely we prove that $\|T_t\| = 1$ holds for all $t > 0$. In 4.2 we showed already $\|T_t\| \leq 1$.

Fix $t > 0$. For $\varepsilon > 0$ there exists $f \in E_+$ with compact support contained in $[t,\infty[$ and satisfying $p_1(f) < \varepsilon$, $p_2(f) = 1$, $(p_3(f) = 0)$. Now obviously $p_3(T_t f) = 0$, $p_2(T_t f) = 1$, and $p_1(T_t f) \leq e^{-2t} p_1(f)$, hence $1 \leq \|T_t f\| \leq 1 + \varepsilon$. Since $\varepsilon > 0$ was arbitrary, the assertion follows (because of $1 \leq \|f\| \leq 1 + \varepsilon$).

4.4 $s(A) < -1$. Obviously for $u \geq -1$ $(u - A)$ is injective since $p_1(e^{u\cdot}) = \infty$. Hence it is enough to show that $E_+ \subset \mathrm{Im}(u - A)$.

For $f \in E_+$ and $u \geq -1$ set

$$F(x) = \int_0^\infty e^{-ut} f(t + x)dt = e^{ux} G(x)$$

where

$$G(x) = \int_x^\infty e^{-us} f(s)ds$$

We show that $F \in E$; obviously then $(u - A)F = f$.

(i) If $a < b$ then using $G'(x) = -e^{-ux} f(x)$ we obtain via integration by part

$$\int_a^b e^{2x} F(x)dx = \frac{1}{2+u} [e^{(2+u)x} G(x)]_a^b + \int_a^b e^{2x} f(x)dx.$$

The second summand converges to $p_1(f)$ (for $a \to -\infty$, $b \to \infty$). Now

$$e^{(2+u)x} G(x) \leq \int_x^\infty e^{2s} f(s)ds,$$

hence the first summand converges, too, and we obtain $p_1(F) \leq 2p_1(f)$.

(ii) For $t > 0$ we have

$$G(t) = \int_t^\infty e^{-us} f(s)ds \leq \int_t^\infty e^{2s} f(s)ds$$

hence for $-1 \leq u < 0$ $\lim_{t\to\infty} F(t) = 0$. The case $u \geq 0$ is obvious.

(iii) $e^{3t} F(t) = e^t \int_t^\infty e^{2t+ut-us} f(s)ds \leq e^t p_1(f)$, thus

$$\lim_{t\to-\infty} e^{3t} F(t) = 0.$$

Thus $s(A) < -1$ is proved.

4.5 Summary and Final Remarks. (i) The group $\mathcal{T}$ of translations on E is exponentially asymptotically stable on the dense ideal generated by D(A), and is strongly asymptotically stable, but *not exponentially asymptotically stable on* E.

(ii) There is no nontrivial closed ideal J on E which is invariant under

$\mathcal{T}$, in other words $\mathcal{T}$ is irreducible (see [10], III.8).

(iii) The following problem remains open: does there exist a group of positive operators on $E = L^2([0,1])$ with property (i) above?

REFERENCES

[1] Angelescu, N. - Protopopescu, V., On a problem in linear transport theory. Rev. Roum. Phys. 22 (1977), 1055-1061.

[2] Derndinger, R. - Nagel, R., Der Generator stark stetiger Verbandshalbgruppen auf C(X) und dessen Spektrum. Math. Annal. 245 (1979), 159-177.

[3] Greiner, G. - Voigt, J. - Wolff, M., On the spectral bound of the generator of semigroups of positive operators. To appear in J. Operator Theory (1981).

[4] Hille, E. - Phillips, R.S., Functional Analysis and Semigroups. 2nd ed., AMS Coll. Publ. XXI, Providence, Rhode Island 1957.

[5] Karlin, S., Positive operators. J. Math. Mech. 8 (1959), 907-937.

[6] Larsen, E.W., The spectrum of the multigroup neutron transport operator for bounded spatial domains. J. Math. Phys. 20 (1979), 1776-1782.

[7] Mil'stein, G.N., Exponential stability of positive semigroups in a linear topological space I. Jzv. vyss. ucebn. Zaved, Mat. 9(160) (1975), 35-42, translated in Soviet Mathematics 19 (1975), 35-42.

[8] Neubrander, F., Stabilität stark stetiger Halbgruppen. Diplomarbeit Tübingen 1980.

[9] Schaefer, H.H., Topological Vector Spaces. 3rd print., Springer Verlag, Berlin/Heidelberg/New-York 1971.

[10] Schaefer, H.H., Banach Lattices and Positive Operators. Springer Verlag, Berlin/Heidelberg/New-York 1974

[11] Schaefer, H.H. - Wolff, M. - Arendt, W., On lattice isomorphisms with positive real spectrum and groups of positive operators. Math. Z. 164 (1978), 115-123.

[12] Simon, B., An abstract Kato's inequality for generators of positivity preserving semigroups. Indiana Univ. Math. J. 26 (1977), 1067-1073.

[13] Slemrod, M., Asymptotic behaviour of C_0-semigroups as determined by the spectrum of the generator. Indiana Univ. Math. J. 25 (1976), 782-792.

[14] Triggiani, R., On the stabilizability problem in Banach space. J. Math. Anal. Appl. 52 (1975), 383-403.

[15] Vidav, I., Existence and uniqueness of nonnegative eigenfunctions of the Boltzmann operators. J. Math. Anal. Appl. 22 (1968), 144-155.

[16] Wolff, M., On C_o-semigroups of lattice homomorphisms on a Banach lattice. Math. Z. 164 (1978), 69-80.

[17] Yang, M.Z. - Zhu, K.T., The spectrum of transport operators with continuous energy in inhomogenious medium with any cavity. Scientia Sinica 21 (1978), 298-304.

[18] Zabczyk, J., A note on C_o-semigroups. Bull. Acad. Pol. Sci. 23 (1975), 895-898.

LOCAL OPERATORS, REGULAR SETS, AND EVOLUTION EQUATIONS OF DIFFUSION TYPE

Gunter Lumer
Institut de Mathématique
Université de l'Etat
Mons, Belgique

The purpose of the present paper is twofold, and correspondingly it is divided into two different but closely related parts.
In Part I, which is expository, we give a very brief and sketchy account of - or merely indications on - some of the developments since around 1975 concerning the evolution equations of diffusion type associated to a local operator A on a locally compact Hausdorff space Ω. We also mention some of the applications to parabolic partial differential equations. While quite incomplete, this account, together with the bibliography at the end of the paper, should be useful in giving the interested reader a first idea and orientation on the mentioned subject.
The local operators A which are considered in the developments mentioned above are assumed to have decisive potential - theoretic properties, i.e. to satisfy a "maximum principle" (local dissipativeness), and to have "enough regular sets" (open sets in Ω, regular with respect to A in some sense related to the usual potential - theoretic meaning of "regular open set").
Part II is not expository. In it we deal with several aspects concerning regularity. In particular, in section 1 of Part II, we discuss relations between restricting of "local" Feller semigroups and evolution equations of diffusion type as treated, respectively, in [21] and [3], and give improved results along such lines (somewhat better suited for applications to partial differential equations).

PART I: LOCAL OPERATORS. SOLVABILITY, AND STUDY OF THE SOLUTIONS, OF ASSOCIATED EVOLUTION EQUATIONS OF DIFFUSION TYPE

In the brief survey below, we can by no means go through a general detailed recalling of definitions, notations, and terminology, but shall refer the reader instead to the appropriate references. However, we recall a few things explicitely to make Part I, as much as possible, directly readable "in a first approximation", and refer for the rest, concerning notions, notations,

and terminology, to [3], [4], [5], unless otherwise mentioned.
Part I deals with work by G. Lumer, L. Paquet, J.P. Roth, and L. Stoica.

1. Local Operators and Associated Evolution Equations.

A local operator A on Ω (Ω a locally compact Hausdorff space satisfying possibly some additional conditions [1]) will play a role somewhat similar to that of a differential operator (on, say, an open set of R^N). We recall that a local operator A on Ω is a family of operators ("operator" meaning "linear operator") A^V, indexed by $V \in \mathcal{O}(\Omega)$ ($\mathcal{O}(\Omega)$ being the set of all non empty open subsets of Ω), with $D(A^V) = D(A,V) \subset C(V)$, $A^V : D(A,V) \to C(V)$, and such that for $V_1, V_2 \in \mathcal{O}(\Omega)$, $V_1 \subset V_2$,

$$
(1) \qquad \begin{aligned} & f \in D(A,V_2) \Rightarrow f \mid V_1 \in D(A,V_1), \\ & (A^{V_2} f) \mid V_1 = A^{V_1}(f \mid V_1). \end{aligned}
$$

(Here, as in [5], we always write $D(A^V) = D(A,V)$, while not assuming a priori A "completed" or "locally closed", see [5]. We shall however assume henceforth that our local operators are "semi - complete" in the sense of [5].)[2]

Given a local operator A on Ω, one can associate to each $V \in \mathcal{O}(\Omega)$, or to each $V \in \mathcal{O}_c(\Omega)$ ($\mathcal{O}_c(\Omega)$ being the collection of all relatively compact $V \in \mathcal{O}(\Omega)$), certain basic evolution equations (initial - value problems with boundary conditions) of the type

$$
(2) \qquad \begin{aligned} & \frac{\partial u}{\partial t} = Au, \quad t > 0, \ x \in V, \qquad (u = u(t,x)), \\ & u(o,x) = f(x), \quad x \in \overline{V}, \\ & u(t,\cdot) \mid \partial V = 0, \quad t > 0 \end{aligned}
$$

or

---

1) To simplify, we shall assume in any case below, that Ω has a countable base.

2) A very similar notion of "local operator" was already introduced by E. Dynkin [2] p. 145, in 1965, in connection with the characteristic operator of a continuous Markov Process.

$$\frac{\partial u}{\partial t} = Au, \quad t > 0, \; x \in V,$$

(3) $$u(o,x) = f(x), \quad x \in \overline{V},$$

$$u(t,\cdot) \mid \partial V = f(\cdot) \mid \partial V, \quad t \geq 0,$$

where in (2) $V \in \mathcal{O}(\Omega)$, $u(t,\cdot) \mid V \in C_o(V)$ (i.e. $u(t,\cdot) \mid V$ tends to 0 at infinity in V) $\forall\, t \geq 0$, and in (3) $V \in \mathcal{O}_c(\Omega)$. These problems, loosely described in (2), (3), are set up precisely (in sup-norm context and with specific uniform convergence behavior) as Banach space Cauchy problems, respectively in $C_o(V)$, $C(\overline{V})$, in the following way:

(2') $$\frac{du}{dt} = A_V u, \quad t \geq 0, \qquad (t \mapsto u(t) \in D(A_V) \subset C_o(V)),$$

$$u(o) = f \qquad (f \in D(A_V)),$$

or

(3') $$\frac{du}{dt} = \tilde{A}_V u, \qquad (u(t) \in D(\tilde{A}_V) \subset C(\overline{V}))$$

$$u(o) = f \qquad (f \in D(\tilde{A}_V)),$$

where the boundary conditions (behavior on ∂V) are now embodied in the way the operators A_V, $\tilde{A}_V$, associated to V, (operating in the Banach spaces $C_o(V)$, $C(\overline{V})$), are defined. (A_V, $\tilde{A}_V$, will be described explicitely in the next section).

By saying that the problem (2'), or (3'), "is solvable", we mean that it is uniformly well posed as a Banach space Cauchy problem, and assuming that A_V, or $\tilde{A}_V$, respectively, is closed, this is equivalent to saying that A_V, or $\tilde{A}_V$, generates a semigroup on, respectively, $C_o(V)$, $C(\overline{V})$.

If T generates the semigroup $(P(t))_{t \geq 0}$ (on some Banach space), we shall often use the symbolic notation $\exp\{tT\}$ instead of $P(t)$, and also say merely "the semigroup $P(t)$" or "the semigroup $\exp\{tT\}$".

Given A on Ω, $V \in \mathcal{O}(\Omega)$, we say that "the Cauchy problem for V (corresponding to A) is solvable" iff (2') is solvable. We abbreviate "Cauchy problem" by "c.p.". Similarly, we say that "the Cauchy problem with continuous boundary values (c.p.c.) is solvable" iff (3') is solvable.

2. Operators Associated to $V \in \mathcal{O}(\Omega)$ (Given a Local Operator A on Ω).

Given as in the previous section A on Ω, the following Banach space operators, in $C_o(V)$, or $C(\overline{V})$, associated to $V \in \mathcal{O}(\Omega)$, or $V \in \mathcal{O}_c(\Omega)$, are of basic importance in the results we are concerned with:

$\forall\ V \in \mathcal{O}(\Omega)$, A_V is defined by

$$\text{(4)} \quad \begin{aligned} D(A_V) &= \{f \in C_o(V) \cap D(A,V) : Af \in C_o(V)\}, \\ A_V f &= Af \text{ in } V, \qquad \text{for } f \in D(A_V). \end{aligned}$$

$\forall\ V \in \mathcal{O}_c(\Omega)$, $\widetilde{A}_V$ is defined by

$$\text{(5)} \quad \begin{aligned} D(\widetilde{A}_V) &= \{f \in C(\overline{V}) : f \mid V \in D(A,V),\ \exists g \in C(\overline{V}) \text{ with } g \mid \partial V = 0,\ g = Af \text{ in } V\}, \\ \widetilde{A}_V f &= g, \qquad \text{for } f \in D(\widetilde{A}_V). \end{aligned}$$

$\forall\ V \in \mathcal{O}_c(\Omega)$, $\widehat{A}_V$ is defined by

$$\text{(6)} \quad \begin{aligned} D(\widehat{A}_V) &= \{f \in C(\overline{V}) : f \mid \partial V = 0,\ f \mid V \in D(A,V),\ \exists\ g \in C(\overline{V}) \text{ with } g = Af \text{ in } V\}, \\ \widehat{A}_V f &= g, \qquad \text{for } f \in D(\widehat{A}_V). \end{aligned}$$

Moreover, given the local operator A, and $\lambda \in \mathbb{C}$ (usually we consider the case $\lambda > 0$), we write A_λ for the local operator $A - \lambda$, and thus may also consider the operators $A_{\lambda V}$, $\widetilde{A}_{\lambda V}$, $\widehat{A}_{\lambda V}$.

As we have seen in Section 1, the operators A_V, $\widetilde{A}_V$, come up in connection with problems of the type (2), (2'), (3), (3'); the operators $\widehat{A}_V$ come up in problems of perturbation (see [6]), and approximation of solutions (see [7]), as well as other related matters.

3. The Potential-Theoretic Assumptions on Local Operators, and the Potential-Theoretic Techniques.

We make essential potential-theoretic assumptions on our local operators A.

A is assumed to be real and locally dissipative (see [3]), and to have "enough regular open sets" (with precision we mean by this, unless otherwise mentioned, the existence of an exhaustive family $\mathcal{R}$ of open A-regular subsets of Ω such as in Theorem 5.4 of [3]). We also assume, until further notice, A to be locally closed (although one can deal adequately with the case of non locally closed local operators satisfying the other assumptions above, as is shown in [5]). Under these circumstances, A has strong potential-theoretic properties; in particular global maximum principles are available[3], and A_λ-superharmonic functions play a fundamental role (see [3], [4]). Such local operators are also intimately connected with the theory of Markov processes (see [23]).

Necessary and sufficient conditions for solvability of the c.p. (or c.p.c.) for general open sets V, can be given in terms of the existence of a "Cauchy barrier" for V (we shall return to such results below). For the notion of Cauchy barrier see [4], Definition 3.1[4].

4. Some Basic Results.

The context and hypothesis are those described above, unless otherwise mentioned.

Concerning solvability of evolution equations of diffusion type, we have

THEOREM 4.1. Given any $V \in \mathcal{O}(\Omega)$, the c.p. for V (corresponding to A) is solvable iff $D(A_V)$ is dense (in $C_o(V)$) and $\exists$ a Cauchy barrier (relative to A) for V. If these conditions are satisfied, the solution u(t,f) corresponding to the initial value $f \in D(A_V)$ is given by $u(t,f) = \exp\{tA_V\}f$ ($u(t,x,f) = (\exp\{tA_V\}f)(x)$, $t \geq 0$, $x \in V$), and $\exp\{tA_V\}$ is a Feller semigroup.

A quite similar result holds also for the c.p.c., see Theorem 1.2 of [8].

Furthermore, it is often necessary to work with non locally closed local operators A. A very useful variant of 4.1 above, using the "closure" $\overline{A}$ of a non locally closed A, is given in [5], Theorem 6. Whether A is assumed to be

3) including "complex variants" of such maximum principles, useful for instance in estimating resolvents $R(\lambda,\cdot)$ for complex λ, and thus studying the holomorphy of solution semigroups; see [14].

4) This is a less restrictive variant of the notion "V is quasi-regular at infinity with respect to 1-A" used earlier in [3]; see (5.1) of [3], and [4] Theorem 2.11.

locally closed or not, Theorem 6 of [5] also gives a less restrictive variant of Theorem 4.1 in another direction, by assuming only (instead of an exhaustive $\mathcal{H}$ as described in the previous section) the existence of an exhaustive family $\mathcal{H}$ of "A - Cauchy regular" open sets (with, correspondingly, the appropriate interpretation of "(A-1) - superharmonic" in the notion of Cauchy barrier).

On the other hand, at least when considering concrete situations with $\Omega \in \mathcal{C}(R^N)$, problems such as loosely described in (2), or (3), can be set up in an L^2 - variational context (i.e. using appropriate Sobolev spaces and variational formulation of the problems) instead of the sup - norm set up considered above. The corresponding "variational problem" is a less stringent one, and the "variational solution" may exist when the sup - norm solution fails to exist (there are simple examples of this in R^3, involving non regular $V \in \mathcal{C}_c(R^3)$ and the Laplacian). Such matters are treated in [9] using both variational and potential - theoretic techniques. Results are obtained first of all in the general context of the previous sections, a measure μ being given on Ω, and a "variational structure" defined on $\mathcal{C}_c(\Omega)$ (to each $V \in \mathcal{C}_c(\Omega)$ is associated a subspace H_V of $L^2(V)$, and a sesquilinear form $a_V : H_V \times H_V \to \mathbb{C}$, satisfying appropriate assumptions - see Section 1 of [9]). A "variational operator $\mathcal{A}_V$ in $L^2(V)$ is then defined, and concerning the "comparison of the L^2 - variational and sup - norm set ups", we have, with the terminology and assumptions described in [9][5].

THEOREM 4.2. *Assume we have a variational structure defined on* $\mathcal{C}_c(\Omega)$, *compatible with* A, *satisfying a coerciveness condition. Then for all* $V \in \mathcal{C}_c(\Omega)$ *we have*

$$A_V \subset \mathcal{A}_V. \tag{7}$$

A useful application of Theorem 4.2 to partial differential equations is described in [9] (Section 2), in which Ω is an open connected (non empty) subset of R^N, and A is the local operator on Ω induced[6] by the differential

5) For the L^2 - variational operator $\mathcal{A}_V$ see [9] p. 551. See also [15] Chap. IV, and [1] p. 63-65, (except for a change of sign in the definition of the operator).

6) In the way explained in [10], Section II.

operator

$$(8)\quad A(x,D) = \sum_{|\alpha|\leq 2} c_\alpha(x)D^\alpha = \sum_{i,j=1}^{N} a_{ij}(x)D_iD_j + \sum_{j=1}^{N} b_j(x)D_j + c(x),$$

$D_i = \partial/\partial x_i$, where one assumes the c_α real, measurable and bounded on Ω for $|\alpha| \leq 1$, with $c_0 \leq 0$ on Ω, and for $|\alpha| = 2$ continuous with distributional derivatives belonging to $L^\infty(\Omega)$. A(x,D) is moreover assumed to be elliptic in Ω. The variational structure is obtained here by taking for $V \in \mathcal{O}_c(\Omega)$,

$$(9)\quad \begin{aligned} H_V &= H_0^1(V), \\ a_V(u,v) &= \sum_{i,j=1}^{N} \int_V a_{ij}D_ju\overline{D_iv}\,dx - \sum_{j=1}^{N} \int_V b_j^*(D_ju)\overline{v}\,dx - \int_V cu\overline{v}\,dx, \end{aligned}$$

where $b_j^* = b_j - \sum_{i=1}^N D_ia_{ij} \in L^\infty(V)$; and for $f \in D(\mathcal{A}_V)$,

$$(10)\quad a_V(f,v) = -(\mathcal{A}_V f,v)_{L^2(V)}, \qquad \forall\, v \in H_0^1(V).$$

Theorem 4.2 is shown to apply yielding $A_V \subset \mathcal{A}_V$, $\forall\, V \in \mathcal{O}_c(\Omega)$. It follows that if the sup - norm Cauchy problem (the c.p.) is solvable for $V \in \mathcal{O}_c(\Omega)$ and we are in the selfadjoint situation, (i.e., a_V is a selfadjoint form), then the sup - norm solution u(t,f), for $f \in D(A_V)$, (considered as an element of $L^2(V)$) can be computed by a spectral expansion convergent in $L^2(V)$, of the form

$$(11)\quad u(t,f) = \sum_{n=1}^{\infty} e^{\lambda_n t} c_n\varphi_n, \qquad c_n = (f,\varphi_n),$$

see [9].

Let us consider again the sup - norm set up only. Using potential - theoretic and semigroup approximation techniques, rather strong results on approximation of solutions (in sup - norm) can be obtained, see [7], both in the general context, and in the classical context. It would be a somewhat lengthy matter to describe these results with any degree of precision, and we thus rather refer the reader to the paper just mentioned. Let us merely say that in the classical context one shows that, roughly speaking, solutions (of Cauchy problems in the sense of Section 1) corresponding to second order elliptic operators with real - valued coefficients having little regularity, posed in regions with "bad boundaries", can be approximated in a strong sense by solutions corresponding to "approximating operators" having C^∞ coefficients and very regular regions (with C^∞ boundaries).

J.P. Roth, [21], has treated evolution equations closely related to those considered above, in the following context (we keep our notations and general conventions from Sections 1 and 2 above[7]; but we mention that the results of Roth also hold without assuming a countable base for Ω): let $P(t)$ be a Feller semigroup on $C_o(\Omega)$ with pregenerator A_o, where A_o is "local" as an operator in $C_o(\Omega)$ (this means that whenever $f \in D(A_o)$, $V \in \mathcal{O}(\Omega)$, $f=0$ in V, we have $A_o f=0$ on V), and satisfies an additional condition on $D(A_o)$ (see [21] p. 55). Interesting results are obtained in [21] concerning the "restriction" of the generator $\overline{A_o}$ to "regular" open subsets of Ω (regular in a certain sense, specified in [21], Chap IV, p. 57[8]). We state now such a result, after introducing some corresponding notations. (We follow directly [21] but adapt everything to the notations specified here above. This translation may cause a bit of trouble to the reader, but still it seems the best procedure. What we call here Ω, A_o, V, $C_{oo}(V)$,..., would be called X, A, Ω, $\mathcal{K}(\Omega)$,..., in [21]). Also in [21], to the pregenerator called there "A", one associates a family of operators A_Ω which we would call here A^V, and which constitute a local operator in the sense of Section 1 above; that local operator we shall call here A. Thus A is the local operator on Ω induced by A_o, via

$$D(A,V) = \{f \in C(V) : \forall x \in V,\ \exists \text{an open neighborhood of } x,\ V_x, \text{ and } g_x \in D(A_o), \text{ with } f = g_x \text{ in } V_x\},$$

and for $f \in D(A,V)$ $(Af)(x) = (A_o g_x)(x)$. Now, $\forall\, V \in \mathcal{O}(\Omega)$, $f \in C_o(V)$, let us denote by $\tilde{f}$ the extension of f to Ω by 0 outside V. One defines $\forall\, V \in \mathcal{O}(\Omega)$, the following two operators, $A_{o,V,1}$, and $A_{o,V,2}$, in $C_o(V)$:

(12)
$$D(A_{o,V,1}) = \{f \in C_o(V) : \exists\, g \in C_o(V) \text{ such that } (P(t)\tilde{f} - \tilde{f})/t \to g \text{ uniformly on compacta of } V, \text{ as } t \to 0\},$$

$$A_{o,V,1} f = g, \qquad \text{for } f \in D(A_{o,V,1});$$

7) Except that in the context of [21] all functions are real-valued, so we shall interpret, while dealing with that context here, $C(V)$ as $C(V,R)$={real-valued functions in $C(V)$}, $C_o(V)$ as $C_o(V,R)$, etc....; see furthermore [3] p. 422 concerning complexification.

8) If V is regular in that sense, Roth says "V satifies the regularity hypothesis $\mathcal{H}$".

$$D(A_{o,V,2}) = D(A,V) \cap C_{oo}(V),$$

(13)

$$A_{o,V,2}f = Af \text{ in } V, \qquad \text{for } f \in D(A_{o,V,2}).$$

One has ([21], Chap. IV)

THEOREM 4.3. <u>Let</u> $V \in \mathcal{O}_c(\Omega)$ <u>satisfy the regularity hypothesis</u> "$\mathcal{R}$" <u>of</u> [21]. <u>Then</u>

(i) $\exists$ <u>a unique Feller semigroup</u> $Q(t) = e^{tB}$ <u>on</u> $C_o(V)$, <u>such that</u> $\forall\ f \in C_o(V)$, K <u>compact</u> $\subset V$, we have (considering restrictions to K)

(14) $$\|P(t)\tilde{f} - Q(t)f\|_{C(K)} = o(t) \qquad \text{(as } t \to 0\text{)};$$

(ii) $B = A_{o,V,1} = \overline{A_{o,V,1}} = \overline{A_{o,V,2}}$, <u>where the closures (of the graph) are taken in</u> $C_o(V) \times C_o(V)$, <u>the first space being provided with the usual sup-norm convergence, the second with uniform convergence on compacta of</u> V.

(iii) $\exists$ <u>constants</u> $c_1, c_2 > 0$, <u>such that</u> $\forall\ f \in C_o(V)$, $\|Q(t)f\| \leq c_1 e^{-c_2 t}\|f\|$.

In connection with Theorem 4.3, an evolution equation of diffusion type, of the type (2) above, is solved; and a Dirichlet problem for V regular in the sense considered in Theorem 4.3 is also solved thereafter in [21] Chap. IV. From these results one can derive useful consequences concerning the above considered c.p., c.p.c., $\bar{A}$-Cauchy regularity, $\bar{\bar{A}}_V^{-1}$, etc. (see Section 1 of Part II, where we consider such direct consequences, and also give improved results in such directions).

Very recently, J.P. Roth, [22], has also proved a quite interesting and useful result on the "patching together of compatible local Feller semigroups", and on "patching together" the corresponding generators.

Also very recently another sort of intertying of local operators, the "connecting of local operators A_i given on the branches Ω_i of a ramified space Ω, via connecting operators" has been taken up in [11], [12], [13], where the results concern essentially "networks" (one-dimensional ramified spaces) except for a brief mention in [13] of the general theory (which is presently being written up). In this sort of intertying, the local operators A_i live on disjoint open subsets of a "ramified space" Ω and the "connecting operators" have their "support" contained in $\Omega\setminus(\cup_i\Omega_i)$ (which is the "ramification space of Ω"). The corresponding evolution equations of diffusion type, with respect to the local operator on Ω obtained by connecting the A_i, constitute a certain type of generalized transmission problems.

Some applications of the results obtained in the general context to partial differential equations have already been mentioned above, in connection with Theorem 4.2 ((8),(11)), and approximation of solutions corresponding to second order elliptic operator problems in which little regularity is assumed. Other applications, to second order elliptic operators having merely continuous coefficients, are given in [10], [16], [17]. Other results concerning the c.p. for degenerate elliptic second order operators with very regular coefficients, and second order operators on manifolds, are also given in the last two references just mentioned. L. Paquet also makes an extensive study of time-dependent local operators (on a "space-time" locally compact space $\Omega_T = \Omega \times [0,\tau[$) and the c.p. in that context; this is then applied to the c.p. corresponding to second order parabolic equations with merely continuous coefficients depending now on time also, as well as to the inhomogeneous Cauchy problem with continuous boundary values depending on time, [16], [18], [19], [20].

Finally, without attempting to go into any detail, we mention again the interesting recent work of L. Stoica, [23], which deals with local operators A (in the sense of Section 1 above, but real-valued), locally closed, locally dissipative (that notion defined slightly differently), having a base of "Dirichlet" (D-) and "Poisson" (P-) regular open sets, (i.e. local operators with strong potential-theoretic properties, closely related to those considered in Section 3 and thereafter), and studies these objects in connection with Markov processes and the potential theory of the "quasi-harmonic spaces" associated to such local operators A. The matter of existence of "enough D- and P-regular open sets" brings up, of course, problems directly related to the c.p. and c.p.c. considered above.

PART II: COMPARSION BETWEEN DIFFERENT TYPES OF REGULARITY, IN RELATION WITH THE CAUCHY PROBLEM FOR LOCAL OPERATORS

Throughout Part II, we use unless otherwise mentioned, the notions, general conventions, notations, and terminology, indicated in Sections 1 and 2 of Part I[9].

1. Cauchy Problems (c.p.) for Local Operators, and Restriction of Feller Semigroups whose Generators are Local.

We consider first the context described in the paragraph containing Theorem I.4.3; we show that under these circumstances the local operator A induced by A_o has a closure $\overline{A}$, and if $V \in \mathcal{O}(\Omega)$ is regular in the sense of Roth (footnote 8)) then the c.p. (corresponding to $\overline{A}$) is solvable for V, and $B = \overline{A}_V$, so $Q(t) = \exp\{t\overline{A}_V\}$. These facts are rather easy to derive from Theorem I.4.3, [21], and [5]. Somewhat deeper and more useful are the facts we establish next, showing that one has similar results but with everything happening in terms of one a priori given local operator A (which is what one wants in applications to partial differential equations), and under weaker hypotheses, applicable for instance to classical diffusion equations in open sets with boundary in R^N (for which the assumptions of Theorem I.4.3 are too restrictive).

Let us thus first consider the already mentioned context of the paragraph containing Theorem I.4.3. We are thus considering in $C_o(\Omega)$, (Ω locally compact Hausdorff with countable base), an operator A_o which is local (i.e. for $f \in D(A_o)$, $V \in \mathcal{O}(\Omega)$, "$f| V=0$" implies "$(A_o f) | V = 0$"), pregenerates a Feller semigroup, and is such that $\forall f \in D(A_o)$, $\varphi \in C_o^\infty(R) = \{\varphi$ real-valued in $C^\infty(R)$: $\varphi(0)=0\}$, $\varphi \circ f \in D(A_o)$. Also as described in the mentioned paragraph of Section I.4, A_o induces a local operator (in the sense of Section I.1) on Ω, A. We have

PROPOSITION 1.1. The local operator A is locally dissipative.

9) To refer to Definition a.b, Theorem a.b, etc., of part I(part II), we say Definition I.a.b (II.a.b), etc.

PROOF. Let $W \in \mathcal{O}_c(\Omega)$, $\partial W \neq \emptyset$, $f \in C(\overline{W})$ with $f \mid W \in D(A,W)$ and

$$(15) \qquad \max_{\partial W} |f| < \sup_{W} |f| .$$

We must show that $\exists\ x_o \in W$ with $|f(x_o)| = \sup_W |f|$ and $(Af)(x_o)f(x_o) \leq 0$. Set $K = \{x \in W: |f(x)| = \sup_W |f|\}$. K is compact by (15). By what is shown in [21] (IV.1.3, Lemma 1 of III.1.4, of [21]), $\exists\ g \in D(A_o)$, g=f near K, say in W_1 open $\subset W$, and $\exists\ \psi \in D(A_o)$, $0 \leq \psi \leq 1$, supp $\psi \subset W_1$, $\psi = 1$ near K. Then $h = \psi g \in D(A_o)$, $\max_{\Omega} |h|$ occurs necessarily on K, and A_o is dissipative as pregenerator of a Feller semigroup, so $\exists\ x_o \in K \subset W$, $|h(x_o)| = |f(x_o)| = \sup_W |f|$, $(A_o h)(x_o)h(x_o) = (Af)(x_o)f(x_o) \leq 0$.

The case $\partial W = \emptyset$ is handled similarly since in that case W is both open and compact.

(We could also alternatively prove our proposition using for f, or -f, a $g \in D(A_o)$ as above, and the local positive maximum principle of [21] p. 55).

LEMMA 1.2. <u>A admits a closure</u> $\overline{A}$ <u>in the sense of</u> [5] (<u>which is again a locally dissipative local operator, locally closed, semi-complete, extending</u> A).

PROOF. Since $D(A_o)$ is dense in $C_o(\Omega)$, and there exist for any K compact in $V \in \mathcal{O}(\Omega)$, some $\psi \in D(A_o)$ $0 \leq \psi \leq 1$, supp $\psi \subset V$, $\psi = 1$ near K, it follows readily that $\exists$ a base for Ω of $W \in \mathcal{O}_c(\Omega)$ such that $D(A_W)$ is dense in $C_o(W)$. In view of this and Proposition II.1.1 above, Theorem 1 of [5] applies to yield our statement.

If a set $V \in \mathcal{O}(\Omega)$ satisfies the regularity hypothesis "$\mathcal{H}$" of Roth (see footnote 8) above), we shall say henceforth, in order to avoid confusion with our own notations and terminology, that "V is regular (R)".

THEOREM 1.3. <u>Let</u> $V \in \mathcal{O}(\Omega)$ <u>be regular</u> (R). <u>Then the "restricted" Feller semigroup</u> Q(t) <u>corresponding to</u> V <u>according to Theorem</u> I.4.3 <u>is equal to</u> $\exp\{t\overline{A}_V\}$, <u>i.e., the generator</u> B <u>of</u> Q(t) <u>is</u> $\overline{A}_V$ (<u>defined as in</u> (4)). <u>Thus the c.p. corresponding to</u> $\overline{A}$ <u>is solvable for</u> V. <u>In particular</u> V <u>is</u> $\overline{A}$-<u>Cauchy regular in the sense of</u> [5].

PROOF. By (13), Theorem I.4.3 (ii), and Lemma II.1.2, we have $B \subset \overline{A}_V$, where the latter is dissipative (and closed). Since B is a generator, we have by maximality $B = \overline{A}_V$.

There are, however, some serious difficulties in applying Theorem I.4.3 as stated here in part I, i.e. in [21][10], or its above consequences, to partial differential equations. Actually, from the point of view of such applications, the natural thing is to look at a local operator given a priori on Ω for which A_Ω (playing the role of the A_0 of Theorem I.4.3) is the pre-generator of a Feller semigroup. However the condition $\varphi \circ f \in D(A_\Omega)$ if $f \in D(A_\Omega)$, $\varphi \in C_0^\infty(R)$, will often not hold (for instance for Ω an open set with boundary in R^N), even if we have $\varphi \circ f \in D(A,V)$ for $f \in D(A,V)$, $\varphi \in C^\infty(V)$; moreover the barrier conditions for "V is regular (R)" are expressed in terms of the local operator A' induced by A_Ω, but should be expressed in terms of A rather than A'. The approach and results below tend to eliminate these difficulties.

We shall now assume for the rest of this section that there is given on Ω a local operator A, completed (see [5] footnote 2)). As above in this section, all functions, function spaces ($C(V)$, $C_0(V)$,...) are real.

(16) $$\Phi = \{\varphi \text{ real-valued} \in C^\infty(R) : 0 \notin \text{supp } \varphi\}.$$

We assume:

(i) A is locally dissipative.

(17) (ii) $\exists$ a base $\mathcal{B}$ for Ω, of $W \in \mathcal{O}_c(\Omega)$ such that $\overline{D(A_W)} = C_0(W)$.

(iii) $\forall\, V \in \mathcal{O}(\Omega)$, $f \in D(A,V)$, $\varphi \in \Phi$, implies $\varphi \circ f \in D(A,V)$.

LEMMA 1.4. <u>Let</u> $V \in \mathcal{O}(\Omega)$, $f \in D(A,V)$, supp $f \subset V$. <u>Then</u> $\tilde{f}$ (<u>extension</u> <u>of</u> f <u>to all</u> <u>of</u> Ω <u>by</u> 0 <u>off</u> supp f) <u>belongs</u> <u>to</u> $D(A,\Omega)$.

PROOF. f belongs locally to $D(A,\cdot)$ in V; $\tilde{f}$ is locally 0, hence belongs to $D(A,\cdot)$, in $\Omega \setminus V$ since supp $f \subset V$. A being completed; we conclude that $\tilde{f} \in D(A,\Omega)$.

---

10) We mention that F. Hirsch has told us recently that - while it is not published - Roth has known for some time that his result, Theorem I.4.3, is valid under weaker assumptions easier to apply to partial differential operators.

LEMMA 1.5. Let $V \in \mathcal{O}(\Omega)$, K compact $\subset V$. Then $\exists\, f \in D(A_\Omega)$, $0 \le f \le 1$, $f = 1$ near K, supp f compact $\subset V$.

PROOF. (a). We assume first that V belongs to the base $\mathcal{B}$; K compact $\subset V$. For this case, the proof goes much like that of IV.1.3. of [21]. $\exists\, \psi \in C_{oo}(V)$, $0 \le \psi \le 1$, supp $\psi \subset V$, $\psi = 1$ near K; and by (17)-(ii) $\exists\, \psi_1 \in D(A_V)$ with $\|\psi - \psi_1\|_{C_o(V)} \le 1/4$. Take $\varphi \in \Phi$ such that $0 \le \varphi \le 1$, $\varphi = 0$ on $]-\infty, 1/4]$, $\varphi = 1$ on $[3/4, +\infty[$. Then by (17)-(iii) $f_1 = \varphi \circ \psi_1 \in D(A,V)$, and $0 \le f_1 \le 1$ on V, $f_1 = 1$ near K, and supp $f_1 \subset$ supp ψ compact $\subset V$. By Lemma II.1.4, and supp f_1 compact, $f = \tilde{f}_1 \in D(A_\Omega)$ and has the required properties.

(b). Given now any $V \in \mathcal{O}(\Omega)$, K compact $\subset V$, then $\forall\, x \in K$, $\exists V_x \in \mathcal{B}$, $x \in V_x \subset \overline{V}_x \subset V$, and $\exists\, f_x \in D(A_\Omega)$ constructed as in (a) corresponding to $\{x\}$ and V_x, with $f_x = 1$ on V^*_x open, $x \in V^*_x \subset \overline{V^*_x} \subset V_x$. Cover K with $V^*_{x_i}$, $i = 1,2,\ldots,N$.

Set $W = \bigcup_i V_{x_i}$, $W^* = \bigcup_i V^*_{x_i}$. Then $W \subset \overline{W} \subset V$, $K \subset W^* \subset \overline{W^*} \subset W$. Set

$$\psi = \sum_{i=1}^{N} f_{x_i} \quad . \tag{18}$$

Then supp $\psi \subset \overline{W} \subset V$, and $\psi \ge 1$ on W^*. Hence, if $\varphi \in \Phi$ is the same as considered in (a) of this proof, $f = \varphi \circ \psi \in D(A,\Omega)$, and moreover $f \in D(A_\Omega)$ since supp $f \subset$ supp $\psi \subset \overline{W}$ compact; supp $f \subset V$, $f = 1$ on W^* hence near K.

REMARK 1.6. If $g \in D(A_\Omega)$, $\varphi \in \Phi$, then $\varphi \circ g \in D(A_\Omega)$. Indeed, $\varphi \circ g \in D(A,\Omega)$. $\exists\, \delta > 0$ such that $\varphi = 0$ on $[-\delta, \delta]$, and since $|g| < \delta$ off some compact $K \subset \Omega$, $\varphi \circ g = 0$ off K, hence $A(\varphi \circ g) \in C_{oo}(\Omega)$, and so $\varphi \circ g \in D(A_\Omega)$.

THEOREM 1.7. If f, $g \in D(A,V)$, $(V \in \mathcal{O}(\Omega))$, then $fg \in D(A,V)$.

PROOF. (a) Let $V \in \mathcal{O}(\Omega)$, $f \in D(A,V)$; let $\varepsilon > 0$ be given. $\exists\, \varphi_\varepsilon \in \Phi$ such that φ_ε coincides with $x \mapsto x^2$ outside $[-\varepsilon, \varepsilon]$. Since $\varphi_\varepsilon \circ f \in D(A,V)$, then if we write V_ε for the open set $\{x \in V: |f(x)| > \varepsilon\}$, we have $f^2 | V_\varepsilon = (\varphi_\varepsilon \circ f) | V_\varepsilon \in D(A, V_\varepsilon)$. Now $\bigcup_{\varepsilon > 0} V_\varepsilon = V_o = \{x \in V: f(x) \neq 0\}$, and therefore, since A is completed, $f^2 | V_o \in D(A, V_o)$.

(b) Again let $V \in \mathcal{O}(\Omega)$, $f \in D(A,V)$. Then $\forall\, x \in V$, by Lemma II.1.5, $\exists\, W_x \in \mathcal{O}_c(\Omega)$, $x \in W_x \subset \overline{W}_x \subset V$, and $\psi \in D(A_\Omega)$, $0 \le \psi \le 1$, $\psi = 1$ on W_x. Let

$\|f|\overline{W}_x\|_{C(\overline{W}_x)} = M$. Then if $M=0$, $f|W_x = 0 \in D(A,W_x)$; otherwise $f+2M\psi \geq M>0$ on W_x, so by (a) of this proof, $(f+2M\psi)^2|W_x \in D(A,W_x)$, and similarly $\psi^2|W_x \in D(A,W_x)$, while $f\psi|W_x = f|W_x \in D(A,W_x)$, hence $f^2|W_x = (f+2M\psi)^2|W_x - 4Mf\psi|W_x - 4M^2\psi^2|W_x \in D(A,W_x)$. Since A is completed we have $f^2 \in D(A,V)$.

(c) $\forall V \in \mathcal{O}(\Omega)$, $f,g \in D(A,V)$, we have by (b) above $fg=(1/2)[(f+g)^2-f^2-g^2] \in D(A,V)$.

COROLLARY 1.8. *If* $f,g \in D(A_\Omega)$, *and* fg *has compact support, then* $fg \in D(A_\Omega)$.

THEOREM 1.9. *Let* K *be compact in* Ω, $\bigcup_{i=1}^N V_i \supset K$ *be an open covering of* K (*i.e.*, $V_i \in \mathcal{O}(\Omega)$, $i=1,2,\dots,N$). *Then* $\exists\ \alpha_i \in D(A_\Omega)$, $\operatorname{supp} \alpha_i \subset V_i$, $0 \leq \alpha_i \leq 1$, *with* $\sum_{i=1}^N \alpha_i = 1$ *near* K.

PROOF. $\exists\ W_i \in \mathcal{O}_c(\Omega)$, $W_i \subset \overline{W}_i \subset V_i$, $K \subset \bigcup_{i=1}^N W_i$, and by Lemma II.1.5, $\exists f_i \in D(A_\Omega)$, $0 \leq f_i \leq 1$, $f_i = 1$ on $\overline{W}_i$, $\operatorname{supp} f_i$ compact $\subset V_i$. $\exists\ \varphi \in \Phi, 0 \leq \varphi \leq 2$, such that φ coincides with $x \mapsto 1/x$ on $[1, +\infty[$. Set $f = \sum_{i=1}^N f_i$, and

$$g = \varphi \circ f; \quad \beta_i = f_i g \qquad \text{for } i=1,2,\dots,N. \tag{19}$$

By Remark II.1.6 and Collary II.1.8, g, $\beta_i \in D(A_\Omega)$. Also, $\exists\ h \in D(A_\Omega)$, $0 \leq h \leq 1$, $\operatorname{supp} h \subset \bigcup_{i=1}^N W_i$, $h=1$ near K. Set finally

$$\alpha_i = h\beta_i = hgf_i. \tag{20}$$

Then $\alpha_i \in D(A_\Omega)$, $\operatorname{supp} \alpha_i \subset V_i$, and on $\bigcup_{i=1}^N W_i$, and in particular on $\operatorname{supp} \alpha_i$, $f \geq 1$, so $g = 1/f$, $\sum \alpha_i = h(1/f)f = 1$ near K, $\alpha_i = hf_i / \sum_{i=1}^N f_i \leq 1$ at all points where $\alpha_i > 0$.

LEMMA 1.10. *Let* $V \in \mathcal{O}(\Omega)$, K *be compact* $\subset V$, *and* $f \in D(A,V)$. *Then* $\exists\ g \in D(A_\Omega)$, supp g *compact* $\subset V$, *with* $g=f$ *near* K.

PROOF. By Lemma II.1.5, $\exists\ h \in D(A_\Omega)$, $0 \leq h \leq 1$, supp h compact $\subset V$, $h=1$ near K. So $g = (hf)^\sim$ will do in view of Lemma II.1.4, and Theorem II.1.7.

LEMMA 1.11. *Suppose* $g \in C_{oo}(\Omega)$, $V \in \mathcal{O}_c(\Omega)$, K *is compact* $\subset V$, *and* $g|V \in D(A,V)$. *Then* $\exists f \in D(A_\Omega)$, $f \geq g$ *on* Ω, $f=g$ *near* K.

PROOF. $\exists\ \psi \in D(A_\Omega)$, $0 \leq \psi \leq 1$, $\psi = 1$ near K, $\operatorname{supp}\psi \subset V$. Thus $g\psi = [(g|V)(\psi|V)]^\sim \in D(A_\Omega)$, and $(g-g\psi)|V \in D(A,V)$ and is 0 near K. Set $h = g - g\psi$. supp h is contained in some V_1 open, $\overline{V} \subset V_1 \subset \overline{V}_1$ compact; $\exists W$ open, so that $K \subset W \subset \overline{W} \subset V$, $h = 0$ on W. Then $\overline{V}_1 \setminus W$ is compact and disjoint from K. So $\exists\, \alpha \in D(A_\Omega)$, $0 \leq \alpha \leq 1$, $\alpha = 1$ near $\overline{V}_1 \setminus W$ and $\alpha = 0$ near K. Set $\|h\|_{C_0(\Omega)} = M$, and $h' = M\alpha$. Then it is easily checked that

$$(21) \qquad h' \geq h \qquad \text{on all of } \Omega.$$

Thus $f = g\psi + h' \in D(A_\Omega)$, $f \geq g$, $f = g$ near K.

Let us next look at the local operator A' on Ω, induced by A_Ω (via $D(A',V) = \{f \in C(V)$: f coincides locally in V with elements $g \in D(A_\Omega)\}$, and if $f \in D(A',V)$, f coincides with g near $x \in V$, then $(A'f)(x) = (A_\Omega g)(x)$). We have now

THEOREM 1.12. $A' = A$.

PROOF. The fact that $A' \subset A$ is immediate from the definition of A' and the fact that A is completed. Conversely, if $f \in D(A,V)$, $V \in \mathcal{O}(\Omega)$, then $\forall\, x \in V$ $\exists\, W_x \in \mathcal{O}_c(\Omega)$, $x \in W_x \subset \overline{W}_x \subset V$, and by Lemma II.1.10 $\exists\, g \in D(A_\Omega)$, $f = g$ near $\overline{W}_x$, so $f|W_x \in D(A',W_x)$. It follows that $f \in D(A',V)$ and $A'f = Af$.

The definition of "V is regular (R)", as given in [21] p.57, is in terms of the local operator induced by the pregenerator A_0 (in the notation of Theorem I.4.3). We shall recall now this definition but stated relative to any a priori given locally dissipative local operator. Thus:

We shall say, given a locally dissipative local operator A on Ω, that $V \in \mathcal{O}(\Omega)$ is "regular (R) relative to A", iff: $V \in \mathcal{O}_c(\Omega)$ and,

(1) V admits a barrier at every $x \in \partial V$ in the following sense: $\exists\, W_x$ open containing x, and a function $h_x \in C(W_x)$ such that

(i) $h_x(x) = 0$

(ii) $h_x > 0$ on $(\overline{V} \cap W_x) \setminus \{x\}$

(iii) $h_x | V \cap W_x \in D(A, V \cap W_x)$, $Ah_x \leq 0$ on $V \cap W_x$,

(2) $\exists$ functions ψ, θ in D(A,W), W some open set containg $\overline{V}$, ψ, θ, > 0 in W, and satisfying $A\psi < 0$, $A\theta > 0$, in W.

We have then the following result

THEOREM 1.13. Let A be a local operator on Ω, which satisfies the hypothesis (17) and is completed. Suppose A_Ω pregenerates a semigroup on $C_o(\Omega)$ (which is then automatically a Feller semigroup). Let $V \in \mathcal{O}(\Omega)$ be regular (R) relative to A. Then the c.p. (corresponding to $\bar{A}$) is solvable for V, i.e., ($\exists$ the closure $\bar{A}$ of A and) $\bar{A}_V$ generates a (Feller) semigroup on $C_o(V)$.

Moreover, $Q(t) = \exp\{t\bar{A}_V\}$ satisfies (and is the unique Feller semigroup on $C_o(V)$ satisfying): $\forall\ f \in C_o(V)$, K compact $\subset V$, $\|\exp\{t\bar{A}_\Omega\}\tilde{f} - Q(t)f\|_{C(K)} = o(t)$ as $t \to 0$, where $\exp\{t\bar{A}_\Omega\} = \exp\{t\overline{A_\Omega}\}$, $\tilde{f}$ being the extension by 0 of f to Ω.[11)]

PROOF. A_Ω plays here the role of A_o of Theorem I.4.3, A' the role of the local operator induced by A_o. Here Φ replaces $C_o^\infty(R)$; but having shown under the present assumptions the validity of the properties II.1.4 to II.1.12 concerning A_Ω, and $A = A'$ (by Theorem II.1.12), the rest of Roth's arguments in [21] Chap. IV needed for the above statement will then work in the present situation yielding a Feller semigroup Q(t) of generator B, satisfying the conclusions of Theorem I.4.3, so that $B \subset \bar{A}'_V = \bar{A}_V$, and by the maximality of dissipative generators among dissipative operators, $B = \bar{A}_V$. (Also by maximality $\overline{A_\Omega} = \bar{A}_\Omega$). The statement thus follows.

Theorem II. 1.13 shows in particular that under the given assumptions on A, A_Ω, regularity (R) relative to A, for a $V \in \mathcal{O}(\Omega)$, implies A - Cauchy regularity of V.

In the next section, we deal with several relations between basic operators associated to a local operator, and the different types of regularity, (relations rather easy to establish, but useful in clarifying the situation).

2. H^V, A_V, A_V^{-1}; A-Regular, Dirichlet Regular, A-Cauchy Regular, and Poisson Regular, Open Sets.

NOTATION 2.1. The context here is the general context of Sections 1 and 2 of Part I. For $V \in \mathcal{O}(\Omega)$, A_V, $\hat{A}_V$, are as defined in the sections just mentioned. "H^V", "A-regular open set" are as defined in [3]; "A-Cauchy regular open set" has the meaning defined in [5]; "Dirichlet-regular open set", "Poisson-regular

11) As in the statement of I.4.3, the terms inside $\| \ \|_{C(K)}$ are to be understood as being the corresponding restrictions to K.

open set", has the meaning defined in [23]. Throughout this section,

(22) A will be a local operator on Ω, real, locally dissipative.

Since in [23] all functions considered are real, we shall when refering to that context - as already done for the context of Theorem I.4.3 - consider our local operator A as restricted to the corresponding real functions.

PROPOSITION 2.2. For $V \in \mathcal{O}(\Omega)$ with $\partial V \neq \emptyset$, the following two are equivalent:

(i) V is P-regular (with respect to A),

(ii) $\exists\ \widehat{A}_V^{-1} \in B(C(\overline{V}))$ and $\overline{D(A_V)} = C_o(V)$.

Again, in another direction, the following two are equivalent, $\lambda > 0$ being given:

(i') V is A-regular and A_λ-regular,

(ii') V is D-regular with respect to A and A_λ.

PROOF. We need only to consider real functions. That (i) implies (ii) is proved using the codissipativeness of $-\widehat{A}_V^{-1}$, and the closed graph theorem in essentially the same manner as one procedes in the proof of 3.1, 3.6, of [3].

Suppose now (ii) holds. Then all we need to check is that for $f \in C(\overline{V})$, $f \geq 0$, we have $-\widehat{A}_V^{-1} f \geq 0$ in the present situation. But by the usual perturbation argument (from $\widehat{A}_V$ to $\widehat{A}_V - \lambda$) we have that $\exists\ \widehat{A}_{\lambda V}^{-1} \in B(C(\overline{V}))$ and $\|\widehat{A}_V^{-1} - \widehat{A}_{\lambda V}^{-1}\| \to 0$ as $\lambda \to 0$; on the other hand $-\widehat{A}_{\lambda V}^{-1} f \geq 0$ follows merely from the local dissipativeness of A, λ being > 0; so we conclude that $-\widehat{A}_V^{-1} f \geq 0$.

Next, in considering the equivalence (i')-(ii'), H_λ^V denotes the same object as H^V but corresponding to A_λ instead of A. For the mentioned equivalence, all that needs really to be checked is that under the present circumstances, if (i') holds, and $f \in C(\partial V)$, $f \geq 0$, then $H^V f \geq 0$. Consider thus such an f, and write $H^V f = u$, $H_\lambda^V f = u_\lambda$. Since A is locally dissipative, the maximum principles for A_λ imply $u_\lambda \geq 0$. Set $w_\lambda = u - u_\lambda$. Then

(23) $$w_\lambda | \partial V = 0,\ A w_\lambda = -\lambda u_\lambda \leq 0 \text{ in } V,$$

so w_λ is A_λ-superharmonic, and hence $w_\lambda \geq 0$ in V, $u \geq u_\lambda \geq 0$ in V.

PROPOSITION 2.3. Suppose that A is locally closed (in addition to (22)). Let $V, V' \in \mathcal{O}_c(\Omega)$, $V \subset \overline{V} \subset V'$. Then, if V and V' are A-Cauchy regular, the following two are equivalent (as additional properties), given any $\lambda > 0$:

(i) V _is_ A_λ - _regular_,

(ii) $\exists\ \widetilde{A}^{-1}_{\lambda V} \in B(C(\overline{V}))$.

PROOF. Suppose (ii) holds. Let $E = \{\varphi \in C(\partial V): \exists\, g \in D(A_{\lambda V'}) = D(A_{V'})$ with $g|\partial V = \varphi\}$. Since V' is A-Cauchy regular,

(24) E is dense in $C(\partial V)$.

Let $\varphi \in E$, and g be as in the definition of E; set $\psi = -A_\lambda g$ in $\overline{V}$. By the assumption (ii) $\exists\, v = \widetilde{A}^{-1}_{\lambda V}\ \psi$. Set $u = v+g$ in $\overline{V}$. Then $u \in C(\overline{V})$, $u|V \in D(A_\lambda, V)$ and

$$A_\lambda u = A_\lambda v + A_\lambda g = \Psi + A_\lambda g = 0 \text{ in } V,$$

$$u|\partial V = v|\partial V + g|\partial V = \varphi,$$

since $v \in D(\widehat{A}_{\lambda V})$. In view of this, (24), and the maximum principle for A_λ-harmonic functions, we see that V is A_λ-regular. Thus (ii) implies (i).

To show that (i) implies (ii), one has by a standard argument (as in 3.1 of [3] for instance) that $I(\widehat{A}_{\lambda V}) = C(\overline{V})$, and the rest goes then as for "(i) → (ii)" of the preceding proposition.

REMARK 2.4. The argument in the preceding Proposition II.2.3 also shows that, (A being as in Proposition II.2.3), if $V \in \mathcal{O}_c(\Omega)$, $\lambda > 0$, $\exists\ \widetilde{A}^{-1}_{\lambda V} \in B(C(\overline{V}))$, then V is A_λ-regular whenever

(25) $E_0 = \{\varphi \in C(\partial V): \exists\ g \in C(\overline{V}),\ g|\partial V = \varphi,\ g|V \in D(A,V),\ Ag$ in V extends continously to $\overline{V}\}$

is dense in $C(\partial V)$.

Finally, let us simply mention that examples can be given where A on Ω is locally dissipative, real, locally closed, $V \in \mathcal{O}_c(\Omega)$ is A-Cauchy regular but not P-regular.

REFERENCES

[1] Caroll, R.W., Abstract Methods in Partial Differential Equations. Harper & Row, New York, 1969.

[2] Dynkin, E., Markov Processes I and II. Springer Verlag, 1965.

[3] Lumer, G., Problème de Cauchy pour opérateurs locaux et "changement de temps". Annales Inst. Fourier, 25 (1975), fasc. 3 et 4, 409-446.

[4] Lumer, G., Problème de Cauchy et Fonctions Surharmoniques. Séminaire de Théorie du Potentiel, Paris No.2, Lect. Notes in Math. vol. 563 p. 202 - 218, Springer Verlag, Berlin, 1976.

[5] Lumer, G., Equations d'évolution pour opérateurs locaux non localement fermés. C.R. Acad. Sci. Paris, 284 (1977), série A, 1361 - 1363.

[6] Lumer, G., Perturbations additives d'operateurs locaux. C.R. Acad. Sci. Paris, 288 (1979), série A, 107 - 110.

[7] Lumer, G., Approximation d'opérateurs locaux et de solutions d'équations d'evolution. Séminaire de Théorie du Potentiel, Paris No. 5, Lect. Notes in Math. vol. 814, p. 166 - 185, Springer-Verlag, Berlin, 1980.

[8] Lumer, G., Problème de Cauchy avec valeurs au bord continues, comportement asymptotique, et applications. Séminaire de Théorie du Potentiel, Paris No.2, Lect. Notes in Math. vol. 563 (1976), p. 193 - 201, Springer-Verlag.

[9] Lumer, G., Evolution equations in sup-norm context and in L^2 variational context. Linear Spaces and Approximation, I.S.N.M. vol. 40 (1978), p. 547 - 558, Birkhäuser-Verlag, Basel.

[10] Lumer, G., Equations d'évolution en norme uniforme pour opérateurs elliptiques. Régularité des solutions. C.R. Acad. Sci. Paris, 284 (1977), sér. A, 1435 - 1437.

[11] Lumer, G., Connecting of local operators and evolution equations on networks. Potential Theory Copenhagen 1979 (Proceedings), Lect. Notes in Math. vol. 787, p. 219 - 234, Springer-Verlag, Berlin, 1980.

[12] Lumer, G., Equations de diffusion sur des réseaux infinis. Séminaire Goulaouic-Schwartz, 1979-1980, Ecole Polytechnique, Palaiseau, p. XVIII.1 - XVIII.9.

[13] Lumer, G., Espaces ramifiés et diffusions sur les réseaux topologiques, C.R. Acad. Sci. Paris, 291 (1980), sér. A, 627 - 630.

[14] Lumer, G.-Paquet, L., Semi-groupes holomorphes, produit tensoriel de semi-groupes et equations d'évolution. Séminaire de Théorie du Potentiel, Paris No. 4, Lect. Notes in Math. vol. 713, p. 156 - 177, Springer Verlag, Berlin, 1979.

[15] Lions, J., Problèmes aux limites dans les équations aux dérivées partielles. 2nd. ed., Les Presses de l'Univ. de Montréal, 1965.

[16] Paquet, L., Sur les équations d'évolution en norme uniforme. Thèse, Université de l'Etat à Mons, 1978.

[17] Paquet, L., Equations d'évolution pour opérateurs locaux et équations aux dérivées partielles. C.R. Acad. Sci. Paris, 286 (1978), sér. A, 215 - 218.

[18] Paquet, L., Semi-groupes généralisés et équations d'évolution. Séminaire de Théorie du Potentiel, Paris No. 4, Lect. Notes in Math. vol. 713, p. 243 - 263, Springer Verlag, Berlin, 1979.

[19] Paquet, L., Opérateurs locaux dépendant du temps et problème de Cauchy. C.R. Acad. Sci. Paris, 286 (1978), sér. A, 613 - 616.

[20] Paquet, L., Problème de Cauchy avec valeurs au bord dépendant du temps et comportement asymptotique. C.R. Acad. Sci. Paris, 286 (1978), sér. A, 819 - 822.

[21] Roth, J.P., Opérateurs dissipatifs et semi-groupes dans les espaces de fonctions continues. Annales Inst. Fourier, 26 (1976), fasc. 4, 1 - 97.

[22] Roth, J.P., Recollement des semi-groupes de Feller locaux. Annales Inst. Fourier, 30 (1980), fasc. 3, 75 - 89.

[23] Stoica, L., Local operators and Markov processes. Lect. Notes in Math. vol. 816, Springer-Verlag, Berlin, 1980.

AN OUTLINE OF THE SPECTRAL THEORY OF PROPAGATORS

P. Masani

Departments of Mathematics

University of Pittsburgh

After indicating recent improvements in the propagator theory of Hilbertian varieties and some applications to Banach algebras, we outline the spectral theory of propagators.

1. Introduction

In this paper

$$(1.1)\quad \left|\begin{array}{l} \Lambda \text{ is a non-void set} \\ W \text{ is a Banach space over } \mathbb{F}\ (\mathbb{F}=\mathbb{R} \text{ or } \mathbb{C}) \\ \mathscr{H} \text{ is a Hilbert space over } \mathbb{F}. \end{array}\right.$$

In many mathematical problems, pure and applied, we have to deal with $\mathscr{H}$-vector-valued functions $x(\cdot)$ on Λ, or with W to $\mathscr{H}$ linear operator-valued functions $X(\cdot)$ on Λ. Propagator theory is concerned with the changes in the functions $x(\cdot)$ and $X(\cdot)$ when a transformation semi-group (s.g.) Γ acts on Λ. Specifically, it is concerned with the case in which the changes are expressible in the form

$$x(t \oplus \lambda) = S(t)\{x(\lambda)\}, \qquad X(t \oplus \lambda) = S(t)\cdot X(\lambda),$$

where $t \in \Gamma$, $\lambda \in \Lambda$, $\oplus$ denotes the action of Γ on Λ, and $S(\cdot)$, called the <u>propagator</u>, is a function on Γ whose values are linear operators from $\mathscr{H}$ to $\mathscr{H}$. The need for a <u>spectral theory</u> of propagators stems from the realization that in many applications Γ is abelian and the $S(t)$, $t \in \Gamma$, form a commuting family of normal operators, and that a "spectral theorem" for the entire family would yield the integral representation encountered in various analytical problems.

To provide the necessary background we shall begin with a resume of propagator theory in the "time domain" so-to-speak as developed by us in

[8,9], but incorporating recent improvements (§2). It transpires that the main theorems in [8,9] are valid even when the involutory s.g. Γ is unitless, and that the Gelfand - Naimark representation theorem for C* algebras is a corollary of our generalized version [8:4.14] of Stinespring's Theorem.

In treating the spectral theory (§3), it is fruitful at the outset to disregard the confines of propagator theory, and taking Γ to be an arbitrary set, to prove a "Kolmogorov extension theorem" for any commuting family $(E_t : t \in \Gamma)$ of spectral measures for $\mathscr{H}$ over $\mathbb{C}$ with compact spectra. In conjunction with the spectral theorem for a single self-adjoint operator, this yields a general spectral theorem 3.6 for any commuting family $(S(t) : t \in \Gamma)$ of normal operators on $\mathscr{H}$ to $\mathscr{H}$. Then assuming, as in propagator theory, that Γ has an abelian algebraic structure and that $S(\cdot)$ is the appropriate morphism on Γ, we show that the spectrum $\sigma(E)$ of the spectral measure $E(\cdot)$ of $(S(t) : t \in \Gamma)$ falls within the class of appropriate "characters" of Γ. For instance, for $\Gamma = \mathbb{A}$, an abelian involutory Banach algebra, and $S(\cdot)$, a $*$ representation of $\mathbb{A}$, we find that $\sigma(E) \subseteq \sigma(\mathbb{A}) \cup \{0\}$, where $\sigma(\mathbb{A})$ is the Gelfand spectrum of $\mathbb{A}$, and that for an abelian C* algebra $\mathbb{A}$, we have $\sigma(E) = \sigma(\mathbb{A}) \cup \{0\}$. The "commutative" version of the Gelfand - Naimark Theorem follows at once from the last equality. For an involutory abelian s.g. Γ with neutral element, our spectral theorem yields the integral representation for a positive-definite function discovered by Lindahl and Maserick [7], and rediscovered by Berg, Christiansen and Ressel [1].

Space will allow the ennuciation of only very basic results, and permit only stray remarks on the proofs. A fuller version of the paper will appear elsewhere.

2. Propagator Theory in the Time-Domain

Since a vector in $\mathscr{H}$ can be regarded as an $\mathbb{F}$-to-$\mathscr{H}$ linear operator, the vectorial case $x(\cdot)$ mentioned at the outset of §1 is subsumed by the operatorial case $X(\cdot)$. We shall accordingly deal only with the latter.

More fully, let $CL(W,\mathscr{H})$ be the space of continuous linear operators on W to $\mathscr{H}$, cf. (1.1); then we are given that

$$(2.1) \quad \begin{cases} \text{(i)} & X(\cdot) \text{ is a function on } \Lambda \text{ to } CL(W,\mathscr{H}) \\ \text{(ii)} & \mathscr{D}_X \overset{=}{_d} \bigcup_{\lambda \in \Lambda} X(\lambda)(W) \subseteq \mathscr{H}. \end{cases}$$

For brevity we refer to such functions $X(\cdot)$ as Hilbertian varieties. The linear manifold in $\mathscr{H}$ spanned by $\mathscr{D}_X$ is denoted by $<\mathscr{D}_X>$, and its closure, called the subspace of $X(\cdot)$, by $\mathscr{S}_X$; thus

$$\mathscr{S}_X = \text{cls.} <\mathscr{D}_X>. \tag{2.2}$$

Also associated with $X(\cdot)$ is its covariance kernel $K_X(\cdot,\cdot)$ defined by

$$K_X(\lambda,\lambda') = X(\lambda')^*\cdot X(\lambda), \qquad \lambda,\lambda' \in \Lambda. \tag{2.3}$$

It is a triviality that

$$K_X(\cdot,\cdot) \text{ is a PD kernel on } \Lambda\times\Lambda \text{ to } CL(W,W^*), \tag{2.4}$$

where W^* is the adjoint (not dual W') of W, and PD means "positive - definite" in the obvious sense as defined in [8:2.5] for instance.[1] Conversely, the Kernel Theorem of Kolmogorov, Aronszajn and Pedrick tells us that given a PD kernel on $\Lambda\times\Lambda$ to $CL(W,W^*)$, there exists a Hilbert space $\mathscr{H}$ and a function $X(\cdot)$ on Λ to $CL(W,\mathscr{H})$ such that $K(\cdot,\cdot) = K_X(\cdot,\cdot)$; moreover $K(\cdot,\cdot)$ determines $X(\cdot)$ up to unitary equivalence, cf. [8:2.10, 2.9]. This theorem is crucial in several applications.

Now let an additive semi - group Γ, possibly non - abelian and unitless, act on Λ in the sense that there is a binary operation $\oplus$ on $\Gamma\times\Lambda$ to Λ such that

$$\begin{cases} \forall\, s,t \in \Lambda \ \&\ \forall\lambda \in \Lambda, & (s+t)\oplus\lambda = s\oplus(t\oplus\lambda), \\ \text{when } \Gamma \text{ has a neutral element } 0, & 0\oplus\lambda = \lambda. \end{cases} \tag{2.5}$$

It is convenient to regard the elements of Γ as moments of a (multidimensional) "time"[2], and to think of $t\oplus\lambda$ as the phase of an evolving system t time - units after its phase is λ. The assumption that Γ is a s.g. satisfying (2.5) then amounts to assuming that our system is deterministic and time - invariant ("temporarily homogeneous") cf. [9:3.4]. As the phase of the system advances from λ to $t\oplus\lambda$, the variety $X(\cdot)$ attached to the system changes from $X(\lambda)$ to $X(t\oplus\lambda)$. When and only when this change is expressible in the form

1 The definition of positive definteness rests on the concepts of a hermitian operator and a non - negative operator on W to W^*, cf. [8:2.3]. Throughout the sequel, "$A \preceq B$" will mean that $B-A$ is non - negative hermitian.

2 Hence our use of the letters s,t, etc. for the elements of Γ.

(2.6) $$X(t \oplus \lambda) = S(t)\cdot X(\lambda), \qquad t \in \Gamma,\ \lambda \in \Lambda,$$

where the S(t) are single - valued linear operators whose domains and ranges contain $<\mathcal{D}_X>$, do we say that the variety X(·) possesses a (linear)[3] propagator S(·).

Whether or not a given variety X(·) possesses a propagator S(·) and whether or not the S(t) have other desirable properties such as continuity depends on the nature of X(·) and therefore on its covariance kernel $K_X(\cdot,\cdot)$, cf. (2.4) et seq. Thus it is natural to seek conditions on $K_X(\cdot,\cdot)$ which ensure the existence of propagators of various sorts. In this paper we shall deal only with involutory semi - groups Γ, i.e. with s.g.'s Γ which admit a one - one function * on Γ onto Γ such that

(2.7) $$\begin{cases} \forall\, s,t \in \Gamma, \quad s^{**} = s, \quad (s+t)^* = t^* + s^*; \\ \text{when } \Gamma \text{ has a neutral element } 0, \quad 0^* = 0. \end{cases}$$

For involutory s.g.'s Γ there are two sets of conditions on the propagator S(·) which are natural and important for the applications:

Condition A. $\forall\, t \in \Gamma$, S(t) is a closed linear operator such that[4]

$$S(t) = \text{cls. Rstr.}_{<\mathcal{D}_X>} S(t) \text{ and } S(t^*) \subseteq S(t)^*.$$

Condition B. $\forall t \in \Gamma$, $S(t) \in CL(\mathcal{S}_X, \mathcal{S}_X)$ and $S(t^*) = S(t)^*$.

The following theorem gives a complete characterization of these cases:

2.8 MAIN THEOREM. Let $\Lambda, W, \mathcal{H}, X(\cdot)$ be as in (1.1), (2.1), and let the involutory s.g. Γ (possibly non - abelian and unitless) act on Λ in the sense of (2.5). Then

(a) X(·) has a propagator S(·) satisfying Condition A, iff. $K_X(\cdot,\cdot)$ has the transfer property:

$$\forall\, t \in \Gamma \text{ and } \forall\, \lambda,\lambda' \in \Lambda, \qquad K_X(t \oplus \lambda, \lambda') = K(\lambda, t^* \oplus \lambda');$$

(b) X(·) has a propagator satisfying Condition B, iff. $K_X(\cdot,\cdot)$ has the transfer property and satisfies the mild translational inequality:

3 The qualification "linear" will be omitted in the sequel, as non - linear propagators will not concern us in this paper.

4 $\text{Rstr.}_S F$ means the restriction of function F to the domain S.

$\exists\ \gamma(\cdot) \in (\mathbb{R}_{0+})^{\Gamma}$ such that

$$\forall\ t \in \Gamma \text{ and } \forall \lambda \in \Lambda, \qquad K_X(t \oplus \lambda, t \oplus \lambda) \preceq \gamma(t) \cdot K_X(\lambda,\lambda).$$

The results 2.8 (a), 2.8 (b) are proved in [8:4.7 and 4.10] under the assumption that Γ has a neutral element 0. But inspection of the proof shows that this assumption is redundant. For the many applications given in [8] , in particular to dilation theory, a neutral element is required. The scope of propagator theory is wider, however, and the admission of unitless Γ is a necessary improvement.

While Thm. 2.8 is deep, it is a triviality that the propagator $S(\cdot)$ has the semi - group property; more precisely

$$(2.9) \quad \begin{cases} \forall\ s,t \in \Gamma, \quad S(s+t) \subseteq \text{cls.}\{S(s)\cdot S(t)\}, \text{ under Condition A.} \\ \qquad\qquad\quad S(s+t) = S(s)\cdot S(t), \text{ under Condition B.} \end{cases}$$

Thus <u>under Condition B, $S(\cdot)$ is a * homomorphism on Γ into the multiplicative * s.g. $CL(\mathscr{S}_X, \mathscr{S}_X)$</u>; furthermore when Γ is abelian, $(S(t) : t \in \Gamma)$ is a commuting family of normal operators.

In many applications $\Lambda = \Gamma$, i.e. the s.g. Γ or Λ acts on itself, the operations $\oplus$ and + being identical, cf. [8: 4.12-] , [9: §5] and Szafraniec [13] , who discovered a new formulation of the Condition B for this case. A significant instance is the following generalized form of a theorem due originally to Stinespring [12] :

2.10 THEOREM (Stinespring). Let

(i) $\mathbb{A}$ be a Banach algebra over $\mathbb{F}$ (possibly non - abelian and unitless) with an isometric involution *,

(ii) $R(\cdot) \in L(\mathbb{A}, CL(W,W^*))$,

(iii) the kernel $K(\cdot,\cdot)$, defined by

$$K(a,b) = R(b^* \cdot a), \quad a,b \in \mathbb{A},$$

be PD on $\mathbb{A} \times \mathbb{A}$ to $CL(W,W^*)$,

(iv) $X(\cdot)$ be the Hilbertian variety with convariance kernel $K(\cdot,\cdot)$.

Then

(a) $X(\cdot)$ possesses a propagator $S(\cdot)$ on $\mathbb{A}$ to $CL(\mathscr{S}_X, \mathscr{S}_X)$;

(b) $S(\cdot)$ is a contractive * homomorphism on $\mathbb{A}$ into the C* algebra

$CL(\mathscr{S}_X, \mathscr{S}_X)$;

(c) when $\mathbb{A}$ has a unit 1, $R(t) = X(1)^*S(t)X(1)$, $t \in \mathbb{A}$.

This theorem follows from its unitized version given in [8: 4.14 and 4.15] by dint of the isometric * isomorphism between $\mathbb{A}$ and its standard unitization.

By associating with $\mathbb{A}$ a canonical Banach space $W_{\mathbb{A}}$ and a canonical function $R_{\mathbb{A}}(\cdot)$ on $\mathbb{A}$ to $CL(W_{\mathbb{A}}, W_{\mathbb{A}}^*)$, we can deduce the Gelfand - Naimark representation theorem from Thm. 2.10. Since $\mathbb{A}$ is trivially isometrically * isomorphic to its standard unitization, and this unitization preserves the C* property, cf. [3: §12, # 19], we may without loss of generality assume that $\mathbb{A}$ has a unit 1 such that $|1| = 1$. The Banach space we associate with $\mathbb{A}$ is the Bochner - Legesgue class

(2.11) $$W_{\mathbb{A}} \underset{d}{=} L_2(\mathscr{S}, 2^{\mathscr{S}}, \text{card}; \mathbb{A}) \underset{d}{=} \ell_2(\mathscr{S}; \mathbb{A})$$

of $\mathbb{A}$ - valued functions on the space $\mathscr{S}$ of <u>normalized states</u> $\emptyset$ of $\mathbb{A}$, i.e. of $\emptyset$ such that

$$\emptyset \in L(\mathbb{A}, \mathbb{F}), \ |\emptyset| = 1, \ \emptyset(a^*) = \overline{\emptyset(a)}, \ \emptyset(a^* \cdot a) \geq 0, \ a \in \mathbb{A}.$$

We define the function $R_{\mathbb{A}}(\cdot)$ by

(2.12) $$[R_{\mathbb{A}}(a)(w_1)](w_2) = \sum_{\emptyset \in \mathscr{S}} \emptyset[w_2(\emptyset)^* \cdot a \cdot w_1(\emptyset)],$$

where $a \in \mathbb{A}$ and $w_1, w_2 \in W_{\mathbb{A}}$. It is then a straightforward exercise to show that $R_{\mathbb{A}}(\cdot)$ is well - defined and fulfills the premises 2.10(ii), (iii), that $R_{\mathbb{A}}$ is a non - negative hermitian contraction, and that for a C*algebra $\mathbb{A}$, $|R_{\mathbb{A}}(t^* \cdot t)| = |t|^2$, $t \in \mathbb{A}$. The conclusions 2.10 (b), (c) yield the following:

2.13 THEOREM (Gelfand - Naimark representation). Let $\mathbb{A}$ be a unital Banach algebra over $\mathbb{F}$ with an isometric involution *. Then

(a) $\exists$ a contractive * homomorphism $S(\cdot)$ on $\mathbb{A}$ into $CL(\mathscr{H}, \mathscr{H})$, where $\mathscr{H}$ is a Hilbert space over $\mathbb{F}$;

(b) when $\mathbb{A}$ is a C* algebra, $S(\cdot)$ is an isometric * isomorphism;

(c) when $\mathbb{A}$ is abelian, the $S(t)$, $t \in \Gamma$, form a commuting family of normal operators.

3. Spectral Theory of Propagators

In the spectral theory of propagators, Γ is an abelian s.g. and $(S(t):t\in\Gamma)$ is a s.g. of normal operators. It is desirable, however, to commence with an arbitrary set Γ and a family of commuting spectral measures $E_t(\cdot)$, $t\in\Gamma$, with compact spetra σ_t, and to seek a single "Kolmogorov" spectral measure which represents the family. Accordingly our initial data will be:

(3.1)

(i) Γ is a non-void set;

(ii) $\forall\, t\in\Gamma$, σ_t is a compact subset of $\mathbb{C}$, τ_t is the $\tau_{\mathbb{C}}$-relative topology for σ_t, where $\tau_{\mathbb{C}}$ is the standard topology for $\mathbb{C}$;

(iii) $$\tilde{\Gamma} = \mathop{\times}_{t\in\Gamma} \sigma_t;$$ $\forall\, t\in\Gamma$, $\mathscr{E}_t$ is "evaluation at t" on $\tilde{\Gamma}$;

(iv) $\mathscr{N} \underset{d}{=} \bigcup_{t\in\Gamma} \mathscr{E}_t^{-1}(\tau_t)$, $\tau \underset{d}{=}$ the topology generated by $\mathscr{N}$;

(v) $\forall\, t\in\Gamma$, $\mathscr{B}_{\tau_t} \underset{d}{=} \sigma\text{-ring}(\tau_t)$, $\mathscr{B}_\tau \underset{d}{=} \sigma\text{-ring}(\tau)$.

Thus τ is the topology of pointwise convergence for $\tilde{\Gamma}$, and by Tychonov's Thm.

(3.2) $(\tilde{\Gamma},\tau)$ is a compact Hausdorff space.

Also, $\mathscr{B}_{\tau_t}$, $\mathscr{B}_\tau$ are the σ-algebras of Borel subsets of the topological spaces (σ_t,τ_t), $(\tilde{\Gamma},\tau)$, respectively. We now assert the following fundamental result:

3.3 THEOREM. (Kolmogorov extension for spectral measures). With the notation (3.1), let

(i) $\mathscr{H}$ be a Hilbert space over $\mathbb{C}$,

(ii) $\forall\, t\in\Gamma$, E_t be a strongly countably additive (s.c.a.) spectral measure for $\mathscr{H}$ on $\mathscr{B}_{\tau_t}$, such that

$$\sigma(E_t) \underset{d}{=} \text{the spectrum of } E_t = \sigma_t,$$

(iii) $\forall\, s,t\in\Gamma$, $\forall\, A\in\mathscr{B}_{\tau_s}$ and $\forall\, B\in\mathscr{B}_{\tau_t}$, $E_s(A)$ and $E_t(B)$ commute.

Then $\exists$ a unique inner regular s.c.a. spectral measure $E(\cdot)$ for $\mathscr{H}$ on $\mathscr{B}_\tau$ such that $\forall$ finite $L \subset \Gamma$ and $\forall\ B_t \in \mathscr{B}_{\tau_t}$, $t \in L$,

$$E[\bigcap_{t \in L} \mathscr{E}_t^{-1}(B_t)] = \prod_{t \in L} E_t(B_t).$$

The proof consists in affecting a Kolmogorov extension of the E_t - family to $\mathscr{B}_{\mathscr{N}} \underset{d}{=} \sigma$ - alg.$(\mathscr{N})$, and then (since in general $\mathscr{B}_{\mathscr{N}} \subset \mathscr{B}_\tau$) a further extension to $\mathscr{B}_\tau$. These extensions are made by applying the classical Kolmogorov and Prokhorov theorems to the families $(|E_L(\cdot)x|^2 : L \text{ finite} \subseteq \Gamma)$, where $E_L = \Pi_{t \in L} E_t$, and $x \in \mathscr{H}$, cf. Kolmogorov [5: p. 29, Fund. Thm.] and Bourbaki [4: Ch. 9, §4, Thm. 1].

We shall call the measure $E(\cdot)$ given by Thm. 3.3 the <u>Kolmogorov measure</u> of the commuting spectral family $(E_t(\cdot) : t \in \Gamma)$. Its spectrum $\sigma(E)$ is obviously a compact set:

(3.4) $$\sigma(E) \subseteq \tilde{\Gamma} = \underset{t \in \Gamma}{\mathrm{X}} \overline{D}(0,r_t), \qquad r_t \underset{d}{=} \max_{z \in \sigma_t} |z|,$$

where $\overline{D}(0,r_t)$ is the closed disk in $\mathbb{C}$ with center 0 and radius r_t. The following simple corollary of Thm. 3.3 plays a central role:

3.5 FUNDAMENTAL COROLLARY. With the notation of Thm. 3.3, let

$$t \in \Gamma, \qquad S(t) \underset{d}{=} \int_{\sigma_t} \lambda E_t(d\lambda), \qquad \mathscr{E}_\sigma(t) \underset{d}{=} \mathrm{Rstr.}_{\sigma(E)} \mathscr{E}_t.$$

Then

(a) $\forall\ t \in \Gamma$, $S(t) \underset{d}{=} \int_{\tilde{\Gamma}} \mathscr{E}_t(f)E(df)$, $\sigma_t = \sigma\{S(t)\} = \mathscr{E}_\sigma(t)\{\sigma(E)\}$;

(b) $\mathscr{E}_\sigma(\Gamma)$ is a $\sigma(E)$ - separating subset of $C(\sigma(E),\mathbb{C})$;

(c) $\mathscr{E}_\sigma \cdot S^{-1}$ is an isometry on $S(\Gamma) \subseteq CL(\mathscr{H},\mathscr{H})$ onto the set $\mathscr{E}_\sigma(\Gamma) \subseteq C(\sigma(E),\mathbb{C})$;

(d) The following conditions are equivalent:
 (α) $\sigma(E)$ is separating on Γ
 (β) $\mathscr{E}_\sigma(\cdot)$ is one - one on Γ to $C(\sigma(E),\mathbb{C})$
 (γ) $S(\cdot)$ is one - one on Γ to $CL(\mathscr{H},\mathscr{H})$;

(e) The following conditions are equivalent:
 (α) $\mathscr{E}_\sigma(\Gamma)$ is uniformly closed in $C(\sigma(E),\mathbb{C})$
 (β) $S(\Gamma)$ is uniformly closed in $CL(\mathscr{H},\mathscr{H})$.

At this stage we have to invoke the spectral theorem for a single continuous self adjoint operator H on $\mathscr{H}$ to $\mathscr{H}$, referring to its direct proof based on the square-root and the explicit exhibition of the spectral measure of H, as given e.g. in [11: pp. 279-280].[5] Now let T be a continuous normal operator on $\mathscr{H}$ to $\mathscr{H}$, $\Gamma = \{1,2\}$ and E_1, E_2 be the spectral measures of the real and imaginary parts of T. Then the premisses of Cor. 3.5 are fulfilled, and from the conclusion 3.5 (a) we readily obtain the spectral theorem for T.

Next, let $(S(t) : t \in \Gamma)$ be a commuting family of continuous normal operators on $\mathscr{H}$ to $\mathscr{H}$, and E_t be the spectral measure of S(t). Then the premisses of Cor. 3.5 are again fulfilled, and we arrive at the following conclusion:

3.6 GENERAL SPECTRAL THEOREM. Let

(i) $(S(t) : t \in \Gamma)$ be a commuting family of continuous normal operators on $\mathscr{H}$ to $\mathscr{H}$,

(ii) $\sigma_t \underset{d}{=} \sigma\{S(t)\}$, $t \in \Gamma$,

(iii) $\tau_t, \tilde{\Gamma}, \mathscr{N}, \tau, \mathscr{E}_t, \mathscr{B}_{\tau_t}, \mathscr{B}_\tau$ be defined as in (3.1).

Then $\exists$ a unique inner regular, s.c.a. spectral measure for $\mathscr{H}$ on $\mathscr{B}_\tau$ such that

$$\sigma(E) \subseteq \tilde{\Gamma} \quad \& \quad S(t) = \int_{\tilde{\Gamma}} \mathscr{E}_t(f)E(df), \qquad t \in \Gamma,$$

and all the conclusions 3.5 (a) - (e) hold.

We shall call $E(\cdot)$, given by 3.6, the <u>spectral measure</u> of the family $(S(t) : t \in \Gamma)$.

An important theorem of Kuratowski asserts that if two complete, separable metric spaces $\mathscr{X}$, $\mathscr{Y}$ have the same cardinality, then there is a one-one function Φ on $\mathscr{X}$ onto $\mathscr{Y}$ such that both Φ and Φ^{-1} are Borel measurable, cf. Parthasarathy [10: p. 14, # 2.12]. The combination of this theorem with Thm. 3.6 immediately yields the following explicit version of a theorem of von Neumann (cf. [11: pp. 358 -]):

---

5 The deep intrinsical nature of this proof is revealed by its adaptibility to the general spectral theorems of H. Freundenthal and U. Krause, cf. G. Birkhoff [2: pp. 362-364] and U. Krause [6: 3.4].

3.7 THEOREM (von Neumann). Let

(i) Γ be a countable set,

(ii) $(S(t) : t \in \Gamma)$ be as in 3.6 (i), and $E(\cdot)$ be its spectral measure,

(iii) Φ be the Kuratowski function on $\sigma(E)$ onto the closed unit disk $\overline{D}$ in $\mathbb{C}$.[6]

Then

$$\forall\, t \in \Gamma, \quad S(t) = \{\mathscr{E}_\sigma(t) \cdot \Phi^{-1}\}(T), \quad T \underset{d}{=} \int_{\sigma(E)} \Phi(f) E(df);$$

i.e. all the $S(t)$ are the values of Borel measurable functions at the same normal operator T.

Thm. 3.6 remains valid of course when, as in propagator theory, Γ has an algebraic structure and $S(\cdot)$ is the corresponding morphism whose values are commuting normal operators. But this additional structure together with the inner regularity of the spetral measure $E(\cdot)$ allows us to infer that $\sigma(E)$ lies within the set of appropriate characters of Γ. There are many such specializations of Thm. 3.6. It will suffice to state just two:

3.8 THEOREM. Let

(i) Γ be an involutory abelian s.g.,

(ii) $S(\cdot)$ be a *homomorphism on Γ into the multiplicative s.g. $CL(\mathscr{H},\mathscr{H})$, where $\mathscr{H}$ is a Hilbert spcce over $\mathbb{C}$,

(iii) $E(\cdot)$ be the spectral measure of $(S(t) : t \in \Gamma)$,

(iv)

$$\hat{\Gamma} \underset{d}{=} \{f : f \in \mathbb{C}^\Gamma \;\&\; \forall\, s,t \in \Gamma,\ f(s+t) = f(s)f(t), \quad f(t^*) = \overline{f(t)}\}.$$

Then

$$\sigma(E) \subseteq \tilde{\Gamma} \cap \hat{\Gamma} \quad \& \quad \forall\, t \in \Gamma, \quad S(t) = \int_{\tilde{\Gamma}} \mathscr{E}_t(f) E(df),$$

and all the conclusions 3.5 (a)-(e) hold.

3.9 THEOREM. Let

(i) $\mathbf{A}$ be an abelian Banach algebra over $\mathbb{C}$, with isometric involution *,

---

6 Since Γ is countable, the compact Hausdorff spaces $\sigma(E)$ and $\overline{D}$ are completely metrizable and separable, and have the same cardinality, viz. c.

(ii) $S(\cdot)$ be a * homomorphism on $\mathbb{A}$ into $CL(\mathscr{H},\mathscr{H})$, where $\mathscr{H}$ is a Hilbert space over $\mathbb{C}$,

(iii) $E(\cdot)$ be the spectral measure of $(S(t) : t \in \Gamma)$,

(iv) $\sigma(\mathbb{A})$ be the Gelfand spectrum of $\mathbb{A}$. Then

(a) $\sigma(E) \subseteq \tilde{\mathbb{A}} \cap \{\sigma(\mathbb{A}) \cup \{0\}\}$ & $\forall\, t \in \mathbb{A}$, $S(t) = \int_{\tilde{\mathbb{A}}} \mathscr{E}_t(f)E(df)$,

and all the conclusions 3.5 (a) - (e) hold;

(b) $\mathscr{E}_\sigma(\cdot)$ is a contractive * homomorphism on $\mathbb{A}$ onto the subalgebra $\mathscr{E}_\sigma(\mathbb{A})$ of $C(\sigma(E),\mathbb{C})$;

(c) when $\mathbb{A}$ is a C* algebra, we have $\sigma(E) = \sigma(\mathbb{A}) \cup \{0\}$, and $\mathscr{E}_\sigma(\cdot)$ is an isometric * isomorphism on $\mathbb{A}$ onto the C* algebra $C(\sigma(\mathbb{A}) \cup \{0\},\mathbb{C})$.

If in Thm. 3.9 we take the $\mathscr{H}$ and the $S(\cdot)$ given by the Gelfand - Naimark Thm. 2.12, then the conclusion (c) gives the so-called "commutative" Gelfand - Naimark Thm., cf. Bonsall & Duncan [3: p. 189, Thms. 4,5].

As an application of Thm. 3.8, consider a bounded $\mathbb{C}$-valued PD function ϕ on an additive abelian involutory s.g. Γ with a neutral element 0. By definition, the kernel $K(\cdot,\cdot)$ such that

$$K(s,t) = \phi(t^* + s), \qquad s,t \in \Gamma$$

is PD on $\Gamma \times \Gamma$ to $\mathbb{C}$, and is therefore the covariance kernel of a vectorial variety $x(\cdot)$ on Γ to $\mathscr{H}$. It follows easily that the conditions of the Main Thm. 2.8 (b) are fulfilled and that $\gamma(t) \leqslant 1$, and consequently that $x(\cdot)$ has a propagator $S(\cdot)$ whose values are normal contractions. Since, cf. (2.9) et seq., $S(\cdot)$ is a *homomorphism on Γ into the multiplicative *s.g. $CL(\mathscr{S}_x, \mathscr{S}_x)$, therefore Thm. 3.8 applies. Thus $S(t) = \int_{\sigma(E)} \mathscr{E}_t(f)E(df)$, and so

$$\phi(t) = K(t,0) = (S(t)x(0),x(0)) = \int_{\sigma(E)} \mathscr{E}_t(f)\mu(df),$$

where $\mu(\cdot) = |E(\cdot)x(0)|^2$. This establishes the result of Lindahl & Maserick, and of Berg et al, mentioned in §1, which they prove by appeal to Choquet theory.

Acknowledgement

This paper bears the impress of fruitful conversations with Professors

C. Berg, W. Hackenbroch, P. Ressel, E. Thomas and G. Vincent - Smith. Their occurence was made possible primarily by an award from the Alexander von Humboldt Stiftung which enabled the writer to spend the academic year 1979-1980 in Germany, as well as by invitations from mathematics departments in Copenhagen, Regensburg, Münster, Groningen and Lausanne. Their help is gratefully acknowledged.

REFERENCES

[1] Berg, C. - Christiansen, J. - Ressel, P., Positive definite functions on abelian semi - groups. Math. Ann. 223 (1976), 253-272.

[2] Birkhoff, G., Lattice Theory. 3rd Ed., Amer. Math. Soc., Providence, R.I., 1979.

[3] Bonsall, F. - Duncan, J., Complete normed algebras. Springer-Verlag, New York, 1973.

[4] Bourbaki, N., Eléments de mathématique. Livre VI, Intégration, Hermann, Paris, 1969.

[5] Kolmogorov, A., Foundations of probability. Chelsea, New York, 1950.

[6] Krause, U., Der Satz von Choquet als ein abstrakter Spektralsatz und vice versa. Math. Ann. 184 (1970), 275-296.

[7] Lindahl, R. - Maserick, P., Positive - definite functions on involution semi - groups. Duke Math. J. 38 (1971), 771-782.

[8] Masani, P., Dilations as propagators of Hilbertian varieties. SIAM J. Anal. 9 (1978), 414-456.

[9] Masani, P., Propagators and dilations. in "Probability theory on vector spaces", edited by A. Weron, Lecture Notes 656, Springer-Verlag, Berlin, 1978, 95-117.

[10] Parthasarathy, K., Probability measures on metric spaces. Acad. Press, New York, 1967.

[11] Riesz, F. - Sz.-Nagy, B., Functional Analysis. Ungar. New York, 1955.

[12] Stinespring, W., Positive functions on C* algebras. Proc. Amer. Math. Soc. 6 (1955), 333-343.

[13] Szafraniec, F., Dilations on involution semi - groups. Proc. Amer. Math. Soc. 66 (1977), 30-32.

ON GENERALIZED INVERSES AND OPERATOR RANGES

M. Z. Nashed
Department of Mathematical Sciences
University of Delaware
Newark, Delaware 19711

Aspects of the theory of operator ranges, factorization and range inclusion are brought to bear on some operator and approximation-theoretic problems for generalized inverses on infinite dimensional Banach and Hilbert spaces. Several criteria are given for an operator to have a bounded outer inverse with infinite rank. It is also shown using one of these criteria that the set of all bounded linear operators with a bounded outer inverse is open. The set of all bounded linear operators with a bounded inner inverse is dense in the space of all bounded linear operators. Comments on related topics in generalized inverse operator theory and some open problems are given.

1. Introduction

A unified approach to the operator theory of generalized inverses has been developed in recent years; see Nashed and Votruba [17]. Within this framework, the algebraic, topological, extremal and proximinal properties have been separately considered and analyzed. Although the *algebraic* theory of generalized inverses is virtually complete, there are still a number of operator-theoretic questions and approximation-perturbation aspects that merit further investigation.

The purpose of this paper is to show that close relationships exist between *operator ranges* (specifically the notions of majorization, factorization, range inclusion, and topological complements) and the operator theory of *generalized inverses* on infinite dimensional spaces (specifically, bounded outer inverses with infinite rank, the structure of all bounded operators with a bounded outer (or inner) inverse). By an *outer inverse* to a linear operator $A: X \to Y$ we shall mean a *nonzero linear* operator $B: Y \to X$ such that $BAB = B$. For other notations and properties of generalized inverses which are used, but not specifically defined or established, see [17].

2. Outer Inverses and Operator Ranges

Let X and Y be (real or complex) Banach spaces and let $L(X,Y)$ be the space of all bounded linear operators on X into Y. The range and null space of $A \in L(X,Y)$ are denoted by $R(A)$ and $N(A)$ respectively. Let $D \supset R(A)$. A linear map $B: D \subset Y \to X$ is called an *inner inverse* of A if $ABA = A$. If B is an inner inverse with domain Y, then

$$X = N(A) \dot{+} R(BA), \quad Y = R(A) \dot{+} N(AB), \tag{2.1}$$

where $\dot{+}$ denotes algebraic direct sum. Similarly, a linear map B is an *outer inverse* of A if $BAB = B$. Each outer inverse induces the direct sum decompositions

$$X = R(B) \dot{+} N(BA) \quad \text{and} \quad Y = N(B) \dot{+} R(AB). \tag{2.2}$$

It is well known that A has a *bounded* inner inverse on Y if and only if $N(A)$ and $R(A)$ have topological complements in X and Y respectively (see, e.g., [17]). The same result holds if A is a closed linear operator with dense domain.

Henceforth by a complement we shall mean a topological complement. A topological direct sum will be denoted by $\oplus$.

REMARK 2.1 If A has a *bounded* outer inverse B then the algebraic decompositions (2.2) are also topological decompositions. Various necessary and sufficient conditions for a linear operator B to be an outer inverse of a given linear operator A are collected in [17; Proposition 1.13].

We are here interested in conditions under which there exists a bounded outer inverse with a given nullspace and a given range. The following two remarks address this question.

REMARK 2.2 Not every closed complemented subspace Y_1 of Y can be the null space of an outer inverse (of A) which has the given range X_1. If $X_1 = R(B)$ then $R(AB) = AX_1$. So Y_1 must be a complement to the given subspace AX_1 in Y.

REMARK 2.3. If X_1 is a closed complemented subspace of X such that $X_1 \cap N(A) = \{0\}$ and Y_1 is a complement to AX_1 then there exists a bounded outer inverse B of A such that $R(B) = X_1$, $N(B) = Y_1$. For $y \in AX_1$ define $By = (A/X_1)^{-1}y$ and extend B linearly to all of Y such that $N(B) = Y_1$. It then follows that B is an outer inverse with the prescribed properties.

Combining Remarks 2.1 - 2.3 we have

THEOREM 2.1. <u>Let</u> X,Y <u>be Banach spaces</u>. $A \in L(X,Y)$ <u>has a bounded outer inverse with given range</u> X_1 <u>and given nullspace</u> Y_1 <u>if and only if the following conditions are satisfied</u>:

a) X_1 <u>is a closed complemented subspace of</u> X <u>and</u> $X_1 \cap N(A) = \{0\}$;

b) Y_1 <u>is a complement for the subspace</u> AX_1.

An excellent survey on operator ranges is given by Fillmore and Williams [7]. They consider a number of elegant but little-known results concerning the ranges of bounded linear operators in Hilbert space. As Fillmore and Williams remark there is reason to believe that the results and techniques of the theory of operator ranges will find increasing applications, for instance in formulating and proving infinite-dimensional versions of finite-dimensional theorems. Here we shall use some results on operator ranges to establish criteria for the existence of bounded outer inverses and related properties. The following result is due to Douglas [4]; since it plays an important role in what follows we include a proof using in part the notation of generalized inverses.

THEOREM 2.2. <u>Let</u> A <u>and</u> T <u>be bounded linear operators on a Hilbert space</u> H. <u>The following statements are equivalent</u>:

a) $R(A) \subset R(T)$.

b) $A = TC$ <u>for some bounded operator</u> C <u>on</u> H.

c) $AA^* \leq \lambda^2\, TT^*$ <u>for some</u> $\lambda \geq 0$.

PROOF. Suppose that (a) holds. Set $C = T^\dagger A$. Then C is bounded and $TC = TT^\dagger A = A$, where $T^\dagger$ is the generalized inverse of T. That (b) implies (a) is trivial. If $A = TC$ then

$$(AA^*x,x) = ||A^*x||^2 = ||C^*T^*x||^2 \leq ||C^*||^2||T^*x||^2 = ||C^*||^2\, (TT^*x,x);$$

thus (b) implies (c). Finally if (c) holds, then $||A^*x|| \leq \lambda||T^*x||$ for all $x \in H$. Therefore, the linear map $D\colon R(T^*) \to R(A^*)$ defined by

$D(T^*x) = A^*x$ is bounded. Extend D to the closure of $R(T^*)$ by continuity and put $D = 0$ on $R(T^*)^\perp = N(T)$, then $DT^* = A^*$, so $A = TD^*$. ∎

If we consider operators A and T with domains being the Hilbert spaces H_1 and H_2, respectively, but with range in a common space, then the operator C in the statement of Theorem 2.2 is defined from H_1 to H_2. In [4] Douglas remarks that the equivalence of statements (a) and (b) in Theorem 2.2 persists in Banach spaces; however this is false since (a) does not imply (b) in Banach spaces. A counterexample (due to Douglas) is published in a paper by M. Embry (Proc. Amer. Math. Soc., 38 (1973), 587-589).

THEOREM 2.3. *Let X,Y be Hilbert spaces. An operator $A \in L(X,Y)$ has a* ***bounded*** *outer inverse of* ***infinite rank*** *(i.e., with infinite dimensional range) if and only if the range of A contains a closed complemented subspace of infinite dimension.*

PROOF. Suppose B is a bounded outer inverse with infinite rank. Then AB is a projector and $R(AB)$ is a closed complemented subspace of infinite dimension which is contained in $R(A)$; compare with (2.2).

Conversely, suppose M is an infinite dimensional closed subspace contained in $R(A)$. Let $Y = M \oplus S$ and let P be the projector on M along S. Since $R(P) \subset R(A)$ it follows from Theorem 2.2 that there is a bounded linear operator C such that $P = AC$. Then $P^2 = P$ implies $CP = CPACP$, so that $B := CP$ is a bounded outer inverse of A of infinite rank. ∎

COROLLARY 2.4. *If A is a bounded linear operator on an infinite-dimensional Hilbert space, then A has a bounded outer inverse of infinite rank if and only if A is not compact.*

PROOF. This follows from Theorem 2.3 and the fact that a bounded linear operator on a Hilbert space is compact if and only if its range contains no closed infinite-dimensional subspaces (see, e.g., [5 ; Corollary 5.10] or [7 ; Theorem 2.5]; a simple proof is given in [8 ; p. 294]). ∎

REMARK 2.4. Let A be an $m\times n$ matrix of rank r. For all integers s,t with $0 \le s \le r$ and $r \le t \le \min(m,n)$, A has outer inverses of rank s and inner inverses of rank t. For operators with infinite rank, we can similarly construct outer inverses with any rank.

EXAMPLE 2.1. Let H_1 and H_2 be Hilbert spaces and let $K\colon H_1 \to H_2$ be compact linear operator with infinite dimensional range. Let $\{\mu_n; u_n, v_n\}$ be a singular system for K, i.e.,

$$u_n = \mu_n K v_n, \qquad v_n = \mu_n K^* u_n$$

where $0 < \mu_1 \le \mu_2 \le \ldots \le \mu_n \le \ldots$ with $\mu_n \to \infty$. We assume that $\{u_n\}_1^\infty$ and $\{v_n\}_1^\infty$ are orthonormal systems. Then (see, e.g., [14])

$$Kx = \sum_{n=1}^{\infty} \mu_n^{-1} (x, v_n)\, u_n$$

and

$$K^{\dagger} y = \sum_{n=1}^{\infty} \mu_n (y, u_n)\, v_n$$

for $y \in \mathcal{D}(K^{\dagger}) := \mathcal{R}(K) + \mathcal{R}(K)^{\perp}$, where $K^{\dagger}$ is the (Moore-Penrose) generalized inverse of K.

Let n be a fixed positive integer and define the operator B_n by

$$B_n y := \sum_{j=1}^{n} \mu_j (y, u_j)\, v_j .$$

It follows that

$$KB_n y = \sum_{j=1}^{n} (y, u_j)\, u_j$$

and

$$B_n K B_n y = \sum_{j=1}^{n} \mu_j \left(\sum_{i=1}^{n} (y, u_i) u_i, u_j \right) v_j$$

$$= \sum_{j=1}^{n} \mu_j (y, u_j)\, v_j = B_n y .$$

Thus for each positive integer, B_n is an outer inverse of K, $\dim \mathcal{R}(B_n) = n$, and $||B_n y|| \le \Gamma_n ||y||$. Also for each $y \in \mathcal{D}(K^{\dagger})$, $||B_n y - K^{\dagger} y|| \to 0$ as $n \to \infty$, but not uniformly, since $K^{\dagger}$ is unbounded. Thus the operators B_n are not uniformly bounded: $||B_n|| \to \infty$ as $n \to \infty$.

COROLLARY 2.5. *A bounded operator* A *on an infinite dimensional Hilbert space has a bounded outer inverse with infinite dimensional range if and only if for each positive integer* n *there is an outer inverse* B_n *with* dim $\mathcal{R}(B_n) = n$ _and the operators_ B_n _are uniformly bounded_: $||B_n|| \le \gamma$ _for all_ n.

REMARK 2.5. In the case of Banach spaces, the existence of an infinite-dimensional closed complemented subspace M contained in $\mathcal{R}(A)$ is necessary for the existence of a bounded outer inverse of infinite rank. In addition, we need that $\mathcal{N}(A)$ is topologically complemented in the Banach space $\{x : Ax \in M\}$. Details and related topics will be discussed elsewhere. A partial result was also given by R. Khalil [9] using the fact that every closed subspace of a Banach space has a b a s i c sequence.

REMARK 2.6. $A \in L(X,Y)$ has a bounded inner inverse if and only if $\mathcal{N}(A)$ and $\mathcal{R}(A)$ are complemented in X and Y, respectively. If X and Y are Hilbert spaces, these conditions are satisfied if and only if $\mathcal{R}(A)$ is closed. Comparing these conditions with the necessary and sufficient conditions for the existence of a bounded outer inverse (Theorem 2.3 and Corollary 2.4) it follows immediately that if an operator has a bounded inner inverse then it has a bounded outer inverse, but not conversely. This last assertion is known and has been established directly. For if A has a bounded inner inverse B, then it follows immediately $B A B$ is a bounded outer inverse (as well as an inner inverse) of A. To prove directly that the converse is false, one has to construct (in view of our criteria for the existence of bounded inner, respectively outer, inverses) an example of a noncompact operator with nonclosed range. Such examples abound, for instance, in the theory of singular integral equations and Fredholm integral equations of the first kind on the whole line. A rather technical example, based on a construction due to E. Asplund, is given in Caradus [2]. Now that the set of all bounded operators which have bounded outer inverses has been characterized in the above simple manner in both Banach and Hilbert spaces, more transparent examples can be given.

REMARK 2.7. Theorem 2.3 is also valid if A is a closed densely defined operator. The modification of the proof is only in the use of the following immediate extension of a part of Theorem 2.2. If $\mathcal{R}(P) \subset \mathcal{R}(A)$, then there exists a densely defined operator C such that $P = AC$ and C is bounded in the graph norm of A. Moreover, if P is bounded, then C is bounded.

3. Topological Properties of the Set of All Operators with Bounded Inner (Outer) Inverses

Let $G_1(X,Y)$ denote the class of all $A \in L(X,Y)$ which have a bounded inner inverse and $G_2(X,Y)$ the set of all $A \in L(X,Y)$ which have a bounded outer inverse. The set of invertible operators (which is a proper subset of $G_1 \cap G_2$) is open. What can we say about G_1 and G_2?

Using Corollary 2.4 we obtain a simple proof that $G_2(X,Y)$ is open in the (uniform) operator topology. This result was first established by the author in [12] for Banach spaces and used in [13] for the stability of inverse mapping theorems when the derivative operator is noninvertible. The analysis in [12], [13] provides also perturbation bounds.

THEOREM 3.1. The set G_2 of all bounded linear operators on an infinite-dimensional Hilbert space with a bounded outer inverse is open.

PROOF. By Corollary 2.4 the set G_2 is the complement in the space $L(X,Y)$ of the set of all compact operators. Now the latter set is closed (see, e.g., [8]), so G_2 is open. ∎

THEOREM 3.2. Let H_1 and H_2 be Hilbert spaces. The set $G_1(H_1,H_2)$ of all bounded linear operators with a bounded inner inverse is dense $L(H_1,H_2)$.

PROOF. Let $A \in L(H_1,H_2)$ and write its polar decomposition $A = VP$ where V is a maximal partial isometry and P is a positive operator (see, e.g., [8]). For any $\varepsilon > 0$ there exists an invertible positive operator Q such that $||Q-P|| < \varepsilon$. Thus $||A-VQ|| = ||VP-VQ|| \leq ||P-Q|| < \varepsilon$. Since V is a partial isometry if and only if $VV^*V = V$ and since V is maximal, it follows that V has a bounded inner inverse. But Q is invertible, so VQ has a bounded inner inverse. This proves that G_1 is dense in $L(H_1,H_2)$. ∎

Graves has shown that the set of all bounded linear operators of X o n t o Y, where X and Y are Banach spaces is open in the Banach space $L(X,Y)$. Dieudonné has shown that the set of all right (left) boundedly invertible operators in $L(X,Y)$ is an open set in $L(X,Y)$. Clearly if A is onto a Banach space, or if A is right (left) boundedly invertible operator then A has a bounded outer inverse (see also [13] for references to the literature).

4. Related Topics, Comments and Problems

4.1 Invariance Properties of Inner, Outer and Generalized Inverses. If A is a linear operator acting between two vector spaces V and W, then every inner inverse of A determines algebraic complements to $N(A)$ in V and $R(A)$ in W, and conversely. What properties of inner inverses (or expressions) are invariants under all choices of inner inverses to a given operator (or equivalently all choices of algebraic complements to $N(A)$ and $R(A)$)? Although fragments of results of this nature are given in several contexts, there does not seem to be a systematic study of invariance properties under choices of projectors or complements, either in the algebraic context or in Banach space. We mention some examples of invariants of inner and generalized inverses.

(i) The transformation $B - BAB$ is invariant under change of projectors (or complements); this transformation is a "measure" of the departure of an inner inverse from being an outer inverse. See [17; Corollary 1.9].

(ii) In the theory of so-called "alternative problems" or operator equations of the form $Fx = Lx$ where F is a nonlinear operator and L is a linear operator with closed range and nontrivial nullspace, some topological complements to subspaces $R(L)$ and $N(L)$ are used to "split" the operator equation into a pair of equations (equivalent to the problem), or to study existence of solutions based on topological degree or coincidence degree. For operators in a certain class, Mawhin has shown that coincidence degree has the invariance properties under choice of different complements to $N(L)$ and $R(L)$. See [11] and references cited therein.

(iii) For a bounded linear operator A on a Banach space, the generalized inverse $A^{\dagger}$ depends on the projects P and Q (see [12]). Continuity of $A^{\dagger}$ is invariant under these projectors. Bounds for the norm of the difference of two generalized inverses of the same linear operator, but corresponding to two different pairs of projectors are given in [12].

4.2 Extremal Characterizations and Operator Ranges. Engl and Nashed [6] have recently established new extremal characterizations of generalized inverses of a closed or bounded linear operator between Hilbert spaces, which generalize the extremal characterization in the Frobenius norm for matrices due to Penrose (see [15] for a comparison of all extremal properties). The generalization utilizes Hermitian order and Schatten norms. For example, for $A \in L(H_1,H_2)$ with closed range, then for any $Y \in L(H_1,H_3)$, where H_i are Hilbert spaces, the set $\{(XA-Y)(XA-Y)^*: X \in L(H_2,H_3)\}$ has a smallest element with respect to the Hermitian order on $L(H_3)$ and the set of all such smallest elements has a unique element which minimizes XX^* with respect to the Hermitian norm; this element is $X = BA^\dagger$. Theorem 2.2 on range inclusion, factorization, and majorization of operators can be used with the results in [6] to provide equivalent "extremal-like" characterizations. For example, with $X = BA^\dagger$, we have

$$R(XA-B) \subset R(ZA-B) \tag{4.1}$$

for all $Z \in L(H_2,H_3)$ and

$$R(BA^\dagger) \subset R(X) \tag{4.2}$$

for all other X that satisfy (4.1).

<u>4.3 A Problem on Drazin Inverse</u>. Find a direct extremal characterization of the Drazin inverse. For definitions and literature on the Drazin inverse, see [16], [2], [17]; some operator-theoretic properties are developed in [16].

<u>4.4 An Operator Equation of the Invariant Subspace Problem</u>. The open question whether every operator on an infinite-dimensional Hilbert space has an invariant subspace other than the zero subspace and the whole space is called the invariant subspace problem. Since this problem deals with an infinite dimensional extension of a problem whose answer is well known in finite dimensional space, operator ranges play an important role in various formulations. The invariant subspace problem can be equivalently formulated in terms of an innocent looking operator equation, namely, a bounded linear operator A on a Hilbert space H into H has a non-trivial invariant subspace if and only if $XAX = AX$ has a solution in $L(H)$ other than zero and the identity operator (see, e.g., [18]). Although this equation is quite different from the equations defining various generalized or approximate inverses, some connections might exist.

4.5 Topological Complements as Operator Ranges. The theory of generalized inverses for a bounded operator acting between Banach spaces X and Y hinges on the existence of topological complements to $N(A)$ and $\overline{R(A)}$, in X and Y, respectively. In the case of Hilbert spaces, such complements always exist, and among them the complements $N(A)^{\perp} = \overline{R(A^*)}$ and $R(A)^{\perp} = N(A^*)$ are especially distinguished. In particular, when $R(A)$ is closed, these complements are, respectively, the range and nullspace of another distinguished operator, A*. No analogous situation exists in Banach spaces. If topological complements can be chosen in Banach spaces so that

$$(4.3) \qquad X = N(A) \oplus R(B) \quad \text{and} \quad Y = R(A) \oplus N(B) ,$$

then the generalized inverse of A relative to complements induced by B can be defined as usual: $A_B^{\dagger}$ is the linear extension of $(A/R(B))^{-1}$ to all of Y such that $N(A_B^{\dagger}) = N(B)$. Of interest is the study of properties of operators B that satisfy (4.3), together with additional restrictions on B so that analogues of results on generalized inverse operator theory in Hilbert space can be immediately constructed in Banach spaces (e.g., iterative methods, spectral approximations, etc.). A restricted attempt is given in [10].

4.6 Quasicomplementation, Quasi-Regularizers and Metric Generalized Inverses. There are still open problems and useful directions for investigations on relationships among these topics; see, e.g., [3], [15].

Acknowledgement

The author would like to thank Professor T. Ando for conversations on operator ranges and for bringing reference [7] to his attention.

REFERENCES

[1] Anselone, P. M.-Nashed, M. Z., Perturbations of outer inverses, in Approximation Theory III, (E. W. Cheney, ed.), pp. 163-169, Academic Press, New York, 1980.

[2] Caradus, S. R., Generalized Inverses and Operator Theory, Queen's Papers in Pure and Applied Mathematics, No. 50, Queen's University, Kingston, Ontario, 1978.

[3] Cross, R. W., Unilateral quasi-regularizers of closed operators, Math. Proc. Camb. Phil. Soc. 87 (1980), 471-480.

[4] Douglas, R. G., On majorization, factorization, and range inclusion of operators on Hilbert space, Proc. Amer. Math. Soc. 17 (1966), 413-415.

[5] Douglas, R. G., Banach Algebra Techniques in Operator Theory, Academic Press, New York, 1972.

[6] Engl, H. W.-Nashed, M. Z., New extremal characterizations of generalized inverses of linear operators, J. Math. Anal. Appl. (1981), to appear.

[7] Fillmore, P. A.-Williams, J. P., On operator ranges, Advances in Math. 7 (1971), 254-281.

[8] Halmos, P. R., A Hilbert Space Problem Book, Van Nostrand, Princeton, N.J., 1967.

[9] Khalil, R., Existence of bounded outer inverses, to appear.

[10] Koliha, J. J., Power convergence and pseudoinverses of operators between Banach spaces, J. Math. Anal. Appl. 48 (1974), 446-469.

[11] Mawhin, J., Topological Degree Methods in Nonlinear Boundary Value Problems, Conference Board of the Mathematical Sciences Regional Conference in Mathematics, No. 40, Amer. Math. Soc., Providence, 1979.

[12] Nashed, M. Z., On the perturbation theory for generalized inverse operators in Banach spaces, in Functional Analysis Methods in Numerical Analysis (M. Z. Nashed, ed.), pp. 180-195, LNM Vol. 701, Springer-Verlag, Berlin/Heidelberg/New York, 1979.

[13] Nashed, M. Z., Generalized inverse mapping theorems and related applications of generalized inverses in Nonlinear Equations in Abstract Spaces (V. Lakshmikantham, ed.), pp. 217-252, Academic Press, New York, 1978.

[14] Nashed, M. Z., Aspects of generalized inverses in analysis and regularization, in Generalized Inverses and Applications (M. Z. Nashed, ed.), pp. 193-252, Academic Press, New York, 1976.

[15] Nashed, M. Z., Best approximation problems arising from generalized inverse operator theory, in Approximation Theory III (E. W. Cheney, ed.), pp. 667-674, Academic Press, New York, 1980.

[16] Nashed, M. Z., ed., Recent Applications of Generalized Inverses, Pitman, London, 1981.

[17] Nashed, M. Z.-Votruba, G. F., A unified operator theory of generalized inverses, in Generalized Inverses and Applications, pp. 1-109, Academic Press, New York, 1976.

[18] Radjavi, H.-Rosenthal, P., Invariant Subspaces, Springer Verlag, Berlin/Heidelberg/New York, 1973.

II Functional Analysis

MODULAR APPROXIMATION BY A FILTERED FAMILY OF LINEAR OPERATORS

Julian Musielak
Institute of Mathematics
A. Mickiewicz University
Poznań

There is introduced the notion of boundedness of a filtered family (T_V) of linear operators in a modular space. This notion is used to get a general theorem on modular convergence $T_V x \to x$. Applications in cases of generalized Orlicz spaces of functions and sequences are given.

1. Introduction

Let X be a real vector space. A functional $\rho: X \to [0,\infty]$ is called a *modular* on X, if $\rho(x) = 0$ iff $x = 0$, $\rho(-x) = \rho(x)$ and $\rho(ax+by) \leqslant \rho(x) + \rho(y)$ for $a,b \geqslant 0$, $a+b=1$, $x,y \in X$. If $\rho(ax+by) \leqslant a\rho(x) + b\rho(y)$ for $a,b \geqslant 0$, $a+b=1$, then ρ is called a *convex modular* on X. The *modular space* X_ρ generated by ρ is defined as $X_\rho = \{x \in X: \rho(ax) \to 0$ as $a \to 0+\}$. The formula $|x|_\rho = \inf\{u > 0: \rho(\frac{x}{u}) \leqslant u\}$ defines an F-norm in X_ρ, and in case of ρ convex, $\|x\|_\rho = \inf\{u > 0: \rho(\frac{x}{u}) \leqslant 1\}$ defines a norm in X_ρ equivalent to $|\ |_\rho$. Convergence $x_n \to 0$ in norm in X_ρ is equivalent to the condition $\rho(ax_n) \to 0$ as $n \to \infty$ for every $a > 0$. Besides this, there is defined in X_ρ a *modular convergence* (*ρ-convergence*) $x_n \xrightarrow{\rho} 0$ by the condition: there exists an $a > 0$ such that $\rho(ax_n) \to 0$ as $n \to \infty$. The *ρ-closure* of a set $S \subset X_\rho$ is defined as the set of all elements $x \in X_\rho$ such that $x_n - x \xrightarrow{\rho} 0$ for a sequence of $x_n \in S$. Obviously, norm convergence implies ρ-convergence but not conversely. Let us remark that if X is a normed space with norm $\|\ \|$, then $\rho(x) = \|x\|$ is a convex modular in X, $X_\rho = X$ and $\|\ \|_\rho = \|\ \|$, ρ-convergence and norm convergence being thus equivalent.

An important example of a modular space is provided by a generalized Orlicz space $L^\varphi = L^\varphi(\Omega,\Sigma,\mu)$, defined as follows. Let (Ω,Σ,μ) be a measure space with a nonnegative, nontrivial σ-finite and complete measure μ, and let X be the space of all extended real-valued, Σ-measurable functions $x = x(\cdot)$ over Ω, finite μ-almost everywhere; two functions equal μ-a.e. will be treated as the same element of X. Let φ be a *φ-function with parameter*, i.e. $\varphi: \Omega \times \mathbb{R} \to \mathbb{R}_+ = [0, \infty)$, $\varphi(t,u)$ is an even, continuous function of u, equal to zero iff $u = 0$ and nondecreasing for $u \geq 0$ for every $t \in \Omega$, and is a measurable function of $t \in \Omega$ for every $u \in \mathbb{R}$. If, moreover, $\varphi(t,u)$ is a convex function of $u \in \mathbb{R}$ for all $t \in \Omega$, then it is called a convex φ-function with parameter. Now, taking

$$(1) \qquad \rho(x) = \int_\Omega \varphi(t, x(t))\, d\mu,$$

ρ is a modular on X (convex modular, if φ is convex). The respective modular space X_ρ is denoted by $L^\varphi(\Omega, \Sigma, \mu)$ (or briefly, L^φ) and is called a *generalized Orlicz space*. In case when Ω is the set of nonnegative integers, Σ is the σ-algebra of all subsets of Ω and $\mu(A)$ is equal to the number of elements of the set A, the respective generalized Orlicz space of sequences $x = (t_i)$ defined by the modular

$$(2) \qquad \rho(x) = \sum_{i=0}^{\infty} \varphi_i(t_i)$$

is denoted by ℓ^φ and is called the *generalized Orlicz sequence space* (for definitions, see e.g. [3]).

We shall be concerned with problems of approximation by singular integrals (convolution operators) and of moduli of smoothness in modular spaces. It is quite natural to consider approximation with respect to ρ-convergence; the respective theorems concerning convergence in norm $|\ |_\rho$ or $\|\ \|_\rho$ are then easily deduced making the number $a > 0$ in the condition $\rho(ax_n) \to 0$ variable. Also, as a norm is a special case of a modular, the results may be interpreted as theorems for normed spaces. Problems of the above type were investigated in [1] and the present paper may be considered as a further contribution in this direction.

In order to put together theorems on convolution operators and on moduli of smoothness, there will be adopted here the technique of filters which makes possible to give a uniform treatment of seemingly different problems.

2. A General Theorem

Let ρ be a modular on a real vector space X and let X_ρ be the respective modular space. Let V be an abstract nonempty set and let $\mathscr{V}$ be a filter of subsets of V. A function $g: V \to \mathbb{R}$ tends to zero with respect to $\mathscr{V}$, $g(v) \xrightarrow{\mathscr{V}} 0$, if for every $\varepsilon > 0$ there is a set $V_o \in \mathscr{V}$ such that $|g(v)| < \varepsilon$ for all $v \in V_o$. A function $G: \mathscr{V} \to \mathbb{R}$ tends to zero with respect to $\mathscr{V}$, $G(V) \xrightarrow{\mathscr{V}} 0$, if for every $\varepsilon > 0$ there is a set $V_\varepsilon \in \mathscr{V}$ such that $|G(V \cap V_\varepsilon)| < \varepsilon$ for every $V \in \mathscr{V}$.

A family $T = (T_v)_{v \in V}$ of linear operators $T_v: X_\rho \to X_\rho$ will be called $\mathscr{V}$ - b o u n d e d (briefly: b o u n d e d), if there exist positive numbers k_1, k_2 and a function $g: V \to \mathbb{R}_+$ such that $g(v) \xrightarrow{\mathscr{V}} 0$ and for every $x \in X_\rho$ there is a set $V_x \in \mathscr{V}$ for which

$$\rho(T_v x) \leqslant k_1 \rho(k_2 x) + g(v)$$

for all $v \in V_x$.

Let us remark that if ρ is convex, then the constant k_1 may be always taken equal to 1 and, moreover, if ρ is convex and linear operators $T_v: X_\rho \to X$ are $\mathscr{V}$-bounded, then $T_v: X_\rho \to X_\rho$ for every $v \in V$.

If X is a normed space with norm $\|\ \|$ and we take $\rho(x) = \|x\|$, then the family $(T_v)_{v \in V}$ of linear operators $T_v: X \to X$ is $\mathscr{V}$-bounded, iff there is a constant $M > 0$ such that for every $x \in X$ there exists a $V \in \mathscr{V}$ for which $\|T_v x\| \leqslant M\|x\|$ for all $v \in V$.

The following theorem is a general tool in various approximation problems:

THEOREM 1. <u>Let $T = (T_v)_{v \in V}$ be a $\mathscr{V}$-bounded family of linear operators $T_v: X_\rho \to X_\rho$ and let $S_o \subset X$ satisfy the following conditions:</u>

(a) <u>for every $x \in S_o$ there is an $a > 0$ such that $\rho(a(T_v x - x)) \xrightarrow{\mathscr{V}} 0$,</u>

(b) <u>X_ρ^S is the ρ - closure in X_ρ of the set of all finite linear combinations of elements of the set S_o.</u>

<u>Then for every $x \in X_\rho^S$ there exists a $b > 0$ such that $\rho(b(T_v x - x)) \xrightarrow{\mathscr{V}} 0$.</u>

PROOF. First, let us remark that the thesis holds for all $x \in S$, since supposing $x = c_1 x_1 + \ldots + c_n x_n$ with $x_i \in S_o$ we have, writing $c = \sum_{i=1}^n |c_i|$,

$$\rho(b(T_v x - x)) \leqslant \sum_{i=1}^{n} \rho(bc(T_v x_i - x_i)) \xrightarrow{\mathscr{V}} 0,$$

if we take $b > 0$ sufficiently small. Now, let $\varepsilon > 0$ be arbitrary and let $x \in X_\rho^S$ be given. Then there exists a $b > 0$ and an element $s \in S$ such that

$$\rho(3bk_2(x-s)) < \frac{\varepsilon}{6k_1} \quad \text{and} \quad \rho(3b(T_v s - s)) \overset{\mathscr{V}}{\to} 0,$$

where we may assume $k_1, k_2 \geq 1$. Let $v \in V_{3b(x-s)}$, the set $V_{3b(x-s)}$ being chosen according to the definition of $\mathscr{V}$-boundedness of $(T_v)_{v \in V}$ corresponding to the element $3b(x-s)$. Then we have

$$\begin{aligned} \rho(b(T_v x - x)) &\leq \rho(3bT_v(x-s)) + \rho(3b(T_v s - s)) + \rho(3b(s-x)) \\ &\leq k_1\rho(3bk_2(x-s)) + g(v) + \rho(3b(T_v s - s)) + \rho(3b(s-x)) \\ &\leq 2k_1\rho(3bk_2(x-s)) + g(v) + \rho(3b(T_v s - s)) \\ &\leq \frac{\varepsilon}{3} + g(v) + \rho(3b(T_v s - s)). \end{aligned}$$

Now, let $V_1, V_2 \in \mathscr{V}$ be so that $g(v) < \varepsilon/3$ for $v \in V_1$ and $\rho(3b(T_v s - s)) < \varepsilon/3$ for $v \in V_2$. Taking $V = V_1 \cap V_2 \cap V_{3b(x-s)}$, we obtain $\rho(b(T_v x - x)) < \varepsilon$ for all $v \in V$. Hence $\rho(b(T_v x - x)) \overset{\mathscr{V}}{\to} 0$.

REMARK. If we assume (a) for every $a > 0$, then the thesis of Theorem 1 holds for every $b > 0$.

One may define the T-modulus of smoothness of an element $x \in X_\rho$ by means of the formula

$$\omega_T(x,V) = \sup_{v \in V} \rho(T_v x - x) \qquad \text{for every } V \in \mathscr{V}.$$

It is easily seen that $\omega_T(x,V) \overset{\mathscr{V}}{\to} 0$, iff $\rho(T_v x - x) \overset{\mathscr{V}}{\to} 0$. Hence Theorem 1 may be reformulated in terms of T-moduli of smoothness in the following way:

THEOREM 2. *Let* $T = (T_v)_{v \in V}$ *be a* $\mathscr{V}$-*bounded family of linear operators* $T_v : X_\rho \to X_\rho$ *and let* S_0 *be a subset of* X. *If for every* $x \in S_0$ *there is an* $a > 0$ *such that* $\omega_T(ax,V) \overset{\mathscr{V}}{\to} 0$, *then the same holds for every* $x \in X_\rho^S$.

These results will be applied below to cases of generalized Orlicz spaces, where S is ρ-dense in X_ρ, i.e., $X_\rho^S = X_\rho$; application in case where ρ is the norm in a normed space X is left to the reader.

3. Application to Generalized Orlicz Function Spaces

In this Section ρ will be given by formula (1), limiting ourselves to the case of Lebesgue measure over an interval $[0,b)$. We shall investigate two families of operators in $X_\rho = L^\varphi$: the translation operator and the convolution operator (singular integral operator).

Let $\Omega = [0,b) \subset \mathbb{R}$, $0 < b < \infty$, μ = Lebesgue measure in the σ-algebra Σ of all Lebesgue measurable subsets of $[0,b)$. The t r a n s l a t i o n o p e r a t o r $\tau_v : X \to X$ will be defined by the equality $\tau_v x(t) = x(t+v)$, where x is extended to the whole $\mathbb{R}$ b-periodically. Also, the φ-function with parameter generating the modular ρ by formula (1) will be extended periodically with respect to the variable $t \in [0,b)$ to the whole $\mathbb{R}$, i.e., $\varphi(t+b,u) = \varphi(t,u)$ for $u,t \in \mathbb{R}$.

We shall say that the function φ is φ - b o u n d e d, if there exist positive constants k_1, k_2 such that

$$\varphi(t-v,u) \leqslant k_1 \varphi(t,k_2 u) + f(t,v) \qquad \text{for } u,v,t \in \mathbb{R},$$

where the function $f: \mathbb{R} \times \mathbb{R} \to \mathbb{R}_+$ is measurable and b-periodic with respect to the first variable and such that writing $h(v) = \int_a^b f(t,v)dt$ for every $v \in \mathbb{R}$, we have $H = \sup_{v \in \mathbb{R}} h(v) < \infty$ and $h(v) \to 0$ as $v \to 0$.

Let us remark that if φ is convex with respect to u, then we may take in the above definition $k_1 = 1$. The above condition was introduced in [2] in connection with the investigation of the translation operator in a generalized Orlicz space. A trivial example of a τ-bounded function is obtained taking φ independent of the parameter t, as in case of usual Orlicz spaces; non-trivial examples are given e.g. in [1].

Now, taking $V = \mathbb{R}$ and denoting by $\mathscr{V}$ the filter of all neighbourhoods of zero in $\mathbb{R}$, we prove first the following statement, writing $\tau = (\tau_v)_{v \in \mathbb{R}}$:

PROPOSITION 1. (a) <u>If φ is τ-bounded, then the family τ of translation operators is $\mathscr{V}$-bounded.</u> (b) <u>If $1 \in L^\varphi$, then the linear combinations of the set S_o of all characteristic functions of Lebesgue measurable subsets of</u> $[0,b)$ <u>form a ρ-dense set in L^φ and for every $x \in S_o$ there is an $a > 0$ such that</u> $\rho(a(T_v x - x)) \xrightarrow{\mathscr{V}} 0$.

PROOF. $\mathcal{V}$ - boundedness of τ follows from the inequality

$$\rho(\tau_v x) = \int_0^b \varphi(t,x(t+v))dt \leq k_1\rho(k_2 x) + g(v),$$

where $g(v) = h(v) = \int_0^b f(t,v)dt$. ρ - density in L^φ of linear combinations of S_o, i.e., of simple functions, is easily proved first for positive functions, applying Lebesgue's dominated convergence theorem, and then for arbitrary $x \in L^\varphi$ by splitting x in positive and negative part. Now, if x is the characteristic function of a set $A \subset [0,b)$, we have for every $v \in \mathbb{R}$,

$$\rho(a(\tau_v x - x)) = \int_{A_v} \varphi(t,a)dt, \quad \text{where} \quad A_v = (A - v) \dot{-} A.$$

But $\int_0^b \varphi(t,a)dt < \infty$ for sufficiently small $a > 0$, since $1 \in L^\varphi$, and $\mu(A_v) \to 0$ as $v \to 0$. Hence $\rho(a(\tau_v x - x)) \to 0$ as $v \to 0$.

From Proposition 1 we deduce easily, applying Theorem 2, the following

THEOREM 3. _If_ φ _is_ τ - _bounded and_ $1 \in L^\varphi$, _then for every_ $x \in L^\varphi$ _there is a_ $c > 0$ _such that_

$$\omega_\tau(cx,\delta) = \sup_{|v| \leq \delta} \int_0^b \varphi[t,c(x(t+v) - x(t))]\,dt \to 0 \qquad \underline{\text{as}}\ \delta \to 0+.$$

Let us still remark that the same holds with respect to the norm of L^φ, if we restrict ourselves to x from the closure E^φ in L^φ of the set of simple functions.

Now, we are going to investigate the convolution operator T_w, where $w \in \mathcal{W}$, $\mathcal{W}$ is an abstract set and $\mathscr{W}$ is a filter in $\mathcal{W}$. Let $K_w : [0,b) \to \mathbb{R}_+$ for $w \in \mathcal{W}$ be integrable in $[0,b)$ and s i n g u l a r, i.e.

$$\sigma(w) = \int_0^b K_w(t)dt \xrightarrow{\mathscr{W}} 1, \quad \sigma_\delta(w) = \int_\delta^{b-\delta} K_w(t)dt \xrightarrow{\mathscr{W}} 0 \qquad \text{for every } 0 < \delta < \tfrac{b}{2},$$

$$\sigma = \sup_{w \in \mathcal{W}} \int_0^b K_w(t)dt < \infty,$$

and let us extend K_w b - periodically to the whole $\mathbb{R}$. Let

$$T_w x(s) = \int_0^b K_w(t - s)x(t)dt. \tag{3}$$

We prove first

PROPOSITION 2. <u>Let</u> φ <u>be a convex,</u> τ <u>- bounded</u> φ <u>- function with a parameter, and let</u> $(K_w)_{w \in \mathcal{W}}$ <u>be singular. Then</u> $T_w : L^\varphi \to L^\varphi$ <u>for every</u> $w \in \mathcal{W}$ <u>and</u> $T = (T_w)_{w \in \mathcal{W}}$ <u>is</u> $\mathcal{W}$ <u>- bounded.</u>

PROOF. It is sufficient to prove that T is $\mathcal{W}$-bounded; henceforth follows that $T_w : L^\varphi \to L^\varphi$. Applying b - periodicity of $\varphi(\cdot, u)$ and $x(\cdot)$, Jensen's inequality and τ - boundedness of φ with $k_1 = 1$, $k_2 = k \geq 1$, we obtain for $x \in L^\varphi$:

$$\begin{aligned} \rho(T_w x) &= \int_0^b \varphi\left(s, \frac{1}{\sigma(w)} \int_0^b K_w(t)\sigma(w)x(s+t)dt\right) ds \\ &\leq \frac{1}{\sigma(w)} \int_0^b \int_0^b K_w(t)\varphi(s, \sigma x(s+t))dt\, ds \\ &= \frac{1}{\sigma(w)} \int_0^b K_w(t) \int_0^b \varphi(u-t, \sigma x(u))\, du\, dt \\ &\leq \rho(k\sigma x) + g(w), \end{aligned}$$

where

$$g(w) = \frac{1}{\sigma(w)} \int_0^b K_w(t)h(t)dt.$$

Splitting the last integral in three integrals over intervals $[a, a+\delta]$, $[a+\delta, b-\delta]$, $[b-\delta, b]$ and applying the usual procedure concerning singular integrals, we obtain $g(w) \xrightarrow{\mathcal{W}} 0$.

Now, we are able to prove the following

THEOREM 4. <u>Let</u> φ <u>be a convex,</u> τ <u>- bounded</u> φ <u>- function with a parameter,</u> $1 \in L^\varphi$, <u>and let</u> $(K_w)_{w \in \mathcal{W}}$ <u>be singular. Then the operators</u> T_w <u>defined by</u> (3) <u>satisfy the condition</u>

$$\rho(a(T_w x - x)) \xrightarrow{\mathcal{W}} 0 \qquad \text{for some } a > 0,$$

<u>for every</u> $x \in L^\varphi$ <u>(with</u> a <u>dependent on</u> x).

PROOF. By Proposition 2, $T_w x \in L^\varphi$ for $x \in L^\varphi$. Let $x \in L^\varphi$, then, by Theorem 3, $\omega_\tau(cx, \delta) \to 0$ as $\delta \to 0+$ for sufficiently small $c > 0$. Now, let us choose $a > 0$ so small that $2a\sigma \leq c$ and $\rho(4\sigma kax) < \infty$, where k is equal to the constant k_2 from the definition of τ - boundedness of φ, k_1 being taken equal to 1; we may suppose $k \geq 1$. We estimate now

$$\rho(a(T_w x - x)) = \int_0^b \varphi\left(s, \frac{1}{\sigma(w)} \int_0^b K_w(t)\sigma(w)ax(s+t)dt - ax(s)\right) ds$$

$$\leq \frac{1}{2} \int_0^b \varphi\left(s, \frac{1}{\sigma(w)} \int_0^b K_w(t)2\sigma(w)a(x(s+t) - x(s))dt\right) ds$$

$$+ \frac{1}{2} \int_0^b \varphi(s, 2a(\sigma(w) - 1)x(s))ds$$

$$\leq \frac{1}{2\sigma(w)} \int_0^b K_w(t) \int_0^b \varphi(s, 2\sigma a(x(s+t) - x(s)))ds\ dt$$

$$+ \frac{1}{2} \rho(2a(\sigma(w) - 1)x).$$

Now, we split the first of the integrals on the right-hand side of this inequality into three integrals over intervals $[0,\delta]$, $[\delta, b-\delta]$ $[b-\delta, b]$, where $0 < \delta < b/2$ is arbitrary. The first integral is estimated as follows:

$$\int_0^\delta K_w(t) \int_0^b \varphi(s, 2\sigma a(x(s+t) - x(s)))ds\ dt$$

$$\leq \int_0^\delta K_w(t)\rho(2\sigma a(\tau_t x - x))dt \leq \sigma(w)\omega_\tau(2\sigma ax, \delta).$$

and the third one, by substitution $t = b-u$,

$$\int_{b-\delta}^b K_w(t) \int_0^b \varphi(s, 2\sigma a(x(s+t) - x(s)))ds\ dt$$

$$\leq \int_0^\delta K_w(b-u)\rho(2\sigma a(\tau_{-u} x - x))du \leq \sigma(w)\omega_\tau(2\sigma ax, \delta).$$

Finally, the second integral

$$\int_\delta^{b-\delta} K_w(t) \int_0^b \varphi(s, 2\sigma a(x(s+t) - x(s)))ds\ dt$$

$$\leq \frac{1}{2} \int_\delta^{b-\delta} K_w(t)[\rho(4\sigma a\tau_t x) + \rho(4\sigma ax)]\,dt$$

$$\leq \frac{1}{2} \int_{\delta}^{b-\delta} K_w(t)[\rho(4\sigma akx) + h(t) + \rho(4\sigma ax)]\,dt$$

$$\leq (\rho(4\sigma akx) + \frac{1}{2} H) \int_{\delta}^{b-\delta} K_w(t)dt.$$

Hence

$$\rho(a(T_w x - x)) \leq \omega_\tau(2\sigma ax, \delta) + \frac{\rho(4\sigma akx) + H/2}{\sigma(w)} \int_{\delta}^{b-\delta} K_w(t)dt + \frac{1}{2}\rho(2a(\sigma(w) - 1)x).$$

Let us take any $\varepsilon > 0$. By Theorem 3, taking $\delta > 0$ sufficiently small, we may make the first term on the right-hand side of the above inequality smaller than $\varepsilon/3$. Then, by singularity of (K_w), the second term may be made less than $\varepsilon/3$, taking $w \in W_1$ with an appropriate $W_1 \in \mathscr{W}$. Since $x \in L^{\varphi}$ and $\sigma(w) \xrightarrow{\mathscr{W}} 1$, the third term becomes less than $\varepsilon/3$ for $w \in W_2$ with some $W_2 \in \mathscr{W}$. Thus, $\rho(a(T_w x - x)) < \varepsilon$ for $w \in W_1 \cap W_2$.

Let us remark that taking as $\mathcal{W}$ the set $\mathbb{N}$ of all nonnegative integers and as $\mathscr{W}$ the filter of all sets of the form $\mathbb{N} \setminus A$ with A finite, $A \subset \mathbb{N}$, we obtain an approximation theorem for a summability method defined by the kernel $(K_n(u))$, $n = 0,1,2,\ldots$ Taking as $\mathcal{W}$ an interval on $\mathbb{R}$ and a point $w_0 \in \overline{\mathcal{W}}$ (may be also ∞) and as $\mathscr{W}$ the filter of all (may be also one-sided) neighbourhoods of w_0, we get an approximation theorem for summability method defined by the kernel $K(u,w)$, where $u \in [0,b)$, $w \in \mathcal{W}$. In the next Section we shall investigate the case of matrix summability methods.

4. Application to Generalized Orlicz Sequence Spaces

We are going now to apply Section 2 to the case of the space X of all sequences $x = (t_j)$ and to a modular of the form (2), where $\varphi = (\varphi_i)$ is a sequence of φ-functions, i.e. $\varphi: \mathbb{N} \times \mathbb{R} \to \mathbb{R}_+$. We shall investigate two families of operators: a sequence of translation operators and a family of convolution operators in the generalized Orlicz sequence space ℓ^{φ}. Here, $\mathcal{V}$ will be the set $\mathbb{N}$ of all nonnegative integers and the filter $\mathscr{V}$ will consist of all sets $V \subset \mathcal{V}$ which are complements of finite sets. The set $\mathcal{W}$ and the filter $\mathscr{W}$ of its subsets will be abstract, as previously.

The *translation operator* τ_m, $m = 0,1,2,\ldots$, will be defined by the formula $\tau_m x = ((\tau_m x)_i)$, where $(\tau_m x)_i = t_i$ for $i \leqslant m$ and $(\tau_m x)_i = t_{i+m}$ for $i > m$, $x = (t_j)$. Hence $\tau_m x - x = (0, \dots, t_{2m+1} - t_{m+1}, t_{2m+2} - t_{m+2}, \dots)$ with zeros on the first m+1 places, and the τ-modulus of smoothness of $x = (t_j)$ is equal to

$$\omega_\tau(x,V) = \sum_{i \in V} \varphi_i(t_{i+m} - t_i);$$

we shall write $\omega_\tau(x,r)$ for $V = \{r+1, r+2, \dots\}$.

In the sequel we shall say that $\varphi = (\varphi_i)_{i=0}^{\infty}$ is τ_-- b o u n d e d, if there exist constants $k_1, k_2 \geqslant 1$ and a double sequence $(\eta_{n,j})$ such that

$$\varphi_n(u) \leqslant k_1 \varphi_{n+j}(k_2 u) + \eta_{n,j} \qquad \text{for } u \in \mathbb{R}, n > j \geqslant 0,$$

where $\eta_{n,j} \geqslant 0$, $\eta_{n,0} = 0$, $\sum_{n=0}^{\infty} \eta_{n,j} < \infty$ uniformly with respect to j. We shall say that φ is τ_+ - b o u n d e d, if there are constants $k_1, k_2 \geqslant 1$ and a double sequence $(\varepsilon_{n,j})$ such that

$$\varphi_{n+j}(u) \leqslant k_1 \varphi_n(k_2 u) + \varepsilon_{n,j} \qquad \text{for } u \in \mathbb{R}; n,j = 0,1,2,\dots,$$

where $\varepsilon_{n,j} \geqslant 0$, $\varepsilon_{n,0} = 0$, $\varepsilon_j = \sum_{n=0}^{\infty} \varepsilon_{n,j} \to 0$ as $j \to \infty$, $s = \sup_{j \in \mathbb{N}} \varepsilon_j < \infty$.

Let us still write $e_\ell = (\delta_{i,\ell})_{i=0}^{\infty}$, where $\delta_{i,\ell}$ is the Kronecker symbol.

PROPOSITION 3. (a) _If φ is τ_- - bounded, then the family $\tau = (\tau_m)_{m=0}^{\infty}$ of translation operators is $\mathscr{V}$-bounded._

(b) _The set of linear combinations of sequences $e_0, e_1, e_2, \dots$ is ρ-dense in ℓ^φ._

(c) $\rho(a(\tau_m e_\ell - e_\ell)) \to 0$ _as_ $m \to \infty$ _for every_ ℓ _and every_ $a > 0$.

PROOF. (a) is obtained, because for $x \in \ell^\varphi$,

$$\rho(\tau_m x) = \sum_{i=0}^{m} \varphi_i(t_i) + \sum_{i=m+1}^{\infty} \varphi_i(t_{i+m}) \leqslant \sum_{i=0}^{m} \varphi_i(t_i) + k_1 \sum_{j=2m+1}^{\infty} \varphi_j(k_2 t_i) + c'_m,$$

where $c'_m = \sum_{i=m+1}^{\infty} \eta_{i,m} \to 0$ as $m \to \infty$. (b) is obvious and (c) follows from the fact that $\tau_m e_\ell - e_\ell = 0$ for $m \geqslant \ell$.

By Proposition 3 and Theorem 2, there holds the following

THEOREM 5. _If φ is τ_--bounded, then for every $x \in \ell^\varphi$ there is an $a > 0$ for_ which $\omega_\tau(ax,r) \to 0$ as $r \to \infty$.

In order to investigate the convolution operator T_w in ℓ^φ, let $\mathcal{W}$ be a filter of subsets of an abstract set $\mathcal{W}$ and let $K_w : \mathbb{N} \to \mathbb{R}_+$ for $w \in \mathcal{W}$ be s i n g u l a r, i.e.

$$\sigma(w) = \sum_{j=0}^{\infty} K_{w,j} \leqslant \sigma < \infty, \quad K_{w,0} \xrightarrow{\mathcal{W}} 1, \quad \frac{K_{w,j}}{\sigma(w)} \xrightarrow{\mathcal{W}} 0 \qquad \text{for } j = 1,2,\dots$$

Let $T_w x = ((T_w x)_i)_{i=0}^{\infty}$, where $(T_w x)_i = \sum_{j=0}^{i} K_{w,i-j} t_j$. We prove first

PROPOSITION 4. *Let $(K_w)_{w \in \mathcal{W}}$ be singular, $\varphi = (\varphi_i)_{i=0}^{\infty}$ τ_+-bounded and let φ_i be convex for $i = 0,1,2,\dots$ Then $T_w : \ell^\varphi \to \ell^\varphi$ for every $w \in \mathcal{W}$ and $T = (T_w)_{w \in \mathcal{W}}$ is $\mathcal{W}$-bounded.*

PROOF. It is enough to show that T is $\mathcal{W}$-bounded. We have for every $x \in \ell^\varphi$

$$\rho(T_w x) = \sum_{i=0}^{\infty} \varphi_i\left(\frac{\sum_{j=0}^{i} K_{w,j}}{\sigma(w)} \, \frac{\sum_{j=0}^{i} K_{w,j}\sigma(w) t_{i+j}}{\sum_{j=1}^{i} K_{w,j}} \right)$$

$$\leqslant \frac{1}{\sigma(w)} \sum_{j=0}^{\infty} \left\{ K_{w,j} \sum_{i=j}^{\infty} \varphi_i(\sigma(w) t_{i-j}) \right\} \leqslant k_1 \rho(k_2 \sigma x) + c(w)$$

where $c(w) = \frac{1}{\sigma(w)} \sum_{j=1}^{\infty} K_{w,j} \varepsilon_j$. But taking any $\eta > 0$ one may choose an index r such that $\sup_{j>r} \varepsilon_j < \eta/2$. Then

$$0 \leqslant c(w) \leqslant \sum_{j=1}^{r} \frac{K_{w,j}}{\sigma(w)} s + \frac{1}{\sigma(w)} \sum_{j=r+1}^{\infty} K_{w,j} \frac{\eta}{2} \leqslant s \sum_{j=1}^{r} \frac{K_{w,j}}{\sigma(w)} + \frac{\eta}{2}.$$

Now, taking $W \in \mathcal{W}$ so that the first term on the right-hand side of the last inequality is less than $\eta/2$ for all $w \in W$, we obtain $c(w) < \eta$ for $w \in W$. Thus, $c(w) \xrightarrow{\mathcal{W}} 0$ and so T is $\mathcal{W}$-bounded.

THEOREM 6. Let $\varphi = (\varphi_i)_{i=0}^{\infty}$ be a τ_+-bounded sequence of convex φ-functions φ_i. Let $(K_w)_{w\in \mathcal{W}}$ be singular, $K_w = (K_{w,j})_{j=0}^{\infty}$, where the family of elements $x_w^{\ell} = (0,\ldots,0,K_{w,1},K_{w,2},\ldots)$ with zeros on the first ℓ places satisfies the condition $\rho(a_{\ell}x_w^{\ell}) \xrightarrow{\mathcal{W}} 0$ for some $a_{\ell} > 0$ for $\ell = 0,1,2,\ldots$ Then for every $x \in \ell^{\varphi}$ there is an $a > 0$ such that $\rho(a(T_w x - x)) \xrightarrow{\mathcal{W}} 0$.

PROOF. By Theorem 1 and Proposition 4, it is sufficient to show the theorem for $x = e_{\ell}, \ell = 0,1,2,\ldots$ However, it is easily calculated that

$$\rho(a(T_w e_{\ell} - e_{\ell})) = \varphi_1(a(K_{w,o} - 1)) + \rho(ax_w^{\ell}) \qquad \text{for } a > 0.$$

Choosing $a > 0$ so small that $\rho(ax_w^{\ell}) \xrightarrow{\mathcal{W}} 0$, we obtain the theorem.

REFERENCES

[1] Hudzik, H. - Musielak, J. - Urbański, R., Linear operators in modular spaces. An application to approximation theory. Proc. of the Conference on Function Spaces and Approximation, Gdańsk 1979, in print.

[2] Kamińska, A., On some compactness criterion for Orlicz subspace $E_{\Psi}(\Omega)$. Commentationes Math. 22 (1980), in print.

[3] Musielak, J. - Orlicz, W., On modular spaces. Studia Math. 18 (1959), 49-65.

INTERPOLATION BETWEEN H^1 AND L^∞

Colin Bennett and Robert Sharpley [1]

Department of Mathematics

University of South Carolina

Columbia

The purpose of this article is to provide a simple proof of the result of N. M. Rivière - Y. Sagher that the real interpolation spaces between H^1 and L^∞ can be identified with the Lorentz L^{pq}-spaces. In contrast to existing proofs, which make heavy use of H^1-structure, the proof given here relies only on the well-known result of E. M. Stein-G. Weiss characterizing the distribution of the Hilbert transform of an arbitrary characteristic function of a set of finite measure, and a simple technique for applying that result due to R. O'Neil-G. Weiss.

1. Introduction

For simplicity only the case of the circle group T will be considered here. When T is equipped with normalized Lebesgue measure, the decreasing rearrangement f^* of a measurable function f on T is the unique nonnegative, decreasing, right-continuous function on the interval (0,1) that is equimeasurable with $|f|$. Recall that the Lorentz space L^{pq} $(1\le p<\infty,\ 1\le q\le\infty)$ consists of all measurable f on T for which

$$(1.1)\qquad ||f||_{pq} = \left(\int_0^\infty [t^{1/p} f^*(t)]^q dt/t\right)^{q}$$

is finite.

The *periodic Hilbert transform*, or *conjugate-function operator*, H is defined on $L^1(T)$ by the principal-value integral

---

[1] The research of both authors is partially supported by National Science Foundation Grant MCS80-01941.

$$(Hf)(x) = \tilde{f}(x) = \frac{1}{2\pi}\int_{-\pi}^{\pi} f(x-t)\cot\frac{t}{2}\,dt.$$

The (real) Hardy space $H^1(T)$ consists of those f in L^1 for which $\tilde{f}$ belongs also to L^1: it is a Banach space under the norm

$$(1.2) \qquad ||f||_{H^1} = ||f||_{L^1} + ||\tilde{f}||_{L^1}.$$

The *Peetre K-functional* $K(f,t,X_0,X_1)$ for a compatible couple (X_0,X_1) of Banach spaces is defined for every f in X_0+X_1 and every $t>0$ by

$$(1.3) \qquad K(f;t;X_0,X_1) = \inf_{f=f_0+f_1} (||f_0||_{X_0} + t||f_1||_{X_1}).$$

The following result is well-known (cf. [2, p. 184]).

THEOREM 1.1. (J. Peetre) For each f in $L^1(T)$,

$$(1.4) \qquad K(f;t;L^1,L^\infty) = \int_0^t f^*(s)ds = tf^{**}(t) \qquad (t>0).$$

The real interpolation space $(X_0,X_1)_{\theta,q}$ between X_0 and X_1 consists of those f in X_0+X_1 for which

$$(1.5) \qquad ||f||_{(X_0,X_1)_{\theta,q}} = \left(\int_0^\infty [t^{-\theta}K(f;t;X_0,X_1)]^q \frac{dt}{t}\right)^{1/q}$$

is finite (cf. [2, p. 167]). Hence, in view of Theorem 1.1 it is a simple matter to use the classical Hardy inequalities to identify the real interpolation spaces between L^1 and L^∞ as follows:

COROLLARY 1.2. If $0<\theta<1$, $1\le q\le\infty$, and $\theta = 1-1/p$, then

$$(1.6) \qquad (L^1,L^\infty)_{\theta,q} = L^{pq}$$

with equivalent norms.

The purpose of this note is to establish by simple methods the same result but with L^1 replaced by H^1. The following well-known result (cf. [4, p. 197]) will be crucial.

THEOREM 1.3. (E. M. Stein-G. Weiss). *Let* E *be an arbitrary measurable subset of* T *and let* χ_E *denote its characteristic function. Then*

$$(1.7)\qquad (\chi_E^{\sim})^*(t) = \frac{1}{\pi}\sinh^{-1}\left(\frac{\sin|E|/2}{\tan \pi t/2}\right) \qquad (0<t<1).$$

2. Interpolation between H^1 and L^∞

Let f be a measurable function on T. For each $t>0$, define the truncates f^t and f_t of f by

$$(2.1)\qquad (f^t)(x) = f(x) - f^*(t)\ \mathrm{sgn}\, f(x)$$

$$(2.2)\qquad f_t = f - f^t.$$

The decreasing rearrangements are given by

$$(2.3)\qquad (f^t)^*(s) = \begin{cases} f^*(s) - f^*(t), & 0<s<t,\\ 0, & t\le s<1,\end{cases}$$

and

$$(2.4)\qquad (f_t)^*(s) = \begin{cases} f^*(t), & 0<s<t,\\ f^*(s), & t\le s<1,\end{cases}$$

so, in particular, for each $t>0$,

$$(2.5)\qquad f^*(s) = (f^t)^*(s) + (f_t)^*(s) \qquad (0<s<1).$$

If f belongs to H^1, then since f_t is bounded and hence belongs to H^1, it is clear that $f^t = f-f_t$ is also in H^1. The H^1-norm may be estimated as follows.

LEMMA 2.1. *If* f *belongs to* H^1, *then*

$$(2.6)\qquad ||(f^t)^{\sim}||_{H^1} \le c(f^{**})^{**}(t) \qquad (0<t<1),$$

where c is a constant independent of f.

PROOF. It follows directly from (2.3) that

$$||f^t||_{L^1} = \int_0^t [f^*(s) - f^*(t)]ds = t[f^{**}(t) - f^*(t)]. \tag{2.7}$$

In order to estimate the L^1-norm of $(f^t)^\sim$ we use a technique employed by R. O'Neil-G. Weiss [4, p. 192]. Let

$$E = \{x: (f^t)^\sim(x) \geq 0\}, \quad F = \{x: (f^t)^\sim(x) < 0\}.$$

Then

$$\begin{aligned}
||(f^t)^\sim||_1 &= \frac{1}{2\pi}\int_E (f^t)^\sim(x)dx - \frac{1}{2\pi}\int_F (f^t)^\sim(x)dx \\
&= \frac{1}{2\pi}\int_{-\pi}^{\pi} (f^t)^\sim(x)\,[\chi_E(x) - \chi_F(x)]dx \\
&= \frac{-1}{2\pi}\int_{-\pi}^{\pi} (f^t)(x)\,[\tilde{\chi}_E - \tilde{\chi}_F](x)dx \\
&\leq \frac{1}{2\chi}\int_0^1 (f^t)^*(s)\,[(\chi_E^\sim)^*(s) + (\chi_F^\sim)^*(s)]ds
\end{aligned}$$

Hence, by (2.3), and the monotonicity of $\sinh^{-1}$,

$$\begin{aligned}
||(f^t)^\sim||_1 &\leq \frac{1}{2\pi}\int_0^t (f^*(s) - f^*(t))\,\frac{2}{\pi}\,[\sinh^{-1}(\frac{\sin(|E|/2) + \sin(|F|/2)}{2\tan \pi s/2})]ds \\
&\leq \frac{1}{\pi^2}\int_0^t (f^*(s) - f^*(t))\sinh^{-1}(\cot \pi s/2)ds \\
&\leq \frac{1}{\pi^2}\int_0^t f^*(s)\sinh^{-1}(\frac{1}{s})ds
\end{aligned}$$

An integration by parts gives

$$\begin{aligned}
||(f^t)^\sim||_1 &\leq \frac{1}{\pi^2}\int_0^t sf^{**}(s)\frac{1}{\sqrt{s^2+1}}\frac{ds}{s} \leq \frac{1}{\pi^2}\int_0^t f^{**}(s)ds \\
&= \frac{1}{\pi^2}\,t(f^{**})^{**}(t).
\end{aligned}$$

Combining this with (2.7) we obtain

$$||(f^t)^{\sim}||_{H^1} = ||f^t||_{L^1} + ||(f^t)^{\sim}||_{L^1}$$

$$\leq ct[f^{**}(t) + (f^{**})^{**}(t)] \leq ct(f^{**})^{**}(t).$$

THEOREM 2.2. *If* $0<\theta<1$, $1\leq q\leq\infty$ *and* $\theta = 1-1/p$, *then*

$$(H^1,L^\infty)_{\theta,q} = L^{pq} \tag{2.8}$$

with equivalent norms.

PROOF. Since the L^1-norm is dominated by the H^1-norm it is clear that $K(f;t;L^1;L^\infty) \leq K(f;t;H^1,L^\infty)$ so by Corollary 1.2,

$$(H^1,L^\infty)_{\theta,q} \subset (L^1,L^\infty)_{\theta,q} = L^{pq}$$

with a continuous embedding. Thus it remains only to show the reverse inclusion.

Fix $t > 0$ and write $f = f^t + f_t$ as in (2.1) and (2.2). Then

$$K(f;t;H^1,L^\infty) \leq ||f^t||_{H^1} + t||f_t||_{L^\infty}$$

so from (2.4) and (2.7) we obtain

$$\begin{aligned} K(f;t;H^1,L^\infty) &\leq ct(f^{**})^{**}(t) + tf^*(t) \\ &\leq ct(f^{**})^{**}(t) \end{aligned} \tag{2.9}$$

Hence from (2.5)

$$||f||_{(H^1,L^\infty)_{\theta,q}} \leq c(\int_0^\infty [t^{1-\theta} (f^{**})^{**}(t)]^q dt/t)^{1/q}$$

whence two applications of Hardy's inequality yields

$$||f||_{(H^1,L^\infty)_{\theta,q}} \leq c(\int_0^\infty [t^{1/p} f^*(t)]^q dt/t)^{1/q} = c||f||_{L^{pq}}$$

This establishes the reverse inclusion $L^{pq} \subset (H^1, L^\infty)_{\theta,q}$ and hence completes the proof.

Rivière and Sagher [5] were the first to establish (2.8). Shortly thereafter Fefferman, Rivière, and Sagher [3] discovered the K-functional for H^p and L^∞ within constants for $0<p<\infty$ by making heavy use of the then newly developed Fefferman-Stein H^p theory. In [1] equation (2.9) was established using $L \log L$ estimates and was incorporated into the framework of weak type inequalities. The proof presented in this paper, although simple, does not extend to $0<p<1$, but does have an easy generalization to $H^1(R^n)$ by using the analogous estimate for Riesz transforms that we stated in Theorem 1.3 [4, p. 193-196].

REFERENCES

[1] Bennett, C. - Sharpley, R., Weak-type inequalities for H^p and BMO. Proc. Symp. Pure Math. 35.1 (1979), 201-229.

[2] Butzer, P. L. - Berens, H., Semigroups of Operators and Approximation. Springer-Verlag, Berlin 1967.

[3] Fefferman, C. - Rivière, N. M. - Sagher, Y., Interpolation between H^p spaces: The real method. Trans. Amer. Math. Soc. 191 (1974), 75-81.

[4] O'Neil, R. - Weiss, G., The Hilbert transform and rearrangement of functions. Studia Math. 23 (1963), 189-198.

[5] Rivière, N. M. - Sagher, Y., Interpolation between L^∞ and H^1, the real method. J. Func. Anal. 14 (1973), 401-409.

THE FRANKLIN ORTHOGONAL SYSTEM AS UNCONDITIONAL BASIS IN $Re H^1$ AND VMO

Zbigniew Ciesielski
Mathematical Institute
Polish Academy of Sciences
Sopot

The aim of this lecture is to present a simplified proof of P. Wojtaszczyk's theorem that the Franklin orthogonal system is an unconditional basis in $Re H^1$.

1. Introduction

The Hardy space H^p, $1 \leqslant p < \infty$, on the unit disc $\{z \in \mathbf{C}: |z| < 1\}$ is a separable Banach space. The question of constructing a basis or unconditional basis in H^p spaces can be considered. According to the celebrated result of M. Riesz [13] (on the boundedness of the Hilbert transform) H^p with $1 < p < \infty$ is linearly isomorphic to L^p, and therefore in this case positive answers to the above questions can be given. The case $p = 1$ is more difficult. P. Billard [1] in 1971 constructed in H^1 a basis by means of the Haar orthogonal system. It was shown in 1976 by S. Kwapień and A. Pełczyński [10] that Billard's basis is not unconditional. In the same paper the authors pose the question of existence of an unconditional basis in H^1. A positive but non-constructive answer to this question was given by P. Maurey [11] in 1979. He simply proved that H^1 is linearly isomorphic to the dyadic (martingale) H^1 in which the Haar system is an unconditional basis. In 1980 L. Carleson [3] constructed an unconditional basis in H^1, and recently P. Wojtaszczyk [14] proved simply that the orthogonal Franklin system is an unconditional basis in $Re H^1$. For the sake of completeness we mention that H^∞ is non-separable and the basis questions make sense only in separable subspaces, e.g. in A, the space of functions analytic in the unit disc and continouus on its boundary. S.V. Bočkariov [2] in 1974 constructed an orthogonal basis in A by means of the Franklin system. On the other hand it is known that A has no unconditional basis (cf. A. Pełczyński [12],

p. 65).

Our contribution concerns the ReH^1 space. We simplify the "most delicate" place in the proof of P. Wojtaszczyk [14], Lemma 2 (see also L. Carleson [3], Lemma 4).

It is apparent (cf. [6] and [7]) that the method of proof is such that it can be extended to obtain simultaneous unconditional bases in H^1_m, i.e. the closure of polynomials with respect to the Sobolev norm in W^m_1 on a given polydisc.

2. Preliminaries

The main tools in this note, as well in Wojtaszczyk [14], are the Franklin orthogonal system and the atomic characterization of the space ReH^1. In both cases we recall the basic properties and the relevant results. Moreover, we shall use elementary properties of the Haar and Schauder bases which we recall below as well.

The dyadic partitions Π_n of $I := \langle 0,1\rangle$ are defined as follows: $\Pi_1 = \{0,1\}$, $\Pi_n = \{s_{n,j},\ j=0,\dots,n\}$ for $n = 2^\mu+k$, $\mu \geq 0$, $1 \leq k \leq 2^\mu$, with

$$s_{n,j} = \begin{cases} \dfrac{j}{2^{\mu+1}}, & j=0,\dots,2k \\[2ex] \dfrac{j-k}{2^\mu}, & j=2k+1,\dots,n. \end{cases}$$

It is also convenient to have the following notation: $t_0 = 0$, $t_1 = 1$, $t_n = \dfrac{2k-1}{2^{\mu+1}}$ and $(n) = \langle (k-1)/2^\mu, k/2^\mu\rangle$. Clearly $\Pi_n = \{t_0,\dots,t_n\}$ and $|(n)| \sim 1/n$. The spaces of all step functions (splines of order 1), say right-continuous, and of all polygonals i.e. piece-wise linear continuous functions (splines of order 2) corresponding to Π_n are denoted by $S^1_n(I)$ and $S^2_n(I)$, respectively. Clearly $S^r_n(I) \subset S^r_{n+1}(I) \subset L^2(I)$ and $\dim S^r_n(I) = n+r-1$, $r = 1,2$.

Using the $L^2(I)$ scalar product

$$(f,g) = \int_I fg$$

we now define, for given $r = 1,2$, an orthonormal system $\{f^{(n)}_n,\ n \geq 2-r\}$ such that: $f^{(r)}_{r-2} = 1$, $f^{(r)}_{n+1} \in S^r_{n+1}(I)$, $f^{(r)}_{n+1}$ is orthogonal to $S^r_n(I)$, and $\|f^{(r)}_n\|_2 = 1$. Now $\{h_n,\ n \geq 1\} := \{f^{(1)}_n,\ n \geq 1\}$ is the orthonormal Haar system and $\{f_n, n \geq 0\} := \{f_n^{(2)}, n \geq 0\}$ is the orthonormal Franklin system.

The properties of the Haar system we are going to use later on are the following:

H.1. $\{h_n, n \geq 1\}$ is a basis in $L^1(I)$.

H.2. For $f \in L^1(I)$ let

$$H_n f = \sum_{j=1}^{n} (f,h_j)h_j.$$

Then

$$H_n f(t) = \frac{1}{s_{n,j} - s_{n,j-1}} \int_{s_{n,j-1}}^{s_{n,j}} f(s)ds \quad \text{for } s_{n,j-1} < t \leq s_{n,j}, \quad j=1,\dots,n.$$

and (cf. [5], Theorem 7)

$$\| f - H_n f \|_1 \leq 6\omega_1^{(1)}(f;\tfrac{1}{n}), \quad n \geq 1;$$

where

$$\omega_1^{(1)}(f;\delta) = \sup_{0<h<\delta} \int_0^{1-h} |f(t+h) - f(t)|dt.$$

H.3. supp $h_n = (n)$.

H.4. If $n = 2^\mu + k$, $\mu \geq 0$, $1 \leq k \leq 2^\mu$, then

$$h_n(t) = \begin{cases} 2^{\mu/2} & \text{for } 2k-2 < 2^{\mu+1}t \leq 2k-1, \\ -2^{\mu/2} & \text{for } 2k-1 < 2^{\mu+1}t \leq 2k. \end{cases}$$

Introducing the integration operator $Gf(t) = \int_0^t f$ we now define the Schauder system as follows:

$$s_o := 1, \quad s_n := Gh_n, \quad n \geq 1.$$

These functions have the following properties:

S.1. $\{s_n, n \geq 0\}$ is a Schauder basis in $C(I)$ and for $f \in C(I)$ we obtain

$$f = f(0)s_o + \sum_{n=1}^{\infty} (\int_I h_n df)s_n.$$

S.2. Let $f \in C(I)$ and let

$$S_n f := f(0)s_o + \sum_{j=1}^{n} (\int_I h_j df)s_j.$$

Then $S_n f(t_j) = f(t_j)$ for $j=0,\dots,n$.

S.3. If f is absolutely continuous on I and D denotes the differentiation operator, then

$$DS_n f = H_n Df.$$

S.4. Let $\{N_{n,j}^{(2)}, j=0,\dots,n\}$ be the set of B-splines of order 2 corresponding to Π_n, i.e. $N_{n,j}^{(2)} \in S_n^2(I)$, supp $N_{n,j}^{(2)} = \langle s_{n,j-1}, s_{n,j}\rangle$ with $s_{n,-1} = 0$, $s_{n,n+1} = 1$ and $N_{n,j}^{(2)}(s_{n,i}) = \delta_{ij}$. Then, for $f \in C(I)$

$$S_n f = \sum_{j=o}^{n} f(s_{n,j})N_{n,j}^{(2)}.$$

It is now time to pass to the F r a n k l i n s y s t e m. For later convenience let, for a given dyadic interval (n) and an interval $J \subset I$, $t \in I$,

$$r(t,(n)) = \frac{\text{dist }(t,(n))}{|(n)|},$$

$$r(J,(n)) = \frac{\text{dist }(J,(n))}{|(n)|}.$$

Now, Theorem 1 of [4] and Theorem 1 of [5] imply:

F.1. $\{f_n, n \geq 0\}$ is a basis in $L^1(I)$.

F.2. There are constants q, $0 < q < 1$, and $C > 0$ such that for $n \geq 1$, $t, t_1, t_2 \in I$ we have

(i) $$|f_n(t)| \leq Cn^{1/2} q^{r(t,(n))},$$

(ii) $$|f_n(t_1) - f_n(t_2)| \leq Cn^{3/2} |t_1 - t_2| q^{r((t_1,t_2),(n))},$$

(iii) $$|Gf_n(t)| \leq Cn^{-1/2} q^{r(t,(n))}.$$

It should be noticed that (i) implies both (ii) and (iii).

In order to construct a basis (unconditional basis) in the Hardy space H^1 over the unit disc it is sufficient to do this in $\mathcal{R}e H^1$ and then by conplexification to pass to H^1. It has been shown by Wojtaszczyk [14] that the Franklin system is an unconditional basis in $\mathcal{R}e H^1$, and consequently the basis constructed by S.V. Bočkariov [2] in the disc algebra A is an unconditional basis in H^1.

Following the work of R.R. Coifman and Guido Weiss [8] we recall their new real variable characterization of ReH^1. A function $a(t)$, $t \in I$, is called an **a t o m** if either $a(t) \equiv 1$ or if it is measurable and such that:

<u>A.(i).</u> supp a is contained in an interval $J \subset I$,

<u>A.(ii).</u> $|a(t)| \leqslant |J|^{-1}$ for $t \in I$,

<u>A.(iii)</u> $(a,1) = 0$.

ReH^1 can be identified with the set of functions in H^1 with imaginary part vanishing at zero. The space ReH^1 with the norm induced from H^1 is a Banach space. This Banach space has the following description:

$f \in ReH^1$ <u>if and only if</u>

$$f = \sum_j \lambda_j a_j, \quad \sum_j |\lambda_j| < \infty, \tag{2.1}$$

<u>where the a_j's are some atoms. Moreover, the infimum of $\sum|\lambda_j|$ taken over all such decompositions defines an equivalent norm in ReH^1 and it is denoted by</u> $\|f\|_{ReH^1}$.

The dual space to ReH^1, i.e. $(ReH^1)^*$, was characterized by C. Fefferman [9] as the space of B o u n d e d M e a n O s c i l l a t i o n (BMO) functions. A function $\ell \in L^1(I)$ is said to be in BMO if

$$\|\ell\|_{BMO} = |(\ell,1)| + \sup_J \frac{1}{|J|} \int_J |\ell - m_J(\ell)| < \infty,$$

where the sup is taken over all subintervals $J \subset I$, and $m_J(\ell) = |J|^{-1}\int_J \ell$. Now, to each $L \in (ReH^1)^*$ there exists a unique $\ell \in BMO$ (the correspondence $L \to \ell$ is linear) such that

$$L(f) = (\ell,f), \qquad |(\ell,f)| \leqslant \|\ell\|_{BMO}\|f\|_{ReH^1}$$

holds for $f \in L^2(I) \subset ReH^1$. The extension of L to ReH^1 is denoted by the same symbol (ℓ,f). Finally we define the space of V a n i s h i n g M e a n O s c i l l a t i o n (VMO) as a subspace of BMO of those ℓ for which

$$\int_J |\ell - m_J(\ell)| = o(|J|) \quad \text{as } |J| \to 0.$$

The norm in VMO is the one induced from BMO. In this setting we have $(VMO)^* = ReH^1$ (cf. [8], Thm.4.1).

3. Characterization of the BMO and VMO spaces.

It is convenient to introduce for a given sequence of real numbers $(a_0, a_1, \dots)$ the quantities

$$A_n := \Big(n \sum_{(m)\subset(n)} |a_m|^2\Big)^{1/2}, \qquad n \geq 2.$$

THEOREM 1. (P. Wojtaszczyk [14]). *Let* $f \in L^1(I)$ *and let*

(3.1) $$f = \sum_{n=o}^{\infty} a_n f_n.$$

Then,

$f \in BMO$ *iff* $A_n = \mathcal{O}(1)$.

$f \in VMO$ *iff* $A_n = o(1)$.

Following Wojtaszczyk we know that this result follows from the following three lemmas.

LEMMA 1. *If* $A_n = \mathcal{O}(1)[o(1)]$ *and* f *is given by* (3.1), *then* $f \in BMO[VMO]$.

LEMMA 2. *For the Franklin functions we have*

$$\|f_n\|_{\mathcal{R}e H^1} = \mathcal{O}(n^{-1/2}).$$

LEMMA 3. *If* $f \in BMO[VMO]$, *then* $A_n = \mathcal{O}(1)[o(1)]$.

The proofs of Lemmas 1 and 3 as given in [14] are simple and they will not be repeated here. We mention only that following Carleson's way [3] of decomposing the sum (3.1) into three parts one finds that Lemma 1 essentially follows from F.2.(ii). Similarly, using Lemma 2 and F.2.(ii), one proves Lemma 3.

PROOF OF LEMMA 2. Let $N = |(n)|^{-1}$ for $n \geq 2$. Then by H.1

(3.2) $$f_n = H_N f_n + \sum_{j=N+1}^{\infty} (f_n, h_j)h_j.$$

Since $(f_n, 1) = 0$ it follows that $Gf_n(0) = Gf_n(1) = 0$. Thus properties S.3 and S.4 give

$$H_N f_n = DS_N Gf_n = D \sum_{j=0}^{N} Gf_n(s_{N,j}) N_{N,j}^{(2)}$$

(3.3)

$$= \sum_{j=1}^{N-1} Gf_n(s_{N,j}) DN_{N,j}^{(2)} = \sum_{j=1}^{N-1} \lambda_j a_j,$$

where $\lambda_j = 2f_n(s_{N,j})$ and $a_j = \frac{1}{2} DN_{N,j}^{(2)}$. These a_j's are atoms. Property F.2.(iii) now implies

$$\sum_{j=1}^{N-1} |\lambda_j| = \mathcal{O}(N^{-1/2}). \tag{3.4}$$

The second sum of the right hand side of (3.2) can be written as

$$\sum_{j=N+1}^{\infty} \lambda_j a_j \tag{3.5}$$

with $\lambda_j = (f_n, h_j)|(j)|^{1/2}$ and $a_j = h_j|(j)|^{-1/2}$. It now follows by H.3 and H.4 that these a_j's are atoms. On the other hand, by H.2, F.2.(i),

$$\sum_{j=N+1}^{\infty} |\lambda_j| = \sum_{\nu=0}^{\infty} \sum_{j=2^\nu N+1}^{2^{\nu+1}N} |\lambda_j|$$

$$= \sum_{\nu=0}^{\infty} \| H_{2^{\nu+1}N} f_n - H_{2^\nu N} f_n \|_1$$

(3.6)

$$\leq C \sum_{\nu=0}^{\infty} \omega_1^{(1)}(f_n; 1/2^\nu N)$$

$$\leq C'(\mathrm{var}\, f_n)/N = \mathcal{O}(N^{-1/2}).$$

To obtain the last but one estimation we have used the inequality $\omega_1^{(1)}(f;\delta) \leq 3\delta \operatorname{var}_I f$. Combining the formulas and estimates (3.2)-(3.6) we complete the proof.

4. Unconditional basis in $\mathcal{Re}H^1$ and in VMO.

THEOREM 2. The Franklin orthonormal system is an unconditional basis both in $\mathcal{Re}H^1$ and in VMO.

PROOF. Using the notation of Section 3 and applying Theorem 1 we obtain uniformly in n

$$\| \sum_{j=0}^{n} \pm b_j f_j \|_{ReH^1} \sim \sup \{ \sum_{j=0}^{n} b_j a_j : |a_o| \leq 1, \ |a_1| \leq 1, \ A_m \leq 1, \ m \geq 2 \}$$

and

$$\| \sum_{j=0}^{n} \pm a_j f_j \|_{BMO} \sim \| \sum_{j=0}^{n} a_j f_j \|_{BMO}.$$

REFERENCES

[1] Billard, P., Bases dans H et bases de sous espaces de dimension finie dans A. Linear Operators and Approximation. (ISNM, vol. 20) Edited by P.L. Butzer, J.-P. Kahane and B.Sz. - Nagy. Proceedings Conf. Oberwolfach, Aug. 14-22, 1971. Birkhäuser Verlag, Basel 1972.

[2] Bočkariov, S.V., Existence of bases in the space of analytic functions, and some properties of the Franklin system. Mat. Sbornik 95 (137), (1974), 3-18 (in Russian).

[3] Carleson, L., An explicit unconditional basis in H^1. Institut Mittag-Leffler. Report No. 2, 1980.

[4] Ciesielski, Z., Properties of the orthonormal Franklin system. Studia Math. 23 (1963), 141-157.

[5] Ciesielski, Z., Properties of the orthonormal Franklin system, II. Studia Math. 27 (1966), 289-323.

[6] Ciesielski, Z., Bases and approximation by splines. Proc. International Congress of Mathematicians. Vancouver 1974, p. 47-51.

[7] Ciesielski, Z., Constructive function theory and spline systems. Studia Math. 53 (1975), 277-302.

[8] Coifman, R.R. and Weiss, G., Extensions of Hardy spcaces and their use in analysis. Bull. Amer. Math. Soc. 83 (1977), 569-645.

[9] Fefferman, C., Characterization of bounded mean oscillation. Bull. Amer. Math. Soc. 77 (1971), 587-588.

[10] Kwapień, S. and Pełczyński, A., Some linear topological properties of the Hardy spaces H^p. Compositio Math. 33 (1976), 261-288.

[11] Maurey, B., paper to appear in Acta Mathematica 1980.

[12] Pełczyński, A., Banach Spaces of Analytic Functions and Absolutely Summing Operators. Conference Board of the Mathematical Sciences. Regional Conference Series in Mathematics No. 30. 1977. Amer. Math. Soc., Providence.

[13] Riesz, M., Sur les fonctions conjuguées. Math. Z. 27 (1927), 218-244.

[14] Wojtaszczyk, P., The Franklin system is an unconditonal basis in H_1. Submitted in 1980 for Ark. Math.

III Abstract Harmonic Analysis

BESSEL POTENTIAL SPACES AND GENERALIZED LIPSCHITZ SPACES ON LOCAL FIELDS

H. Ombe and C. W. Onneweer[1]
Department of Mathematics
University of New Mexico
Albuquerque, NM 87131

In this paper we prove an embedding theorem for Bessel potential spaces and generalized Lipschitz spaces in $L_r(K)$, $2 < r < \infty$, where K is a local field. This theorem complements a result of the second author who has proved a similar embedding theorem for such spaces in $L_r(K)$ when $1 < r \leq 2$.

1. Notation and Definitions

In this paper N, Z and R will denote the natural numbers, the integers and the real numbers, respectively. Let K be a local field, that is, K is a locally compact, non-discrete, totally disconnected topological field. Let dx denote a Haar measure on K^+, K considered as an additive group. For each $a \in K$ with $a \neq 0$ the measure $d(ax)$ is again a Haar measure on K^+. Thus $d(ax) = \|a\|\, dx$ for some $\|a\| \in R$. If $\|0\|$ is defined by $\|0\| = 0$ then it can be shown that the function $a \to \|a\|$ from K to R defines a (non-archimedean) norm on K. This norm has the properties that $\|ab\| = \|a\| \cdot \|b\|$ and $\|a+b\| \leq \max\{\|a\|, \|b\|\}$ for all $a,b \in K$.

Next, let $P_0 = \{x \in K; \|x\| \leq 1\}$ and $P_1 = \{x \in K; \|x\| < 1\}$. Then P_0 is a ring in K, P_1 is a maximal ideal in P_0 and $P_0/P_1 \cong GF(q)$, the finite field of q elements, where q is a power of some prime number p. For each $k \in Z$ let

$$P_k = \{x \in k; \|x\| \leq q^{-k}\} .$$

[1]The research of the second author was partially supported by NSF grants MCS 79-01957 and MCS 80-01870.

Then (i) each P_k is a compact open subgroup of K, (ii) $P_{k+1} \subsetneq P_k$, (iii) $\bigcap_{k\in Z} P_k = \{0\}$ and $\bigcup_{k\in Z} P_k = K$, (iv) if m is the Haar measure on K^+ normalized by $m(P_0) = 1$ then $m(P_k) = q^{-k}$ for each $k \in Z$. From here on m or dx will denote this particular Haar measure on K^+.

To describe the dual group $\hat{K}$ of K^+, choose a character $\chi \in \hat{K}$ so that $\chi(x) = 1$ for $x \in P_0$ and $\chi(x) \neq 1$ for some $x \in P_{-1}$. Then $\hat{K} = \{\chi_y; y \in K\}$, where $\chi_y(x) = \chi(yx)$. If $\hat{f}(\chi_y)$ is defined the notation $\hat{f}(y)$ will be used for $\hat{f}(\chi_y)$. For each $k \in Z$ let the function Δ_k on K be defined by

$$\Delta_k(x) = \begin{cases} q^k & \text{if } x \in P_k, \\ 0 & \text{if } x \notin P_k. \end{cases}$$

Then $\|\Delta_k\|_1 = 1$ and

$$(\Delta_k)\hat{}\,(y) = \begin{cases} 1 & \text{if } y \in P_{-k}, \\ 0 & \text{if } y \notin P_{-k}. \end{cases}$$

We now present the definitions of two spaces of functions which were given first by Taibleson in [3] and [4], respectively. These definitions can also be found in [5]. We first observe that for each $\alpha > 0$ there exists a function $G_\alpha \in L_1(K)$ such that $\|G_\alpha\|_1 = 1$ and

$$(G_\alpha)\hat{}\,(y) = \begin{cases} 1 & \text{if } y \in P_0, \\ \|y\|^{-\alpha} & \text{if } y \notin P_0. \end{cases}$$

DEFINITION 1. <u>For</u> $\alpha > 0$ <u>and</u> $1 \leq r < \infty$ <u>the Bessel potential space</u> $L(r,\alpha)$ <u>is defined by</u>

$$L(r,\alpha) = \{f \in L_r(K); f = G_\alpha * g \ \underline{\text{for some}}\ g \in L_r(K)\}.$$

If we set $\|f\|_{L(r,\alpha)} = \|g\|_r$ when $f = G_\alpha * g$ then $L(r,\alpha)$ is a Banach space with respect to the norm $\|\cdot\|_{L(r,\alpha)}$.

In order to study the smoothness properties of the functions in $L(r,\alpha)$ we introduce the generalized Lipschitz spaces (or Besov spaces) $\Lambda(\alpha,r,s)$ on K.

DEFINITION 2. <u>Let</u> $1 \leq r \leq \infty$, $1 \leq s < \infty$ <u>and</u> $\alpha > 0$. <u>Then</u>

$$\Lambda(\alpha,r,s) = \{f \in L_r(K) ; {}_K\int (\|y\|^{-\alpha} \|f_y - f\|_r)^s \|y\|^{-1} dy < \infty\} .$$

Here f_y denotes a translate of f : $f_y(x) = f(x-y)$. In [4] Taibleson proved that the following two expressions define equivalent norms on $\Lambda(\alpha,r,s)$:

(a) $\|f\|_r + ({}_K\int (\|y\|^{-\alpha}\|f_y - f\|_r)^s \|y\|^{-1} dy)^{1/s}$,

(b) $\|f\|_r + (\sum_{k=-\infty}^{\infty} \|q^{k\alpha}(\Delta_k - \Delta_{k+1}) * f\|_r^s)^{1/s}$.

We shall denote the second of these norms by $\|\cdot\|_{\Lambda(\alpha,r,s)}$.

2. The Embedding Theorem.

In [1, Theorem 7] we proved that for $1 < r \le 2$ we have

$$\Lambda(\alpha,r,r) \subset L(r,\alpha) \subset \Lambda(\alpha,r,2) ,$$

where $\subset$ denotes a continuous embedding mapping. In this paper we will prove the following complement of this result.

THEOREM. If $\alpha > 0$ and $2 < r < \infty$ then

$$\Lambda(\alpha,r,2) \subset L(r,\alpha) \subset \Lambda(\alpha,r,r) .$$

Before giving a proof of the theorem we review some results needed later on. We begin by stating the relevant facts about the generalized Littlewood-Paley function $G_r(f)$ of a function $f \in L_r(K)$. For $f \in L_{1,loc}(K)$ and $1 \le r \le \infty$ let

$$G_r(f)(x) = (\sum_{k=-\infty}^{\infty} |(\Delta_k - \Delta_{k+1}) * f(x)|^r)^{1/r} .$$

Then we have the following.

(1) If $f \in L_r(K)$, $1 < r < \infty$, then $G_2(f) \in L_r(K)$ and the norms $\|f\|_r$ and $\|G_2(f)\|_r$ are equivalent.

(2) If $f \in L_r(K)$, $2 \le r < \infty$, then $G_r(f) \in L_r(K)$ and $\|G_r(f)\|_r \le C\|f\|_r$.

A proof of (1) and (2) can be found in [4] or [5].

We now prove that the Bessel potential spaces $L(r,\alpha)$ can be identified with the spaces $\mathcal{D}(D_r^{[\alpha]})$ of strongly differentiable functions in $L_r(K)$ of order $\alpha > 0$. We repeat here two definitions that can be found in [1] and [2], respectively.

DEFINITION 3. For $f \in L_r(K)$, $1 \le r < \infty$, $\alpha > 0$, $m \in N$ and $x \in K$ let

$$E_{m,\alpha}f(x) = \sum_{\ell=-\infty}^{m-1} (q^{(\ell+1)\alpha} - q^{\ell\alpha})(f - \Delta_\ell * f)(x) .$$

If $\lim_{m\to\infty} E_{m,\alpha}f$ exists in $L_r(K)$ the limit is called the strong derivative of order α of f, the limit will be denoted by $D_r^{[\alpha]}f$. Also, we set $\mathcal{D}(D_r^{[\alpha]}) = \{f \in L_r(K); D_r^{[\alpha]}f \text{ exists}\}$.

For later reference we now state some results that were proved in [1].

(3) If $1 < r < \infty$ then $\mathcal{D}(D_r^{[\alpha]})$ is dense in $L_r(K)$.

(4) If $1 \le r \le 2$ then $f \in \mathcal{D}(D_r^{[\alpha]})$ if and only if there exists a $g \in L_r(K)$ such that $\hat{g}(y) = \|y\|^\alpha \hat{f}(y)$ a.e.; moreover $g = D_r^{[\alpha]}f$.

(5) If $1 \le r \le 2$ then $D_r^{[\alpha]}$ is a closed linear operator and, hence, $\mathcal{D}(D_r^{[\alpha]})$ is a Banach space with respect to the norm

$$\|f\|_{D(r,\alpha)} = \|f\|_r + \|D_r^{[\alpha]}f\|_r .$$

DEFINITION 4. If for $f \in L_r(K)$, $1 \le r < \infty$, and $\alpha > 0$ and r' satisfying $r + r' = rr'$ there exists a $g \in L_r(K)$ so that for all $\phi \in \mathcal{D}(D_{r'}^{[\alpha]})$ we have

$$\int_K f(x) D_{r'}^{[\alpha]}\phi(x)dx = \int_K g(x)\phi(x)dx ,$$

we say that f is differentiable of order α in $L_r(K)$ in the weak sense. We call g the weak derivative of f, denoted by $g = w\text{-}D_r^{[\alpha]}f$. Also, we set $\mathcal{D}(w\text{-}D_r^{[\alpha]}) = \{f \in L_r(K); w\text{-}D_r^{[\alpha]}f \text{ exists}\}$.

In [2] we proved that if $\alpha > 0$ and $2 \le r < \infty$ then $\mathcal{D}(D_r^{[\alpha]}) = \mathcal{D}(w\text{-}D_r^{[\alpha]})$. Moreover, if $f \in \mathcal{D}(D_r^{[\alpha]})$ then its weak and strong derivatives of order α are equal a.e. As a simple consequence we can prove that (5) holds for all r with $1 \le r < \infty$.

LEMMA. Let $\alpha > 0$ and $1 \le r < \infty$. Then $\mathcal{D}(D_r^{[\alpha]}) = L(r,\alpha)$ and the norms in these spaces are equivalent.

PROOF. For $1 \le r \le 2$ the lemma was proved in [1, Theorem 6]. So we shall assume that $2 < r < \infty$. Let $f \in L(r,\alpha)$ and assume $f = G_\alpha * f_\alpha$ for some $f_\alpha \in L_r(K)$. For any $\phi \in \mathcal{D}(D_{r'}^{[\alpha]})$, where $r + r' = rr'$, we have $\phi \in L(r',\alpha)$ and, hence, $\phi = G_\alpha * \phi_\alpha$ for some $\phi_\alpha \in L_{r'}(K)$. Also, in [1, page 161] it is shown that $D_{r'}^{[\alpha]}\phi = \phi_\alpha * \mu_\alpha$, where $\mu_\alpha = \delta_0 - \Delta_0 + D_1^{[\alpha]}\Delta_0$ and δ_0 is the Dirac δ-measure concentrated at $0 \in K$. Therefore, since both G_α and μ_α are inversion-invariant, we have

$$\int_K f(x)D_{r'}^{[\alpha]}\phi(x)dx = \int_K G_\alpha * f_\alpha(x)\phi_\alpha * \mu_\alpha(x)dx$$

$$= \int_K G_\alpha * \phi_\alpha(x)f_\alpha * \mu_\alpha(x)dx$$

$$= \int_K \phi(x)f_\alpha * \mu_\alpha(x)dx .$$

Since $f_\alpha * \mu_\alpha \in L_r(K)$, we see that $f \in \mathcal{D}(w\text{-}D_r^{[\alpha]}) = \mathcal{D}(D_r^{[\alpha]})$ and $D_r^{[\alpha]}f = f_\alpha * \mu_\alpha$. In addition,

$$\|f\|_{D(r,\alpha)} = \|f\|_r + \|D_r^{[\alpha]}f\|_r$$

$$= \|G_\alpha * f_\alpha\|_r + \|f_\alpha * \mu_\alpha\|_r$$

$$\le C\,\|f_\alpha\|_r = C\,\|f\|_{L(r,\alpha)} .$$

Conversely, let $f \in \mathcal{D}(D_r^{[\alpha]})$. Define the function h by

$$h = D_r^{[\alpha]}f + (\Delta_0 - D_1^{[\alpha]}\Delta_0) * f .$$

Clearly, $h \in L_r(K)$. We shall show that $f \in L(r,\alpha)$ by proving that $f = G_\alpha * h$. Take any $\phi \in \mathcal{D}(D_{r'}^{[\alpha]})$. Using the characterization for $\mathcal{D}(D_{r'}^{[\alpha]})$ given in (4) we can easily show that $G_\alpha * \phi \in \mathcal{D}(D_{r'}^{[\alpha]})$. Furthermore, we have

$$\int_K G_\alpha * h(x)\phi(x)dx$$

$$= \int_K h(x)G_\alpha * \phi(x)dx$$

$$= \int_K D_r^{[\alpha]}f(x)G_\alpha * \phi(x)dx + \int_K (\Delta_0 - D_1^{[\alpha]}\Delta_0) * f(x)G_\alpha * \phi(x)dx$$

$$= \int_K f(x)D_{r'}^{[\alpha]}(G_\alpha*\phi)(x)dx + \int_K f(x)(\Delta_0 - D_1^{[\alpha]}\Delta_0) * G_\alpha * \phi(x)dx .$$

A computation of the Fourier transform, in which we use (4), and an application of the Uniqueness Theorem shows that

$$D_{r'}^{[\alpha]}(G_\alpha * \phi) + (\Delta_0 - D_1^{[\alpha]}\Delta_0) * G_\alpha * \phi = \phi .$$

Therefore,

$$\int_K G_\alpha * h(x)\phi(x)dx = \int_K f(x)\phi(x)dx \ .$$

Since, according to (3), $\mathcal{D}(D_{r'}^{[\alpha]})$ is dense in $L_{r'}(K)$, we may conclude that $G_\alpha * h(x) = f(x)$ a.e., that is, $f \in L(r,\alpha)$. Finally, we observe that

$$\begin{aligned}\|f\|_{L(r,\alpha)} = \|h\|_r &\le \|D_r^{[\alpha]}f\|_r + \|(\Delta_0 - D_1^{[\alpha]}\Delta_0) * f\|_r \\ &\le \|D_r^{[\alpha]}f\|_r + \|\Delta_0 - D_1^{[\alpha]}\Delta_0\|_1 \ \|f\|_r \\ &\le C\,\|f\|_{D(r,\alpha)} \ .\end{aligned}$$

This completes the proof of the lemma.

PROOF OF THE THEOREM. Let $f \in \Lambda(\alpha,r,2)$. For each $n \in Z$ define f_n by $f_n = f * \Delta_n$. Then $\lim_{n\to\infty} f_n = f$ in $L_r(K)$. Also, according to [1, Theorem 1(b)], $f_n \in \mathcal{D}(D_r^{[\alpha]})$ and

$$D_r^{[\alpha]}f_n = \sum_{\ell=-\infty}^{n} q^{\ell\alpha}(\Delta_\ell - \Delta_{\ell-1}) * f \ .$$

For $m < n$ let $f_{m,n} = D_r^{[\alpha]}f_n - D_r^{[\alpha]}f_m$. Then, according to (1), we have

$$\|f_{m,n}\|_r \le C\|\mathcal{G}_2(f_{m,n})\|_r \ .$$

Also, a simple computation, compare [1, page 163], shows that

$$\mathcal{G}_2(f_{m,n}) = \Big(\sum_{\ell=m}^{n-1} |q^{(\ell+1)\alpha}(\Delta_\ell - \Delta_{\ell+1}) * f|^2\Big)^{1/2} \ .$$

Therefore, it follows from the generalized Minkowski inequality that

$$\begin{aligned}\|\mathcal{G}_2(f_{m,n})\|_r &= \Big(\int_K \Big(\sum_{\ell=m}^{n-1} |q^{(\ell+1)\alpha}(\Delta_\ell - \Delta_{\ell+1}) * f(x)|^2\Big)^{r/2} dx\Big)^r \\ &\le \Big\{\sum_{\ell=m}^{n-1}\Big(\int_K \big(|q^{(\ell+1)\alpha}(\Delta_\ell - \Delta_{\ell+1}) * f(x)|^2\big)^{r/2}\Big)^{2/r}\Big\}^{1/2} \\ &= \Big(\sum_{\ell=m}^{n-1} \|q^{(\ell+1)\alpha}(\Delta_\ell - \Delta_{\ell+1}) * f\|_r^2\Big)^{1/2} \ .\end{aligned}$$

Since $f \in \Lambda(\alpha,r,2)$, we see that for each $\varepsilon > 0$ there exists an $M \in N$ so that for all $m,n > N$ we have $\|\mathcal{G}_2(f_{m,n})\|_r < \varepsilon$. Hence

$\{D_r^{[\alpha]} f_m\}_{m\varepsilon N}$ is a Cauchy sequence in $L_r(K)$. Consequently, $\lim_{m\to\infty} D_r^{[\alpha]} f_m$ exists in $L_r(K)$ and, because $D_r^{[\alpha]}$ is a closed operator, we may conclude that $f \varepsilon \mathcal{D}(D_r^{[\alpha]})$. Furthermore, since

$$D_r^{[\alpha]} f_n = \lim_{m\to-\infty} f_{m,n} ,$$

we see that for each $n \varepsilon Z$

$$\|D_r^{[\alpha]} f_n\|_r \le C\Big(\sum_{\ell=-\infty}^{n-1} \| q^{(\ell+1)\alpha}(\Delta_\ell - \Delta_{\ell+1}) * f\|_r^2\Big)^{1/2}$$

$$\le C(\|f\|_{\Lambda(\alpha,r,2)} - \|f\|_r) .$$

Therefore,

$$\|D_r^{[\alpha]} f\|_r \le C(\|f\|_{\Lambda(\alpha,r,2)} - \|f\|_r) .$$

and, hence, $\|f\|_{L(r,\alpha)} \le C\,\|f\|_{\Lambda(\alpha,r,2)}$.

Conversely, assume $f \varepsilon L(r,\alpha)$. We first show that for each $k \varepsilon Z$ we have

(6) $$q^{k\alpha}(\Delta_k - \Delta_{k+1}) * f = q^{-\alpha}(\Delta_k - \Delta_{k+1}) * w\text{-}D_r^{[\alpha]} f .$$

Take any $\phi \varepsilon \mathcal{D}(D_{r'}^{[\alpha]})$. A comparison of the Fourier transforms shows that

$$q^{k\alpha}(\Delta_k - \Delta_{k+1}) * \phi = q^{-\alpha}(\Delta_k - \Delta_{k+1}) * D_{r'}^{[\alpha]} \phi .$$

Therefore,

$$\int_K q^{k\alpha}(\Delta_k - \Delta_{k+1}) * f(x)\phi(x)dx$$

$$= \int_K f(x) q^{k\alpha}(\Delta_k - \Delta_{k+1}) * \phi(x)dx$$

$$= \int_K f(x) q^{-\alpha}(\Delta_k - \Delta_{k+1}) * D_{r'}^{[\alpha]}\phi(x)dx$$

$$= \int_K f(x) D_{r'}^{[\alpha]}(q^{-\alpha}(\Delta_k - \Delta_{k+1}) * \phi)(x)dx$$

$$= \int_K w\text{-}D_r^{[\alpha]} f(x) q^{-\alpha}(\Delta_k - \Delta_{k+1}) * \phi(x)dx$$

$$= \int_K q^{-\alpha}(\Delta_k - \Delta_{k+1}) * w\text{-}D_r^{[\alpha]} f(x)\phi(x)dx .$$

An application of the Hahn-Banach theorem implies that (6) holds. Next, applying (6), Fubini's Theorem and (2), respectively, we see that

$$\begin{aligned}\|f\|_{\Lambda(\alpha,r,r)} &= \|f\|_r + \Big(\sum_{\ell=-\infty}^{\infty} \|q^{\ell\alpha}(\Delta_\ell-\Delta_{\ell+1}) * f\|_r^r\Big)^{1/r} \\ &= \|f\|_r + q^{-\alpha}\Big(\sum_{\ell=-\infty}^{\infty} \int_K |(\Delta_\ell-\Delta_{\ell+1}) * \text{w-}D_r^{[\alpha]}f(x)|^r dx\Big)^{1/r} \\ &= \|f\|_r + q^{-\alpha}\Big(\int_K |\sum_{\ell=-\infty}^{\infty} (\Delta_\ell-\Delta_{\ell+1}) * \text{w-}D_r^{[\alpha]}f(x)|^r dx\Big)^{1/r} \\ &= \|f\|_r + q^{-\alpha}\| G_r(\text{w-}D_r^{[\alpha]}f)\|_r \\ &\le \|f\|_r + C\,\|\text{w-}D_r^{[\alpha]}f\|_r \\ &\le C\,\|f\|_{D(r,\alpha)} \le C\,\|f\|_{L(r,\alpha)} < \infty\ .\end{aligned}$$

Thus $f \in \Lambda(\alpha,r,r)$. This completes the proof of the theorem.

REMARK. The first author recently proved that the inclusion relations stated in the main theorem are sharp. For each $k \in Z$ let $y(k)$ denote the element p^{-k} in K where p is a fixed element of $P_1\backslash P_2$. The following holds.

(i) Assume $2 \le r < \infty$ and $\alpha > 0$. Let f be defined by

$$f(x) = \Delta_0(x) \sum_{\ell=1}^{\infty} q^{-\ell\alpha}\ell^{-1/2}\chi_{y(\ell)}(x)\ .$$

Then $f \in \Lambda(\alpha,r,s)$ for $s > 2$, but $f \notin L(r,\alpha)$. Thus

$$\Lambda(\alpha,r,s) \not\subset L(r,\alpha) \quad \text{for} \quad s > 2\ .$$

(ii) Assume $2 \le r < \infty$ and $\alpha > 0$. Let g be defined by

$$g(x) = \sum_{\ell=1}^{\infty} q^{-\ell(\alpha+1-r^{-1})}\ell^{-s^{-1}}(\Delta_\ell-\Delta_{\ell-1})(x)\ .$$

Then $g \in L(r,\alpha)$ and $g \notin \Lambda(\alpha,r,s)$ for $s < r$. Thus $L(r,\alpha) \not\subset \Lambda(\alpha,r,s)$ if $s < r$.

A detailed proof of (i) and (ii) will appear elsewhere.

REFERENCES

[1] Onneweer, C.W., Fractional differentiation and Lipschitz spaces on local fields. Trans. Amer. Math. Soc. 258 (1980), 155-165.

[2] Onneweer, C.W., Saturation results for operators defining fractional derivatives on local fields. To appear in the Proceedings of the Conference on Functions, Series, Operators, Budapest 1980.

[3] Taibleson, M.H., Harmonic analysis on n-dimensional vector spaces over local fields I. Math. Ann. 176 (1968), 191-207.

[4] Taibleson, M.H., Harmonic analysis on n-dimensional vector spaces over local fields II. Math. Ann. 186 (1970), 1-19.

[5] Taibleson, M.H., Fourier analysis on local fields. Mathematical Notes, Princeton University Press, Princeton, N.J. 1975.

ON THE MINIMUM NORM PROPERTY OF THE FOURIER PROJECTION IN L^1-SPACES

P.V. Lambert
Department of Mathematics
Limburgs Universitair Centrum
B - 3610 Diepenbeek

Let G be a compact abelian group, $\hat{G}$ its dual, N a finite part of $\hat{G}$, and E_N the (complex) linear hull of the characters e_γ, $\gamma \in N$. The Fourier projection $x \to x * k$, where k is the Dirichlet kernel $\sum_{\gamma \in N} e_\gamma$, has minimum norm among all projections $L^1(G) \to E_N$. We proved in [4] that the Fourier projection is the unique minimum norm projection $L^1(G) \to E_N$, whenever the kernel k is determined up to a constant factor as an element of E_N by its roots in G. Hence if G is the circle group T and E_N the space of trignometric polynomials $\sum_{j=-n}^{n} c_j e^{ijt}$, $t \in T$, the Fourier projection is characterized by its minimum norm. On the other side we also showed there that the convex set C_k^1 of minimum norm projections $L^1(G) \to E_N$ can have arbitrarily large dimension by suitable choices of G and E_N. In this paper we prove a partial converse to those results: if the kernel k is real and if the Fourier projection $L^1(G) \to E_N$ is characterized by its minimum norm, then the kernel k is continuously determined up to a constant factor as an element of E_N by its roots in G; moreover, when the real kernel k does not satisfy this condition, we give a lower bound for dimension (C_k^1), which can reach the power of the continuum. These results as those in [4] are valid and written for more classes of operators then only the class of projectors.

1. General Setting and Generalization of D.L. Berman's Relation

DEFINITION 1.1. <u>We say that the compact abelian</u> T_o<u>-group</u> G <u>operators continuously on the Banach space</u> E, <u>when a mapping</u>:

$$(1.1) \qquad \varphi : (g,x) \to gx : G \times E \to E$$

is given, which is separately continuous in g *and* x *and such that* $g \to (x \to gx)$ *is a representation of* G *in* Lin Aut (E). *If furthermore for each* g *the mapping* $\varphi_g : x \to gx$ *is an isometry, we say that* G *operates continuously and isometrically on* E.

REMARK 1.1. It is a corollary of the Banach - Steinhaus theorem that these assumptions imply the continuity of $(g,x) \to gx$.

DEFINITION 1.2. *For every* $\gamma \in \hat{G}$, *where* $\hat{G}$ *is the dual of* G, *the subspace* E_γ *of* E *is defined by:*

$$(1.2) \qquad E_\gamma = \{x \in E : gx = (-g,\gamma)x \text{ for every } g \in G\}.$$

The mapping $S_\gamma : E \to E_\gamma$ *is then defined by:*

$$(1.3) \qquad S_\gamma x := x_\gamma := x * e_\gamma := \int_G (gx)(g,\gamma)dm(g)$$

strongly, where m *is the normalized Haar measure of* G *and* e_γ *is the character* γ *considered as a mapping* $G \to \mathbb{C}$, i.e., $e_\gamma(g) = (g,\gamma)$.

REMARK 1.2. $x_\gamma \in E_\gamma$: first of all the integral converges strongly, furthermore

$$hx_\gamma = h \int_G (gx)(g,\gamma)dm(g) = \int_G (hgx)(g,\gamma)dm(g)$$

$$= \int_G (fx)(f-h,\gamma)dm(f) = (-h,\gamma)x_\gamma \qquad \text{if } h \in G.$$

DEFINITION 1.3. *For a finite subset* N *of* $\hat{G}$ *and a finite subset* $\{c_\gamma : \gamma \in N\}$ *of* $\mathbb{C} \setminus \{0\}$ *let* $E_N = \sum_{\gamma \in N} E_\gamma$ *and* $k = \sum_{\gamma \in N} c_\gamma e_\gamma$. *We then define the mapping* $S_k : E \to E_N$ *by*

$$(1.4) \qquad S_k = \sum_{\gamma \in N} c_\gamma S_\gamma,$$

i.e.

$$\forall x \in E : S_k x = \sum_{\gamma \in N} c_\gamma x_\gamma = x * k = \int_G (gx)k(g)dm(g)$$

and denote $S_k|_{E_N}$ *by* s_k.

REMARK 1.3. The spaces E_γ are linearly independent, i.e. the sum $\sum_{\gamma \in N} E_j$ is a direct sum: the S_γ are projections with ranges E_γ, and S_γ vanishes on $E_{\gamma'}$, for $\gamma' \neq \gamma$.

THEOREM 1.1. *If the C.A. group G operates continuously on the Banach space E, then any continuous linear extension $S : E \to E_N$ of the transformation $s_k : x \to x * k$ of E_N (see Def. 1.3.) satisfies the relation:*

$$\forall\, x \in E : S_k x = \int_G (g^{-1} S g x)\, dm(g), \tag{1.5}$$

strongly.

COROLLARY. *If furthermore G operates isometrically on E, then the norm of S_k is minimum among those of the continuous linear extensions $E \to E_N$ of s_k.*

PROOF OF THEOREM 1.1. AND OF ITS COROLLARY. The relation (1.5) is first proved when $x \in E_\gamma$. The relation (1.5) will then be true if we can show that $\oplus E_\gamma$ is dense in E.

The complex linear hull $>\hat{G}< := >\{e_\gamma : \gamma \in \hat{G}\}<$ of the characters is a self-adjoint complex algebra, which vanishes nowhere in G and which separates the points of G. Hence by the Stone - Weierstrass theorem $>\hat{G}<$ is dense in $C_o(G) = C(G)$. It follows thus

(*) $>\hat{G}<$ contains a continuous approximation of the identity, i.e., a family of positive continuous functions y_α, with $\int_G y_\alpha dm = 1$, and such that for every neighbourhood U of the origin in G, and every $\varepsilon > 0$, a y_α can be found such that $|y_\alpha(g)| < \varepsilon$ for $g \notin U$.

Let U denote the set of all neighbourhoods of $0 \in G$, $A = \{(U,\varepsilon) : U \in U, \varepsilon > 0\}$. A is a directed set with respect to the partial ordering

$$(U,\varepsilon) < (V,\varepsilon') \Leftrightarrow V \subset U \text{ and } \varepsilon' \leq \varepsilon.$$

For each $\alpha = (U,\varepsilon) \in A$ let

$$F_\alpha = \{y : y \in >\hat{G}< \text{ and } y \text{ meets all the conditions of (*) with respect to } (U,\varepsilon)\}.$$

Then $F' = \{F_\alpha : \alpha \in A\}$ is a filterbase in $>\hat{G}<$, which approximates the identity. One obtains a directed family associated with F' by choosing precisely one y_α in each F_α and setting $y_\alpha < y_\beta \Leftrightarrow \alpha < \beta$.

Let now $x \in E$, $\alpha \in A$ and $y \in F_\alpha$. It follows then from the definition (1.3):

$$x * y = x * \Big(\sum_{\text{finite}} a_\gamma e_\gamma \Big) = \sum_{\text{finite}} a_\gamma x * e_\gamma = \sum_{\text{finite}} a_\gamma S_\gamma x \in \oplus E_\gamma .$$

Hence for every $x \in E$ $H_x := \{H_\alpha^x : \alpha \in A\}$, with $H_\alpha^x = \{x * y : y \in F_\alpha\}$ for every $\alpha \in A$, is a filterbase in $\oplus E_\gamma$. One then ends the proof of Thm. 1.1 by showing

(**) If G operates continuously on E then for every $x \in E$ the filterbase H_x converges to x in E.

We now prove the corollary.

If $x \to gx$ is an isometry it is also a bijective isometry of E. Hence $x \to Sgx$ has the same norm as S and $S_{(g)} := x \to g^{-1}Sgx$ has the same norm as S_g, i.e. as S. Hence from

$$S_k x = \int_G S_{(g)} x \, dm(g)$$

strongly, it follows

$$\| S_k x \| \leqslant \int_G \| S_{(g)} x \| \, dm(g)$$

$$\leqslant \int_G \| S_{(g)} \| \; \| x \| \, dm(g) = \int_G \| S \| \; \| x \| \, dm(g) = \| S \| \; \| x \| ;$$

hence $\| S_k \| \leqslant \| S \|$.

Comments

1. If $c_\gamma = 1$ for each $\gamma \in N$, then $k = d_N = \sum_{\gamma \in N} e_\gamma$ is the Dirichlet kernel, s_{d_N} is the identity transformation of E_N, and its natural extension $S_{d_N} : x \to x * d_N$ to $E \to E_N$ is the Fourier projector:

(1.6) $$F_N : x \to x * d_N, \text{ where } x * d_N = \sum_{\gamma \in N} \int_G (gx)(g,\gamma) dm(g).$$

The continuous linear extensions of s_{d_N} are the projectors $E \to E_N$.

2. In the future we shall be concerned with the case where G operates on $C(G)$ or on $L^p(G)$, $1 \leqslant p < \infty$, in the following way:

(1.7) $$(hx)(g) = x(g-h) .$$

The operation is continuous, the representation operators are isometric. Theorem 1.1, its corollary, and Comment 1 are applicable. The convolution considered above is the usual convolution of functions. Furthermore, for any $\gamma \in \hat{G}$, the space $E_\gamma = \{x \in E : gx = (-g,\gamma)x$ for every $g \in G\}$ is the one-dimensional space spanned by the character e_γ: if, for every $g \in G$, $gx = (-g,\gamma)x$, i.e., if, for every $(g,h) \in G^2$, $(-g,\gamma)x(h) = (gx)(h) = x(h-g)$, then it follows in particular for $h = 0$ that $(-g,\gamma)x(0) = x(-g)$, i.e. $x = x(0)\, e_\gamma$.

3. Michael Golomb in his lectures on Theory of Approximation at the Summer School on Numerical Analysis in "Le Bréau-sans-nappe" France in 1963 mentioned the original D.L. Berman's relation in the special case of $C(T)$ and $L^p(T)$, $1 \leq p \leq \infty$, where T is the circlegroup, and where S_k is the Fourier projection F_N, in addition to which E_N is the linear span of the classical set of characters $\{e^{ikt} : -n \leq k \leq n\}$ of T. He mentioned also the then open problem: it was not known whether F_N is the unique projection of minimum norm $C(T) \to E_N$, resp. $L^p(T) \to E_N$, i.e. whether other continuous projections, having the same norm as F_N, might not also exist.

Of course the case $p = 2$ is trivial.

The general case $S_k : C(G) \to E_N$ has been most satisfactorily settled in 1969 in the one side by a group of five mathematicians: E.W. Cheney, C.R. Hobby, P.D. Morris, E. Schurer and D.E. Wulbert, and in the other side by our previous work (see § 2). The purpose of this paper is to settle the general case $S_k : L^1(G) \to E_N$. In 1969 (see § 3) we gave a sufficient condition in order that S_k should be the unique minimum norm extension $L^1(G) \to F_N$ of s_k. In § 3 of this paper we shall prove that if this condition is slightly weakened, it is also necessary. The case $L^p(G) \to E_N$ with $1 < p < 2$ or $2 < p < \infty$ is still an open problem.

4. Let us assume that G operates continuously and isometrically on E. The set of minimum norm extensions $E \to E_N$ of s_k is then convex. It is a facet of the sphere with radius $\|S_k\|$ of the complex normed space $L(E;E_N)$: this facet consists of the common points of this sphere with the complex affine subspace:

$$V_k = \{S : S = S_k + R,\ R \in L(E;E_N),\ R(E_N) = \{0\}\}$$

of $L(E;E_N)$, i.e., the complex affine subspace of continuous linear extension $E \to E_N$ of s_k. We denote this facet by C_k, and $\dim(C_k)$ will be the complex dimension of the complex affine subspace of V_k generated by C_k.

2. The Case of Spaces of Continuous Functions

We first consider again the circle group T and for any $n \in \mathbb{N}$ the linear span E_n of the classical subset of characters $\{t \to e^{ikt} : -n \leqslant k \leqslant n\}$ of T: It was proved in an outstanding work in 1969 by E.W. Cheney, C.R. Hobbey, P.D. Morris, F. Schurer and D.E. Wulbert [1], that F_n is the unique minimum norm projection $C_{\mathbb{R}}(T) \to E_n$. The proof of this uniqueness is based on the peculiar form of the Dirichlet kernel d_n, i.e. is based on the facts that

a) a trigonometric polynomial vanishing at the 2n alternating points of d_n in T is determined up to a constant factor,

b) the Fourier coefficients

$$(|d_n|)_k = \frac{1}{2\pi} \int_{-\pi}^{\pi} |d_n|(t) e^{-ikt} \, dt$$

are $\neq 0$ for $k \notin \{-2n, -(2n-1), \ldots, 0, \ldots, (2n-1), 2n\}$. The proof that these Fourier coefficients are indeed $\neq 0$ is difficult.

Let us now return to the general case $E \to E_N$ of § 1 when $E = C(G)$, G being any compact abelian group. Simultaneously and independently of the above result we proved in 1969 [3] the following general theorem.

THEOREM 2.1. *Let A_k be the symmetric set $\{\chi \in \hat{G} : \chi \in N - N$ and $(|k|)_\chi = 0\}$. Then $\dim(C_k) \geqslant \mathrm{card}(A_k)$. More precisely the real parts and the non-zero imaginary parts of the characters e_χ, $\chi \in A_k$, yield linearly independent mappings $R_\chi : x \to x(\mathrm{Re}\; e_\chi) * k$ and $R_{-\chi} : x \to x(\mathrm{Im}\; e_\gamma) * k$, $R_\chi(E_N) = R_{-\chi}(E_N) = \{0\}$, such that the mappings $S_\chi = S_k + R_\chi$, $S_{-\chi} = S_k + R_{-\chi}$ are all minimum norm extensions $C(G) \to E_N$ of s_k.*

For examples see [3].

We now specify $G = T$ in order to be able to combine Theorem 2.1 with a slight extension of results of [1]. This combination leads to a criterion for uniqueness of the minimum norm extension $C(T) \to E_N$ of s_k, whenever the kernel k has a special form.

DEFINITION 2.1. *The point $g \in T$ will be called an alternating point of a continuous real function $x : T \to \mathbb{R}$ iff x vanishes and changes sign at g.*

DEFINITION 2.2. <u>We say that an element</u> x <u>of</u> E_N <u>is determined, up to a constant factor, by a subset</u> H <u>of</u> $\{g \in G : x(g) = 0\}$ <u>if and only if every element</u> $y \in E_N$, <u>which vanishes at every point of</u> H <u>satisfies</u>, $y = cx$ <u>for some</u> $c \in \mathbb{C}$.

COROLLARY OF THEOREM 2.1. <u>Assume</u> $G = T$ <u>and the kernel</u> k <u>real and determined, up to a constant factor, by the set of its alternating points. Then</u> $S_k : x \to x * k$ <u>is the unique minimum norm extension</u> $C(T) \to E_N$ <u>of its restriction</u> s_k <u>to</u> E_N <u>if and only if the symmetric set</u> $A_k = \{\chi \in \hat{T} : \chi \in N - N$ <u>and</u> $(|k|)_\chi = 0\}$ <u>is empty. Furthermore</u> $\dim(C_k) \geq \mathrm{card}(A_k)$.

REMARK 2.2. (Connection with approximation theory.) Our C.A. groups G are at least T_0-spaces, so they are also T_4-spaces (normal and separated). Let now P be the set of all continuous linear projections $P : C(G) \to E_N$. We look for a projection $P_0 \in P$, which minimizes the maximal normalized approximation error of any $x \in C(G)$ by its projection $P_0 x$ in E_N, i.e., we look for a $P_0 \in P$ such that

$$\|I-P_0\| = \inf_{P \in P} \|I-P\| = \inf_{p \in P} \sup_{\substack{x \in C(G) \\ \|x\|_\infty = 1}} \|(I-P)(x)\| . \tag{2.1}$$

The solutions are given by the minimum norm projections because

$$\forall\ P \in P : \|I-P\| = 1 + \|P\| ,$$

if either G has no isolated point (Thm. of Daugavet-Arens, using the fact that G is T_4 and that the linear operators P are of finite rank) or G is metric (Thm. of Krasnosel'skii, using the fact that the operators P are compact linear operators).

3. The Case of L^1-Spaces

We now consider $S_k : x \to x * k : L^1(G) \to E_N$ and the bounded linear extensions $S : L^1(G) \to E_N$ of $s_k := S_k|E_N$. In 1969 we proved (see [4]) that if k is determined, up to a constant factor, by the set of its roots in G, (see Def. 2.2), then S_k is the unique minimum norm extension $L^1(G) \to E_N$ of its restriction s_k to E_N. We shall now first weaken this condition.

DEFINITION 3.1. <u>We say that an element</u> x <u>of</u> E_N <u>is continuously determined, up</u>

<u>to a constant factor, by the set of its roots in</u> G, <u>iff</u>:

$$\forall\, y \in E_N : \frac{y}{x} \text{ continuous on } G \Rightarrow y \in \mathbb{C}\, x.$$

REMARK 3.1. Of course if $x \in E_N$ is determined, up to a constant factor, by the set of its roots in G (see Def. 2.2), then it is also continuously determined, up to a constant factor, by this set.

LEMMA 3.1. <u>If</u> $x \in E_N \setminus \{0\}$ <u>is not continuously determined, up to a constant factor, by the set of its roots in</u> G, <u>then there is an</u> $y \in E_N$ <u>such that</u> y/x <u>is continuous on</u> G, $\dim_{\mathbb{C}}(>x,y<) = 2$ <u>and</u> $\int_G y(g)(\operatorname{sgn}\bar{x})(g)\,dm(g) = 0$. <u>Moreover, if</u> x <u>is a real function and if</u> N <u>is symmetric, the function</u> y <u>may be assumed to be real</u>.

LEMMA 3.2. <u>The kernel</u> $k = \sum_{\gamma \in N} c_\gamma e_\gamma$, $\forall\, \gamma \in N : c_\gamma \neq 0$, <u>is real if and only if</u> N <u>is symmetric and</u> $\forall\, \gamma \in N : c_{-\gamma} = \overline{c_\gamma}$.

LEMMA 3.3. <u>Any continuous linear mapping of finite rank</u> $S : L^1(G) \to L^1(G)$ <u>of the special form</u> $S = \sum_{i=1}^n x_i' \otimes y_i \in C(G) \otimes L^1(G)$, $n \in \mathbb{N}$, <u>satisfies</u>

$$\|S\|_1 = \sup_{h \in G} \|K_S(h,\cdot)\|_1$$

<u>where</u> $K_S(h,g) := \sum_{i=1}^n x_i'(h) y_i(g)$ <u>for all</u> h <u>and almost all</u> g <u>and</u> $K_S(h,\cdot)$ <u>is the</u> L^1<u>-function</u> $g \to K_S(h,g)$.

PROOF. We know that $C(G) \subseteq L^\infty(G) = (L^1(G))'$ and we use the known inclusions: $L^\infty(G) \otimes_\varepsilon L^1(G) \subseteq L(L^1(G);\, L^1(G)) \subseteq (L^1(G) \otimes_\pi L^\infty(G))'$ and the fact that the unit ball $B_1(G)$ of $L^1(G)$ is $\sigma(M(G), C(G))$ dense in the unit ball $B_M(G)$ of $M(G)$. We also use the following:

Let e be the neutral element of G and δ the Dirac-measure on G i.e., $\delta(E) = 1$, if $e \in E$, and $\delta(E) = 0$ if $e \notin E$, for any Borel set E of G. The Borel measure $g\delta$, $g \in G$, on G is then defined by $g\delta(E) = \delta(E-g)$ for any Borel set E of G. Then it is known that $\{cg\delta : g \in G, c \in T\}$ are the extremal points of the unit ball $B_M(G)$.

This, together with Krein - Milman, yields:

$$\|S\|_1 = \sup_{x \in B_1(G)} \Big\| \sum_{i=1}^n \langle x_i', x\rangle y_i \Big\|_{L^1(G)} = \sup_{x \in B_M(G)} \Big\| \sum_{i=1}^n \langle x_i', x\rangle y_i \Big\|_{L^1(G)} =$$

$$= \sup_{h \in G, c \in T} \| c \sum_{i=1}^{n} \langle x_i', h\delta \rangle y_i \|_{L^1(G)} = \sup_{h \in G} \| \sum_{i=1}^{n} x_i'(h) y_i \|_{L^1(G)}$$

$$= \sup_{h \in G} \| K_S(h, \bullet) \|_1 .$$

THEOREM 3.1. <u>If the kernel</u> k <u>is real and if</u> S_k <u>is the unique minimum norm extension</u> $L^1(G) \to E_N$ <u>of its restriction</u> s_k <u>to</u> E_N, <u>and if</u> $\hat{G} \setminus (N-N) \neq \emptyset$, <u>then</u> k <u>is continously determined, up to a constant factor, by the set of its roots in</u> G.

PROOF. Assume that $\hat{G} \setminus (N-N) \neq \emptyset$ and that $k = \sum_{\gamma \in N} c_\gamma e_\gamma$, $c_\gamma \neq 0$, is real and not continuously determined, up to a constant factor, by the set of its roots in G. Then N is symmetric (Lemma 3.2) and there is a real $y \in E_N$ such that y/k is continuous in G, $\dim_{\mathbb{C}}(\rangle k, y \langle) = 2$ and $\int_G y \operatorname{sgn} k = 0$, (Lemma 3.1). Consequently $a_y := \| y/k \|_\infty$ is defined and > 0, and of course:

$$\forall \alpha : 0 \leqslant \alpha < \frac{1}{a_y} \Rightarrow \operatorname{sgn}(k + \alpha y) = \operatorname{sgn} k \quad . \tag{3.1}$$

Consider now a function $u \in C_{\mathbb{R}}(G)$ such that $u \neq 0$ and $\forall \gamma \in N - N$: $\int_G u(h)(h,\gamma) dm(h) = 0$, i.e. a non-zero continuous real function on G such that (since N − N is symmetric) its Fourier-coefficient $\hat{u}(\gamma)$ vanishes whenever $\gamma \in N - N$. Such a function u of course exists and there is no restriction in assuming that

$$0 < \| u \|_\infty < \frac{1}{a_y} \quad . \tag{3.2}$$

The function y has the form $y = \sum_{\gamma \in N} d_\gamma e_\gamma$, $d_\gamma \in \mathbb{C}$; consider then the kernel $H(h,g) = u(h) y(g-h) = \sum_{\gamma \in N} d_\gamma u(h)(-h,\gamma)(g,\gamma)$ and the continuous linear mappings

$$R_u : \begin{cases} L^1(G) \to E_N \\ x \to \int_G x(h) H(h, \bullet) dm(h) = \sum_{\gamma \in N} d_\gamma \langle u e_\gamma', x \rangle e_\gamma , \end{cases}$$

where e_γ' is the character $e_{-\gamma}$ considered as an element of the dual space $L^\infty(G)$ of $L^1(G)$, and

$$S_u = S_k + R_u : L^1(G) \to E_N .$$

According to Lemma 3.3 and to (3.2) and (3.1) this yields

$$\|S_u\| = \sup_{h \in G} \|(k+H)(h,\bullet)\|_1 = \sup_{h \in G} \int_G |k(g-k) + u(h)y(g-k)|\,dm(g)$$

$$= \sup_{h \in G} \int_G [k(g-h) + u(h)y(g-h)](\text{sgn } k)(g-h)\,dm(g)$$

$$= \sup_{h \in G} \int_G |k(g-h)|\,dm(g) = \|S_k\| = \|k\|_1 .$$

So S_u has minimum norm. It remains to show that

$$S_{u|E_N} = s_k, \text{ i.e. that } R_u(E_N) = \{0\},$$

and that

$$S_u \neq S_k, \text{ i.e. that } R_u \neq 0.$$

First according to the choice of u, we have

$$\forall\, \chi \in N\ \forall\, g \in G : (R_u(e_\chi))(g) = \int_G \sum_{\gamma \in N} d_\gamma u(h)(h,\chi-\gamma)(g,\gamma)\,dm(h) = 0.$$

Hence $R_u(E_N) = \{0\}$.

Not let h_o be a point of G where $u(h_o) \neq 0$, and consider the extension $\overline{R}_u$ to $M(G) \to E_N$ or R_u. Then

$$\overline{R}_u(h_o\delta) = u(h_o) \sum_{\gamma \in N} d_\gamma(h_o,-\gamma)e_\gamma \neq 0 \in E_N ,$$

since there is a $d_\gamma \neq 0$ and since the e_γ's are linearly independent. Then Lemma 3.3 yields

$$\|R_u\| = \sup_{h \in G} \|H(h,\bullet)\|_1 \geqslant \|H(h_o,\bullet)\|_1 = \|\overline{R}_u(h_o\delta)\|_1 > 0.$$

Let now C_k be again the convex facet of the sphere with radius $\|S_k\|$ of $L^1(G)$ consisting of the minimum norm extensions $L^1(G) \to E_N$ of s_k. Then we have

COROLLARY. <u>Assume that the real kernel</u> k <u>is not continuously determined, up to a constant factor, by the set of its roots in</u> G. <u>Then we have</u>:

$$\dim_{\mathbb{C}}(C_k) \geqslant \text{card}(\widehat{G} \setminus (N-N)).$$

PROOF. We keep the notations of the proof of Theorem 3.1. Let first $A_k := \hat{G} \setminus (N-N)$ and $A_r = \{\gamma \in A_k : e_\gamma \text{ is real}\}$. Then $A_k \setminus A_r$ can be written as a disjoint union $A_c \cup (-A_c)$ i.e., we use the disjoint union $A_k = A_r \cup A_c \cup (-A_c)$. Choose then a fixed $a \in \mathbb{R}$ s.t. $0 < a < 1/a_y$ and let for each $\chi \in A' := A_r \cup A_c$:

$$\alpha_\chi := \operatorname{Re}(e_\chi) = \frac{1}{2}(e_\chi + e_{-\chi}), \quad \alpha_{-\chi} := \operatorname{Im}(e_\chi) = \frac{1}{2i}(e_\chi - e_{-\chi})$$

$$R_\chi : \left| \begin{array}{l} L^1(G) \to E_N \\ x \to (a\alpha_\chi x) * y \end{array} \right. , \quad R_{-\chi} : \left| \begin{array}{l} L^1(G) \to E_N \\ x \to (a\alpha_{-\chi} x) * y \end{array} \right.$$

$$S_\chi = S_k + R, \quad S_{-\chi} = S_k + R_{-\chi} .$$

It follows from the proof of Theorem 3.1 that both S_χ and $S_{-\chi}$ are minimum norm extensions $L^1(G) \to E_N$ of s_k. It remains to show that the elements of

$$\{R_\chi : \chi \in A'\} \cup \{R_{-\chi} : \chi \in A_c\}$$

are linearly independent. In order to show this let N_1 be a finite subset of A' such that

$$\sum_{\chi \in N_1} (\lambda_\chi R_\chi + \lambda_{-\chi} R_{-\chi}) = 0, \text{ where } \lambda_\chi, \lambda_{-\chi} \in \mathbb{C} \text{ and } \lambda_{-\chi} = 0 \text{ if } \chi \in A_r. \tag{3.3}$$

We put again $y = \sum_{\gamma \in N} d_\gamma e_\gamma$. Then (3.3) yields

$$a \sum_{\chi \in N_1} [\lambda_\chi \sum_{\gamma \in N} (\alpha_\chi e_{-\gamma} \otimes d_\gamma e_\gamma) + \lambda_{-\chi} \sum_{\gamma \in N} (\alpha_{-\chi} e_{-\gamma} \otimes d_\gamma e_\gamma)] = 0$$

which is equivalent to:

$$\frac{a}{2} \sum_{\gamma \in N} [\sum_{\chi \in N_1} [\lambda_\chi (e_{-(\gamma-\chi)} + e_{-(\gamma+\chi)}) - i\lambda_{-\chi}(e_{-(\gamma-\chi)} - e_{-(\gamma+\chi)})]] \otimes d_\gamma e_\gamma = 0 . \tag{3.4}$$

Let $N_0 := \{\gamma \in N : d_\gamma \neq 0\}$. We know that $N_0 \neq \emptyset$. Then, since the e_γ's are linearly independent, (3.4) is equivalent to

(3.5) $$\forall\, \gamma \in N_o : \sum_{\chi \in N_1} [\,(\lambda_\chi - i\lambda_{-\chi})e_{-(\gamma-\chi)} + (\lambda_\chi + i\lambda_{-\chi})e_{-(\gamma+\chi)}] = 0,$$

i.e., since $\lambda_{-\chi} = 0$ for $\chi \in A_r$,

(3.6) $$\forall\, \gamma \in N_o : \sum_{\chi \in N_1 \cap A_r} (\lambda_\chi e_{-(\gamma-\chi)} + \lambda_\chi e_{-(\gamma+\chi)}) + \sum_{\chi \in N_1 \setminus A_r} [\,(\lambda_\chi - i\lambda_{-\chi})e_{-(\gamma-\chi)} + (\lambda_\chi + i\lambda_{-\chi})e_{-(\gamma+\chi)}] = 0\,.$$

Of course $\gamma-\chi_1 \neq \gamma-\chi_2$ and $\gamma+\chi_1 \neq \gamma+\chi_2$, whenever $\chi_1 \neq \chi_2$. But also, by the definition of A' it can not occur that $\gamma-\chi_1 = \chi+\chi_2$, i.e. that $\chi_2+\chi_2 = 0$, whenever $\chi_1 \neq \chi_2$. Hence it follows from (3.6) and the linear independence of the characters that $\lambda_\chi = 0$, whenever $\chi \in N_1 \cap A_r$, and $\lambda_\chi - i\lambda_{-\chi} = 0 = \lambda_\chi + i\lambda_{-\chi}$, i.e., $\lambda_\chi = \lambda_{-\chi} = 0$, whenever $\chi \in N_1 \setminus A_r$.

This achieves the proof, since then we have

$$\dim_{\mathbb{C}}(C_k) = \dim_{\mathbb{C}}(>C_k - S_k<)$$

$$\geqslant \dim_{\mathbb{C}}(>\{R_\chi : \chi \in A'\} \cup \{R_{-x} : \chi \in A_c\}<) = \mathrm{Card}(\hat{G} \setminus (N-N)).$$

REMARK 3.2. If the set A_k is infinite, i.e. if G is infinite, one can of course find a set of algebraically linearly independent vectors α_ν, which has the power of the continuum and which is in the closed convex hull of $\{ae_\chi : \chi \in A_k\}$, $0 < a < 1/a_y$. If follows easily from the preceeding proof that these vectors α_ν can be chosen such as to define linearly independent elements of the facet C_k, the dimension of which has hence the power of the continuum. Furthermore it is easily seen that S_k is the center of the facet C_k.

EXAMPLE 3.1. (Uniqueness)

a) (This is the first example of [4]) . Let G be the circle group $T \cong \mathbb{R}/_{2\pi\mathbb{Z}}$ and N the classical part $\{-n,-(n-1),\dots,0,\dots,(n-1),n\}$ of $\hat{T} = \mathbb{Z}$. The Dirichlet kernel d_N is then given by:

$$d_N(t) = \sum_{k=-n}^{+n} e^{ikt} = \begin{cases} \dfrac{\sin(2n+1)t/2}{\sin t/2}, & \text{if } 0 < t < 2\pi \\ 1 + 2n, & \text{if } t = 0 \end{cases}$$

and has the 2n distinct roots $x_j = 2j\pi/(2n+1)$, $j = 1,2,\ldots,2n$, in T. It is easy to show that any element $y \in E_N$ vanishing at these roots satisfies $y = (y(0)/d_N(0))d_N$. Hence the Fourier projection

$$x \to x * d_N : L^1(G) \to E_N$$

is the unique minimum norm projection $L^1(G) \to E_N$.

b) Let again $G = T$ and for any $n, k \in \mathbb{N}$, $k \geq 1$, let N be as in a) and $M = kN$. For any $y \in E_M$ there is precisely one $y^* \in E_N$ with $\forall\, t \in T : y(t) = y^*(kt)$, i.e., $d_M^* = d_N$.

Hence, if y vanishes at the roots of d_M, then y^* vanishes at the roots of $d_M^* = d_N$. It follows by a) that $\exists\, c \in \mathbb{C} : y^* = cd_N$, i.e. $y = cd_M$. Hence the Fourier projection is the unique minimum norm projection $L^1(G) \to E_M$.

c) Other examples of uniqueness are given in [4].

EXAMPLE 3.2. (Non-uniqueness) Let again G be the circle group T and $N = \{-4,-3,-2,0,2,3,4\}$. Putting $\alpha = \cos t$, we know that $t \to \cos kt$ is also a function of α for each $k \in \mathbb{N}$. We denote this function by T_k (= the kth Tchebyshev polynomial of the first kind).

It follows that $t \to d_N(t)$ is also a function of α, i.e.

$$d_N(t) = 1 + 2[\sum_{k=2}^{4} T_k(\alpha)] = 16\alpha^4 + 8\alpha^3 - 12\alpha^2 - 6\alpha + 1 =: P(\alpha).$$

Some study of the signs of P shows that P has only two distinct roots α_1 and α_2 in $[-1,1]$. A study of the derivative P' shows then that these roots are simple. Hence, since $V := >T_0, T_2, T_3, T_4<$ is four-dimensional, it is not difficult to find a real $Q \in V$ such that $\dim_{\mathbb{C}}(>P,Q<) = 2$ and $Q(\alpha_1) = Q(\alpha_2) = 0$. Since the roots α_1 and α_2 of P are simple, the function Q/P is continuous in $[-1,1]$. The element y of E_N, defined by $\forall\, t \in T : y(t) = Q(\cos t)$, satisfies then: $\dim_{\mathbb{C}}(>d_N, y<) = 2$ and y/d_N continuous on T. The corollary of Thm. 3.1 then yields $\dim_{\mathbb{C}}(C_{d_N}) = \infty$.

REFERENCES

[1] Cheney, E.W. - Hobby, C.A. - Morris, P.D. - Schurer, F. - Wulbert, D.E., On the minimal property of the Fourier projection. Trans. A.M.S. 143 (1969), 249-258.

[2] Lambert, P.V., Réalité des projecteurs de norme minimum sur certains espaces de Banach. Bull. de la Classe des Sciences. Acad. Royale de Belgique, 5 - LIII, (1967-2), 91-100.

[3] Lambert, P.V., Minimum norm property of the Fourier projection in spaces of continuous functions. Bull. de la Soc. Math. de Belgique, 21 (1969), 359-369.

[4] Lambert, P.V., On the minimum norm property of the Fourier projection in L^1-spaces. Bull. de la Soc. Math. de Belgique 21 (1969), 370-391.

[5] Lambert, P.V., On the minimum norm property of the Fourier projection in L^1-spaces and in spaces of continuous functions. Bull. A.M.S., 76 (1970), 798-804.

BANACH SPACES OF DISTRIBUTIONS OF WIENER'S TYPE AND INTERPOLATION

Hans G. Feichtinger
Institut für Mathematik
Universität Wien
Wien

In the parallel paper [9] we have introduced "spaces of Wiener's type", a family of Banach spaces of (classes of) measurable functions, measures or distributions on locally compact groups. The elements of these spaces are characterized by - what we call - the global behaviour of certain of their local properties. In the present paper it is to be shown that interpolation methods can be applied to these spaces in a very natural way. Using the results on interpolation it is not difficult to extend various theorems of analysis to the setting of Wiener-type spaces. As illustration we present a version of the Hausdorff - Young inequality for locally compact abelian groups. As a consequence, one obtains a sharpened version of Sobolev's embedding theorem.

1. Definitions and Basic Properties

Throughout G will be a locally compact group with left Haar measure dx. We shall mainly be interested in non-discrete, non-compact groups (e.g. $G = \mathbb{R}^m$). $\mathcal{K}(G)$ denotes the space of all continuous, complex-valued functions on G with compact support (supp), endowed with its natural inductive limit topology. $(L^p, \| \ \|_p)$, $1 \leqq p \leqq \infty$, denotes the usual Lebesgue spaces. Given a subset $M \subseteqq G$ we write c_M for its characteristic function. The space $L^1_{loc}(G)$ consists of all (classes of) measurable functions f on G such that $fc_K \in L^1(G)$ for any compact subset $K \subseteqq G$. It is a topological vector space with the family of seminorms $f \to \|fc_K\|$. A BF-space on G is a Banach space $(B, \| \ \|_B)$ which is continuously embedded into $L^1_{loc}(G)$. As usual we shall speak of "functions" in such spaces, identifying two measurable functions in B, if they are equal locally almost everywhere (l.a.e.). A BF-space is called solid if any measurable function g, for which there exists $f \in B$ such that $|g(x)| \leqq |f(x)|$ l.a.e. belongs to B, with $\|g\|_B \leqq \|f\|_B$. A BF-space B is called left translation invariant (translation invariant) if the left (left and right) translation operators, given by

$$L_y f(x) := f(y^{-1}x), \; A_y f(x) := f(xy)$$

act boundedly on B. Their operator norm is written as $|||\ \ |||_B$. Corresponding terminology is applied to spaces of measures or distributions, to which the translation operators are extended by transposition. A left invariant BF-space will be called a homogeneous Banach space on G if G acts (by left translations) isometrically on B, and if translation is continuous in B, i.e. if $\lim_{y \to e} \|L_{y_1} f - f\|_B = 0$ for all $f \in B$. The homogeneous Banach spaces which are dense in $L^1(G)$ are exactly the Segal algebras in the sense of Reiter ([15]).

A triple (B^1, B^2, B^3) will be called a Banach convolution triple (BCT), if convolution, given by

$$f^1 * f^2(x) := \int_G f^1(y^{-1}x) f^2(y)\,dy \quad \text{for} \quad f^i \in \mathcal{K}(G) \quad B^i,\ i = 1,2,$$

extends to a bounded, bilinear map (of norm 1) from $B^1 \times B^2$ into B^3. Clearly (A,A,A) is a BCT for some $A \subseteq L^1(G)$ iff A is a Banach convolution algebra. Any weighted L^1-space

$$L^1_w(G) = \{ f \mid fw \in L^1(G) \},\ \|f\|_{1,w} := \|fw\|_1$$

is a BCA, called Beurling algebra, if w is a continuous function satisfying $w(x) \leq 1$, and $w(xy) \leq Cw(x)w(y)$ for all $x,y \in G$. (cf. [15]). Such functions are called weight functions. A Banach space B is a (left) Banach convolution module over the Banach algebra A iff (A,B,B) is a BCT, and a (left) Banach ideal in A if furthermore $B \subseteq A$. Any homogeneous Banach space is known to be a left $L^1(G)$ Banach convolution module. Constants without importance will be denoted by $C, C_1, \dots$.

<u>General Hypothesis</u>

As a standing assumption we suppose throughout this paper that for any Banach space B used below there exists some "nice" Banach algebra A acting on B by "pointwise" multiplication.

More precisely, we suppose that there exists a homogeneous Banach space $(A, \|\ \ \|_A)$, continuously embedded into the Banach algebra with respect to pointwise multiplication (i.e. separating points from closed sets), and which is closed under complex conjugation, and that B is a Banach module over A with respect to "pointwise" multiplication, i.e.

$$\| hf \|_B \leq \| h \|_A \| f \|_B \quad \text{for all} \quad h \in A,\ f \in B.$$

Here some comment concerning the term "pointwise" multiplication is in order. Of course, there is no problem of interpretation, if B happens to be a BF-space on G (which covers the most important examples). In this case the pointwise product of a continuous function with a locally integrable function is to be taken in the ordinary sense. In order to cover more general situations (which occur naturally in the investigations) we assume in the sequel that the following situation is given:

B is continuously embedded into the topological dual A_o' of $A_o := A \cap \mathcal{K}(G)$ (endowed with its natural inductive limit topology). On A_o' an action of A by "pointwise multiplication" is given in a natural way, i.e. by transposition of the operation of A on A_o by ordinary multiplication (remember the definition of a "pointwise product" of a test function and a distribution). Since the assumptions imply that A_o is always a dense subspace of $\mathcal{K}(G)$, R(G) (the space of all Radon measures on G) and in particular $L^1_{loc}(G)$ (identified with the closed subspace of all absolutely continuos measures) is alway continuously embedded into A_o' in a natural way. Since the action of A on a subspace of L^1_{loc} defined in the way just mentioned coincides of course with the natural action mentioned above we gain flexibility in adopting our assumptions concerning the definition of pointwise products. We define B_{loc} to be the space of all elements σ of A_o' such that $h\sigma \in B$ for all $h \in A_o$. (Otherwise we would have to restrict our attention to spaces of locally integrable function, which would sometimes be a quite unnatural restriction).

EXAMPLES. The most important examples of algebras A which are defined for arbitrary locally compact groups are the spaces $(C^o(G), \| \ \|_\infty)$ of continuous functions vanishing at infinity, and Eymard's Fourier A(G), which coincides with $\mathcal{F}L^1(G) := \{\mathcal{F}f \mid f \in L^1(\hat{G})\}$ if G is a locally compact abelian group with dual group $\hat{G}$ ($\hat{\hat{G}}$ is identified with G). Therefore any solid BF-space B on G, in particular the spaces $L^p(G)$, $1 \leq p \leq \infty$, is included in our consideration (considered as $C^o(G)$-module), but may take $B = C^o(G)$ itself. If G is abelian, one may consider $B = \mathcal{F} L^p(G)$, $1 \leq p \leq \infty$ (the Fourier transform being taken in the sense of tempered distributions or as a quasimeasure, cf. [8] or [12]) as a module over A(G). As further examples we only mention here the spaces of Besov-Hardy-Sobolev type $B^s_{p,q}$ and $F^s_{p,q}$, $s \in \mathbb{R}$, $1 \leq p,q \leq \infty$, as considered by H. Triebel (see [18], [19]) including Lipschitz and Bessel potential spaces (cf. [16]). For further examples cf. [9]. The Wiener type spaces W(B,C) are now defined as follows:

DEFINITION 1.1. *Let B satisfy the general hypothesis, and let C be a solid, translation invariant BF-space on G. Given any open subset Q of G with compact closure and* $f \in B_{\ell oc}$ *we set:* $F := F_f\colon z \mapsto \|f\|_{B(zQ)}$, *with*
$\|f\|_{B(zQ)} := \inf\{\|g\|_B \mid g \in B,\ g$ *coincides with f on zQ, i.e.*
$hf = hg$ *for all* $h \in A_0$ *with* $\operatorname{supp} h \subseteq zQ\}$.
The Wiener-type space W(B,C) with local component B and global component C is then defined by

$$(1.1) \qquad W(B,C) := \{ f \mid f \in B_{\ell oc},\ F \in C \}.$$

The natural norm on W(B,C) is given by

$$(1.2) \qquad \|f\|_{W(B,C)} := \|F\|_C .$$

THEOREM 1.1. *Let B,C be as in Definition 1.1. Then W(B,C) is a Banach space, continuously embedded into* $B_{\ell oc}$. *It does not depend on the particular choice Q, i.e. two different open subsets of G with compact closure define the same space and equivalent norms.*

It should be mentioned here that good examples of solid translation invariant BF-spaces are weighted L^p-spaces $L^p_w(G) = \{ f \mid fw \in L^p(G) \}$, $\|f\|_{p,w} := = \|fw\|_p$, for w being a continuous weight function on G.

In the present paper we shall consider mainly spaces of the form $W(L^p,L^q)$ or $W(\mathcal{F} L^p,L^q)$, $1 \leq p,q \leq \infty$. Practically all spaces of Wiener's type that have been considered in a number of mostly recent papers (only to mention [2-4,6-11,13,15,17,20]) arise as special cases of the above families, most of them are even of the first kind. In order to give the reader some orientation concerning inclusions among these spaces we state the following lemma:

LEMMA 1.2. Let $1 \leq p, p_1, p_2, q_1, q_2 \leq \infty$ be given. Then

i) $W(L^p, L^p) = L^p(G)$;

ii) $W(L^{p_1}, L^{q_1}) \subseteq W(L^{p_2}, L^{q_2})$ if $p_1 \geq p_2$, $q_1 \leq q_2$;

iii) $W(\mathcal{F}L^{p_1}, L^{q_1}) \subseteq W(\mathcal{F}L^{p_2}, L^{q_2})$ if $p_1 \leq p_2$, $q_1 \leq q_2$;

iv) $W(L^p, L^q) \subseteq W(\mathcal{F}L^{p'}, L^q)$ for $1 \leq p \leq 2$, and

$W(\mathcal{F}L^{p'}, L^q) \subseteq W(L^p, L^q)$ for $2 \leq p \leq \infty$, for all $q, 1 \leq q \leq \infty$

and $1/p' := 1 - 1/p$.

REMARK 1.1. If G is nondiscrete and noncompact it can be shown that equality holds in ii) and iii) only for $p_1 = p_2$ and $q_1 = q_2$, and in iv) only for $p = 2$.

REMARK 1.2. One also has $W(M(G), L^q) \subseteq W(\mathcal{F}L^\infty, L^q)$ for all q (here $M(G) = (C^0(G))'$ denotes the space of bounded measures on G). The spaces $W(M(G), L^q)$, $q > 1$, arise as dual spaces of the spaces $W(C^0(G), L^{q'})$ (cf. [10, 11, 13, 17]).

2. The Abstract Main Result

The following theorem is the basic result of this paper:

THEOREM 2.1. Let A,B,C be as in Definition 1.1. Assume furthermore that C is a left Banach convolution module over some Beurling algebra $L^1_w(G)$. Then $W(B,C)$ is a retract of the vector-valued function space $C(B)$, i.e. there exist bounded linear operators $T: W(B,C) \mapsto C(B)$ and $S: C(B) \mapsto W(B,C)$ such that $S \circ T = Id_{W(B,C)}$.

REMARK 2.1. It can be shown that C satisfies the above condition for a solid translation invariant BF-space containing $\mathcal{K}(G)$ as a dense subspace, or if C is of the form $C = L^p_w(G)$, $1 \leq p \leq \infty$, for some weight function w.

PROOF. The proof is given in four steps.

Step 1. In order to define a mapping T in a suitable way we choose some $g \in A_0$

satisfying $\int_G g(x^{-1})dx = 1$, and supp $g \subseteq Q$. Then we set:

$$\text{(2.1)} \qquad Tf(z) := (L_z g)f, \quad z \in G.$$

We show first that $z \to (L_z g)f$ defines a continuous mapping from G into B, for g as above and for every $f \in W(B,C) \subseteq B_{\ell oc}$. In fact, let $x \in G$ and some relatively compact neighbourhood V of x be given. Then there exists $h \in A_o$ such that $h(x) = 1$ on V(supp g). This implies for $x,y \in V$:

$$\text{(2.2)} \quad \|(L_y g)f - (L_x g)f\|_B = \|(L_y g - L_x g)fh\|_B \leqq \|L_y g - L_x g\|_A \|fh\|_B \to 0$$

for $y \to x$ (in V), since translation is continuous in A. It therefore remains to give an estimate of $z \mapsto \|(L_z g)f\|_B$ in the space C. Making use of the following inequality

$$\text{(2.3)} \qquad \|(L_z g)f\|_B \leqq \|L_z g\|_A \|f\|_{B(zQ)} \text{ for any } z \in G$$

we obtain

$$\text{(2.4)} \qquad \|Tf\|_{C(B)} \leqq \|g\|_A \|f\|_{W(B,C)} \text{ for all } f \in W(B,C).$$

This completes the proof of step 1.

Step 2. Having defined T as above we are now looking for the corresponding operator $S: C(B) \mapsto W(B,C)$. Choosing $g^1 \in A_o$, satisfying $g^1(x) \equiv 1$ on supp g (g as above) we shall define SF (at first formally) by

$$\text{(2.5)} \qquad SF := \int_G (L_z g^1)F(z)dz \text{ for } f \in C(B).$$

Before we can verify that S satisfies all requirements we have to make (2.5) precise: At a first stage we claim that it makes sense to interprete SF as the element of A_o', given by

$$\text{(2.6)} \qquad \langle SF,h\rangle := \int_G \langle (L_z g^1)F(z),h\rangle dz, \quad h \in A_o.$$

We have now to verify that the right hand expression is well defined as an element of A_o' (i.e. as a measure, quasimeasure or distribution in our applications). In order to show continuity of the functional defined in (2.6) let some compact subset K of G, and any $h \in A_o$ with supp $h \subseteq K$ be given. Writing K_1 for supp g^1 and using the fact that $(L_z g^1)h = 0$ for $z \notin KK_1^{-1}$ first, and the continuous embeddings $B \hookrightarrow A_o'$ and $C \hookrightarrow L^1_{\ell oc}(G)$ then, we obtain

$$(2.7)\qquad \left| \int_G \langle (L_z g^1)F(z), h\rangle\, dz \right| \leq \int_{KK_1^{-1}} |\langle (L_z g^1)F(z), h\rangle|\, dz \leq$$

$$\leq C_1 \|h\|_A \int_{KK_1^{-1}} \|(L_z g)F(z)\|_B dz \leq C_2 \|g^1\|_A \|h\|_A \|F(z)\|_{C(B)},$$

where C_2 denotes a constant depending on the space C and on K (and K_1) only.

Step 3. We intend to prove now the boundedness of S as a mapping from C(B) into W(B,C). That SF belongs to $B_{\ell oc}$, i.e. that $h(SF) \in B$ for all $h \in A_o$ can be shown as follows: Since multiplication of $h \in A_o$ with $SF \in A_o'$ is to be understood in the usual sense, i.e. as being defined by $\langle h(SF), h_1\rangle = \langle SF, hh_1\rangle$ for all $h_1 \in A_o$, one has $h(SF) = \int_G h(L_z g^1)F(z)dz$. But the last integral is convergent in B, since the integrand is an integrable function on G with values in B and compact support (recall $C \to L^1_{\ell oc}(G)$!).

In order to show that SF belongs to W(B,C) let us look for an estimate for $y \mapsto \|SF\|_{B(zQ)}$, for $F \in C(B)$. Let $g^2 \in A_o$ be choosen such that $g^2(x) \equiv 1$ on Q. Then one has (as in step 2) for any $y \in G$:

$$(2.8)\qquad \|SF\|_{B(yQ)} \leq \|(L_y g) \int_G (L_z g^1)F(z)dz\|_B \leq \|g^2\|_A \|g^1\|_A \int_{yN} \|F(z)\|_B dz,$$

if we set $N := (\mathrm{supp}\, g^2)(\mathrm{supp}\, g^1)^{-1}$. Noting that the function $\phi: z \to \|F(z)\|_B$ belongs to C, and that C is a left Banach convolution module over some Beurling algebra $L^1_w(G)$ we obtain, as a continuation of (2.8)

$$(2.9)\qquad \|SF\|_{B(yQ)} \leq \|g^2\|_A \|g^1\|_A \|c_N\|_{1,w} \|\phi\|_C.$$

Combining (2.8) and (2.9) we arrive at

$$(2.10)\qquad \|SF\|_{W(B,C)} \leq C_3 \|\phi\|_C = C_3 \|F\|_{C(B)} \text{ for all } F \in W(B,C).$$

Step 4. In this last step it is shown that under the assumptions made the relation $S \circ T(f) = f$ holds true for all $f \in W(B,C)$. Since W(B,C) is continuously embedded into $B_{\ell oc}$, and hence into A_o', it will be sufficient to verify that this identity holds in A_o'. Given any $h \in A_o$ one has (using the identity $g^1 = g^2 g^1$ and applying Fubini's theorem):

(2.11) $$<S(Tf),h> = <\int_G (L_z g^2)(L_z g^1)f,h> = \int_G <(L_z g^1)f,h>\,dz =$$

$$= \int_G\int_G g^1(z^{-1}y)f(y)h(y)\,dz\,dy = \left(\int_G g^1(x^{-1})\,dx\right)<f,h> = <f,h>, \text{q.e.d.}$$

This completes the proof of Theorem 2.1.

REMARK 2.2. There is also a more elementary, but somewhat longer proof showing that the spaces W(B,C) can be represented as retracts of vectorvalued sequence spaces. In this case one makes use of the characterization of W(B,C) by means of uniform, bounded partitions of unity (cf. [9], Theorem 2).

THEOREM 2.2. <u>Suppose that the same algebra</u> A <u>acts on</u> B^1 <u>and</u> B^2, <u>and assume that</u> C^1 <u>or</u> C^2 <u>has absolutely continuous norm</u> (<u>i.e. that</u> $f_n(x) \to 0$ <u>for</u> $n \to \infty$ <u>and</u> $|f_n(x)| \leq |f(x)|$ a.e. <u>implies</u> $\|f_n\|_C \to 0$). <u>Then one has for</u> $\theta \in (0,1)$:

$$\left(W(B^1,C^1),\ W(B^2,C^2)\right)_{[\theta]} = W\left((B^1,B^2)_{[\theta]}, (C^1,C^2)_{[\theta]}\right).$$

PROOF. As a consequence of Theorem 2.1 and general interpolation principles the interpolation results follow from the corresponding interpolation results for the vector-valued function spaces $C^i(B^i)$ (cf. [1], § 6.4). The needed "complex" result is then found in § 13/6 of Calderon's paper ([5]).

COROLLARY 2.3. <u>For</u> $\theta \in (0,1)$, $1 \leq p_1,p_2,q_1,q_2 \leq \infty$, $q_2 < \infty$ <u>one has</u>

$$\left(W(L^{p_1},L^{q_1}_{w_1}), W(L^{p_2},L^{q_2}_{w_2})\right)_{[\theta]} = W(L^p,L^q_w),$$

<u>and</u>

$$\left(W(\mathcal{F}L^{p_1},L^{q_1}_{w_1}), W(\mathcal{F}L^{p_1},L^{q_2}_{w_2})\right)_{[\theta]} = W(\mathcal{F}L^p,L^q_w),$$

<u>with</u> $1/p = (1-\theta)/p_1 + \theta/p_2$, $1/q = (1-\theta)/q_1 + \theta/q_2$, $w = w_1^{1-\theta}w_2^{\theta}$.

REMARK 2.3. There are of course corresponding results for real interpolation spaces, based on the real interpolation results for - say - weighted vector-valued L^p-spaces (cf. [1], [18]). Since we do not need these results here we leave it to the reader to combine known results to new explicit statements, if they should be useful to him.

3. Applications

As in related fields interpolation results for a family of functional spaces imply a number of results concerning operators (convolution operators, Fourier transform, etc.) on these spaces. A typical application of this kind is the following one:

THEOREM 3.1. *Let $f \in L^1_{loc}(G)$ be given, such that $\sup_{y \in G} \int_{yQ} |f(x^{-1})| dx \leq C$ for some open set Q with compact closure. If $T_f: k \to k * f$ acts boundely from $(\mathcal{K}(G), \| \ \|_p)$ into $L^p(G)$, then T_f defines a bounded operator from $W(L^r, L^s)$ into $L^s(G)$ for any $r \in [1,p]$, with $s = r'(p-1)$.*

PROOF. It is easily checked that the first assumption implies $\|L_y h * f\|_\infty \leq$ $\leq C \|h\|_\infty$ for any $y \in G$ and any $h \in L^\infty(G)$ with $\operatorname{supp} h \subseteq Q$. Consequently T_f is bounded operator from $W(L^\infty, L^1)$ into $L^\infty = W(L^\infty, L^\infty)$. Complex interpolation between $W(L^\infty, L^1)$ and $L^p(G)$ yields just the spaces $W(L^r, L^s)$ with $s = r'(p-1)$, while interpolation with the same parameter $\theta \in (0,1)$ between L^∞ and L^p gives exactly $L^s(G)$.

The following result is an extension of the usual Hausdorff-Young inequality:

THEOREM 3.2. *Let G be a locally compact abelian group. For $1 \leq r \leq p \leq \infty$ the Fourier transform defines a bounded linear mapping from $W(\mathcal{F}L^p, L^r)$ into $W(\mathcal{F}L^r, L^p)$. In particular, $W(\mathcal{F}L^p, L^p)$ on G is mapped onto the corresponding space on $\hat{G}$ by the Fourier transform.*

The theorem will follow essentially by means of complex interpolation from the following proposition, which is of interest for itself.

PROPOSITION 3.3. *For $1 \leq p \leq \infty$ the Fourier transform maps $W(\mathcal{F}L^p, L^1)$ into $W(\mathcal{F}L^1, L^p)$.*

PROOF. It is known (see [9], Theorem 2, cf. also [7]) that there exists some compact set $K \subseteq G$ and $C > 0$ such that any $f \in W(\mathcal{F}L^p, L^1)$ has a representation of the form $f = \sum_1^\infty a_n L_{y_n} f_n$, with $\sum_1^\infty |a_n| \leq C \|f\|_{W(\mathcal{F}L^p, L^1)}$, $\operatorname{supp} f_n \subseteq K$ and $\|f_n\|_{\mathcal{F}L^p} \leq 1$ for all n.

Applying Theorem 5 of [9] to $\mathcal{F}f_n$ (take $B = \mathcal{F}L^1(\hat{G})$, $C = L^p(\hat{G})$ there) one obtains

$$\|f_n\|_{W(\mathcal{F}L^1(\hat{G}),L^p)} \leq C_1 \|\mathcal{F}f_n\|_{L^p} = C_1 \|f_n\|_{\mathcal{F}L^p}$$

and

$$\|f\|_{W(\mathcal{F}L^1(\hat{G}),L^p)} \leq C_1 \sum_{k=1}^{\infty} |a_n| \|(L_{y_n} f_n)\|_{\mathcal{F}L^p} \leq C_2 \|f\|_{W(\mathcal{F}L^p,L^1)}.$$

PROOF (of Theorem 3.2). We first consider the case $r = p$. By Proposition 3.3 $\mathcal{F}$ (and also $\mathcal{F}^{-1}$) map $W(\mathcal{F}L^1,L^1)$ onto the corresponding space on the dual group (cf. also [8], Theorem A2 i), $W(\mathcal{F}L^1,L^1) = S_0(G)$!). By Plancherel's theorem the same assertion is true for $W(\mathcal{F}L^2,L^2) = L^2(G)$, hence for all $p \in [1,2]$ by complex interpolation. For $p \geq 2$ it can be proved by transposition (i.e. as in the case of tempered distributions, as we shall prove in detail elsewhere one has $W(\mathcal{F}L^r,L^s)' = W(\mathcal{F}L^{r'},L^{s'})$ for $1 \leq r,s < \infty$). The general case is then derived by means of further complex interpolation between the "diagonal" case and the result of Proposition 3.3.

REMARK 3.1. The above result is in various direction best possible. We shall show below that the Fourier transform does not map $W(\mathcal{F}L^1,L^p)$ (which is contained in $W(L^r,L^p)$ and $W(\mathcal{F}L^r,L^p)$ for any $r \geq 1$) into $W(\mathcal{F}L^q,L^\infty)$ nor into $W(\mathcal{F}L^\infty,L^q)$ for any $q < p$. In particular, the assertions of Theorem 3.2 break down for $r < p$. It also follows therefrom that the Fourier transform is never surjective in Theorem 3.2 for $r \neq p$.

REMARK 3.2. Combining Theorem 3.2 with Lemma 2.2 one obtains the main result of [3], which has been proved by F. Holland for the case $G = \mathbb{R}$. Theorems 3.4, 3.5 and 4.2 of [17] (cf. Remark 1.2) also arise as consequences of our result.

PROOF (of Remark 3.1). It will be sufficient to show that for any $p < \infty$, and $q < p$ there is a bounded sequence $(f_n)_{n=1}^{\infty}$ in $W(\mathcal{F}L^1(G),L^p)$ for which $(\mathcal{F}f_n)_{n=1}^{\infty}$ is unbounded in $W(\mathcal{F}L^\infty,L^q)$ or $W(\mathcal{F}L^q,L^\infty)$ respectively. Given any $f_0 \neq 0$, $f_0 \in W(\mathcal{F}L^1,L^1)$ let us consider expressions of the form $g_n = \sum_1^n L_{y_k} M_{t_k} f_0$ (recall that $M_t, t \in \hat{G}$, denote the operator of pointwise multiplication with the character t).

Since $\mathcal{K}(G) \cap W(\mathcal{F}L^1, L^p)$ is a dense subspace of $W(\mathcal{F}L^1, L^p)$ for $p < \infty$ it is possible to choose $(y_k)_{k=1}^n$ ("sufficiently large") such that

$$\|g_n\|_{W(\mathcal{F}L^1,L^p)} \leq 2\, n^{1/p} \|f_o\|_{W(\mathcal{F}L^1,L^p)} \quad \text{and} \quad \|g_n\|_q \geq (1/2) n^{1/q} \|f_o\|_q$$

for an arbitrary sequence $(t_k)_{k=1}^n \subseteq \hat{G}$. On the other hand one has

$\mathcal{F} g_n = \sum_1^n M_{z_k} L_{t_k} \mathcal{F} f_o$, which implies $\|\mathcal{F} g_n\|_{W(\mathcal{F}L ,L^q)} \geq (1/2) n^{1/q} \|f_o\|_{W(\mathcal{F}L ,L^q)}$

for an appropriate choice of $(t_k)_{k=1}^n \in \hat{G}$. Hence $f_n := n^{-1/p} g_n$ is a suitable sequence for our first assertion. If $\mathcal{F} f_o$ has suitable compact support, then the second assertion follows if $t_k = t_o$ for all k, because then

$$\|\mathcal{F} f_n\|_{W(\mathcal{F}L^q, L^\infty)} = \|\mathcal{F} f_n\|_{\mathcal{F}L^q} = \|f_n\|_q \geq (1/2) n^{1/q-1/p} \|f_o\|_q .$$

As the last application to be mentioned here we give a version of Sovolev's embedding theorem (cf. [16] Chap. V, § 2.2) for the (fractional) potential spaces $\mathcal{L}_s^p$ in the setting of Wiener type spaces:

THEOREM 3.4. i) <u>For</u> $s > m/2$ <u>one has the following continuous embeddings</u>:

$$\mathcal{L}_s^2(\mathbb{R}^m) \hookrightarrow W(\mathcal{F}L^1, L^2) \hookrightarrow W(C^o, L^2) \hookrightarrow C^o(\mathbb{R}^m).$$

ii) <u>More generally, one has for</u> $p \in [1,2]$ <u>and</u> $s > m(1/q-1/p) \geq 0$ <u>the embedding</u>

$$\mathcal{L}_s^p(\mathbb{R}^m) \hookrightarrow W(\;\; L^{q'}, L^p).$$

PROOF. (i) By definition one has $\mathcal{F}\mathcal{L}_s^2 = L^2_{w_2}(\mathbb{R}^m) := \{ h \mid hw_s \in L^s_{w_s}\}$, with $w_s(x) := (1+|x|^2)^{s/2}$. Since $w_s^{-1} \in L^2(\mathbb{R}^m)$ for $s > m/2$, Hölder's inequality implies $L^2_{w_s} = W(L^2, L^2_{w_s}) \hookrightarrow W(L^2, L^1)$. Assertion (i) follows now from 3.3.

(ii) We apply complex interpolation to the pair of inclusions given by (i) and $L^p \to W(\mathcal{F}L^{p'}, L^p)$ (cf. Lemma 2.2). Using the fact that $(\mathcal{L}_s^p, \mathcal{L}_t^r)_{[\theta]} = \mathcal{L}_u^s$ for $\theta \in (0,1)$, $1/s = (1-\theta)/r + \theta/r$ and $u = (1-\theta)s + \theta t$. (cf. [14], Chap. 5, Theorem 5).

Further results concerning Wiener-type spaces, in particular on their multiplier spaces, Tauberian theorems, as well as a characterization of the Banach dual of W(B,C) will be given in subsequent papers.

REFERENCES

[1] Bergh, J. - Löfström, J., Interpolation Spaces. Grundl. d. math. Wiss., 223, Springer, Berlin, 1976.

[2] Bertrandias, J.P. - Datry, C. - Dupuis, C., Unions et intersections d'espaces L^p invariantes par translation ou convolution. Ann. Inst. Fourier 28/2 (1978), 53-84.

[3] Bertrandias, J.P. - Dupuis, C., Transformation de Fourier sur les espaces $l^p(L^p)$. Ann. Inst. Fourier 29/1 (1979), 189-206.

[4] Busby, R.C. - Smith, H.A., Product - convolution operators and mixed norm spaces. Trans. Amer. Math. Soc., to appear.

[5] Calderón, A.P., Intermediate spaces and interpolation, the complex method. Studia Math. 24 (1964), 113-190.

[6] Feichtinger, H.G., A characterization of Wiener's algebra on locally compact groups. Arch. Math. (Basel) 29 (1977), 136-140.

[7] Feichtinger, H.G., A characterization of minimal homogeneous Banach spaces. Proc. Amer. Math. Soc. (1980), to appear.

[8] Feichtinger, H.G., Un espace de distributions tempérées sur les groupes localement compactes abélians. C. R. Acad. Sci. Paris Sér. A 290 (1980), 791-794.

[9] Feichtinger, H.G., Banach convolution algebras of Wiener's type. In "Functions, Series, Operators" Proc. Conf., Budapest (August 1980), to appear.

[10] Goldberg, R., On a space of functions of Wiener. Duke Math. J. 34 (1967), 683-691.

[11] Holland, F., Harmonic analysis on amalgams of L^p and l^q. J. London Math. Soc. (2) 10 (1975), 295-305.

[12] Larsen, R., An Introduction to the Theory of Multipliers. Grundl. d. math. Wiss., 175, Springer, Berlin, 1971.

[13] Liu, T.S. - van Rooij, A. - Wang, J.K., On some group algebra modules related to Wiener's algebra M_1. Pacific J. Math. 55 (1974), 507-520.

[14] Peetre, J., New Thoughts on Besov Spaces. Duke Univ. Math. Series, Vol. 1, Durham, 1976.

[15] Reiter, H., Classical Harmonic Analysis and Locally Compact Groups. Oxford Univ. Press, 1968.

[16] Stein, E.M., Singular Integrals and Differentiability Properties of Functions. Princeton Math. Series, Vol. 30, Princeton Univ. Press, 1970.

[17] Stewart, J., Fourier transforms of unbounded measures. Canad. J. Math. 31 (1979), 1281–1292.

[18] Triebel, H., Interpolation theory, function spaces, differential operators. North Holland Math. Library, Vol. 18, Amsterdam - New York - Oxford, 1978.

[19] Triebel, H., Spaces of Besov - Hardy - Sobolev Type. Teubner Texte zur Mathematik, Leipzig, 1978.

[20] Wiener, N., The Fourier Integral and certain of its Applications. Cambridge Univ. Press, 1933.

APPROXIMATION THEORY ON THE COMPACT SOLENOID

Walter R. Bloom
School of Mathematical and Physical Sciences
Murdoch University
Perth

The compact solenoid Σ is the $\mathfrak{a}$-adic solenoid with $\mathfrak{a} = (2,3,\ldots)$. It is a compact connected metrisable abelian group with dual the group of rational numbers. We give an analogue of the M. Riesz theorem on the boundedness of partial sums of the Fourier series of functions in $L^p(\Sigma)$, and use this to characterize the Lipschitz functions on Σ in terms of the rate of convergence of their Fourier series. In addition we prove a factorization theorem for these functions.

1. Introduction

We write $\mathbb{R}$, $\mathbb{T}$, $\mathbb{Q}$, $\mathbb{Z}$ and Δ for the groups of reals, complex numbers of modulus one, rationals, integers and $\mathfrak{a}$-adic integers respectively, where $\mathfrak{a} = (2,3,\ldots)$. For $\mathfrak{u} = (1,0,\ldots)$ let B denote the cyclic subgroup of $\mathbb{R}\times \Delta$ generated by $(1,\mathfrak{u})$, and put $\Sigma = (\mathbb{R}\times \Delta)/B$. Then Σ is the $\mathfrak{a}$-adic solenoid described in [9], (10.12). It is a compact connected metrisable divisible torsion-free abelian group with character group isomorphic to $\mathbb{Q}$; to each rational number of the form $m/n!$, where $m \in \mathbb{Z}$ and n is a non-negative integer, there corresponds a character $\gamma_{m,n}$ of Σ given by

$$\gamma_{m,n}((\xi,\mathrm{x}) + B) = \exp \left[2\pi i \frac{m}{n!}(\xi-(x_0 + 2!\, x_1 + \ldots + (n-1)!\, x_{n-2})) \right]$$

for $\xi \in \mathbb{R}$ and $\mathrm{x} = (x_0,x_1,\ldots) \in \Delta$.

A metric on Σ will be given as follows. Write $\Lambda_0 = \Delta$ and, for $n = 1,2,\ldots$, put

$$\Lambda_n = \{x \in \Delta: x_k = 0 \quad \text{for } k<n\}.$$

Then (Λ_n) is a neighbourhood basis at zero consisting of a strictly decreasing sequence of compact open subgroups of Δ. Let (β_n) be any strictly decreasing sequence of positive numbers tending to zero, and define d' on $\Delta \times \Delta$ by $d'(x,x) = 0$ and

$$(1.1) \qquad d'(x,y) = \beta_{n+1}, \quad x-y \in \Lambda_n \setminus \Lambda_{n+1}.$$

Then d' is a translation-invariant metric on Δ compatible with the given topology. The real line will be given its Euclidean metric, and then a (translation-invariant) metric d on Σ will be specified by

$$d(\ (\xi,x)+B,B\ ) = \inf \{\max \{|\eta|,d(y,0)\}: (\eta,y) \in (\xi,x)+B\};$$

this is just the metric assigned in the usual way to products and quotients.

We are interested here in how the classical approximation theorems carry over to the solenoid, and in particular the properties of Lipschitz functions on Σ. Some results in this direction have been obtained already by Walker [11] and Bloom [1], [2] and [3]. In Section 2 we give an analogue of the M. Riesz theorem on the uniform boundedness of partial sums of the Fourier series of a pth-integrable function, $1 < p < \infty$. Section 3 will be concerned with the characterization of Lipschitz functions on Σ by the rate of convergence of their Fourier series, and in Section 4 we consider their factorization properties.

2. M. Riesz Theorem for Σ

The classical theorem of M. Riesz holds for R, T, Z and finite products of these three groups; see [6], Chapter 6. To extend the result to Σ we define for positive integers ℓ,n the (ℓ,n)th partial sum $S_{\ell,n}f$ of the Fourier series of $f \in L^1(\Sigma)$ by

$$S_{\ell,n}f = \sum\{\ \hat{f}(\gamma)\gamma: \gamma \in T_{\ell,n}\},$$

where $T_{\ell,n} = \{\gamma_{m,n}: |m/n!| \leq \ell\}$. For each $p \in (1,\infty)$ the operators $S_{\ell,n}$ will be shown to be uniformly bounded on $L^p(\Sigma)$, the proof using results from multiplier theory on locally compact abelian groups.

Let G be any locally compact abelian group, with character group Γ_G. Given $p \in [1,2]$ a function ϕ on Γ_G will be called a multiplier of $L^p(G)$ if for every $f \in L^p(G)$ there exists $T_\phi f \in L^p(G)$ with $T_\phi f\hat{} = \phi\hat{f}$. The smallest admissible K for which $\|T_\phi f\|_p \leq K\|f\|_p$ for all $f \in L^p(G)$ will be denoted by $\|\phi\|_{p,p}$, and termed the multiplier norm of ϕ.

A bounded measurable function h on G is called regulated if there exists an approximate unit (k_ι) in $L^1(G)$ such that $\|k_\iota\|_1 = 1$ and $\lim k_\iota * h = h$ pointwise. Finally, given any non-empty set $E \subset G$, ξ_E will denote its characteristic function and $A(\Gamma_G,E)$ its annihilator in Γ_G; for the latter see [9], (23.23).

THEOREM 2.1. <u>For each</u> $p \in (1,\infty)$ <u>there exists a constant</u> K_p <u>such that</u> $\|S_{\ell,n}f\|_p \leq K_p\|f\|_p$ <u>for all</u> $f \in L^p(\Sigma)$.

PROOF. First consider $p \in (1,2]$. The M. Riesz theorem for R shows that there exists K_p' independent of ℓ such that $\|\xi_{[-\ell,\ell]}\|_{p,p} = K_p'$; see [6], Theorem 6.2.2. Then

$$\tilde{\xi}_{[-\ell,\ell]} = \xi_{[-\ell,\ell]} - \tfrac{1}{2}\xi_{\{-\ell\}} - \tfrac{1}{2}\xi_{\{\ell\}} \tag{2.2}$$

is regulated and $\|\tilde{\xi}_{[-\ell,\ell]}\|_{p,p} = \|\xi_{[-\ell,\ell]}\|_{p,p}$. Furthermore $(\lambda(\Lambda_n)^{-1}\xi_{\Lambda_n})\hat{} = \xi_{L_n}$, where $L_n = A(\Gamma_\Delta,\Lambda_n)$, from which it follows that $\xi_{L_n} \in M_p(\Gamma_\Delta)$ and $\|\xi_{L_n}\|_{p,p} = 1$. Note also that since Γ_Δ is discrete, ξ_{L_n} is regulated.

Define $\psi_{\ell,n}$ on $R \times \Gamma_\Delta$ by $\psi_{\ell,n} = \tilde{\xi}_{[-\ell,\ell]} \otimes \xi_{L_{n-1}}$ **(the tensor product)**. Then $\psi_{\ell,n} \in M_p(R \times \Gamma_\Delta)$ ([5], Lemma 1, p. 375) and $\psi_{\ell,n}$ is regulated. Hence, appealing to [10], Corollary 4.6, the restriction $\phi_{\ell,n}$ of $\psi_{\ell,n}$ to $A(R \times \Gamma_\Delta,B)$ satisfies

$$\|\phi_{\ell,n}\|_{p,p} \leq \|\psi_{\ell,n}\|_{p,p} = \|\tilde{\xi}_{[-\ell,\ell]}\|_{p,p}\,\|\xi_{L_{n-1}}\|_{p,p} = K_p'.$$

Identifying $A(R \times \Gamma_\Delta,B)$ with Γ_Σ we have that $\phi_{\ell,n} \in M_p(\Gamma_\Sigma)$.

Under this identification we can write

$$([-\ell,\ell]\times L_{n-1}) \cap A(R \times \Gamma_\Delta, B) = \{\gamma_{m,n} : |m/n!| \leq \ell\};$$

see [9], (25.3). Furthermore $(\{-\ell,\ell\} \times L_{n-1}) \cap A(R \times \Gamma_\Delta, B) = \{-\ell,\ell\}$ so that, by (2.2), $\phi_{\ell,n} = \xi_{T_{\ell,n}} - 1/2\ \xi_{\{-\ell,\ell\}}$. It follows that for any $f \in L^p(\Sigma)$,

$$\begin{aligned}\|S_{\ell,n}f\|_p &= \|T_{\phi_{\ell,n}} f + 1/2\ \hat{f}(-\ell)\gamma_{-\ell,1} + 1/2\ \hat{f}(\ell)\gamma_{\ell,1}\|_p \\ &\leq (K_p' + 1)\|f\|_p.\end{aligned}$$

This takes care of the case $p \in (1,2]$. A standard duality argument gives the same result for $q \in [2,\infty)$ with constant $K_p' + 1$, where $p^{-1} + q^{-1} = 1$.

COROLLARY 2.3. <u>For</u> $p \in (1,\infty)$ <u>the Fourier series of</u> $f \in L^p(\Sigma)$ <u>converges in the sense that</u> $S_{\ell,n}f \to f$ <u>in</u> $L^p(\Sigma)$ <u>as</u> $\ell,n \to \infty$.

For $p = 1$ or ∞ the convergence no longer holds. This is a standard result once the unboundedness of the Lebesgue constants $\|D_{\ell,n}\|_1$ is established, where $\hat{D}_{\ell,n} = \xi_{T_{\ell,n}}$; see also Hawley [8].

THEOREM 2.4. $$\|D_{\ell,n}\|_1 \sim 4\pi^{-2} \log(n!\ell).$$

PROOF. Let $i: Z \to Q$ denote the inclusion map and define ρ_n on Q by $\rho_n(r) = (n!)^{-1}r$. Then $D_{\ell,n} = D'_{\ell,n} \circ \tilde{i} \circ \tilde{\rho}_n$, where $D'_{\ell,n}$ is the Dirichlet polynomial on T of order $n!\ell$ and $\tilde{i}$, $\tilde{\rho}_n$ are the adjoints ([9], (24.37)) of i, ρ_n respectively. The result now follows by appealing to [9], (28.54) (v).

COROLLARY 2.5. <u>There exist functions in</u> $L^1(\Sigma)$ <u>and</u> $C(\Sigma)$ <u>whose Fourier series do not converge in norm</u>.

3. Lipschitz Spaces

For $p \in [1,\infty]$ and $\alpha \in (0,1)$ the Lipschitz space $\mathrm{Lip}(\alpha;p)$ is defined by

$$\mathrm{Lip}(\alpha;p) = \{f \in L^p(\Sigma): \|{}_af - f\|_p = O(d(a,0)^\alpha),\ a \to 0\} ,$$

where ${}_af: x \to f(x - a)$; when $p = \infty$ the members of $\mathrm{Lip}(\alpha;p)$ are taken to be continuous. It is known ([2], Theorem 5) that for certain choices of (β_n) in (1.1) the members of $\mathrm{Lip}(\alpha;p)$ can be characterized by the rate of decay of

$$E_{\ell,n}(p;f) = \inf\{\|f - t\|_p : \mathrm{supp}\,(\hat{t}) \subset T_{\ell,n}\} ,$$

the best approximation in $L^p(\Sigma)$ of f by trigonometric polynomials of degree (ℓ,n). Important in this characterization is the following analogue of the classical approximation theorem of Jackson (for a proof see [1], Theorem 4).

THEOREM 3.1. *The Banach algebra* $L^1(\Sigma)$ *admits a bounded positive approximate unit* $(k_{\ell,n})$ *such that for each* ℓ,n, $k_{\ell,n} \in C(\Sigma)$, $\hat{k}_{\ell,n}(0) = 1$, $\mathrm{supp}(\hat{k}_{\ell,n}) \subset T_{\ell,n}$ *and*

$$\|k_{\ell,n} * f - f\|_p \leq K \sup\{\|{}_af - f\|_p : a \in \pi_B((-\ell^{-1},\ell^{-1}) \times \Lambda_{n-1})\}$$

for every $f \in L^p(\Sigma)$ *if* $p \in [1,\infty)$, *or for every continuous* f *if* $p = \infty$. *Here* π_B *denotes the natural homomorphism of* $R \times \Delta$ *onto* Σ *and* K *is a constant.*

In particular if $f \in \mathrm{Lip}(\alpha;p)$ and β_n^{-1} is an integer then $E_{\beta_n^{-1},n}(p;f) = O(\beta_n^\alpha)$. The converse of this result holds for $\beta_n = 2^{-n}$ (see [2], Theorem 5). Using Theorem 2.1 we can give the characterization in terms of the partial sums of the Fourier series of f.

THEOREM 3.2. *Take* $\beta_n = 2^{-n}$ *and let* $p \in (1,\infty)$. *Then* $f \in \mathrm{Lip}(\alpha;p)$ *if and only if* $\|S_{2^n,n}f - f\|_p = O(2^{-n\alpha})$.

PROOF. Since $\mathrm{supp}(S_{2^n,n}f)\hat{} \subset T_{2^n,n}$, one implication follows immediately from [2], Theorem 1. Conversely if $f \in \mathrm{Lip}(\alpha;p)$ then, by the remark following Theorem 3.1, $E_{2^n,n}(p;f) = O(2^{-n\alpha})$. Now let t be any trigonometric polynomial with $\mathrm{supp}(\hat{t}) \subset T_{\ell,n}$. Then $S_{\ell,n}t = t$ and

$$\|S_{\ell,n}f - f\|_p \leq \|S_{\ell,n}f - S_{\ell,n}t\|_p + \|f - t\|_p \leq (K_p + 1)\|f - t\|_p,$$

where K_p is the constant of Theorem 2.1. Since t was chosen arbitrarily we have

$$\|S_{\ell,n}f - f\|_p \leq (K_p + 1)E_{\ell,n}(p;f), \tag{3.3}$$

and the result follows on putting $\ell = 2^n$.

4. Factorization of Lipschitz Functions

The problem of factorizing Lipschitz functions on Euclidean space or the torus was first considered by L.-S. Hahn [7]. More recently Bloom [4] has given a factorization theorem for Lipschitz functions on an arbitrary locally compact metrisable zero dimensional group; see also the references cited there for other results in this direction.

THEOREM 4.1. *Take* $\beta_n = 2^{-n^2}$ *and let* $p \in (1,2]$. *There exists* $g \in L^p(\Sigma)$ *such that for all* $f \in \mathrm{Lip}(\alpha;q)$ *with* $\alpha > q^{-1}$ *there corresponds* $h \in L^q(\Sigma)$ *with* $f = g * h$, *where* $p^{-1} + q^{-1} = 1$.

PROOF. Choose $\beta \in (q^{-1},\alpha)$ and put

$$g = D_1 + \sum_{n=1}^{\infty} 2^{-n^2\beta}(D_{n+1} - D_n),$$

where $D_n = D_{2^{n^2},n}$. Now, from Theorem 2.4, $\|D_n\|_1 \sim 4\pi^{-2}\log(n!2^{n^2})$ and Plancherel's theorem gives $\|D_n\|_2 = (n!2^{n^2+1} + 1)^{1/2}$. Using Hölder's inequality we obtain for some constant K

$$\|g\|_p \leq \|D_1\|_p + K\sum_{n=1}^{\infty} 2^{-n^2\beta}(\log((n+1)!2^{n^2+2n+1}))^{(2-p)/p}((n+1)!2^{n^2+2n+2}+1)^{(p-1)/p},$$

which is finite for $\beta > (p-1)/p = q^{-1}$. Thus $g \in L^p(\Sigma)$. Write

$$h = D_1 * f + \sum_{n=1}^{\infty} 2^{n^2\beta}(D_{n+1} - D_n) * f.$$

As the Fourier transforms of the $(D_{n+1} - D_n) * f$ are pairwise disjoint, we have

$$g * h = D_1 * f + \sum_{n=1}^{\infty} (D_{n+1} - D_n) * f = \lim_{n\to\infty} D_n * f = f,$$

the last equality following from Corollary 2.3 since $S_{2^{n^2},n} f = D_n * f$. Also, by (3.3) and the remark following Theorem 1,

$$\begin{aligned}\|h\|_q &\leq \|D_1 * f\|_q + \sum_{n=1}^{\infty} 2^{n^2\beta}\|D_{n+1} * f - D_n * f\|_q \\ &\leq \|D_1 * f\|_q + \sum_{n=1}^{\infty} 2^{n^2\beta}(K_p + 1)K2^{-n^2\alpha+1} < \infty\end{aligned}$$

since $\alpha > \beta$, so that $h \in L^q(\Sigma)$.

It should be noted that a version of Theorem 4.1 holds also when $p = 1$ since in this case for $\alpha > 0$,

$$\mathrm{Lip}(\alpha;\infty) \subset C(\Sigma) = L^1(\Sigma) * L^{\infty}(\Sigma)$$

by [9], (32.45)(b).

COROLLARY 4.2. _Take_ $\beta_n = 2^{-n^2}$ _and_ $p \in [1,2]$. _Then for_ $\alpha > q^{-1}$, $\mathrm{Lip}(\alpha;q)^{\wedge} \subset \ell^r$, _where_ $r = 2p/(3p - 2)$.

The proof of Corollary 4.2 just uses Theorem 4.1, the Hausdorff-Young theorem and Hölder's inequality. This result has been obtained previously ([3], Theorem 3 and the remarks following it), where it was also shown that the range of values of α could not be extended. In particular the same is true of Theorem 4.1. Corollary 4.2 is also given in [11], Theorem 1 in the case $p = 2$, but for the smaller Lipschitz space obtained by taking $\beta_n = e^{-(n+1)!}$.

REFERENCES

[1] Bloom, W.R., _Jackson's theorem for finite products and homomorphic_

images of locally compact abelian groups. Bull. Austral. Math. Soc. 12 (1975), 301-309.

[2] Bloom, W.R., A characterisation of Lipschitz classes on finite dimensional groups. Proc. Amer. Math. Soc. 59 (1976), 297-304.

[3] Bloom, W.R., Absolute convergence of Fourier series on finite dimensional groups. Colloq. Math. (to appear).

[4] Bloom, W.R., Factorisation of Lipschitz functions on zero dimensional groups. Bull. Austral. Math. Soc. (to appear).

[5] Bonami, A., Étüde des coefficients de Fourier des fonctions de $L^p(G)$. Ann. Inst. Fourier (Grenoble) 20 (1970), 335-402.

[6] Edwards, R.E. and Gaudry, G.I., Littlewood-Paley and multiplier theory. Ergebnisse der Mathematik und ihrer Grenzgebiete, Band 90, Springer-Verlag, Berlin/Heidelberg/New York 1977.

[7] Hahn, L.-S., On multipliers of p-integrable functions. Trans. Amer. Math. Soc. 128 (1967), 321-335.

[8] Hawley, D.N., Fourier analysis on a-adic solenoids, Doctoral Dissertation, University of Oregon, Eugene, Oregon, 1969.

[9] Hewitt, E. and Ross, K.A., Abstract harmonic analysis, vols. I, II. Die Grundlehren der mathematischen Wissenschaften, Bände 115, 152, Springer-Verlag, Berlin/Heidelberg/New York 1963, 1970.

[10] Saeki, S., Translation invariant operators on groups. Tôhoku Math. J. 22 (1970), 409-419.

[11] Walker, P.L., Lipschitz classes on finite dimensional groups. Proc. Cambridge Philos. Soc. 66 (1969), 31-38.

IV Fourier Analysis and Integral Transforms

BERNSTEIN AND MARKOV TYPE ESTIMATES FOR THE DERIVATIVE OF A POLYNOMIAL WITH REAL ZEROS

József Szabados
Mathematical Institute
of the Hungarian Academy of Sciences
Budapest

Starting from an old result of P. Erdös [1], we give Bernstein and Markov type estimates for the derivative of algebraic and trigonometric polynomials with real zeros. As for the order of magnitude, in some cases these estimates turn out to be optimal.

1. The Algebraic Case

Denote by $P(n,k)$ $(0 \leq k \leq n;\ n = 1,2,\ldots)$ the set of those algebraic polynomials $p(x)$ of degree n which have only real roots, k of them in the interval $(-1,1)$, and for which $\max_{|x| \leq 1} |p(x)| \leq 1$. P. Erdös [1] proved that if $p(x) \in P(n,0)$ then

$$|p'(x)| < \frac{1}{2} en \qquad (|x| \leq 1);$$

and this is the best possible estimate in the sense that there exists a sequence of polynomials $p_n(x) \in P(n,0)$ $(n = 1,2,\ldots)$ such that $\lim_{n\to\infty} n\, p_n'(1) = e/2$. In a joint paper with A.K. Varma [3] we generalized this result by showing that if $p(x) \in P(n,1)$ then

$$|p'(x)| \leq c_1 n \qquad (|x| \leq 1)$$

with an absolute constant $c_1 > 0$ [1]. Later I was able to further extend

---

[1] In what follows, $c_1, c_2, \ldots$ will denote absolute positive constants.

this to

$$|p'(x)| \leq \left(e^{c_2 k}\right) n \qquad (|x| \leq 1)$$

whenever $p(x) \in P(n,k)$. Nevertheless, I did not publish this result because meanwhile A. Máté [2] has shown that [2)]

$$(1) \qquad |p'(x)| \leq \left(e^{c_3\sqrt{k}}\right) n \qquad (|x| \leq 1)$$

provided $p(x) \in P(n,k)$. First I would like to state the following

PROBLEM 1. *Is it true that*

$$(2) \qquad |p'(x)| \leq c_4\, k\, n \qquad (|x| \leq 1),$$

if $p(x) \in P(n,k)$?

I think the answer is yes, but even the ingenious method of Máté cannot give (2). (He used a result of D. Newman on rational approximation of $|x|$ which cannot be further improved.) Being rather far from the best estimate, I can only show that (2) already cannot be sharpened, by the following

Example 1: Let (denoting by $P_k^{(\alpha,\beta)}(x)$ the kth Jacobi polynomial with parameters α, β)

$$p(x) = \left(\frac{1-x}{2}\right)^{n-k} P_k^{(2n-2k-\frac{1}{2},0)}(x).$$

Then by Szegö [4], (7.21.2), $p(x) \in P(n,k)$. Further, by Szegö [4], (4.21.7) and (4.1.4)

$$p'(-1) = (n-k)\left(-\frac{1}{2}\right)(-1)^k + \frac{3n-2k+\frac{1}{2}}{2}(-1)^{k-1} k$$

$$= (-1)^{k+1}\left(n\,\frac{3k+1}{2} - k\,\frac{4k+1}{4}\right)$$

---

2) Actually, he proved (1) even under somewhat weaker restrictions for the roots of $p(x)$, and extended the result for higher derivatives and L_p-metric.

i.e.,

$$\frac{1}{2} \leqslant \frac{\max_{|x|\leqslant 1} |p'(x)|}{nk} \leqslant 2 \qquad (0 < k \leqslant n).$$

To finish this section, I mention a problem concerning the pointwise estimate of derivatives of polynomials from the class $P(n,0)$.

PROBLEM 2. _Is it true that_

$$|p'(x)| \leqslant c_5 \sqrt{\frac{n}{1-x^2}} \qquad (|x| < 1)$$

whenever $p(x) \in P(n,0)$?

A slightly different form of this inequality (when $p(x)$ has no root in the unit circle, and $\sqrt{1-x^2}$ is replaced by $(1-x^2)^2$) has been proved in the cited paper of Erdös [1].

2. The Trigonometric Case.

It is easily seen that the obvious transformation $x' = \cos x$ reducing the trigonometric case to the algebraic one does not work in our case. Therefore we make a direct approach to the problem similar to the method of Erdös [1].

Denote by T_n the set of all trigonometric polynomials of degree n, and by $T_n(\omega)$ ($\subset T_n$, $0 \leqslant \omega < \pi$) that subset which contains those trigonometric polynomials of degree n which have only real roots, outside of the interval $(-\omega, \omega)$ 3). Generalizing the classical Bernstein's inequality

$$\max_{-\infty<x<\infty} |t'(x)| \leqslant n\Big(\max_{-\infty<x<\infty} |t(x)|\Big) \qquad (t(x) \in T_n),$$

3) For $\omega = 0$, $T_n(0)$ denotes the set of trigonometric polynomials of degree n which have only real roots.

V.S. Videnskiĭ [5] proved that

$$(3) \qquad \max_{|x|\leq\omega} |t'(x)| \leq 2n^2 \cot\frac{\omega}{2} \max_{|x|\leq\omega} |t(x)| \qquad (t(x)\in T_n, 0<\omega<\pi)$$

and

$$(4) \qquad |t'(x)| \leq n\sqrt{\frac{1+\cos x}{\cos x-\cos\omega}} \max_{|y|\leq\omega} |t(y)| \qquad (t(x)\in T_n, |x|<\omega<\pi).$$

The polynomial

$$\cos\left(2n \arccos\left(\frac{\sin\frac{x}{2}}{\sin\frac{\omega}{2}}\right)\right)$$

shows that these inequalities are sharp. It is our purpose to show that for the class $T_n(\omega)$, the order of magnitude in (3) and (4) can be essentially improved.

THEOREM 1. <u>If</u> $t(x)\in T_n(\omega)$ <u>then</u>

$$\max_{|x|\leq\omega} |t'(x)| \leq c_6\left(\frac{n}{\sin\frac{\omega}{2}}\right) \max_{|x|\leq\omega} |t(x)|.$$

For the proof we need some lemmas.

LEMMA 1. <u>If</u> $0<\beta-\alpha\leq\pi, t(x)\in T_n(0), t(x)t'(x)>0$ <u>in</u> (α,β) <u>and</u>

$$(5) \qquad x_1 = \sup\{x : t(x)=0,\ x\leq\alpha\}$$

<u>then</u>

$$|t(x)| \leq e\,\frac{\sin\frac{x-x_1}{2}\left(\cos\frac{\beta-x}{2}\right)^{2n-1}}{\sin\frac{\beta-x_1}{2}}\,|t(\beta)| \qquad (\alpha\leq x\leq\beta).$$

Of course, similar statement holds when $t(x)t'(x)<0$ in (α,β). Then denoting

$$x_2 = \inf\{x : t(x) = 0,\ x \geqslant \beta\}$$

we have

$$|t(x)| \leqslant e\,\frac{\sin\dfrac{x_2-x}{2}\left(\cos\dfrac{x-\alpha}{2}\right)^{2n-1}}{\sin\dfrac{x_2-\alpha}{2}}\,|t(\alpha)| \qquad (\alpha \leqslant x \leqslant x_2).$$

PROOF OF LEMMA 1. Let

$$t(x) = c\prod_{k=1}^{2n}\sin\frac{x-x_k}{2}\,; \tag{6}$$

then

$$0 \leqslant \frac{t(x)}{t(\beta)} = \prod_{k=1}^{2n}\left\{\frac{\sin\dfrac{x-x_k}{2}}{\sin\dfrac{\beta-x_k}{2}}\right\} = \frac{\sin\dfrac{x-x_1}{2}}{\sin\dfrac{\beta-x_1}{2}}\left(\cos\frac{\beta-x}{2}\right)^{2n-1}\prod_{k=2}^{2n}\left(1-\tan\frac{\beta-x}{2}\cot\frac{\beta-x_k}{2}\right).$$

Here, using that $1-u \leqslant e^{-u}$ and

$$2\,\frac{t'(\beta)}{t(\beta)} = \sum_{k=1}^{2n}\cot\frac{\beta-x_k}{2} \geqslant 0,$$

we get

$$\prod_{k=2}^{2n}\left(1-\tan\frac{\beta-x}{2}\cot\frac{\beta-x_k}{2}\right) \leqslant \exp\left\{-\tan\frac{\beta-x}{2}\sum_{k=2}^{2n}\cot\frac{\beta-x_k}{2}\right\}$$

$$\leqslant \exp\left\{\tan\frac{\beta-x}{2}\cot\frac{\beta-x_1}{2}\right\} \leqslant e \qquad (x_1 \leqslant x \leqslant \beta)$$

which proves the lemma.

LEMMA 2. <u>If</u> $t(x) \in T_n(\omega)$ $(0<\omega<\pi/2)$ <u>and</u> $x_o \in (-\omega,\omega)$ <u>is such that</u> $t'(x_o)=0$ <u>then with the notation</u> (6) <u>we have</u>

$$\sum_{k=1}^{2n} \left|\cot \frac{x_o - x_k}{2}\right| \leqslant 2n \cot\frac{\omega}{2}.$$

PROOF. $t'(x_o) = 0$ implies

$$\sum_{k=1}^{2n} \cot \frac{x_o - x_k}{2} = 0, \tag{7}$$

i.e.,

$$A \overset{\text{def}}{=} \sum_{\cot \frac{x_o - x_k}{2} \geqslant 0} \cot \frac{x_o - x_k}{2} = \sum_{\cot \frac{x_k - x_o}{2} > 0} \cot \frac{x_k - x_o}{2}.$$

We have

$$A \leqslant \max_{1 \leqslant \ell \leqslant 2n} \min\left(\ell \cot \frac{x_o + \omega}{2}, (2n-\ell) \cot \frac{\omega - x_o}{2}\right).$$

Here the maximum is attained when ℓ is one of the two integers nearest

$$\frac{2n \cot \frac{\omega - x_o}{2}}{\cot \frac{\omega + x_o}{2} + \cot \frac{\omega - x_o}{2}} = \frac{2n \sin \frac{\omega + x_o}{2} \cos \frac{\omega - x_o}{2}}{\sin \omega}.$$

Thus

$$A \leqslant 2n \frac{\cos \frac{\omega + x_o}{2} \cos \frac{\omega - x_o}{2}}{\sin \omega} \leqslant n \cot \frac{\omega}{2}.$$

Q.E.D.

PROOF OF THEOREM 1. First we prove the theorem when $\omega \leqslant \pi/8$. We distinguish two cases.

Case 1: $t'(x) \neq 0$ in $(-\omega,\omega)$; say $t(x)t'(x) > 0$. Let x_1 be defined as in (5), provided $-\alpha = \beta = \omega$. Then evidently

$$(8)\qquad 0 \leqslant 2\,\frac{t'(x)}{t(x)} = \sum_{k=1}^{2n} \cot \frac{x-x_k}{2}$$

$$\leqslant \sum_{x-\pi \leqslant x_k \leqslant x_1} \cot \frac{x-x_k}{2} \leqslant \frac{2n}{\sin\frac{x-x_1}{2}} \qquad (|x| \leqslant \omega).$$

Thus we have by Lemma 1

$$|t'(x)| \leqslant |t(x)| \frac{n}{\sin\frac{x-x_1}{2}} \leqslant e\, \frac{\sin\frac{x-x_1}{2}\left(\cos\frac{\omega-x}{2}\right)^{2n-1}}{\sin\frac{\omega-x_1}{2}}\, |t(\omega)| \frac{n}{\sin\frac{x-x_1}{2}}$$

$$\leqslant \frac{en}{\sin\frac{\omega}{2}} \max_{|x| \leqslant \omega} |t(x)| .$$

The proof is similar when $t(x)t'(x) < 0$ in $(-\omega,\omega)$.

Case 2: $t'(x_0) = 0, x_0 \in (-\omega,\omega)$. We may assume that $t(x) > 0$ in $(-\omega,\omega)$. Apply Lemma 1 with $\alpha = -\omega, \beta = x_0$:

$$(9)\qquad 0 \leqslant t(x) \leqslant e\, \frac{\sin\frac{x-x_1}{2}}{\sin\frac{x_0-x_1}{2}}\, t(x_0) \qquad (-\omega \leqslant x \leqslant x_0).$$

We now distinguish two subcases.

Subcase 2a: $x_0 - x_1 \geqslant \omega$. Then similarly as in (8), we get from (9)

$$0 \leqslant t'(x) \leqslant \frac{2en}{\sin\frac{\omega}{2}} \max_{|x| \leqslant \omega} |t(x)| .$$

Subcase 2b: $x_0 - x_1 < \omega$. The function

$$\varphi(u) = \frac{\cot\frac{x-u}{2}}{\cot\frac{x_o-u}{2}} \qquad (-\omega \leq x \leq x_o)$$

being monotone increasing for $\left|\frac{x_o+x}{2} - u\right| \leq \frac{\pi}{2}$, we get by Lemma 2

$$\sum_{k=1}^{2n} \cot\frac{x-x_k}{2} \leq \sum_{x-\pi\leq x_k\leq x_1} \cot\frac{x-x_k}{2}$$

$$= \left(\sum_{x-\pi\leq x_k\leq\omega-\pi/2} + \sum_{\omega-\pi/2<x_k\leq-\omega}\right)\cot\frac{x-x_k}{2}$$

$$\leq 2n\cot\frac{\pi}{8} + \frac{\cot\frac{x-x_1}{2}}{\cot\frac{x_o-x_1}{2}}\sum_{k=1}^{2n}\left|\cot\frac{x_o-x_k}{2}\right|$$

$$\leq 2n\left(\cot\frac{\pi}{8} + \frac{\sin\frac{x_o-x_1}{2}}{\cos\frac{\omega}{2}\sin\frac{x-x_1}{2}}\cot\frac{\omega}{2}\right).$$

Thus by (9)

$$0 \leq t'(x) = \frac{1}{2}t(x)\sum_{k=1}^{2n}\cot\frac{x-x_k}{2} \leq n\left(t(x)\cot\frac{\pi}{8} + \frac{e}{\sin\frac{\omega}{2}}t(x_o)\right)$$

$$\leq c_7\frac{n}{\sin\frac{\omega}{2}}\max_{|x|\leq\omega}|t(x)|.$$

The interval $[x_o,\omega]$ can be treated analogously.

Finally, if $\omega > \pi/8$ then the interval $(-\omega,\omega)$ can be divided into subintervals of length $< \pi/8$ and repeated application of the just proved part of the theorem gives the desired result. Q.E.D.

Apart from the constant c_5 in Theorem 1, the estimate given there is asymptotically best possible when $\omega\to 0$ or $n\to\infty$. This can be seen from

Example 2: Let

$$t_1(x) = (\sin\omega - \sin x)^{n-1}\ (\sin\omega + \sin x) \qquad (0 < \omega < \tfrac{\pi}{2}).$$

Indeed, then $t_1(x) \in T_n(\omega)$ and

$$\max_{|x|\leqslant\omega} |t_1(x)| \cong \frac{2^n}{en}\sin^n\omega, \qquad |t_1'(-\omega)| \cong 2^{n-1}\sin^{n-1}\omega\cos\omega,$$

thus

$$\frac{\max_{|x|\leqslant\omega}|t_1'(x)|}{\max_{|x|\leqslant\omega}|t_1(x)|} \geqslant c_8\frac{n}{\sin\omega}\cos\omega.$$

The following problem remains open.

PROBLEM 3. <u>What is the best constant</u> c_6 <u>in Theorem</u> 1 ?

So far we have not used the $\left(\cos\frac{\beta-1}{2}\right)^{2n-1}$ factor in the estimate of Lemma 1. This will be done in the proof of the next pointwise estimate.

THEOREM 2. <u>If</u> $t(x) \in T_n(\omega)$ <u>then</u>

$$|t'(x)| \leqslant c_9\frac{\sqrt{n}\cot(\omega/4)}{\sin\frac{\omega-x}{2}\sin\frac{\omega+x}{2}}\max_{|y|\leqslant\omega}|t(y)| \qquad (|x| < \omega).$$

PROOF. First we prove the statement when $\omega \leqslant \pi/2$. Just like in the proof of Theorem 1, we distinguish two cases.

<u>Case 1</u>: $t'(x) \neq 0$ in $(-\omega,\omega)$; say $t(x)t'(x) > 0$. Then applying Lemma 1 with $-\alpha = \beta = \omega$ and using

$$\sin\frac{\omega-x}{2}\left(\cos\frac{\omega-x}{2}\right)^{2n-1} \leqslant \frac{c_{10}}{\sqrt{n}} \tag{10}$$

we obtain

$$|t(x)| \leqslant \frac{ec_{10}\sin\frac{x-x_1}{2}}{\sqrt{n}\sin\omega\sin\frac{\omega-x}{2}}|t(\omega)| \qquad (-\omega<x<\omega).$$

Hence by (8),

$$0 < t'(x) \leqslant t(x)\frac{n}{\sin\frac{x-x_1}{2}} \leqslant \frac{c_{11}\sqrt{n}}{\sin\frac{\omega-x}{2}}\frac{|t(\omega)|}{\sin\omega} \qquad (|x|<\omega).$$

Case 2: There exists an $x_o \in (-\omega,\omega)$ such that $t'(x_o)=0$. We may assume $t(x)>0, x\in(-\omega,\omega)$. Using (7) and Lemma 2 we get with $|x_o-x_k|\leqslant\pi$

$$0 < \sum_{k=1}^{2n}\cot\frac{x-x_k}{2} = \sum_{k=1}^{2n}\left(\cot\frac{x-x_k}{2}-\cot\frac{x_o-x_k}{2}\right)$$

$$= \sin\frac{x_o-x}{2}\sum_{k=1}^{2n}\left(\sin\frac{x-x_k}{2}\sin\frac{x_o-x_k}{2}\right)^{-1}$$

$$= \sin\frac{x_o-x}{2}\left\{\left(\sin\frac{\omega+x}{2}\right)^{-1}\sum_{x_k\leqslant-\omega}\left(\sin\frac{x_o-x_k}{2}\right)^{-1}\right.$$

$$\left.+\left(\sin\frac{\omega-x}{2}\right)^{-1}\sum_{x_k\geqslant\omega}\left(\sin\frac{x_k-x_o}{2}\right)^{-1}\right\}$$

$$\leqslant c_{12}\frac{\sin\frac{x_o-x}{2}}{\sin\frac{\omega-x}{2}\sin\frac{\omega+x}{2}}\sum_{k=1}^{2n}\left|\sin\frac{x_o-x_k}{2}\right|^{-1}$$

$$\leqslant c_{13}\frac{\sin\frac{x_o-x}{2}}{\sin\frac{\omega-x}{2}\sin\frac{\omega+x}{2}}\left\{\sum_{|x_o-x_k|<\pi/2}\left|\cot\frac{x_o-x_k}{2}\right| + \sum_{\pi/2\leqslant|x_o-x_k|\leqslant\pi}(\sqrt{2})\right\}$$

$$\leqslant c_{14}\frac{n\sin\frac{x_o-x}{2}\cot\frac{\omega}{2}}{\sin\frac{\omega-x}{2}\sin\frac{\omega+x}{2}} \qquad (-\omega<x\leqslant x_o).$$

Thus Lemma 1 with $\alpha=-\omega$, $\beta=x_o$ yields

$$0 \leqslant t'(x) = \frac{1}{2}\, t(x) \sum_{k=1}^{2n} \cot \frac{x-x_k}{2}$$

$$\leqslant \frac{e}{2}\left(\cos\frac{x_o-x}{2}\right)^{2n-1} c_{14} \frac{n \sin\frac{x_o-x}{2}\cot\frac{\omega}{2}}{\sin\frac{\omega-x}{2}\sin\frac{\omega+x}{2}} |t(x_o)|$$

$$\leqslant c_{15} \frac{\sqrt{n}\cot(\omega/2)}{\sin\frac{\omega-x}{2}\sin\frac{\omega+x}{2}} \max_{|y|\leqslant\omega} |t(y)| \qquad (-\omega<x\leqslant x_o).$$

The proof for the interval $x_o \leqslant x<\omega$ is analogous.

It remains to settle the case $\pi/2<\omega<\pi$. If $t(x)\in T_n(\omega)$ then $t(2x)\in T_{2n}(\omega/2)$. Applying the just proved statement for $t(2x)$ we get

$$|t'(2x)| \leqslant c_{16} \frac{\sqrt{n}\cot(\omega/4)}{\sin\frac{\omega/2-x}{2}\sin\frac{\omega/2+x}{2}} \max_{|y|\leqslant\omega/2} |t(2y)| \qquad (|x|<\omega/2)$$

i.e.,

$$|t'(x)| \leqslant c_{17} \frac{\sqrt{n}\cot(\omega/4)}{\sin\frac{\omega-x}{4}\sin\frac{\omega+x}{4}} \max_{|y|\leqslant\omega} |t(y)|$$

$$\leqslant c_{18} \frac{\sqrt{n}\cot(\omega/4)}{\sin\frac{\omega-x}{2}\sin\frac{\omega+x}{2}} \max_{|y|\leqslant\omega} |t(y)| \qquad (|x|<\omega).$$

Thus Theorem 2 is completely proved.

The following example shows that Theorem 2 cannot be essentially improved.

Example 3: Let

$$t_2(x) = \frac{\left(\sin\frac{\omega-x}{2}\sin\frac{\omega+x}{2}\right)^n}{\sin^{2n}(\omega/2)} .$$

Then $t_2(x) \in T_n(\omega)$ and $\max_{|x|\le\omega} |t_2(x)| = 1$. We have

$$\sin\frac{\omega-x}{2}\sin\frac{\omega+x}{2}t'(x) = n\,\frac{\left(\sin\frac{\omega-x}{2}\sin\frac{\omega+x}{2}\right)^n}{2\sin^{2n}(\omega/2)}\sin x = \frac{n}{2}\left(\frac{\cos x-\cos\omega}{1-\cos\omega}\right)^n \sin x.$$

Put here $x = y_0$ defined by

$$\cos y_0 = 1 - \frac{2\sin^2(\omega/2)}{n}, \quad \sin y_0 \sim \frac{2\sin(\omega/2)}{\sqrt{n}},$$

then

$$\sin\frac{\omega-y_0}{2}\sin\frac{\omega+y_0}{2}t'(y_0) \sim \sqrt{n}\sin\frac{\omega}{2}.$$

REFERENCES

[1] Erdös, P., On extremal properties of the derivatives of polynomials. Ann. of Math. 41 (1940), 310 - 313.

[2] Máté, A., Inequalities for derivatives of polynomials with restricted zeros. (to appear).

[3] Szabados, J. - Varma, A.K., Inequalities for derivatives of polynomials having real zeros. (to appear).

[4] Szegö, G., Orthogonal Polynomials. Coll. Publ. Amer. Math. Soc. vol. 23, Providence, 1975.

[5] Videnskiĭ, V.S., Extremal estimates for the derivative of a trigonometric polynomial on an interval shorter than the period (Russ.). Dokl. Akad. Nauk SSSR 130 (1) (1960), 13 - 16.

PROJECTIONS WITH NORMS SMALLER THAN THOSE OF THE ULTRASPHERICAL AND LAGUERRE PARTIAL SUMS

E. Görlich and C. Markett*)

Lehrstuhl A für Mathematik

Rheinisch - Westfälische Technische Hochschule

Aachen

Norm estimates from above and below for partial sum operators of ultraspherical and Laguerre expansions on a class of weighted Lebesgue spaces are established, using ultraspherical and Laguerre weights with parameters different from the parameters of the orthogonal expansions. It turns out that a suitable shifting of the parameters leads to a considerable reduction of the rate of growth of the operator norms. In this way projection operators on weighted Lebesgue spaces can be constructed, the norms of which are smaller than those of the corresponding partial sums. Thus first upper estimates for the minimal projections in these spaces are obtained.

1. Introduction and Main Results

As is well known, the Fourier partial sums are the minimal projections from $C_{2\pi}$ onto the trigonometric polynomials, but the Chebyshev partial sums $S_n^{-1/2}$ do not have the corresponding property with respect to $C[-1,1]$ and the algebraic polynomials. The latter fact has been established by Cheney and Rivlin [3] for each n by showing that the Lebesgue function of $S_n^{-1/2}$ attains its maximum at the two end points of the interval only, a fact which contradicts a necessary condition for minimal projections due to Morris and Cheney [10].

In the present paper it will be shown that a similar negative statement holds for the ultraspherical partial sum operator S_n^{α} for $\alpha > -1/2$ as well as for the Laguerre partial sums S_n^{α} for $\alpha > -1/3$. In both cases we explicitely give projection operators on the corresponding spaces with norms smaller than those of the partial sums. Indeed, it will be shown that, on a fixed space,

*) This author was supported by a DFG grant (Ne 171/4) which is gratefully acknowledged.

such "better" projections consist e.g. in partial sum operators corresponding
to a weight with parameter shifted. The main purpose of this paper is to give
a quantitative description of this effect. Concerning the convergence of partial sums on weighted L^p - spaces, Muckenhoupt already tried to enlarge the
p - interval of convergence by a variation of the weight parameters. In the
Jacobi case [11] he succeeded, while in the Laguerre case [12,I] he could
prove that the p - interval cannot be enlarged this way.

Let P_n be the set of algebraic polynomials of degree $\leqslant n$, $n \in \mathbb{P} = \{0,1,2,...\}$,
$\mathbb{N}$ the set of naturals. By M_n we always denote a m i n i m a l   p r o j e c -
t i o n from the given space onto P_n. By $L^p_{w(\alpha)}$ and $L^p_{\mathcal{w}(\alpha)}$ we mean the Lebesgue spaces with ultraspherical or Laguerre weight, respectively, as indicated
below (cf. (2.3), (2.11)). Denoting further by C a positive constant which
may have different values at each occurrence and writing $a_n \sim b_n$ for two se-
quences $\{a_n\},\{b_n\}$ with the property that $a_n = \mathcal{O}(b_n)$ and $b_n = \mathcal{O}(a_n)$ as $n \to \infty$, our
main results are as follows:

THEOREM 1. (Ultraspherical case) Let $a > -1$. For each $n \in \mathbb{N}$ there exists a projection operator $P_n : L^1_{w(a)} \to P_n$ such that

(1.1) $$\|M_n\|_{[L^1_{w(a)}]} \leqslant \|P_n\|_{[L^1_{w(a)}]} \leqslant C \log(n+1).$$

In particular, for $a > -1/2$ the $\|M_n\|_{[L^1_{w(a)}]}$ are asymptotically smaller than
the Lebesgue constants of the partial sums $\|S^a_n\|_{[L^1_{w(a)}]} \sim n^{a+1/2}$, $n \to \infty$.

THEOREM 2. (Laguerre case) Let $a > -1$. For each $n \in \mathbb{N}$ there exists a projection operator $P_n : L^1_{\mathcal{w}(a)} \to P_n$ such that

(1.2) $$\|M_n\|_{[L^1_{\mathcal{w}(a)}]} \leqslant \|P_n\|_{[L^1_{\mathcal{w}(a)}]} \leqslant C n^{1/6}.$$

In particular, for $a > -1/3$ the $\|M_n\|_{[L^1_{\mathcal{w}(a)}]}$ are asymptotically smaller than the
Lebesgue constants $\|S^a_n\|_{[L^1_{\mathcal{w}(a)}]} \sim n^{a+1/2}$, $n \to \infty$.

These results will be obtained as corollaries of Theorems 3 and 4 below which
describe the asymptotical norm behaviour of the partial sums with parameters

$\alpha \geqslant a$. The latter two theorems generalize results of Rau [13] and Lorch [7] in the ultraspherical case and of the authors [5] , [8] , [9] in the Laguerre case, respectively.

REMARKS. i) Theorem 1 was formulated for convenience only for the ultraspherical case, but it can be extended to general Jacobi weights. Moreover, Theorems 1 and 2 may be extended to $L^p_{w(a)}$ - and $L^p_{\omega(a)}$ - spaces for $p > 1$ (cf. the remark following Thm. 4).

ii) In contrast to the ultraspherical case, the Laguerre results are given here in terms of spaces $L^p_{\omega(a)}$ which have not been customary so far. But these spaces appear to be particulary suited for Laguerre expansions under several aspects which will be discussed in a subsequent paper. In particular, they lead to a marked similarity between the statements in the ultraspherical and the Laguerre case.

iii) Besides the upper bounds of the $\|M_n\|$ given, it would of course be of interest to have lower bounds, too. In this context let us only mention that the usual tool for lower bounds of minimal projections, namely a Berman - Marcinkiewicz - type identity, does not yield any new information here. In fact, there exist generalizations of this identity to Jacobi and Laguerre expansions. Instead of the ordinary translation operator, they contain the generalized translation which corresponds to the respective orthogonal system (see [2] , [4] for the Jacobi case and [6] ,[9] for the Laguerre case). But for reasons of normalization an additional multiplier operator appears, so that a straightforward generalization of the argument used in the trigonometric case only yields

$$\|S_n^{\alpha,\beta}\|_{[L^1_{w(\alpha,\beta)}]} \leqslant C\, n^{2\alpha+1} \|M_n\|_{[L^1_{w(\alpha,\beta)}]} \qquad (n \in \mathbb{N}), \tag{1.3}$$

where $S_n^{\alpha,\beta}$ are the Jacobi partial sums, $\alpha \geqslant \beta \geqslant -1/2$, $\alpha > -1/2$, and

$$\|\mathcal{S}_n^{\alpha}\|_{[L^1_{\omega(\alpha)}]} \leqslant C\, n^{2\alpha+1} \|M_n\|_{[L^1_{\omega(\alpha)}]} \qquad (n \in \mathbb{N}), \tag{1.4}$$

where $\mathcal{S}_n^{\alpha}$ are the Laguerre partial sums, $\alpha \geqslant 0$. In both inequalities, however, the left hand sides behave like $n^{\alpha+1/2}$ as $n \to \infty$ (cf. Thms. 3 and 4 below), so that one is still far from obtaining a non - trivial lower bound for $\|M_n\|$.

More sophisticated adaptations of this device yield minor improvements only.

2. Preliminaries

The following definitions and formulas will be used (cf. [14]). Denoting the Jacobi polynomials by

$$(2.1)\qquad P_n^{\alpha,\beta}(x) = \sum_{k=0}^{n} \binom{n+\alpha}{n-k}\binom{n+\beta}{k}\left(\frac{x-1}{2}\right)^k\left(\frac{x+1}{2}\right)^{n-k}$$

where $\alpha,\beta > -1$, $x \in [-1,1]$, $n \in \mathbb{P}$, the partial sums of the Jacobi expansion of a function f are defined by

$$(2.2)\qquad \begin{aligned} S_n^{\alpha,\beta}(f;x) &= \int_{-1}^{1} f(t)\, K_n^{\alpha,\beta}(x,t)\, w^{\alpha,\beta}(t)dt, \\ w^{\alpha,\beta}(x) &= (1-x)^\alpha(1+x)^\beta, \\ K_n^{\alpha,\beta}(x,t) &= \sum_{k=0}^{n} (h_k^{\alpha,\beta})^{-1} P_k^{\alpha,\beta}(x)P_k^{\alpha,\beta}(t), \end{aligned}$$

$$h_k^{\alpha,\beta} = \int_{-1}^{1} [P_k^{\alpha,\beta}(x)]^2\, w^{\alpha,\beta}(x)dx = \frac{2^{\alpha+\beta+1}\Gamma(k+\alpha+1)\Gamma(k+\beta+1)}{(2k+\alpha+\beta+1)\Gamma(k+1)\Gamma(k+\alpha+\beta+1)}.$$

Here f is supposed to belong to one of the spaces

$$(2.3)\qquad L^p_{w(a,b)} = \begin{cases} \{f;\{\int_{-1}^{1} |f(x)|^p w^{a,b}(x)dx\}^{1/p} < \infty\}, & 1 \leq p < \infty \\ \{f;\ \operatorname*{ess\,sup}_{-1\leq x\leq 1} |f(x)| < \infty\}, & p = \infty \end{cases}$$

where $a,b > -1$. In particular, $\alpha \neq a$, $\beta \neq b$ are admitted, as far as (2.2) makes sense for such f (further restrictions will be made in Thm. 3). One of the two parameters will be dropped in order to denote the ultraspherical case of (2.2) and (2.3), thus $S_n^\alpha = S_n^{\alpha,\alpha}$, $L^p_{w(a)} = L^p_{w(a,a)}$, etc. According to [14; (4.1.3), (7.32.5), (8.21.18), (7.34.1)], the Jacobi polynomials satisfy

$$(2.4)\qquad P_n^{\alpha,\beta}(x) = (-1)^n P_n^{\beta,\alpha}(-x) \qquad (-1\leq x\leq 1,\ n \in \mathbb{P}),$$

$$(2.5)\qquad |P_n^{\alpha,\beta}(x)| \leq C \begin{cases} n^{-1/2}(1-x)^{-\alpha/2-1/4}, & 0 \leq x \leq 1-n^{-2} \\ n^\alpha, & 1-n^{-2} < x \leq 1 \end{cases} \qquad (n \in \mathbb{N}),$$

(2.6)
$$\max_{0\leqslant x\leqslant 1} |P_n^{\alpha,\beta}(x)(1-x)^{\mu}| \sim \begin{cases} n^{\alpha-2\mu}, & 2\mu<\alpha+1/2 \\ n^{-1/2}, & 2\mu\geqslant\alpha+1/2 \quad (\mu\geqslant 0,\ n\to\infty), \end{cases}$$

(2.7)
$$\int_0^1 |P_n^{\alpha,\beta}(x)|(1-x)^{\mu}dx \sim \begin{cases} n^{\alpha-2\mu-2}, & 2\mu<\alpha-3/2 \\ n^{-1/2}\log n, & 2\mu=\alpha-3/2 \\ n^{-1/2}, & 2\mu>\alpha-3/2 \quad (\mu>-1,\ n\to\infty). \end{cases}$$

The Laguerre polynomials and - functions will be written as

(2.8)
$$L_n^{\alpha}(x) = \sum_{k=0}^{n} \binom{n+\alpha}{n-k} \frac{(-x)^k}{k!},$$

(2.9)
$$\mathcal{L}_n^{\alpha}(x) = (n!/\Gamma(n+\alpha+1))^{1/2} e^{-x/2} x^{\alpha/2} L_n^{\alpha}(x),$$

respectively, where $\alpha>-1$, $x\geqslant 0$, $n\in\mathbb{P}$, and the partial sums of the Laguerre expansion of an f are defined by

(2.10)
$$S_n^{\alpha}(f;x) = \int_0^{\infty} f(t)K_n^{\alpha}(x,t)e^{-t}t^{\alpha}dt,$$
$$K_n^{\alpha}(x,t) = \sum_{k=0}^{n} \frac{k!}{\Gamma(k+\alpha+1)} L_k^{\alpha}(x)L_k^{\alpha}(t).$$

Here f is supposed to be a member of one of the spaces

(2.11)
$$L^p_{w(a)} = \begin{cases} \{f;\{\int_0^{\infty} |f(x)e^{-x/2}|^p x^a dx\}^{1/p}<\infty\}, & 1\leqslant p<\infty \\ \{f;\ \operatorname*{ess\,sup}_{x\geqslant 0} |f(x)e^{-x/2}|<\infty\}, & p=\infty \end{cases}$$

as far as (2.10) makes sense. For properties of the Laguerre polynomials to be used we refer to [9].

3. Norm Estimates for Partial Sums, Proof of Theorems 1 and 2

The following theorem describes the asymptotic behaviour of the ultraspherical partial sums S_n^{α} as operators on $L^1_{w(a)}$ for $\alpha\geqslant a>-1$.

THEOREM 3. (Ultraspherical case) For each $\alpha \geq a > -1$ one has

$$(3.1)\qquad \|S_n^\alpha\|_{[L^1_{w(a)}]} \sim \begin{cases} n^{2a-\alpha+1/2}, & a \leq \alpha < 2a+1/2 \\ \log n, & 2a+1/2 \leq \alpha \leq 2a+3/2 \\ n^{\alpha-2a-3/2}, & \alpha > 2a+3/2 \end{cases} \qquad (n \to \infty).$$

In particular, for $\alpha = a$ this covers the known rates of increase of the Lebesgue constants obtained by Szegö, Rau, and Lorch (cf. [13], [7]), namely

$$(3.2)\qquad \|S_n^a\|_{[L^1_{w(a)}]} = \begin{cases} C_a\, n^{a+1/2} + o(n^{a+1/2}), & a > -1/2 \\ (4/\pi^2)\log n + O(1), & a = -1/2 \end{cases} \qquad (n \to \infty).$$

The counterpart of Thm. 3 for the Laguerre system is

THEOREM 4. (Laguerre case) For each $\alpha \geq a > -1$ one has

$$(3.3)\qquad \|S_n^\alpha\|_{[L^1_{w(a)}]} \sim \begin{cases} n^{2a-\alpha+1/2}, & a \leq \alpha < 2a+1/3 \\ n^{1/6}, & 2a+1/3 \leq \alpha < 2a+3/2 \\ n^{1/6}\log n, & \alpha = 2a+3/2 \\ n^{\alpha-2a-4/3}, & \alpha > 2a+3/2 \end{cases} \qquad (n \to \infty).$$

The similarity between the ultraspherical and Laguerre cases mentioned above, which is due to the particular norm $L^1_{w(a)}$ chosen here, can be read off from the exponents in the first lines of Thms. 3 and 4. Thm. 4 also includes an estimate for the usual type of norm, as employed e.g. by Askey and Wainger [1], i.e., the case $a = \alpha/2 \geq 0$, namely

$$(3.4)\qquad \|S_n^\alpha\|_{[L^1_{w(\alpha/2)}]} \sim n^{1/2} \qquad (n \to \infty)$$

(cf. [8]). The particular case $a = \alpha/2 - 1/4 \geq -1/2$ has been treated in [9]; here

$$(3.5)\qquad \|S_n^\alpha\|_{[L^1_{w(\alpha/2-1/4)}]} \sim n^{1/6} \qquad (n \to \infty).$$

The analogues of Thms. 3 and 4 for L^p-spaces, $p > 1$, are easily derived by the familiar interpolation and duality methods, by making use of Muckenhoupt's results on norm convergence [11], [12,II].

PROOF OF THEOREM 3. We use the representation of the operator norm via the Lebesgue function $\Lambda_n(t)$:

$$(3.6)\qquad \begin{aligned} \|S_n^\alpha\|_{[L^1_{w(a)}]} &= \sup_{-1\leqslant t\leqslant 1} \Lambda_n(t), \\ \Lambda_n(t) = \Lambda_n(t;\alpha,a) &= \int_{-1}^{1} |K_n^\alpha(x,t)|(1-x^2)^a\,dx\,(1-t^2)^{\alpha-a}. \end{aligned}$$

In case $\alpha = a \geqslant -1/2$, as a consequence of the positivity of the Jacobi translation operator [4], the supremum is attained at $t=\pm 1$. Hence, in view of [14; (4.5.3)],

$$(3.7)\qquad \|S_n^a\|_{[L^1_{w(a)}]} = \Lambda(1) = 2^{-2a-1}\,\frac{\Gamma(n+2a+2)}{\Gamma(a+1)\Gamma(n+a+1)}\,\|P_n^{a+1,a}\|_{L^1_{w(a)}},$$

and by (2.7) an evaluation of the norm of $P_n^{a+1,a}$ yields (3.2).

In the remaining cases we need an estimate of $\Lambda_n(t)$ for all $t\in[-1,1]$. For the u p p e r e s t i m a t e we use (cf. (2.4))

$$(3.8)\qquad \sup_{-1\leqslant t\leqslant 1} \Lambda_n(t) \leqslant \sup_{-1\leqslant t\leqslant 1} 2\int_0^1 |K_n^\alpha(x,t)|(1-x^2)^a\,dx\,(1-t^2)^{\alpha-a}$$

and represent the kernel by means of the Christoffel-Darboux formula [11; (2.6-10)] as

$$(3.9)\qquad \begin{aligned} K_n^\alpha(x,t) &= a_n h_1(n,x,t) + b_n[h_2(n,x,t)+h_3(n,x,t)], \\ h_1(n,x,t) &= (n+1)\,P_n^\alpha(x)\,P_n^\alpha(t), \\ h_2(n,x,t) &= n(1-t^2)\,P_n^\alpha(x)\,P_{n-1}^{\alpha+1}(t)\,(x-t)^{-1}, \\ h_3(n,x,t) &= h_2(n,t,x), \end{aligned}$$

where the a_n, b_n are uniformly bounded in n. Since h_2 and h_3 contain singularities at $x=t$, we split up the integral into

$$(3.10)\qquad \int_0^1 |K_n^\alpha(x,t)|(1-x^2)^a(1-t^2)^{\alpha-a}\,dx = \Big\{\int_{U_\varepsilon(t)} + \int_{\complement U_\varepsilon(t)}\Big\}\ldots dx$$

where

$$U_\varepsilon(t) = \begin{cases} \{x \geqslant 0;\ |x-t| < \varepsilon\}, & t \in [-1, 1-n^{-2}) \\ (t-\varepsilon, 1] & ,\quad t \in [1-n^{-2}, 1], \end{cases}$$

$$\complement U_\varepsilon(t) = [0,1] \setminus U_\varepsilon(t), \qquad \varepsilon = \varepsilon(n) = \frac{1}{2} n^{-2}.$$

The first integral in (3.10) is now easily seen to be uniformly bounded with respect to t, in view of (2.5-7). Using (3.9), the second integral in (3.10) may be estimated by

$$\int_{\complement U_\varepsilon(t)} \ldots\, dx \leqslant C \sum_{j=1}^{3} \int_{\complement U_\varepsilon(t)} |h_j(n,x,t)|(1-x^2)^a (1-t^2)^{\alpha-a}\, dx$$
$$= C \sum_{j=1}^{3} I_j,$$

say, the first term of which having the upper bound

$$I_1 \leqslant C \begin{cases} n^{2a-\alpha+1/2}, & a \leqslant \alpha < 2a+1/2 \\ 1, & 2a+1/2 \leqslant \alpha < 2a+3/2 \\ \log (n+1), & \alpha = 2a+3/2 \\ n^{\alpha-2a-3/2}, & \alpha > 2a+3/2, \end{cases}$$

uniformly in t, $t \in [-1,1]$. A careful estimation of I_2 and I_3, carried out by means of (2.5-7), separately on the t - invervals $[-1,-1/2]$, $(-1/2,\ 1-n^{-2})$, $[1-n^{-2},1]$, then yields the same bound as obtained for I_1, except for the fact that the number 1 in case $2a+1/2 \leqslant \alpha < 2a+3/2$ has now to be replaced by $\log (n+1)$.

For the l o w e r e s t i m a t e the inequality

$$\|S_n^\alpha\|_{[L^1_{w(a)}]} \geqslant \|S_{[n/2]}^{\alpha,-1/2}\|_{[L^1_{w(a,-1/2)}]} \tag{3.11}$$

is used, which may be established as follows. Setting $f(x) = g(2x^2-1)$, $x \in [-1,1]$, for some $g \in L^1_{w(a,-1/2)}$ one has $f \in L^1_{w(a)}$ and, by the first identity from [14; Thm. 4.1], $S_n^\alpha(f;x) = S_{[n/2]}^{\alpha,-1/2}(g;2x^2-1)$. Then

$$\|S_{[n/2]}^{\alpha,-1/2} g\|_{L^1_{w(a,-1/2)}} \leqslant \|S_n^\alpha\|_{[L^1_{w(a)}]}\ 2^{a+1/2} \|f\|_{L^1_{w(a)}}$$

$$= \| S_n^{\alpha} \|_{[L^1_{w(a)}]} \| g \|_{[L^1_{w(a,-1/2)}]} ,$$

which proves (3.11). The right hand side of (3.11) can now be estimated from below in several different ways. In order to verify the middle line of (3.1) we show that the right hand side of (3.11) is always bounded from below by $C \cdot \log (n+1)$. Indeed,

$$\| S^{\alpha,-1/2}_{[n/2]} \|_{[L^1_{w(a,-1/2)}]} = \| S^{-1/2,\alpha}_{[n/2]} \|_{[L^1_{w(-1/2,a)}]}$$

$$= \sup_{-1 \leqslant t \leqslant 1} \int_{-1}^{1} |K^{-1/2,\alpha}_{[n/2]}(x,t)| (1-x)^{-1/2} (1+x)^{a}\, dx\, (1+t)^{\alpha-a}$$

$$\geqslant \int_{-1}^{1} |K^{-1/2,\alpha}_{[n/2]}(x,1)| (1-x)^{-1/2} (1+x)^{a}\, dx\, 2^{\alpha-a}$$

(3.12)

$$\geqslant \int_{-1}^{1} |K^{-1/2,\alpha}_{[n/2]}(x,1)| (1-x)^{-1/2} (1+x)^{\alpha}\, dx$$

$$= 2^{-\alpha-1/2} \frac{\Gamma([n/2]+\alpha+3/2)}{\Gamma(1/2)\Gamma([n/2]+\alpha+1)} \| P^{1/2,\alpha}_{[n/2]} \|_{[L^1_{w(-1/2,\alpha)}]}$$

$$= (4/\pi^2) \log n + C_\alpha + O\left(\frac{\log n}{n}\right) + O(n^{-\alpha-3/2}) \qquad (n \to \infty)$$

where in the last step an asymptotic expansion due to Lorch [7,II] has been used.

The first line of (3.1) is obtained by an application of the partial sum operators $S_n^{\alpha,-1/2}$ to the functions

$$(3.13) \qquad f^{\mu}_{2n}(x) = \frac{\Gamma(2n+\alpha+\mu+3/2)}{\Gamma(2n+1/2)} P^{\alpha+\mu+1,-1/2}_{2n}(x) \qquad (\mu \in \mathbb{N}),$$

which, according to [14; (9.4.3)], may also be written as

$$f^{\mu}_{2n}(x) = \sum_{k=0}^{2n} \frac{(2n+k+\alpha+\mu+3/2)}{(2n+k+\alpha+3/2)} A^{\mu}_{2n-k} (2k+\alpha+1/2) \frac{\Gamma(k+\alpha+1/2)}{\Gamma(k+1/2)} P^{\alpha,-1/2}_{k}(x),$$

where $A^k_n = \binom{n+k}{n}$. After a μ fold partial summation one obtains

$$(3.14) \quad S^{\alpha,-1/2}_n(f^{\mu}_{2n};x) = \sum_{j=0}^{\mu} \frac{\Gamma(3n-j+\alpha+\mu+3/2)}{\Gamma(3n+\alpha+3/2)} A^{\mu-j}_{n+j} \frac{\Gamma(n+\alpha+3/2)}{\Gamma(n-j+1/2)} P^{\alpha+j+1,-1/2}_{n-j}(x).$$

Setting $\mu=[2a-\alpha+1/2]+1$, it can be shown that the term for $j=0$ is the principal one. So, in view of (2.7),

$$\|S_n^{\alpha,-1/2}\|_{[L^1_{w(a,-1/2)}]} \geq \|S_n^{\alpha,-1/2} f_{2n}^{\mu}\| \,/\, \|f_{2n}^{\mu}\|$$

$$(3.15) \qquad \geq C\, \frac{\Gamma(3n+\alpha+\mu+3/2)}{\Gamma(3n+\alpha+3/2)}\, A_n^{\mu}\, \frac{\Gamma(n+\alpha+3/2)}{\Gamma(n+1/2)}\, \|P_n^{\alpha+1,-1/2}\| \,/\, \|f_{2n}^{\mu}\|$$

$$\geq C\, n^{2a-\alpha+1/2} \qquad (a\leq\alpha<2a+1/2,\ n\to\infty).$$

The third entry in (3.1) is obtained in a similar way, using the test functions $f_{2n}^{\mu}(x)$, $\mu=[\alpha-2a-3/2]+1$, and the dual norm

$$\|S_n^{\alpha,-1/2}\|_{[L^1_{w(a,-1/2)}]} = \sup_{f\neq 0} \frac{\|S_n^{\alpha,-1/2}(f;x)(1-x)^{\alpha-a}\|_{L^{\infty}}}{\|f(x)(1-x)^{\alpha-a}\|_{L^{\infty}}}$$

$$(3.16) \qquad \geq \|S_n^{\alpha,-1/2}(f_{2n}^{\mu};x)(1-x)^{\alpha-a}\|_{L^{\infty}} \,/\, \|f_{2n}^{\mu}(x)(1-x)^{\alpha-a}\|_{L^{\infty}}$$

$$\geq C\, n^{\alpha-2a-3/2} \qquad (\alpha>2a+3/2,\ n\to\infty).$$

This completes the proof of Theorem 3.

PROOF OF THEOREM 4. Proceeding as in the ultraspherical case, we start with the representation of the operator norm by means of the Lebesgue function, which we denote by $\Lambda_n(t)$ again, thus

$$(3.17) \qquad \|S_n^{\alpha}\|_{[L^1_{w(a)}]} = \sup_{t\geq 0} \Lambda_n(t),$$

$$\Lambda_n(t) = \Lambda_n(t;\alpha,a) = \int_0^{\infty} |K_n^{\alpha}(x,t)|\, e^{-x/2} x^a dx\, e^{-t/2} t^{\alpha-a}.$$

As for the ultraspherical system, the case when the parameters α and a coincide ($\alpha\geq 0$) is exceptional in the sense that the Lebesgue function attains its supremum at the end point $t=0$ of the interval. (This is one of the properties to be proved in the forthcoming paper mentioned.) Hence

$$(3.18) \qquad \|S_n^{a}\|_{[L^1_{w(a)}]} = \Lambda_n(0) = \frac{1}{\Gamma(a+1)}\, \|L_n^{a+1}\|_{L^1_{w(a)}}.$$

Using [9; (2.9)] for the rate of increase of the latter term, the assertion for $\alpha = a$ follows.

In the general case, we have to proceed as in [9; Thms. 1 and 3] where the particular case $a = \alpha/2 - 1/4$, $\alpha \geq -1/2$, has been treated (note that $L^1_{u(a)} = L^1_{w(a/2)}$). We indicate the main steps only. The Lebesgue function in (3.17) may be written as

$$(3.19)\qquad \Lambda_n(t) = \int_0^\infty \Big| \sum_{k=0}^{n} \mathcal{L}_k^\alpha(x)\, \mathcal{L}_k^\alpha(t) \Big| (x/t)^{a-\alpha/2}\, dx.$$

In order to deduce an upper bound of $\Lambda_n(t)$ for each $t \geq 0$ we use the Christoffel - Darboux formula for the kernel $\sum_{k=0}^{n} \mathcal{L}_k^\alpha(x)\, \mathcal{L}_k^\alpha(t)$, as well as estimates of $|\mathcal{L}_n^\alpha(x)|$ and $|\mathcal{L}_{n+1}^\alpha(x) - \mathcal{L}_{n-1}^\alpha(x)|$ and of their norms, which can be found in [9; (2.11), (2.5-6), (2.9-10)] (cf. also [12,II]).

For $t > 3\nu/2$, $\nu = 4n + 2\alpha + 2$ one immediately obtains

$$\Lambda_n(t) \leq \sum_{k=0}^{n} \| \mathcal{L}_k^\alpha(x) x^{a-\alpha/2} \|_{L^1} \cdot \sup_{t > 3\nu/2} | \mathcal{L}_k^\alpha(t) t^{\alpha/2 - a} | \leq C.$$

For $0 \leq t \leq 3\nu/2$, in view of the singularity at $x = t$ in two terms of the Christoffel - Darboux formula, we make the decomposition

$$(3.20)\qquad \Lambda_n(t) = \Big\{ \int_{U_\varepsilon(t)} + \int_{\complement U_\varepsilon(t)} \Big\} \ldots\, dx = L_1(t) + L_2(t),$$

say, where

$$U_\varepsilon(t) = \begin{cases} \{x \geq 0;\ |x - t| < \varepsilon\}, & t > 1/\nu \\ [0,\ t + \varepsilon), & 0 \leq t \leq 1/\nu, \end{cases}$$

$$\complement U_\varepsilon(t) = [0,\infty) \setminus U_\varepsilon(t), \qquad \varepsilon = 1/(2\nu).$$

Now a rough estimation shows that $L_1(t)$ is uniformly bounded for $t \in [0, 3\nu/2]$. In $L_2(t)$, we represent the kernel by the Christoffel - Darboux formula and make estimates for the resulting three terms, the first of which already furnishes the final upper bound as given in (3.3), by (2.6-7). The other two terms have to be treated separately for $t \in [0, 1/\nu]$, $(1/\nu, \nu/2]$, $(\nu/2, 3\nu/2]$. Since the Laguerre functions show a different behaviour on each of these intervals, also the integrals have to be split up accordingly. The upper estimate given in Thm. 4 then follows by carefully estimating the various terms obtained.

As to the lower estimate, the second and third entry of (3.3) can be obtained as in [9; (5.9)] by estimating the Lebesgue function in (3.19) at the particular point $t=\nu(\alpha)$, and using asymptotic expansions of the Laguerre functions. The first entry in (3.3) follows by an application of S_n^α to the test functions

$$f_{2n}^{\mu}(x) = L_{2n}^{\alpha+\mu+1}(x), \qquad (\mu \in \mathbb{N}), \tag{3.21}$$

by observing that (cf. (3.13-14))

$$S_n^\alpha(f_{2n}^\mu;x) = \sum_{k=0}^{n} A_{2n-k}^{\mu} L_k^\alpha(x) = \sum_{j=0}^{\mu} A_{n+j}^{\mu-j} L_{n-j}^{\alpha+j+1}(x). \tag{3.22}$$

For $\mu=[2a-\alpha+1/2]+1$, the first term is the principal one again, and thus with [9; La. 1] it follows that

$$\begin{aligned} \|S_n^\alpha\|_{[L^1_{w(a)}]} &\geqslant \|S_n^\alpha f_{2n}^\mu\|_{L^1_{w(a)}} / \|f_{2n}^\mu\|_{L^1_{w(a)}} \\ &\geqslant C\, A_n^\mu \|L_n^{\alpha+1}\|_{L^1_{w(a)}} / \|L_{2n}^{\alpha+\mu+1}\|_{L^1_{w(a)}} \\ &\geqslant C\, n^{2a-\alpha+1/2} \qquad (a\leqslant\alpha<2a+1/2,\ n\to\infty). \end{aligned} \tag{3.23}$$

The last entry in (3.3) is obtained by estimating the dual norm from below by means of the test functions f_{2n}^μ, $\mu=[\alpha-2a-4/3]+1$, as in (3.22-23):

$$\begin{aligned} \|S_n^\alpha\|_{[L^1_{w(a)}]} &= \sup_{f\neq 0} \frac{\|S_n^\alpha(f;x)e^{-x/2}x^{\alpha-a}\|_{L^\infty}}{\|f(x)e^{-x/2}x^{\alpha-a}\|_{L^\infty}} \\ &\geqslant \|S_n^\alpha(L_{2n}^{\alpha+\mu+1};x)e^{-x/2}x^{\alpha-a}\|_{L^\infty} / \|L_{2n}^{\alpha+\mu+1}(x)e^{-x/2}x^{\alpha-a}\|_{L^\infty} \\ &\geqslant C\, n^{\alpha-2a-4/3} \qquad (\alpha>2a+4/3,\ n\to\infty). \end{aligned} \tag{3.24}$$

PROOF OF THEOREM 1. If in Theorem 3 the parameter a of the space $L^1_{w(a)}$ is fixed, the S_n^α, for the various α admitted, form a particular set of projections, containing several elements which lie much closer to the minimal projection than the S_n^a. For example, choosing $\alpha=2a+1$ for $a>-1$, Thm. 3 gives

$$\|S_n^{2a+1}\|_{[L^1_{w(a)}]} \leqslant C \log (n+1) \qquad (n \in \mathbb{N}),$$

which, for $\alpha > -1/2$, increases less rapidly than the Lebesgue constants $\|S_n^a\|_{[L^1_{w(a)}]}$ (cf. (3.2)).

PROOF OF THEOREM 2. Choosing $P_n = S_n^{2a+1}$ for some $a > -1$, assertion (1.2) follows immediately by Theorem 4. By Theorem 4 again, the behaviour of the Lebesgue constants is

$$\|S_n^a\|_{[L^1_{\omega(a)}]} \sim \begin{cases} n^{a+1/2}, & a > -1/3 \\ n^{1/6}, & -1 < a \leqslant -1/3 \end{cases} \qquad (n \to \infty)$$

which increases more rapidly than $\|S_n^{2a+1}\|_{[L^1_{\omega(a)}]}$, provided $a > -1/3$.

REFERENCES

[1] Askey, R. - Wainger, S., Mean convergence of expansions in Laguerre and Hermite series. Amer. J. Math. 87 (1965), 695-708.

[2] Askey, R. - Wainger, S., A convolution structure for Jacobi series. Amer. J. Math. 91 (1969), 463-485.

[3] Cheney, E.W. - Rivlin, T.J., Some polynomial approximation operators. Math. Z. 145 (1975), 33-42.

[4] Gasper, G., Positivity and the convolution structure for Jacobi series. Annals of Math. 93 (1971), 112-118.

[5] Görlich, E.- Markett, C., Mean Cesàro summability and operator norms for Laguerre expansions. Comment. Math. Prace Mat. Tomus specialis II (1979), 139-148.

[6] Görlich, E. - Markett, C., Estimates for the norm of the Laguerre translation operator. Numer. Funct. Anal. Optim. 1 (1979), 203-222.

[7] Lorch, L., The Lebesgue constants for Jacobi series I, II. Proc. Amer. Math. Soc. 10 (1959), 756-761, Amer. J. Math. 81 (1959), 875-888.

[8] Markett, C., Norm estimates for Cesàro means of Laguerre expansions. In: Approximation and Function Spaces (Proc. Conf., Gdańsk 1979). (to appear)

[9] Markett, C., Mean Cesàro summability of Laguerre expansions and norm estimates with shifted parameter. (to appear)

[10] Morris, P.D. - Cheney, E.W., On the existence and characterization of minimal projections. J. Reine Angew. Math. 270 (1974), 61-76.

[11] Muckenhoupt, B., Mean convergence of Jacobi series. Proc. Amer. Math. Soc. 23 (1969), 306-310.

[12] Muckenhoupt, B., Mean convergence of Hermite and Laguerre series I,II. Trans. Amer. Math. Soc. 147 (1970), 419-431, 433-460.

[13] Rau, H., Über die Lebesgueschen Konstanten der Reihenentwicklungen nach Jacobischen Polynomen. J. Reine Angew. Math. 161 (1929), 237-254.

[14] Szegö, G., Orthogonal Polynomials. 3 rd. ed., Amer. Math. Soc. Colloq. Publ. 23, Providence, R.I. 1967.

THE REGULAR CONVERGENCE OF MULTIPLE SERIES

Ferenc Móricz
Bolyai Institute
University of Szeged
Szeged

Denote Z_+^d the set of d-tuples $\underset{\sim}{k} = (k_1,\ldots,k_d)$ with positive integers for coordinates. A d-multiple series $\sum u_{\underset{\sim}{k}} = \sum\{u_{\underset{\sim}{k}}: \underset{\sim}{k} \in Z_+^d\}$, where the summation is extended over $\underset{\sim}{k} \in Z_+^d$, is said to converge regularly if for every positive η there exists a number $N = N(\eta)$ so that $|\sum\{u_{\underset{\sim}{k}}: \underset{\sim}{k} \in R\}| < \eta$ for every rectangle $R = \{\underset{\sim}{k} \in Z_+^d: \underset{\sim}{\ell} \leqslant \underset{\sim}{k} \leqslant \underset{\sim}{m}\}$ provided $\max(\ell_1,\ldots,\ell_d) > N$ and $\underset{\sim}{m} \geqslant \underset{\sim}{\ell}$. Convergence in Pringsheim's sense follows from regular convergence, but the converse implication is not true in case $d \geqslant 2$. A benefit of the notion of regular convergence is that it makes possible to extend the validity of Kronecker's lemmas from single series to multiple series and these extensions meet a number of applications, among others, in the theory of multiple orthogonal series and of random fields.

1. The Notion of Regular Convergence

Consider a single numerical series $\sum_{i=1}^{\infty} u_i$. The statement that it converges to a finite number s, roughly speaking means the following:

(i) The partial sums $s_m = \sum_{i=1}^{m} u_i$ are as close to s as we wish if m is large enough;

(ii) The remainder sums $\sum_{i=1}^{n} u_i$ $(= s_n - s_{m-1})$ are as small as we wish if n and m are large enough, $n \geqslant m$.

It is well-known that (i) and (ii) are equivalent to each other. But the situation is different in the case of multiple series.

Let Z_+^d be the set of d-tuples $\underset{\sim}{k} = (k_1,\ldots,k_d)$ with positive integers for coordinates, where d is a fixed positive integer. As usual, we write $\underset{\sim}{k} \pm \underset{\sim}{m} = (k_1 \pm m_1,\ldots,k_d \pm m_d)$, $\underset{\sim}{k} \leqslant \underset{\sim}{m}$ iff $k_j \leqslant m_j$ for each j, and $\underset{\sim}{N} = (N,\ldots,N)$ for $N = 0,1,\ldots$. Finally, we set $k^* = \max_{1 \leqslant j \leqslant d} k_j$ and $k_* = \min_{1 \leqslant j \leqslant d} k_j$.

We shall consider the d-multiple numerical series

$$(1) \qquad \sum_{\underset{\sim}{k} \in Z_+^d} u_{\underset{\sim}{k}} = \sum_{k_1=1}^{\infty} \cdots \sum_{k_d=1}^{\infty} u_{k_1,\dots,k_d}$$

with the rectangular partial sums

$$s_{\underset{\sim}{m}} = \sum_{\underset{\sim}{1} \leq \underset{\sim}{k} \leq \underset{\sim}{m}} u_{\underset{\sim}{k}} = \sum_{k_1=1}^{m_1} \cdots \sum_{k_d=1}^{m_d} u_{k_1,\dots,k_d} \qquad (\underset{\sim}{m} \in Z_+^d).$$

More generally, given a rectangle $R = \{\underset{\sim}{k} \in Z_+^d : \underset{\sim}{\ell} \leq \underset{\sim}{k} \leq \underset{\sim}{m}\}$, set

$$s(R) = \sum_{\underset{\sim}{k} \in R} u_{\underset{\sim}{k}} \qquad (\underset{\sim}{\ell}, \underset{\sim}{m} \in Z_+^d;\ \underset{\sim}{\ell} \leq \underset{\sim}{m}).$$

It is clear that $s(R) = s_{\underset{\sim}{m}}$ in case $\underset{\sim}{\ell} = 1$, further, $s(R)$ can be considered as a remainder sum of series (1) in case ℓ^* is large enough.

We remind that the multiple series (1) is said to be *convergent in Pringsheim's sense* to the sum s if for every positive number η there exists a number $N = N(\eta)$ so that

$$|s_{\underset{\sim}{m}} - s| < \eta \qquad \text{whenever } m_* > N,$$

or equivalently, if

$$|s_{\underset{\sim}{m}} - s_{\underset{\sim}{n}}| < \eta \qquad \text{whenever } m_* > N \text{ and } n_* > N.$$

In other words, convergence in Pringsheim's sense means that the rectangular partial sums $s_{\underset{\sim}{m}}$ are as close to s as we wish if each coordinate of $\underset{\sim}{m}$ is large enough.

We shall say that the multiple series (1) *regularly converges* if for every positive number η there exists a number $N = N(\eta)$ so that

$$|s(R)| < \eta \qquad \text{whenever } \ell^* > N \text{ and } \underset{\sim}{m} \geq \underset{\sim}{\ell}.$$

We recall that $\underset{\sim}{\ell}$ is the bottom left-hand corner of the rectangle $R = \{\underset{\sim}{k} \in Z_+^d : \underset{\sim}{\ell} \leq \underset{\sim}{k} \leq \underset{\sim}{m}\}$, while $\underset{\sim}{m}$ is its top right-hand corner. Thus, regular convergence means that the remainder sums $s(R)$ are as small as we wish if at least one of the coordinates of the bottom left-hand corner $\underset{\sim}{\ell}$ of the rectangle R is large enough.

It is not hard to see that convergence in Pringsheim's sense follows from regular convergence. The converse statement is not true in general. For example, the double series indicated in Fig. 1 converges to 0 in Pringsheim's sense, since its rectangular partial sums $s_{mn} = \sum_{k=1}^{m} \sum_{\ell=1}^{n} u_{k\ell} = 0$ if $m \geqslant 2$ and $n \geqslant 2$; but it fails to converge regularly, even its terms are not bounded. We note that if the terms $u_{\underset{\sim}{k}}$ of series (1) are of constant sign, then these two notions of convergence coincide.

ℓ				
.	.	.	.	.
.	.	.	.	.
.	.	.	.	.
3	−3	0	0	0 ...
2	−2	0	0	0 ...
1	−1	0	0	0 ...
0	0	−1	−2	−3 ...
0	0	1	2	3 ...
				k

Fig. 1: $u_{k\ell}$ $(k,\ell = 1,2,\ldots)$

The definition of regular convergence is due to Hardy [3] in case $d = 2$, and to the present author [5] in case $d \geqslant 2$. The former paper, unfortunately, had escaped the attention of the present author, and this is the reason why this kind of convergence of multiple series was rediscovered and called in [5] convergence in a restricted sense.

We remark that in [3] regular convergence is defined by an equivalent condition which is true only for $d = 2$, namely: "A (double) series is said to be regularly convergent if it is convergent in the ordinary sense (i.e. in Pringsheim's sense) and all its rows and columns are also convergent." The treatment of the case $d \geqslant 3$ is not clear from here. In fact, the triple series $\sum_{k=1}^{\infty} \sum_{\ell=1}^{\infty} \sum_{m=1}^{\infty} u_{k\ell m}$ whose terms $u_{k\ell m}$ for $m = 1,2$ are indicated in Fig. 2 and $u_{k\ell m} = 0$ for $m = 3,4,\ldots$ is such that it converges to 0 in Pringsheim's sense and all the single series $\sum_{k=1}^{\infty} u_{k\ell m}$ (for each $\ell, m = 1,2,\ldots$), $\sum_{\ell=1}^{\infty} u_{k\ell m}$ (for each $k,m = 1,2,\ldots$), and $\sum_{m=1}^{\infty} u_{k\ell m}$(for each $k,\ell = 1,2,\ldots$) converge, but the triple series in question fails to converge regularly.

The reason why this triple series does not converge regularly is that the double series $\sum_{k=1}^{\infty} \sum_{\ell=1}^{\infty} u_{k\ell m}$ does not converge even in Pringsheim's sense for $m = 1$ and 2. Indeed, the following theorem holds.

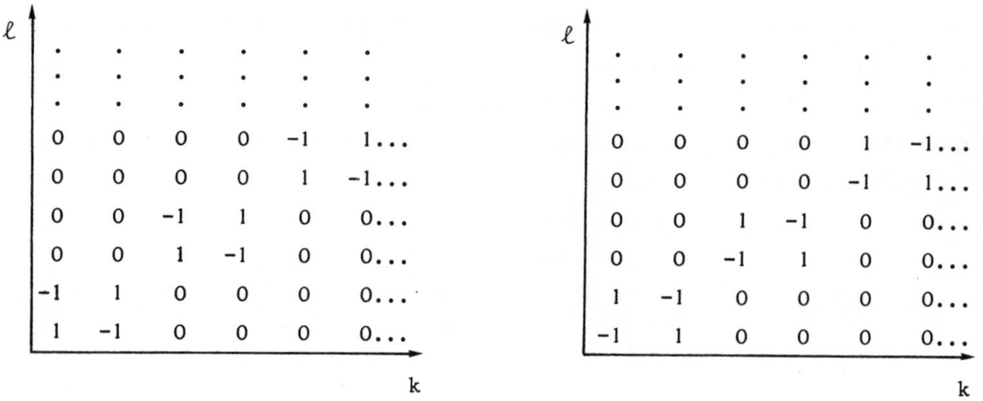

Fig.2:
m=1: $u_{k\ell 1}$ $(k,\ell=1,2,\ldots)$ m=2: $u_{k\ell 2}$ $(k,\ell=1,2,\ldots)$

THEOREM 1. *The* d-*multiple series* (1) *regularly converges if and only if*

(i) *it converges in Pringsheim's sense, and*

(ii) *the* $(d-1)$-*multiple series*

$$\sum_{k_1=1}^{\infty}\cdots\sum_{k_{j-1}=1}^{\infty}\sum_{k_{j+1}=1}^{\infty}\cdots\sum_{k_d=1}^{\infty} u_{k_1,\ldots,k_{j-1},k_j,k_{j+1},\ldots,k_d}$$

regularly converges for each fixed value of $k_j=1,2,\ldots$ *and for each* $j=1,2,\ldots,d$.

The following corollary hence follows by induction.

COROLLARY 1. *Let* $U=\{j_1,\ldots,j_e\}$ *and* $V=\{\ell_{e+1},\ldots,\ell_d\}$ *be two disjoint subsets of* $\{1,\ldots,d\}$, *where* $1\leqslant j_1<\ldots<j_e\leqslant d$ *and* $1\leqslant \ell_{e+1}<\ldots<\ell_d\leqslant d$ *with* $1\leqslant e\leqslant d$ (V *is empty in case* $e=d$). *The* d-*multiple series* (1) *regularly converges if and only if the* e-*multiple series*

$$\sum_{k_{j_1}=1}^{\infty}\cdots\sum_{k_{j_e}=1}^{\infty} u_{k_1,\ldots,k_d}$$

converges in Pringsheim's sense for each choice of U *with* $1\leqslant e\leqslant d$ *and for each fixed value of* $k_{\ell_{e+1}}=1,2,\ldots;\ldots;\ k_{\ell_d}=1,2,\ldots$.

In addition, if series (1) *regularly converges and*

$$\sum_{k_{j_1}=1}^{\infty} \cdots \sum_{k_{j_e}=1}^{\infty} u_{k_1,\ldots,k_d} = s_{k_{\ell_{e+1}},\ldots,k_{\ell_d}},$$

then

$$\sum_{k_{\ell_{e+1}}=1}^{\infty} \cdots \sum_{k_{\ell_d}=1}^{\infty} s_{k_{\ell_{e+1}},\ldots,k_{\ell_d}} = \sum_{\underset{\sim}{k} \in Z_+^d}^{\infty} u_{\underset{\sim}{k}} \qquad (1 \leq e \leq d-1).$$

The main goal of the present paper is to point out that the notion of regular convergence is more appropriate in the study of convergence properties of multiple series than the notion of convergence in Pringsheim's sense.

2. Kronecker's Lemmas for Multiple Series

Beside series (1) we shall consider the tranformed series

(2)
$$\sum_{\underset{\sim}{k} \in Z_+^d} \frac{u_{\underset{\sim}{k}}}{\lambda_{\underset{\sim}{k}}}$$

with the rectangular partial sums

$$s_{\underset{\sim}{m}} = \sum_{\underset{\sim}{1} \leq \underset{\sim}{k} \leq \underset{\sim}{m}} \frac{u_{\underset{\sim}{k}}}{\lambda_{\underset{\sim}{k}}} \qquad (\underset{\sim}{m} \in Z_+^d),$$

where $\lambda = \{\lambda_{\underset{\sim}{k}} : \underset{\sim}{k} \in Z_+^d\}$ is a given d-multiple sequence of positive numbers.

As usual, the finite differences $\Delta_{\underset{\sim}{\varepsilon}} \lambda_{\underset{\sim}{k}}$ and $\nabla_{\underset{\sim}{\varepsilon}} \lambda_{\underset{\sim}{k}}$ are defined as follows, where $\underset{\sim}{\varepsilon} = (\varepsilon_1, \ldots, \varepsilon_d)$ is a d-tuple with $\varepsilon_j = 0$ or 1 for coordinates. In case $\underset{\sim}{\varepsilon} = \underset{\sim}{0}$ set

$$\Delta_{\underset{\sim}{0}} \lambda_{\underset{\sim}{k}} = \lambda_{\underset{\sim}{k}} \qquad (\underset{\sim}{k} \in Z_+^d),$$

while in case $\underset{\sim}{\varepsilon} \neq \underset{\sim}{0}$ let $\varepsilon_j = 1$ iff $j = j_1, \ldots, j_e$ with $1 \leq e \leq d$ and set

$$\Delta_{\underset{\sim}{\varepsilon}} \lambda_{\underset{\sim}{k}} = \delta_{j_1}(\delta_{j_2}(\ldots(\delta_{j_e} \lambda_{\underset{\sim}{k}})\ldots)) \qquad (\underset{\sim}{k} \in Z_+^d),$$

where

$$\delta_j \lambda_{\underset{\sim}{k}} = \lambda_{k_1,\ldots,k_{j-1},k_j+1,k_{j+1},\ldots,k_d} - \lambda_{k_1,\ldots,k_{j-1},k_j,k_{j+1},\ldots,k_d};$$

finally set

$$\nabla_{\underset{\sim}{\varepsilon}} \lambda_{\underset{\sim}{k}} = \Delta_{\underset{\sim}{\varepsilon}} \lambda_{\underset{\sim}{k}-\underset{\sim}{\varepsilon}} \qquad (\underset{\sim}{k} \in Z_+^d;\ k_j \geq 2 \text{ for } j = j_1, \ldots, j_e).$$

Observe that the order of succession of the "operators" $\delta_{j_1},\dots,\delta_{j_e}$ is indifferent.

The use of the forward and backward Abel transformation formulas leads to expressions for the rectangular partial sums $s_{\underset{\sim}{m}}$ of series (1) in terms of the remainder sums

$$S(R) = S(\underset{\sim}{\ell},\underset{\sim}{m}) = \sum_{\underset{\sim}{\ell}\le\underset{\sim}{k}\le\underset{\sim}{m}} \frac{u_{\underset{\sim}{k}}}{\lambda_{\underset{\sim}{k}}} \qquad (\underset{\sim}{\ell},\underset{\sim}{m}\in Z_+^d,\ \underset{\sim}{\ell}\le\underset{\sim}{m})$$

of series (2), in particular, in terms of the rectangular partial sums $S_{\underset{\sim}{m}}$, and in terms of the differences of the sequence λ.

Indeed, the forward Abel transformation formula can be given as

$$s_{\underset{\sim}{m}} = \sum_{\underset{\sim}{\varepsilon}} (-1)^{\varepsilon_1+\dots+\varepsilon_d} \sum_{\underset{\sim}{k}}{}^{(\underset{\sim}{\varepsilon})}\, S_{\underset{\sim}{k}}\Delta_{\underset{\sim}{\varepsilon}}\lambda_{\underset{\sim}{k}}, \tag{3}$$

where the sum $\sum_{\underset{\sim}{\varepsilon}}$ is extended over all 2^d possible choices of $\underset{\sim}{\varepsilon}$ with $\varepsilon_j=0$ or 1 for coordinates; $\sum_{\underset{\sim}{k}}^{(\underset{\sim}{\varepsilon})}$ means the single term $S_{\underset{\sim}{m}}\lambda_{\underset{\sim}{m}}$ in case $\underset{\sim}{\varepsilon}=\underset{\sim}{0}$, while in case $\underset{\sim}{\varepsilon}\ne\underset{\sim}{0}$ with $\varepsilon_j=1$ iff $j=j_1,\dots,j_e$ it means the $\tilde{\varepsilon}$-fold sum

$$\sum_{k_{j_1}=1}^{m_{j_1}-1}\dots\sum_{k_{j_e}=1}^{m_{j_e}-1} S_{\underset{\sim}{k}}\Delta_{\underset{\sim}{\varepsilon}}\lambda_{\underset{\sim}{k}} \quad \text{with } k_\ell=m_\ell \text{ for } \ell\ne j_1,\dots,j_e.$$

Ij $m_j=1$ for at least one $j=j_1,\dots,j_e$, then this e-fold sum is neglected. For instance, in case $d=2$

$$s_{mn} = \sum_{k=1}^{m-1}\sum_{\ell=1}^{n-1} S_{k\ell}(\lambda_{k+1,\ell+1}-\lambda_{k+1,\ell}-\lambda_{k,\ell+1}+\lambda_{k\ell})$$
$$-\sum_{k=1}^{m-1} S_{kn}(\lambda_{k+1,n}-\lambda_{kn}) - \sum_{\ell=1}^{n-1} S_{m\ell}(\lambda_{m,\ell+1}-\lambda_{m\ell}) + S_{mn}\lambda_{mn}$$

provided that $m\ge 2$ and $n\ge 2$, while if, e.g., $m\ge 2$ and $n=1$, then

$$s_{m1} = -\sum_{k=1}^{m-1} S_{k1}(\lambda_{k+1,1}-\lambda_{k1}) + S_{m1}\lambda_{m1}.$$

The backward Abel transformation formula is the following:

$$s_{\underset{\sim}{m}} = \sum_{\underset{\sim}{\varepsilon}} {}^{(\underset{\sim}{\varepsilon})}\sum_{\underset{\sim}{k}} S(\underset{\sim}{k},\underset{\sim}{m})\nabla_{\underset{\sim}{\varepsilon}}\lambda_{\underset{\sim}{k}}, \tag{4}$$

where the sum $\sum_{\underset{\sim}{\varepsilon}}$ is again extended over all 2^d possible choices of $\underset{\sim}{\varepsilon}$ with $\varepsilon_j=0$ or 1 for coordinates; ${}^{(\underset{\sim}{\varepsilon})}\sum_{\underset{\sim}{k}}$ means the single term $S(\underset{\sim}{1},\underset{\sim}{m})\lambda_{\underset{\sim}{1}}$ in case $\underset{\sim}{\varepsilon}=\underset{\sim}{0}$, while in case $\underset{\sim}{\varepsilon} \neq \underset{\sim}{0}$ with $\varepsilon_j = 1$ iff $j = j_1, \ldots, j_e$ it means the ε-fold sum

$$\sum_{k_{j_1}=2}^{m_{j_1}} \cdots \sum_{k_{j_e}=2}^{m_{j_e}} S(\underset{\sim}{k}, \underset{\sim}{m}) \nabla_{\underset{\sim}{\varepsilon}} \lambda_{\underset{\sim}{k}} \text{ with } k_\ell = 1 \text{ for } \ell \neq j_1, \ldots, j_e .$$

If $m_j = 1$ for at least one $j = j_1, \ldots, j_e$, then this e-fold sum is also neglected.

These Abel transformation formulas in case $d = 1$ are wellknown (see, e.g., [1, p.71]), and their various forms in case $d \geq 2$ have been used by a lot of authors. We only mention here that formula (3) in another notation occurs in [2] , while in this form it is in [6]. As to formula (4), see also [6].

After these preliminaries we turn to the Kronecker lemmas. A benefit of the notion of regular convergence is that it makes possible to extend the validity of Kronecker's lemmas from single series to multiple series and these extensions meet a number of applications, among others, in the theory of multiple orthogonal series and in probability theory (see [6]).

One of the Kronecker lemmas in case $d = 1$ states that if $\{\lambda_i : i = 1, 2, \ldots\}$ is a non-decreasing sequence of positive numbers, tending to infinity, then the convergence of the series $\sum_{i=1}^{\infty} u_i/\lambda_i$ implies the estimate $s_m = \sum_{i=1}^{m} u_i = o(\lambda_m)$ as $m \to \infty$ (see, e.g., [1, p. 72]). The generalization of this lemma whose proof is based on (4) reads as follows.

THEOREM 2. Let λ be a d-multiple sequence of positive numbers such that for each $\underset{\sim}{\varepsilon} \neq \underset{\sim}{0}$ with $\varepsilon_j = 0$ or 1 for coordinates, $\varepsilon_j = 1$ iff $j = j_1, \ldots, j_e$ where $1 \leq e \leq d$, we have

$$\Delta_{\underset{\sim}{\varepsilon}} \lambda_{\underset{\sim}{k}} \text{ is } \begin{cases} \text{non-negative if } e = 1, \\ \text{of constant sign in } \underset{\sim}{k} \text{ if } e \geq 2, \end{cases} \tag{5}$$

and

$$\lambda_{\underset{\sim}{k}} \to \infty \qquad \text{as } k^* \to \infty \text{ (or } k_* \to \infty). \tag{6}$$

If the d-multiple series (2) regularly converges, then

$$s_{\underset{\sim}{m}} = \sum_{\underset{\sim}{1} \leq \underset{\sim}{k} \leq \underset{\sim}{m}} u_{\underset{\sim}{k}} = o(\lambda_{\underset{\sim}{m}}) \qquad \text{as } m^* \to \infty \text{ (or } m_* \to \infty). \tag{7}$$

Another Kronecker lemma in case $d=1$ asserts that if $\{\lambda_i\}$ and $\{\nu_i\}$ are two non-decreasing sequences of positive numbers, λ_i tending to infinity as $i\to\infty$, then the estimate $S_m=\sum_{i=1}^m u_i/\lambda_i=o(\nu_m)$ implies the estimate $s_m=\sum_{i=1}^m u_i = =o(\lambda_m\nu_m)$ as $m\to\infty$. Making use of (3) this lemma can be generalized in the following form.

THEOREM 3. *Let* $\lambda=\{\lambda_{\underset{\sim}{k}}:\underset{\sim}{k}\in Z_+^d\}$ *and* $\nu=\{\nu_{\underset{\sim}{k}}:\underset{\sim}{k}\in Z_+^d\}$ *be two* d*-multiple sequences of positive numbers such that for* λ *conditions* (5) *and* (6) *are satisfied, and for* ν *we have*

$$\nu_{\underset{\sim}{k}}\leq\nu_{\underset{\sim}{m}} \qquad \textit{whenever } \underset{\sim}{k}\leq\underset{\sim}{m}.$$

If

$$(8)\qquad S_{\underset{\sim}{m}}=\sum_{\underset{\sim}{1}\leq\underset{\sim}{k}\leq\underset{\sim}{m}}\frac{u_{\underset{\sim}{k}}}{\lambda_{\underset{\sim}{k}}}=o(\nu_{\underset{\sim}{m}}) \qquad \textit{as } m^*\to\infty,$$

then

$$(9)\qquad s_{\underset{\sim}{m}}=\sum_{\underset{\sim}{1}\leq\underset{\sim}{k}\leq\underset{\sim}{m}}u_{\underset{\sim}{k}}=o(\lambda_{\underset{\sim}{m}}\nu_{\underset{\sim}{m}}) \qquad \textit{as } m^*\to\infty \ (\textit{or } m_*\to\infty).$$

Theorems 2 and 3 seem to be new and a detailed proof of them will appear in a forthcoming paper [6] of the present author.

We are going to make a few remarks. Conditions (5) are obviously satisfied, among others, if $\lambda_{\underset{\sim}{k}}=\Pi_{j=1}^d\lambda_{k_j}^{(j)}$ or $\lambda_{\underset{\sim}{k}}=\mu_{k_*}$ or $\lambda_{\underset{\sim}{k}}=\mu_{k^*}$, where each $\{\lambda_{k_j}^{(j)}:k_j=1,2,\ldots\}$ and $\{\mu_i:1,2,\ldots\}$ are non-decreasing sequences of positive numbers.

It is somewhat striking that Theorem 2 is no longer true if series (2) converges in Pringsheim's sense only. This is illustrated by the following example. Let $d=2$ and

$$\lambda_{k\ell}=2^{[k/2]+[\ell/2]} \qquad (k,\ell=1,2,\ldots),$$

where $[\cdot]$ means the integral part. Conditions (5) and (6) are clearly fulfilled. Let

$$u_{k\ell}=\begin{cases}(-1)^{k+\ell}\,2^{2[(k-1)/2]}\lambda_{k\ell} & \text{for } \ell=1,2;\ k=1,2,\ldots;\\ 0 & \text{for } \ell=3,4,\ldots;\ k=1,2,\ldots\,.\end{cases}$$

On the one hand,

$$S_{mn} = \sum_{k=1}^{m} \sum_{\ell=1}^{n} \frac{u_{k\ell}}{\lambda_{k\ell}} = 0 \qquad \text{for } m=1,2,\ldots;\ n=2,3,\ldots$$

(see Fig. 3). Consequently, the double series $\sum_{k=1}^{\infty} \sum_{\ell=1}^{\infty} u_{k\ell}/\lambda_{k\ell}$ converges in Pringsheim's sense. On the other hand,

$$s_{2m,n} = \sum_{k=1}^{2m} \sum_{\ell=1}^{n} u_{k\ell} = \frac{1}{7}(8^m - 1) \qquad \text{for } m=1,2,\ldots;\ n=2,3,\ldots$$

(see Fig. 4). Thus

ℓ						
	.	.	.	.	.	.
	.	.	.	.	.	.
	.	.	.	.	.	.
	0	0	0	0	0	0...
	0	0	0	0	0	0...
	-1	1	-4	4	-16	16...
	1	-1	4	-4	16	-16...
						k

Fig.3 : $u_{k\ell}/\lambda_{k\ell}$

ℓ						
	.	.	.	.	.	.
	.	.	.	.	.	.
	.	.	.	.	.	.
	0	0	0	0	0	0...
	0	0	0	0	0	0...
	-2	4	-16	32	-128	256...
	1	-2	8	-16	64	-128...
						k

Fig.4 : $u_{k\ell}$

$$\lim_{m\to\infty} \frac{s_{2m,2m}}{\lambda_{2m,2m}} = \lim_{m\to\infty} \frac{8^m-1}{7\cdot 4^m} = \infty,$$

i.e. statement (7) does not hold even in the special case $m=n$.

The same example shows that Theorem 3 becomes also false if condition (8) is required in the less restricted case when $m_* \to \infty$. To be more concrete, since $S_{mn}=0$ for $n \geqslant 2$, we have (8) with $\nu_{mn} \equiv 1$ as $\min(m,n) \to \infty$. On the other hand, even with $\nu'_{mn} = \lambda_{mn}^{1/3}$

$$\lim_{m\to\infty} \frac{s_{2m,2m}}{\lambda_{2m,2m}\nu'_{2m,2m}} = \lim_{m\to\infty} \frac{8^m-1}{7\cdot 4^m \cdot 4^{m/3}} = \infty ,$$

which is opposite to statement (9).

3. Regular Convergence of Multiple Orthogonal Series

Let $\varphi = \{\varphi_{\underset{\sim}{k}}(x) : \underset{\sim}{k} \in Z_+^d\}$ be a multiple orthonormal system (in abbreviation: ONS) on $I^d = [0,1]^d$, the unit cube in the d-dimensional Euclidean space, i.e.,

$$\int_{I^d} \varphi_{\underset{\sim}{k}}(x)\varphi_{\underset{\sim}{m}}(x)dx = \begin{cases} 0 & \text{if } \underset{\sim}{k} \neq \underset{\sim}{m}, \\ 1 & \text{if } \underset{\sim}{k} = \underset{\sim}{m} \end{cases} \qquad (\underset{\sim}{k}, \underset{\sim}{m} \in Z_+^d).$$

We shall consider the d-multiple orthogonal series

$$\sum_{\underset{\sim}{k} \in Z_+^d} a_{\underset{\sim}{k}} \varphi_{\underset{\sim}{k}}(x), \tag{10}$$

where $a = \{a_{\underset{\sim}{k}} : \underset{\sim}{k} \in Z_+^d\}$ is a d-multiple sequence of numbers (coefficients). The rectangular partial sums of (10) will be denoted by

$$s_{\underset{\sim}{m}}(x) = \sum_{\underset{\sim}{1} \leq \underset{\sim}{k} \leq \underset{\sim}{m}} a_{\underset{\sim}{k}} \varphi_{\underset{\sim}{k}}(x) \qquad (\underset{\sim}{m} \in Z_+^d),$$

and more generally, if R is a finite rectangle in Z_+^d, then set

$$s(R;x) = \sum_{\underset{\sim}{k} \in R} a_{\underset{\sim}{k}} \varphi_{\underset{\sim}{k}}(x).$$

In this section we follow the definitions and arguments due to Tandori [9], [10] in the special case $d = 1$. Denote by M the class of those d-multiple sequences a for which series (10) regularly converges a.e. for every d-multiple ONS φ on I^d. The set of measure zero of the divergence points may vary with every φ.

The embedding $M \subset \ell^2$ follows from the fact that the d-multiple Rademacher system $\{r_{\underset{\sim}{k}}(x) = \Pi_{j=1}^d r_{k_j}(x_j),\ x = (x_1, \ldots, x_d)\}$ is such that the series $\sum_{\underset{\sim}{k} \in Z_+^d} a_{\underset{\sim}{k}} r_{\underset{\sim}{k}}(x)$ diverges in Pringsheim's sense a.e. for every a with $\sum_{\underset{\sim}{k} \in Z_+^d} a_{\underset{\sim}{k}}^2 = \infty$.

For any given d-multiple sequence a of coefficients set

$$\|a\|^2 = \sup_R \sup_\varphi \int_{I^d} (\max_{Q \subset R} |s(Q;x)|)^2 dx \qquad (\leq \infty), \tag{11}$$

the first supremum being taken over all finite rectangles R in Z_+^d, the second supremum over all ONS φ on I^d, and the maximum over all rectangles Q contained in R.

The main result which is proved in [7] reads as follows.

THEOREM 4. (i) $a \in M$ <u>if and only if</u> $\|a\| < \infty$, <u>and</u>

(ii) M <u>endowed with the norm</u> $\|\cdot\|$ <u>is a separable Banach space</u>.

Part (i) of this theorem says, roughly speaking, that the a.e. regular convergence of series (10) for every ONS on I^d is equivalent to the following "boundedness" property: the sums $s(R;x)$ are majorized by some square integrable function on I^d, the square integral of which depends only on the sequence a of coefficients.

Using the d-dimensional generalization of the famous Rademacher - Menšov inequality (see, e.g., [5, Corollary 2]), it is not hard to give an upper bound for $\|a\|$. Namely, for arbitrary a we have

$$\|a\| \leq C_1 \Big(\sum_{\underset{\sim}{k} \in Z_+^d} a_{\underset{\sim}{k}}^2 \prod_{j=1}^{d} (\log 2k_j)^2 \Big)^{1/2}, \tag{12}$$

where C_1 is a constant depending only on d.

An exact lower bound for $\|a\|$ is not known in general. But in the special case when $\{|a_{\underset{\sim}{k}}| : \underset{\sim}{k} \in Z_+^d\}$ is non-increasing in the sense that $|a_{\underset{\sim}{k}}| \geq |a_{\underset{\sim}{m}}|$ whenever $\underset{\sim}{k} \leq \underset{\sim}{m}$, an opposite inequality to (12) is true:

$$\|a\| \geq C_2 \Big(\sum_{\underset{\sim}{k} \in Z_+^d} a_{\underset{\sim}{k}}^2 \prod_{j=1}^{d} (\log 2k_j)^2 \Big)^{1/2},$$

where C_2 is a positive constant also depending on d. This lower estimate follows from the results of [8] in a routine way.

This approaching method which uses the notion of Banach space in the study of convergence of orthogonal series, makes it possible to deduce the following theorems (for the case $d = 1$, see also Tandori [10]).

THEOREM 5. <u>Let</u> $a = \{a_{\underset{\sim}{k}} : \underset{\sim}{k} \in Z_+^d\}$ <u>and</u> $b = \{b_{\underset{\sim}{k}} : \underset{\sim}{k} \in Z_+^d\}$ <u>be two</u> d-<u>multiple sequences of numbers for which</u> $|a_{\underset{\sim}{k}}| \leq |b_{\underset{\sim}{k}}|$, $\underset{\sim}{k} \in Z_+^d$. <u>If</u> $b \in M$, <u>then</u> $a \in M$ <u>and</u> $\|a\| \leq \|b\|$.

THEOREM 6. (i) <u>If</u> $a \in M$, <u>then there exists a</u> d-<u>multiple sequence</u> $\mu = \{\mu_{\underset{\sim}{k}} : \underset{\sim}{k} \in Z_+^d\}$ <u>of positive numbers</u>, $\mu_{\underset{\sim}{k}}$ <u>tends to infinity as</u> $k^* \to \infty$, <u>such that</u> $\{a_{\underset{\sim}{k}} \mu_{\underset{\sim}{k}} : \underset{\sim}{k} \in Z_+^d\} \in M$.

(ii) *If* $a \notin M$, *then there exists a sequence* μ *with the same properties as in* (i) *and such that* $\{a_{\underset{\sim}{k}}/\mu_{\underset{\sim}{k}} : \underset{\sim}{k} \in Z_+^d\} \notin M$.

It is a remarkable thing that Theorem 4 remains valid if regular convergence is replaced by convergence in Pringsheim's sense in it. This can be simply motivated by the fact that the norm defined by (11) is equivalent to the following one:

$$\|a\|_*^2 = \sup_{\underset{\sim}{m} \in Z_+^d} \sup_{\varphi} \int_{I^d} \left(\max_{\underset{\sim}{1} \leq \underset{\sim}{k} \leq \underset{\sim}{m}} |s_{\underset{\sim}{k}}(x)| \right)^2 dx,$$

where

$$\max_{\underset{\sim}{1} \leq \underset{\sim}{k} \leq \underset{\sim}{m}} |s_{\underset{\sim}{k}}(x)| = \max_{1 \leq k_1 \leq m_1} \cdots \max_{1 \leq k_d \leq m_d} |s_{k_1,\ldots,k_d}(x)|.$$

Thus we can obtain the following

COROLLARY 2. *Let a* d-*multiple sequence* a *is given. If the a.e. convergence of series* (10) *is considered for every ONS on* I^d, *then regular convergence and convergence in Pringsheim's sense are equivalent.*

For individual ONS the notions of a.e. regular convergence and a.e. convergence in Pringsheim's sense may essentially differ from each other. We present a simple example in case $d = 2$. Let $\{r_i(x) : i = 1,2,\ldots\}$ be the Rademacher system and divide it into two disjoint infinite subsystems: $\{r_{i_k}(x) : k = 1,2,\ldots\}$ and $\{r_{j_p}(x) : p = 1,2,\ldots\}$. It is well-known that the series $\sum_{i=1}^\infty a_i r_i(x)$ converges a.e. whenever $\sum_{i=1}^\infty a_i^2 < \infty$. It is clear that every subsystem $\{r_{i_k}(x)\}$ also possesses this property. Further, let $\{\psi_i(x) : i = 1,2,\ldots\}$ be an ONS such that there exists a sequence $\{A_i : i = 1,2,\ldots\}$ of coefficients in ℓ^2 such that the series $\sum_{i=1}^\infty A_i \psi_i(x)$ diverges a.e. on I. Then we set for $k = 1,2,\ldots$

$$\varphi_{k1}(x,y) = \begin{cases} r_{i_k}(2x) & \text{for } 0 \leq x \leq 1/2, \\ \psi_k(2x-1) & \text{for } 1/2 < x \leq 1; \end{cases}$$

$$\varphi_{k2}(x,y) = \begin{cases} r_{i_k}(2x) & \text{for } 0 \leq x \leq 1/2, \\ -\psi_k(2x-1) & \text{for } 1/2 < x \leq 1; \end{cases}$$

and for $\ell = 3,4,\ldots$

$$\varphi_{k\ell}(x,y) = \begin{cases} \sqrt{2}\, r_{j_{p(k,\ell)}}(2x) & \text{for } 0 \leq x \leq 1/2, \\ 0 & \text{for } 1/2 < x \leq 1, \end{cases}$$

where $p = p(k,\ell)$ is a one-to-one mapping of $\{(k,\ell) : k = 1,2,\ldots;\ \ell = 3,4,\ldots\}$ onto $\{p : p = 1,2,\ldots\}$. It is easy to check that $\{\varphi_{k\ell}(x,y) :\ k,\ell = 1,2,\ldots\}$ is an ONS on I^2. If we set

$$a_{k1} = a_{k2} = A_k \quad (k = 1,2,\ldots) \quad \text{and } a_{k\ell} = 0 \quad \text{else},$$

then the double series $\sum_{k=1}^{\infty} \sum_{\ell=1}^{\infty} a_{k\ell}\varphi_{k\ell}(x,y)$ converges a.e. in Pringsheim's sense, but does not converge regularly on a set of measure at least 1/2. It is only a difficulty of technical character to modify this example so as the resulting orthogonal series converge a.e. in Pringsheim's sense and do not converge regularly a.e.

This phenomenon cannot occur in the case of double Fourier series of functions from $L^2(I^2)$. In fact, if $f(x,y) \in L^2(I^2)$ and

$$f(x,y) \sim \sum_{k=-\infty}^{\infty} \sum_{\ell=-\infty}^{\infty} a_{k\ell} e^{-2\pi i(kx+\ell y)} \tag{13}$$

is its Fourier series (for convenience we use complex notation), then $\sum_{k=-\infty}^{\infty} \sum_{\ell=-\infty}^{\infty} |a_{k\ell}|^2 < \infty$. Therefore, by the celebrated result of L. Carleson, all rows and columns of the double series on the right of (13) converge a.e.

It is an open problem whether the a.e. regular convergence and the a.e. convergence in Pringsheim's sense are equivalent to each other or not for the multiple Fourier series of functions $f(x_1,\ldots,x_d) \in L^2(I^d)$ in case $d \geq 3$.

Finally, we remark that for double Fourier series of functions $f(x,y) \in L(I^2)$ the above two kinds of convergence no longer coincide. Let us take two functions: $g(x)$ and $h(y)$, $g(x)$ is drawn in Fig. 5, while $h(y) \in L(I)$ is such a function that its Fourier series boundedly diverges a.e. (see, e.g., [11, p. 308]). Then $f(x,y) = g(x)h(y) \in L(I^2)$, whose double Fourier series (13) converges to 0 in Pringsheim's sense a.e. on $(1/4,3/4) \times (0,1)$, but the columns of (13) diverge a.e. on I^2.

It is a further open question what is the situation in connection with the double Fourier series of functions $f(x,y) \in L^p(I^2)$ in case $1 < p < 2$.

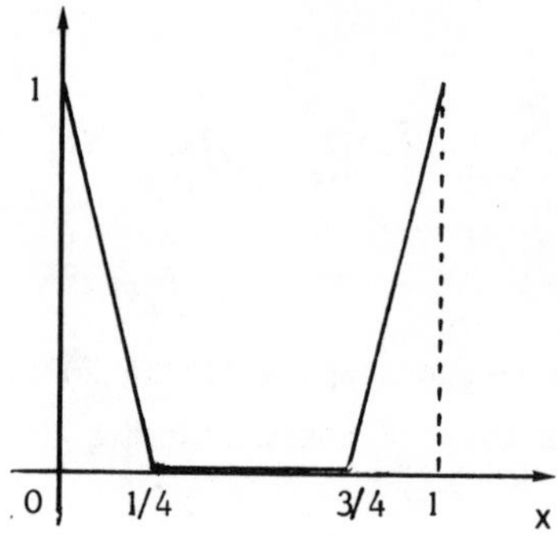

Fig. 5 : g(x)

4. Moore - Smith Convergence and Regular Convergence

In this concluding Section we briefly sketch a possible definition of the notion of regular convergence in the case when the index set is a general directed one (instead of Z_+^d). We begin with the repetition of the definitions of the notion of directed set and Moore - Smith convergence (see, e.g., [4, Ch.2]).

A binary relation "$\leqslant$" *directs* a set D if D is non-void and

(i) $\leqslant$ is transitive on D,

(ii) $\leqslant$ is reflexive on D,

(iii) if k and m are members of D, then there is an element p in D such that $k \leqslant p$ and $m \leqslant p$.

A *directed set* is a pair $(D, \leqslant)$ such that $\leqslant$ directs D. A *net* is a pair $(S, \leqslant)$ such that S is a function and $\leqslant$ directs the domain of S. If S is a function whose domain contains D and D is directed by $\leqslant$, then $\{S_m, m \in D, \leqslant\}$ is the net $(S|D, \leqslant)$ where $S|D$ is S restricted to D.

A net $\{S_m, m \in D, \leqslant\}$ is *eventually* in a set V if there is an element p of D such that if $m \in D$ and $p \leqslant m$, then $S_m \in V$. A net $(S, \leqslant)$ in a topological space (X,J) *converges* to s relative to J (in the Moore - Smith sense) if it is eventually in each J-neighbourhood of s.

Now let (X,+) be an Abel group endowed with a topology (X,J). Given a formal series

$$\sum_{k \in D} a_k \qquad (a_k \in X), \tag{14}$$

consider its all possible partial sums

$$s_m = \sum_{k \in D, k \leq m} a_k \qquad (m \in D)$$

for which the number of those a_k with $k \leq m$ which differ from 0, the neutral element of X with respect to "+", is finite. Denote by D' the subset of D for which this is the case. If D' is non-void and the net $\{s_m : m \in D', \leq\}$ converges to s relative to J, then we may say that series (14) converges and its sum is equal to s.

After these preliminaries, our proposed definition of regular convergence reads as follows. Series (14) is said to be *regularly convergent* if for each neighbourhood V of 0 there exists an element p of D such that for every $m,n \in D$ for which $m \not\leq p$, $m \leq n$ and the number of those a_k with $m \leq k \leq n$ which differ from 0 is finite, we have $\sum_{k \in D, m \leq k \leq n} a_k \in V$. It may happen, of course, that a series (14) converges regularly, but it does not converge in the Moore - Smith sense, and so we cannot attribute any sum s to it.

There are a lot of natural questions arising in connection with these two very general kinds of convergence of series. For instance, for which directed sets $(D, \leq)$ and for which Abel groups (S,+) endowed with a topology (X,J) the following statements hold:

(i) regular convergence implies Moore - Smith convergence,

(ii) Moore - Smith convergence implies regular convergence,

(iii) these two kinds of convergence are equivalent, and

(iv) they are incomparable.

REFERENCES

[1] Alexits, G., Convergence problems of orthogonal series. Pergamon Press, New York - Oxford - Paris 1961.

[2] Hardy, G.H., On the convergence of certain multiple series. Proc. London Math. Soc., Ser. 2, 1 (1903-1904), 124-128.

[3] Hardy, G.H., On the convergence of certain multiple series. Proc. Cambridge Philosoph. Soc. 19 (1916-1919), 86-95.

[4] Kelley, J.L., General topology. Van Nostrand Co, Inc., Princeton 1955.

[5] Móricz, F., On the convergence in a restricted sense of multiple series. Analysis Math. 5 (1979), 135-147.

[6] Móricz, F., The Kronecker lemmas for multiple series and some applications. Acta Math. Acad. Sci. Hungar. 36 (1981) (to appear).

[7] Móricz, F., On the convergence of multiple orthogonal series. Acta Sci. Math. (Szeged) 43 (1981) (to appear).

[8] Móricz, F., - Tandori, K., On the divergence of multiple orthogonal series. Acta Sci. Math. (Szeged) 42 (1980), 133-142.

[9] Tandori, K., Über die Konvergenz der Orthogonalreihen. Acta Sci. Math. (Szeged) 24 (1963), 139-151.

[10] Tandori, K., Über die Konvergenz der Orthogonalreihen II. Acta Sci. Math. (Szeged) 25 (1964), 219-232.

[11] Zygmund, A., Trigonometric series I. University Press, Cambridge 1959.

NORM INEQUALITIES RELATING THE HILBERT TRANSFORM TO THE HARDY-LITTLEWOOD MAXIMAL FUNCTION

Benjamin Muckenhoupt[1]
Department of Mathematics
Rutgers University
New Brunswick, New Jersey

R. Coifman and C. Fefferman have shown for $1 < p < \infty$ that the weighted L^p norm of the Hilbert transform is bounded by the weighted L^p norm of the Hardy-Littlewood maximal function if the weight function satisfies the condition A_∞. It is shown in the first part of this paper that A_∞ is not a necessary condition by deriving a large class of weight functions not in A_∞ for which the norm inequality holds. The rest of the paper consists of the derivation of a necessary condition for the norm inequality; this condition closely resembles the A_∞ condition.

1. Introduction

The problem considered here is the determination of all non-negative functions $W(x)$ such that

$$(1.1)\qquad \int_{-\infty}^{\infty} |\tilde{f}(x)|^p W(x)\,dx \le A \int_{-\infty}^{\infty} [f^*(x)]^p W(x)\,dx,$$

where A is independent of f,

$$\tilde{f}(x) = \lim_{\varepsilon\to 0^+} \int_{|y|>\varepsilon} \frac{f(x-y)}{y}\,dy$$

is the Hilbert transform of f,

1) Supported in part by N.S.F. Grant MCS 80-03098.

$$f^*(x) = \sup_{y\neq x} \frac{1}{y-x}\int_x^y |f(t)|dt$$

is the Hardy-Littlewood maximal function of f and p is a fixed number satisfying $1 < p < \infty$. The principal known result concerning this problem is one by R. Coifman and C. Fefferman that appeared in 1974 in [3]. Theorem III of [3] proves (1.1) for any p satisfying $1 < p < \infty$ provided $W(x)$ is non-negative and there are positive constants C and ε such that $\int_E W dx \leq C(|E|/|I|)^\varepsilon \int_I W dx$ for every interval I and subset E of I. This condition on W, known as the A_∞ condition, has been a popular condition on weight functions since that time. It has been used in various norm inequalities between Littlewood-Paley functions and the Lusin area function and in the theory of weighted H^p spaces.

Coifman and Fefferman did not consider whether A_∞ was a necessary condition for (1.1); in fact A_∞ has not been shown to be a necessary condition for any of the norm inequalities for which it has been shown sufficient. It is shown here that A_∞ is not a necessary condition for (1.1). Theorem 2.1 in §2 describes a large class of W's not in A_∞ for which (1.1) holds; in particular, $\chi_{[0,\infty)}$ is such a function.

The rest of this paper consists of the derivation of a necessary condition for (1.1). This is done by proving the following theorem.

THEOREM 1.2. <u>If</u> $1 < p < \infty$, $W(x)$ <u>is non-negative and</u> $W(x)$ <u>satisfies</u> (1.1), <u>then there are positive constants</u> C <u>and</u> ε <u>such that for every interval</u> I <u>and every subset</u> E <u>of</u> I

$$\int_E W(x)dx \leq C\left[\frac{|E|}{|I|}\right]^\varepsilon \int_{-\infty}^{\infty} \frac{|I|^p W(x)dx}{|I|^p+|x-x_I|^p}.$$

The necessary condition of Theorem 1.2 will be referred to as the C_p condition. It does, of course, resemble the A_∞ condition and is clearly weaker since the integral on the right is larger. We conjecture that the C_p condition is also a sufficient condition for (1.1).

The proof of Theorem 1.2 is fairly long and is broken into several lemmas that are discussed and proved in §§3-5. The proof is completed in §6.

The following notation will be used throughout this paper. For a set E, $|E|$ will denote the Lebesgue measure of E and χ_E the characteristic function of E. If $a > 0$ and I is an interval, aI will denote the interval with the same center and with $|aI| = a|I|$. If $1 < p < \infty$, p' will denote the number such that $p^{-1}+(p')^{-1} = 1$. The letter C will denote constants, not necessarily the same at each occurrence.

2. A_∞ Is Not Necessary for (1.1)

Here we show that A_∞ is not a necessary condition by deriving functions W that satisfy (1.1) but are not in A_∞. Since translations, reflections and sums of weight functions satisfying (1.1) also satisfy (1.1), a great many weight functions can be generated by use of theorem 2.1.

We will need the following definition. If $1 < p < \infty$, then a non-negative function $U(x)$ is in A_p if for every interval I

$$\left[\frac{1}{|I|}\int_I U(x)dx\right]\left[\frac{1}{|I|}\int_I [U(x)]^{-1/(p-1)}dx\right]^{p-1} \leq C,$$

where C is independent of I.

THEOREM 2.1. <u>If</u> $1 < p < \infty$ <u>and</u> $W(x) = U(x)\chi_{[0,\infty)}(x)$, <u>where</u> $U(x)$ is <u>in</u> A_p, <u>then</u> (1.1) <u>holds</u>.

Except for the case $U(x) = 0$ almost everywhere, the functions W of Theorem 2.1 are not in A_∞ as is shown by the following reasoning. First, observe that $\int_{-N}^{N} U dx > 0$ for sufficiently large N. Since U is in A_p, then $U^{-1/(p-1)}$ must be locally integrable. Therefore, $U(x) > 0$ almost everywhere and $\int_0^h U dx > 0$ for all $h > 0$. Then with $h > 0$, $I = [-1,h]$ and $E = [0,h]$, the definition of A_∞ for W would require that $1 \leq C[h/(1+h)]^\varepsilon$ with C and ε independent of h. Since this is impossible, W is not in A_∞.

The proof of Theorem 2.1 will use various weighted norm inequalities. By taking $U(x) \equiv 1$, the better known unweighted versions can be used and this is, of course, sufficient to show that (1.1) does not imply that W is in A_∞. To prove Theorem 2.1, it is sufficient to show that

$$(2.2)\qquad \int_0^\infty |(f(x)\chi_{[0,\infty)}(x))^\sim|^p U(x)dx \le C\int_0^\infty [f^*(x)]^p U(x)dx$$

and

$$(2.3)\qquad \int_0^\infty |(f(x)\chi_{(-\infty,0]}(x))^\sim|^p U(x)dx \le C\int_0^\infty [f^*(x)]^p U(x)dx.$$

By theorem 9, p. 247 of [4], the left side of (2.2) has the bound $C\int_{-\infty}^\infty |f\chi_{[0,\infty)}|^p U dx$. This is bounded by the right side of (2.2) since $|f(x)| \le f^*(x)$ almost everywhere.

To prove (2.3), use the definition of the Hilbert transform to show that the left side is bounded by

$$C\int_0^\infty \left|\int_{-x}^0 \frac{f(t)}{x-t}\,dt\right|^p U(x)dx + C\int_0^\infty \left|\int_{-\infty}^{-x} \frac{f(t)}{x-t}dt\right|^p U(x)dx.$$

Since x and t have opposite signs, this is bounded by the sum of

$$(2.4)\qquad C\int_0^\infty \left[\frac{1}{x}\int_{-x}^0 |f(t)|dt\right]^p U(x)dx$$

and

$$(2.5)\qquad C\int_0^\infty \left[\int_{-\infty}^{-x} \frac{|f(t)|}{|t|}\,dt\right]^p U(x)dx.$$

Now (2.4) is bounded by

$$C\int_0^\infty \left[\frac{1}{2x}\int_{-x}^x |f(t)|dt\right]^p U(x)dx \le C\int_0^\infty [f^*(x)]^p U(x)dx$$

by the definition of $f^*(x)$. To estimate (2.5), it is sufficient to show that (2.5) is bounded by (2.4). With $g(t) = f(-t)$, this is equivalent to showing that

(2.6) $$\int_0^\infty \left[\int_x^\infty \frac{|g(t)|}{t}\right]^p U(x)dx \le C\int_0^\infty \left|\frac{1}{x}\int_0^x |g(t)|dt\right|^p U(x)dx.$$

With $U(x) = x^a$ for $x > 0$ and $-1 < a < p-1$, (2.6) is a result of Boas [1]. To prove it for all U in A_p, observe that the left side of (2.6) is bounded by

(2.7) $$\int_0^\infty \left[\int_x^\infty (\frac{1}{u^2}\int_0^u |g(t)|dt)du\right]^p U(x)dx.$$

Boas' result now follows by an application of Hardy's inequality, see [7], vol. I, p. 20. For general U in A_p, observe that for $r > 0$

(2.8) $$\int_r^\infty \frac{U(x)^{-1/(p-1)}dx}{x^{p'}} \le \frac{C}{r^{p'}}\int_0^r U(x)^{-1/p-1)}dx$$

by use of Lemma 1 p. 232 of [4] since $U(x)^{-1/(p-1)}$ is in $A_{p'}$. By (2.8) and the definition of A_p we see that for $r > 0$.

$$\left[\int_0^r U(x)dx\right]\left[\int_r^\infty [x^pU(x)]^{-1/(p-1)}dx\right]^{p-1} \le C$$

with C independent of r. By Theorem 2, p. 32 of [5], we have (2.7) bounded by

$$\int_0^\infty \left[\frac{1}{u^2}\int_0^u |g(t)|dt\right]^p [u^pU(u)]du.$$

This completes the proof of (2.6), and, thereby, of Theorem 2.1.

3. A Basic Lemma

In Lemma 3, p. 268 of [6], Stein and Weiss showed that if D is a finite union of disjoint intervals, $\alpha > 0$ and E is the set where $|\tilde{\chi}_D(x)| > \alpha$, then $|E| = 2|D|/\sinh \alpha$. Here we treat the inverse problem; given a set E and $\alpha > 0$ we want to find a corresponding set D so that

$|\tilde{\chi}_D(x)| > \alpha$ on E and $|D| \le |E|\sinh\alpha$. This can be done as shown by the following lemma. Like the proof in [6], this proof is based on the fact that the sum of the roots of a monic polynomial equals minus one times the second coefficient.

LEMMA 3.1. _If_ $E = \bigcup_{i=1}^{n}(c_i,d_i)$ _is a finite union of open intervals with disjoint closures and_ $\alpha > 0$, _then there exists a finite disjoint union of open intervals_ $D = \bigcup_{i=1}^{n}(a_i,b_i)$ _such that_ $|D| = |E|\sinh\alpha$, $b_i \in (c_i,d_i)$ _and_ $|\tilde{\chi}_D(x)| > \alpha$ _for_ $x \in E$.

We may assume that the intervals (c_i,d_i) are in their natural order so that $d_i < c_{i+1}$.

The polynomial

$$a(x) = \frac{1+e^{\alpha}}{2}\prod_{i=1}^{n}(x-c_i) + \frac{1-e^{\alpha}}{2}\prod_{i=1}^{n}(x-d_i)$$

has $(-1)^{n-k}a(d_k) > 0$, $(-1)^{n-k-1}a(c_{k+1}) > 0$ and $\lim_{x\to-\infty}(-1)^n a(x) > 0$.

Therefore, $a(x)$ has one root in each interval (d_k,c_{k+1}) and one in $(-\infty,c_1)$. Call the root in $(-\infty,c_1)$, a_1. Call the root in (d_k,c_{k+1}), a_{k+1}. Similarly, the polynomial

$$b(x) = \frac{1+e^{-\alpha}}{2}\prod_{i=1}^{n}(x-c_i) + \frac{1-e^{-\alpha}}{2}\prod_{i=1}^{n}(x-d_i)$$

has $(-1)^{n-k}b(c_k) < 0$ and $(-1)^{n-k}b(d_k) > 0$. Therefore, $b(x)$ has one root in each interval (c_k,d_k); call this root b_k.

Let $D = \bigcup_{k=1}^{n}(a_k,b_k)$ and note that the intervals (a_k,b_k) are disjoint since $(a_k,b_k) \subset (d_{k-1},d_k)$, where d_{-1} is defined to be $-\infty$. Since $a(x) = \Pi(x-a_i)$,

$$\sum_{i=1}^{n} a_i = \frac{1+e^{\alpha}}{2}\sum_{i=1}^{n} c_i + \frac{1-e^{\alpha}}{2}\sum_{i=1}^{n} d_i$$

because the sum of the roots of a monic nth degree polynomial is minus the coefficient of x^{n-1}. Similarly,

$$\sum_{i=1}^{n} b_i = \frac{1+e^{-\alpha}}{2} \sum_{i=1}^{n} c_i + \frac{1-e^{-\alpha}}{2} \sum_{i=1}^{n} d_i .$$

Combining these we get

$$|D| = \sum_{i=1}^{n} (b_i - a_i) = \frac{e^{\alpha} - e^{-\alpha}}{2} \sum_{i=1}^{n} (d_i - c_i) = |E| \sinh \alpha .$$

Now

$$|\tilde{\chi}_D(x)| = \left| \sum_{i=1}^{n} \log \left| \frac{x-a_i}{x-b_i} \right| \right| = \left| \log \left| \frac{b(x)}{a(x)} \right| \right| .$$

If $x \in (c_k, d_k)$ for some k, then $\Pi \frac{x-d_i}{x-c_i} < 0$ and

$$\left| \frac{b(x)}{a(x)} \right| = e^{-\alpha} \left| -1 + \frac{2(1+e^{\alpha})}{1+e^{\alpha} - (e^{\alpha}-1)\Pi \frac{x-d_i}{x-c_i}} \right| < e^{-\alpha} .$$

Therefore, if $x \in (c_i, d_i)$, $|\tilde{\chi}_D(x)| > \alpha$. This completes the proof of Lemma 3.1.

4. A Preliminary Necessary Condition

The condition in the theorem of this section is in fact equivalent to the C_p condition in Theorem 1.2. The equivalence is proved in §§5-6 and makes no use of the equation (1.1). Theorem 4.1 is a straightforward application of Lemma 3.1 to the condition (1.1).

THEOREM 4.1. If $1 < p < \infty$, (1.1) holds, I is an interval with center x_I and E is a subset of I, then

$$\int_E W(x)\,dx \le \frac{3^p A}{\left[\sinh^{-1}\left(\frac{|I|}{|E|}\right)\right]^p} \int_{-\infty}^{\infty} \frac{|I|^p W(x)\,dx}{(|I| + |x - x_I|)^p} . \tag{4.2}$$

If W is not integrable on I, the result is trivial; therefore, assume that $\int_I W dx < \infty$. Given $\varepsilon > 0$, there is a finite union of disjoint open intervals E_1 such that $E_1 \subset I$, $|E_1| < |E| + \varepsilon$ and $\int_E W dx \le \varepsilon + \int_{E_1} W dx$. By Lemma 3.1 with $\alpha = \sinh^{-1}\left[\frac{|I|}{|E_1|}\right]$, there is a finite union of intervals D such that $|D| = |I|$, each interval in D intersects E_1, and $|\tilde{\chi}_D(x)| > \sinh^{-1}(|I|/|E_1|)$ for x in E_1. Therefore

$$\int_E W(x)dx \le \varepsilon + \int_{E_1} W(x)dx \le \varepsilon + \left[\sinh^{-1} \frac{|I|}{|E_1|}\right]^{-p} \int_{E_1} |\tilde{\chi}_D(x)|^p W(x)dx.$$

An application of (1.1) then shows that

$$\int_E W(x)dx \le \varepsilon + A\left[\sinh^{-1} \frac{|I|}{|E_1|}\right]^{-p} \int_{-\infty}^{\infty} |\chi_D^*(x)|^p W(x)dx.$$

Since each interval in D intersects E_1 and $E_1 \subset I$, it follows that $D \subset 3I$ and $\chi_D^*(x) \le \frac{3|I|}{|I|+|x-x_I|}$. Therefore,

$$\int_E W(x)dx \le \varepsilon + A\left[\sinh^{-1}\left(\frac{|I|}{|E|+\varepsilon}\right)\right] \int_{-\infty}^{\infty} \frac{3^p|I|^p W(x)dx}{(|I|+|x-x_I|)^p}.$$

Since ε is arbitrary, (4.2) follows from this.

5. A Stronger Necessary Condition

The condition C_p does not follow immediately from Theorem 4.1 since $\sinh^{-1}y \approx \log 2y$ for large y instead of $|y|^\varepsilon$ for some $\varepsilon > 0$. To prove the C_p condition, we will need a form of the necessity condition that can be used repeatedly. The conclusion of Lemma 5.1 has the desired form; it has similar expressions on both sides of the inequality. Note that Lemma 5.1 does not follow directly from the proof of Theorem 4.1 since the inequality $\Delta(x) \le c|\tilde{\chi}_E(x)|^p$ is not true; $\tilde{\chi}_E$ has zeros while $\Delta(x)$

does not.

LEMMA 5.1. <u>If</u> $1 < p < \infty$, W <u>is non-negative</u>, W <u>satisfies</u> (4.2) <u>for every interval</u> I <u>and subset</u> E <u>of</u> I, $\{I_k\}_{k=1}^n$ <u>is a set of disjoint subintervals of an interval</u> I, x_k <u>is the center of</u> I_k, x_I <u>is the center of</u> I <u>and</u>

$$\Delta(x) = \sum_{k=1}^{n} \frac{|I_k|^p}{|I_k|^p + |x-x_I|^p}, \tag{5.2}$$

<u>then</u>

$$\int_{-\infty}^{\infty} \Delta(x)W(x)dx \le C\left[\log\left(\frac{|I|}{\sum|I_k|}\right)\right]^{1-p} \int_{-\infty}^{\infty} \frac{|I|^p W(x)dx}{|I|^p + |x-x_I|^p}, \tag{5.3}$$

<u>where</u> C <u>depends only on</u> A <u>and</u> p.

To prove this observe first that since $\sum|I_k|^p \le (\sum|I_k|)^p$, then for x not in $3I$

$$\Delta(x) \le C\left[\frac{\sum|I_k|}{|I|}\right]^p \frac{|I|^p}{|I|^p + |x-x_I|^p}.$$

Therefore, $\int_{(3I)^c} \Delta x W(x)dx$ is bounded by the right side of (5.3), and the proof can be completed by estimating

$$\int_{3I} \Delta(x)W(x)dx. \tag{5.4}$$

To estimate (5.4) we will need a Lemma of Carleson, Lemma 5 p. 140 of [2], that asserts the existence of constants B and D; depending only on p, such that for $a > 0$

$$|\{\Delta(x) > a\}| \le Be^{-Da}|I|, \tag{5.5}$$

where $\{I_k\}$ is any subdivision of the interval I into subintervals. Since any disjoint collection $\{I_k\}$ of subintervals of I can have intervals added to it to become a subdivision of I and since adding subintervals increases $\Delta(x)$, (5.5) is valid with the same constants for any disjoint collection $\{I_k\}$ of subintervals of I. Since (5.5) remains true if D is decreased or B is increased, we may also assume that $B \geq 1$ and $0 < D < 1$.

Now let j be the least integer greater than $\log(\Sigma|I_k|/|I|)$, and let J be the least integer greater than $\log\left[\frac{1}{D}\log\left(\frac{e|I|}{\sum|I_k|}\right)\right]$, where D is the constant in (5.5). Note that since $j \leq 0$ and $J > 0$, that $j < J$. Let Q, S and T be respectively the intersection of 3I with the set where $\Delta(x) \leq e^j$, the set where $e^j < \Delta(x) \leq e^J$ and the set where $e^J < \Delta(x)$. We will estimate (5.4) by estimating the integral of $\Delta(x)W(x)$ over the sets Q, S and T separately.

First, we have from the definition of Q that

$$\int_Q \Delta(x)W(x)dx \leq e\frac{\Sigma|I_k|}{|I|}\int_{3I} W(x)dx.$$

The right side is bounded by the right side of (5.3). Next

$$\int_S \Delta x W(x)dx \leq \sum_{i=j}^{J} \int_{\{\Delta x > e^i\} \cap 3I} e^{i+1}W(x)dx. \tag{5.6}$$

Now since $\int_{-\infty}^{\infty} \Delta x dx \leq \frac{2p}{p-1}\sum|I_k|$, we have $|\{\Delta x > e^i\}| \leq \frac{2p}{p-1}e^{-i}\sum|I_k|$. Using this fact and (4.2) shows that the right side of (5.6) is bounded by the product of

$$\sum_{i=j}^{J} Ce^{i+1}\left[\sinh^{-1}\left(\frac{3(p-1)|I|}{2pe^{-i}\Sigma|I_k|}\right)\right]^{-p} \tag{5.7}$$

and

$$\int_{-\infty}^{\infty} \frac{|I|^p W(x)dx}{(|I|+|x-x_I|)^p}. \tag{5.8}$$

To complete this part we need to show that (5.7) has the bound

$$C\left[\log\frac{|I|}{\Sigma|I_k|}\right]^{1-p}. \tag{5.9}$$

To estimate (5.7) use the fact that the argument of the $\sinh^{-1}$ in

(5.7) is bounded below by $\frac{p-1}{p} e^{i-j} \geq \frac{p-1}{p}$. From this we obtain

$$\sum_{i=j}^{J} Ce^{i+1}\left[\log e^{i-j+1}\right]^{-p}$$

as an upper bound for (5.7). The change of variables $i = m+j$ gives the estimate

$$Ce^{j+1}\sum_{m=0}^{J-j} \frac{e^{m}}{(m+1)^{p}} \leq Ce^{j+1} \frac{e^{J-j}}{(J-j+1)^{p}}.$$

Since $J \geq 1$ and $j \leq 0$, this is bounded by $Ce^{J}(1-j)^{-p}$. Using the definition of j and J shows that this is bounded by (5.9) and completes the estimation of the integral of $\Delta(x)W(x)$ over S.

Finally, we have

$$\int_{T} \Delta x W(x)dx \leq \sum_{i=J}^{\infty} \int_{\{\Delta x>e^{i}\}\cap 3I} e^{i+1}W(x)dx.$$

The estimates (5.5) and (4.2) show that the right side is bounded by the product of (5.8) and

$$C\sum_{i=J}^{\infty}\left[\sinh^{-1}\left(\frac{e^{De^{i}}}{B}\right)\right]^{-p}. \tag{5.10}$$

As in the estimation of (5.7) we use the fact that the argument of the $\sinh^{-1}$ in (5.10) has a positive lower bound, $D > 0$ and $J \geq 1$ to show that (5.10) is bounded by

$$C\sum_{i=J}^{\infty}\left[\log(e^{De^{i}})\right]^{-p}.$$

This geometric series is easily estimated; with the definition of J we get the upper bound $C[\log(e|I|/\Sigma|I_k|)]^{-p}$ which is bounded by (5.9). This completes the proof of Lemma 5.1.

6. Proof of Theorem 1.2

We can now complete the proof that C_p is a necessary condition for (1.1). To do this use Theorem 4.1 and Lemma 5.1 to choose a $\delta > 0$ such that if I is an interval and $\{I_k\}$ is a collection of disjoint subintervals of I with $\sum|I_k| < 2\delta|I|$, then

(6.1) $$\int_{-\infty}^{\infty} \Delta(x)W(x)dx \le \tfrac{1}{2}\int_{-\infty}^{\infty} \frac{|I|^p W(x)dx}{|I|^p+|x-x_I|^p}.$$

Furthermore, δ should be chosen small enough that

(6.2) $$3^p A[\sinh^{-1}(1/2\delta)]^{-p} \le \tfrac{1}{2}.$$

Now given $E\subset I$, let n be the least integer such that $\delta^n|I| < |E|$. Define $E_0 = E$ and $E_j = \{\chi_E^* > \delta^j\}$ for $1 \le j \le n$. Define $\Delta_j(x)$ for $1 \le j \le n$ to be the function (5.2) based on the component intervals of E_j.

If $1 \le j \le n$, H is a component interval of E_j and x is an endpoint of H, then $\chi_E^*(x) \le \delta^j$. Therefore,

(6.3) $$|H \cap E| \le \delta^j|H|,$$

and if J is an interval containing an endpoint of H, then $|J\cap E| \le \delta^j|J|$. Consequently, if x is in $H\cap E_{j-1}$, there is an interval $J\subset H$ such that x is in J and $|J\cap E| > |J|\delta^{j-1}$. Therefore, $H\cap E_{j-1} = \{\chi_{E\cap H}^* > \delta^{j-1}\}$. By a covering lemma argument we then have $|H\cap E_{j-1}| \le 2\delta^{1-j}|E\cap H|$. Combining this with (6.3) shows that $|H\cap E_{j-1}| \le 2\delta|H|$. Now let Q be the set of component intervals in $H\cap E_{j-1}$ and let Δ_Q be the corresponding function as defined in (5.2) with $\{I_k\} = Q$. Then by the definition of δ we have

$$\int_{-\infty}^{\infty} \Delta_Q(x)W(x)dx \le \tfrac{1}{2}\int_{-\infty}^{\infty} \frac{|H|^p W(x)dx}{|H|^p+|x-x_H|^p}.$$

Adding these inequalities for all the components H of E_j then gives

(6.4) $$\int_{-\infty}^{\infty} \Delta_{j-1}(x)W(x)dx \le \tfrac{1}{2}\int_{-\infty}^{\infty} \Delta_j(x)W(x)dx$$

for $2 \le j \le n$, where $\Delta_j(x)$ is the function of (5.2) for the collection of component intervals of E_j. Similarly, using (6.2) and Theorem 4.1, we have

$$(6.5)\qquad \int_{-\infty}^{\infty} \chi_E(x)W(x)\,dx \le \tfrac{1}{2}\int_{-\infty}^{\infty} \Delta_1(x)W(x)\,dx.$$

Combining (6.5) and (6.4) for $2 \le j \le n$ shows that

$$(6.6)\qquad \int_E W(x)\,dx \le 2^{-n}\int \Delta_n(x)W(x)\,dx.$$

Since $\delta^{n-1} \ge |E|/|I| > \delta^n$, E_n consists of one interval that contains I and is contained in $2\delta^{-1}I$. Therefore,

$$(6.7)\qquad \Delta_n(x) \le \frac{C|I|^p}{|I|^p+|x-x_I|^p}.$$

Furthermore, since $n > (\log \frac{|E|}{|I|})/\log \delta$, we have $2^{-n} < \left[\frac{|E|}{|I|}\right]^a$, where $a = \frac{-\log 2}{\log \delta}$. Combining this with (6.6) and (6.7) then completes the proof of Theorem 1.2.

REFERENCES

[1] Boas, R., Some integral inequalities related to Hardy's inequality. J. Analyse Math. 23 (1970), 53-63.

[2] Carleson, L., On convergence and growth of partial sums of Fourier series. Acta Math. 116 (1966), 135-157.

[3] Coifman, R.-Fefferman, C., Weighted norm inequalities for maximal functions and singular integrals. Studia Math. 51 (1974), 241-250.

[4] Hunt, R.-Muckenhoupt, B.-Wheeden, R., Weighted norm inequalities for the conjugate function and Hilbert transform. Trans. Amer. Math. Soc. 176 (1973), 227-251.

[5] Muckenhoupt, B., Hardy's inequality with weights. Studia. Math. 44 (1972), 31-38.

[6] Stein, E.-Weiss, G., An extension of a theorem of Marcinkiewicz and some of its applications. J. Math. Mech. 8 (1959), 263-284.

[7] Zygmund, A., Trigonometric Series. Vols. I, II, Second Edition, Cambridge Univ. Press, New York, 1959.

BEST APPROXIMATION ON THE UNIT SPHERE IN $\mathbb{R}^k$

Matthias Wehrens
Lehrstuhl A für Mathematik
Rheinisch-Westfälische Technische Hochschule
Aachen

The translation of a function f on the surface of the unit sphere in k-dimensional Euclidean space is defined by the integral means of f over the circle $\langle x,y\rangle = h$ on the sphere. Via this translation there are introduced the strong Laplace-Beltrami differential operator and the r-th modulus of continuity of functions defined on the sphere. The rate of best approximation by sums of spherical harmonics of degree $\leq n$ is then completely characterized by higher order Lipschitz conditions and differentiability properties.

1. Introduction

The aim of this paper is the characterization of the rate of best approximation of functions defined on the surface of the unit sphere in $\mathbb{R}^k$, $k = 3,4,\ldots$, by Lipschitz conditions and differentiability properties.

To illustrate the problem we recall to mind the corresponding well-known related results on the approximation of continuous, 2π-periodic functions due to Jackson, Bernstein (1911/12) and Zygmund (1945):

THEOREM 0. *Denoting by Π_n, $n \in \mathbb{P} = \{0,1,\ldots\}$, the set of all trigonometric polynomials of degree $\leq n$, one has for $F \in C_{2\pi}$, $r \in \mathbb{P}$ and $0 < \alpha \leq 1$:*

$$E_n(F;C_{2\pi}) := \inf_{t_n \in \Pi_n} \|F - t_n\|_{C_{2\pi}} = \mathcal{O}(n^{-r-\alpha}) \qquad (n \to \infty)$$

$$\Longleftrightarrow F^{(r)} \in \begin{cases} \mathrm{Lip}_1(\alpha;C_{2\pi}), & 0 < \alpha < 1 \\ \mathrm{Lip}_2(1;C_{2\pi}), & \alpha = 1 . \end{cases}$$

This result can also be understood as the characterization of best approximation on the unit circle. This case, namely $k = 2$, could also be treated by the same methods used here; but that would lead to a characterization of best approximation by means of derivatives and Lipschitz conditions of e v e n order. This means that the results obtained by our methods for $k = 2$ are somewhat weaker than those of Theorem 0. So we will only consider the unit sphere in the Euclidean space with dimension ≥ 3.

2. Basic Concepts

Let us denote by S^k, $k = 3,4,\ldots$, the surface of the unit sphere in $\mathbb{R}^k$,

$$S^k := \{x \in \mathbb{R}^k;\ |x| = (\sum_{j=1}^{k} |x_j|^2)^{1/2} = 1\},$$

by $C(S^k)$ the set of all continuous functions f defined on S^k, endowed with the norm

$$\|f\|_C := \max_{x \in S^k} |f(x)|,$$

and by $L^p(S^k)$, $1 \leq p < \infty$, the set of all measurable functions f defined on S^k for which the norm

$$\|f\|_p := \left(\frac{1}{\Omega_k} \int_{S^k} |f(x)|^p \, ds(x) \right)^{1/p}$$

is finite, $ds(x)$ being the $(k-1)$- dimensional surface element of S^k, and $\Omega_k = 2\pi^{\lambda+1}/\Gamma(\lambda+1)$, $\lambda = (k-2)/2$, the surface of S^k. In the following, X is one of these Banach spaces, and $\|\cdot\|_X$ the corresponding norm. Furthermore, L^1_λ is the Banach space of all functions χ measurable on $(-1,1)$ having finite norm

$$\|\chi\|_{1,\lambda} := \frac{\Omega_{k-1}}{\Omega_k} \int_{-1}^{1} |\chi(t)| (1-t^2)^{\lambda-1/2} \, dt.$$

Now, the appropriate fundamental set for our desired approximation theorem is the set of all "spherical harmonics" of degree $n \in \mathbb{P}$. A spherical harmonic Y_n is defined as the restriction to the unit sphere S^k of an harmonic, homogeneous polynomial Q_n of degree n defined on R^k. The spherical harmonics form an orthogonal fundamental system in X, i.e.,

$$\int_{S^k} Y_n(x)\, Y_m(x)\, ds(x) = 0 \qquad (n \neq m),$$

$$\overline{\operatorname{span}\{Y_n\}}^X = X .$$

They are closely connected with the Gegenbauer (or ultraspherical) polynomials

$$P_n^\lambda(t) := \frac{(-2)^n\, \Gamma(n+\lambda)\, \Gamma(n+2\lambda)}{n!\, \Gamma(\lambda)\, \Gamma(2n+2\lambda)} (1-t^2)^{1/2-\lambda} \frac{d^n}{dt^n}(1-t^2)^{n+\lambda-1/2}.$$

(For these and other properties of spherical harmonics see e.g. [12] or [13].)

By means of the normalized Gegenbauer polynomials

$$R_n^\lambda(t) := P_n^\lambda(t)/P_n^\lambda(1), \quad P_n^\lambda(1) = \binom{n+2\lambda-1}{n},$$

we associate to each function $f \in X$ a sequence of spherical harmonics, namely its "spherical Fourier coefficients"

$$\text{(2.1)} \qquad Y_n(f;x) := \frac{1}{\Omega_k} \int_{S^k} R_n^\lambda(\langle x,y\rangle)\, f(y)\, ds(y) \qquad (n \in \mathbb{P}; x \in S^k)$$

($\langle\cdot,\cdot\rangle$ = inner product in $\mathbb{R}^k$).

For this transform one has a uniqueness theorem: If $f \in X$, then

$$Y_n(f;x) = 0 \quad (n \in \mathbb{P}, x \in S^k) \iff f(x) = 0 \qquad \text{(a.e.)}.$$

Whereas the classical translation operator T_h of functions defined on $\mathbb{R}^1$ consists in a simple shifting of the variable, it is here more complicated. For in defining T_h one only has to decide whether the variable is to be shifted a given distance h to the left or to the right; which direction is taken is only a problem of convention and of no mathematical significance. But on the sphere S^k one has infinitely many possible directions for shifting the variable a certain distance h, none of them with any preference. So we define the "spherical translation" τ_h of $f \in X$ as the integral means of such shiftings, namely,

$$(2.2) \qquad \tau_h f(x) := \frac{1}{\Omega_{k-1}(1-h^2)^\lambda} \int_{<x,y>=h} f(y)\, dt(y)$$

where $h \in (-1,1)$, $x \in S^k$ and $dt(y)$ is the "curve" element of the "circle" $<x,y> = h$ on S^k.

This definition has the disadvantage that τ_h does not possess the semi-group property anymore as does T_h $(T_u\, T_v = T_{u+v})$, and it cannot be inverted $(T_h\, T_{-h} = \text{identity})$. Nevertheless, τ_h is a positive, linear operator mapping X into X, with operator norm $\|\tau_h\|_{[X]} = 1$ and

$$\lim_{h\to 1-} \|f-\tau_h f\|_X = 0 \qquad (f \in X).$$

Furthermore, for τ_h there hold the product formulae

$$\tau_h\, Y_n(x) = R_n^\lambda(h)\, Y_n(x) ,$$

$$Y_n(\tau_h f;x) = R_n^\lambda(h)\, Y_n(f;x) .$$

These are typical for a translation operator and correspond to the product formulae

$$T_h\, e^{int} := e^{in(t+h)} = e^{inh}\, e^{int} ,$$

$$\int_0^{2\pi} T_h f(t)\, e^{-int}\, dt = e^{inh} \int_0^{2\pi} f(t) e^{-int}\, dt$$

for the trigonometric fundamental system $\{e^{int}\}_{n=-\infty}^{+\infty}$.

This concept of translation allows one to define the "spherical convolution product" of a function $\chi \in L_\lambda^1$ and $f \in X$, namely

$$(2.3) \qquad (\chi * f)(x) := \frac{\Omega_{k-1}}{\Omega_k} \int_{-1}^{1} \chi(t)\, \tau_t f(x)(1-t^2)^{\lambda-1/2}\, dt \qquad (x \in S^k) ;$$

which may also be rewritten in the more classical form

$$(2.4)\qquad (\chi * f)(x) = \frac{1}{\Omega_k} \int_{S^k} \chi(\langle x,y\rangle) f(y)\, ds(y) \qquad \text{(a.e.)}.$$

As long as 72 years ago L. Fejér [9] wrote the second arithmetical means of the Laplace series on S^3 in this form.

Now, this convolution product has the same properties as does the usual convolution product of periodic functions (cf. [4]), namely:

$$(2.5)\qquad \chi * f \in X,\ \|\chi * f\|_X \leq \|\chi\|_{1,\lambda} \|f\|_X \qquad (\chi \in L^1_\lambda; f \in X),$$

$$(2.6)\qquad Y_n(\chi * f; x) = \chi^\wedge(n)\, Y_n(f;x) \qquad (n \in \mathbb{P}; x \in S^k; \chi \in L^1_\lambda; f \in X),$$

where $\chi^\wedge(n)$ are the "Fourier - Gegenbauer coefficients":

$$(2.7)\qquad \chi^\wedge(n) := \frac{\Omega_{k-1}}{\Omega_k} \int_{-1}^{1} \chi(t)\, R_n^\lambda(t)(1-t^2)^{\lambda-1/2}\, dt .$$

(For the properties of translation and convolution cited here see e.g. [2].)

3. The Strong Laplace - Beltrami Derivative and Integral

Since the spherical harmonics Y_n are derived from harmonic polynomials Q_n, i.e. $\nabla^2 Q_n \equiv 0$, ∇^2 being the usual Laplace differential operator in k dimensions, they are eigenfunctions of the Laplace - Beltrami differential operator ∇_*^2 with eigenvalues $-n(n+2\lambda)$ (cf. [12] or [13]):

$$\nabla_*^2\, Y_n(x) = -n(n+2\lambda)\, Y_n(x) \qquad (n \in \mathbb{P}; x \in S^k).$$

Here ∇_*^2 is a pointwise differential operator. The corresponding strong operator, i.e. when the limit is considered in the norm, can be defined - up to a constant factor $-1/(2\lambda+1)$ - by means of the spherical difference

$$\Delta_h f(x) := f(x) - \tau_h f(x) \qquad (-1<h<1; x \in S^k; f \in X).$$

Indeed, if for $f \in X$ there exists a function $Df \in X$ such that

$$\text{(3.1)} \qquad \lim_{h\to 1-} \left\| \frac{\Delta_h f}{1-h} - Df \right\|_X = 0 ,$$

then $Df(=D^1 f)$ is called the (first) "strong Laplace - Beltrami derivative" of f. Higher derivatives are defined inductively by

$$D^r f := D^1(D^{r-1} f) \qquad (r \in \{2,3,\dots\}) .$$

The sets of functions differentiable in this sense are denoted by

$$W_X^r := \{f \in X; D^r f \in X\} \qquad (r \in \mathbb{N} = \{1,2,\dots\}) .$$

In this respect, we have for all spherical harmonics Y_n, $n \in \mathbb{P}$:

$$Y_n \in W_X^1,\ DY_n(x) = \frac{n(n+2\lambda)}{2\lambda+1} Y_n(x) \qquad (x \in S^k) .$$

There also exists an inverse operator to D^r, defined by the "spherical integral" of $f \in X$:

$$\text{(3.2)} \qquad \begin{cases} (J^1 f)(x) := (\xi * f)(x) & (x \in S^k) \\ (J^r f)(x) := J^1(J^{r-1} f)(x) & (r \in \{2,3,\dots\}; x \in S^k) , \end{cases}$$

where

$$\xi(t) := (2\lambda+1) \int_{-1}^{t} (1-s^2)^{-\lambda-1/2} \int_{-1}^{s} (1-u^2)^{\lambda-1/2}\, du\, ds \qquad (-1 < t < 1).$$

Since $\xi \in L_\lambda^1$, we have in view of (2.5)

$$J^r f \in X \qquad (r \in \mathbb{N}; f \in X) .$$

The fundamental result for the strong Laplace - Beltrami derivative now reads.

THEOREM 1. <u>The following three assertions are equivalent to another for</u> $f \in X$, $r \in \mathbb{N}$:

i) $f \in W_X^r$;

ii) <u>there exists</u> $g_1 \in X$ <u>such that</u>

$$Y_n(g_1;x) = \left(\frac{n(n+2\lambda)}{2\lambda+1}\right)^r Y_n(f;x) \qquad (n \in \mathbb{P}; x \in S^k) ;$$

iii) <u>there exists</u> $g_2 \in X$ <u>such that</u>

$$f(x) = (J^r g_2)(x) \qquad \text{(a.e.)} .$$

<u>The functions</u> g_1, g_2 <u>are uniquely determined:</u>

(3.3) $$D^r f(x) = g_1(x) = g_2(x) - Y_0(f;x) \qquad \text{(a.e.)} .$$

So, in analogy to the classical fundamental theorems of the differential and integral calculus, the integral of the derivative of a differentiable function f is equal to f (except for an additive constant) and, on the other hand, the integral of a function f is differentiable and its derivative is equal to f (up to an additive constant).

Note that it follows from this theorem that the operator D^r is closed on W_X^r.

4. Moduli of Smoothness

For describing the smoothness of a function $f \in X$ it is near at hand to use the difference $\Delta_h f$. For $f \in X$, $0 < \delta < 2$ the (first) "spherical modulus of continuity" is defined by

$$\omega_1^S(\delta;f;X) := \sup_{1-\delta<h<1} \|\Delta_h f\|_X .$$

Virtually one would expect to introduce higher moduli of continuity by replacing the difference $\Delta_h f$ by the r-th difference Δ_h^r, defined by applying Δ_h r-times. But such a definition would not lead to the desired result (namely to the equivalence of this modulus with an appropriate K-functional, and so to a characterization of the best approximation by higher Lipschitz conditions).

So for $r \in \mathbb{N}$ we define the r-th spherical modulus of continuity of $f \in X$ by

$$(4.1)\qquad \omega_r^S(\delta;f;X) := \sup_{\substack{1-\delta<h_j<1\\ j=1,2,\dots,r}} \|\Delta_{h_1}\Delta_{h_2}\dots\Delta_{h_r} f\|_X \qquad (0<\delta<2),$$

and the corresponding spherical Lipschitz classes by

$$\mathrm{Lip}_r^S(\alpha;X) := \{f \in X;\ \omega_r^S(\delta;f;X) = \mathcal{O}(\delta^\alpha),\ \delta\to 0+\} \qquad (\alpha>0).$$

The seminorm ω_r^S has all the properties which are to be expected of a modulus of continuity. Indeed,

LEMMA 1. <u>For</u> $f \in X$, $r \in \mathbb{N}$, <u>there holds:</u>

$$(4.2)\qquad \lim_{\delta\to 0+} \omega_r^S(\delta;f;X) = 0\ ;$$

$$(4.3)\qquad \omega_r^S(\delta_1;f;X) \leqslant \omega_r^S(\delta_2;f;X) \qquad (0<\delta_1<\delta_2<2)\ ;$$

$$(4.4)\qquad \omega_{r+q}^S(\delta;f;X) \leqslant 2^q \omega_r^S(\delta;f;X) \qquad (q\in\mathbb{N}; 0<\delta<2)\ ;$$

$$(4.5)\qquad \lim_{\delta\to 0+} \omega_r^S(\delta;f;X)\,\delta^r = 0 \iff f(x) = \text{const.} \quad \text{(a.e.)}.$$

<u>If furthermore</u> $f \in W_X^q$, $0<q<r$, <u>then</u>

$$(4.6)\qquad \omega_r^S(\delta;f;X) \leqslant M\ \delta^q \omega_{r-q}^S(\delta;D^q f;X) \qquad (0<\delta<2)$$

<u>and</u>

$$(4.7)\qquad \omega_q^S(\delta;f;X) \leqslant M\ \delta^q \|D^q f\|_X \qquad (0<\delta<2)$$

<u>for a positive constant</u> $M(=M(q))$.

Another concept for describing the smoothness of a function $f \in X$ is given by the "K-functional"

$$(4.8) \qquad K(t,f;X,W_X^r) := \inf_{g \in W_X^r} \{\|f-g\|_X + t\|D_g^r\|_X\} \qquad (t>0; r \in \mathbb{N}) .$$

This seminorm is equivalent to the modulus of continuity in the following sense:

LEMMA 2. <u>For</u> $r \in \mathbb{N}$ <u>there exist constants</u> $0 < m \leqslant M < \infty$ <u>such that for</u> $f \in X$, $0 < \delta < 2$:

$$(4.9) \qquad m\,\omega_r^S(\delta;f;X) \leqslant K(\delta^r;f;X,W_X^r) \leqslant M\,\omega_r^S(\delta;f;X) .$$

This result yields a further property of ω_r^S: if $f \in X$, $r \in \mathbb{N}$, and $0 < \delta_1, \delta_2 < 2$, then

$$\omega_r^S(\delta_1;f;X) \leqslant \max\{1,(\delta_1/\delta_2)^r\}\,\omega_r^S(\delta_2;f;X) .$$

5. Best Approximation

Denoting by P_n, $n \in \mathbb{P}$, the set of polynomials of the form $P_n(x) = \sum_{k=o}^{n} Y_k(x)$ for some spherical harmonics Y_k of degree k, $k = 0,1,\dots,n$, we define the best approximation of $f \in X$ with respect to P_n by

$$(5.1) \qquad E_n(f;X) := \inf_{P_n \in P_n} \|f-P_n\|_X \qquad (n \in \mathbb{P}) .$$

For $f \in X$, $n \in \mathbb{P}$ there always exists a polynomial of best approximation $P_n^* \in P_n$ (cf. [11, p. 17]), i.e.,

$$E_n(f;X) = \|f-P_n^*\|_X .$$

Using the kernel of Fejér - Korovkin type (cf. [1] or [6])

$$q_{2n}(t) := c_n \left(\frac{R_n^\lambda(t)}{t-t_n} \right)^2 \qquad (n \in \mathbb{P}; t \in (-1,1)) ,$$

where

$$c_n^{-1} = \left\| \left(\frac{R_n^\lambda(t)}{t-t_n} \right)^2 \right\|_{1,\lambda} \qquad (n \in \mathbf{P}) ,$$

t_n denoting the largest root of R_n^λ, one can show as in e.g. [15], that for $r \in \mathbb{N}$, $f \in W_X^r$ one has the Jackson type inequality:

$$(5.2) \qquad E_n(f;X) \leq M\, n^{-2r} \|D^r f\|_X \qquad (n \in \mathbb{N}) .$$

Furthermore, from this inequality and (4.9) one can derive the following Jackson type theorem:

LEMMA 3. <u>Let</u> $f \in X$, $r \in \mathbb{N}$; <u>then there exists a positive constant</u> $M(=M(r))$ <u>such that</u>

$$(5.3) \qquad E_n(f;X) \leq M\, \omega_r^S(n^{-2};f;X) \qquad (n \in \mathbb{N}) .$$

On the other hand, the corresponding Bernstein type inequality is known (see e.g. [15] or [3]):

$$(5.4) \qquad \|D^r P_n\|_X \leq M_r\, n^{2r} \|P_n\|_X \qquad (r \in \mathbb{N}; P_n \in \mathcal{P}_n; n \in \mathbb{P}) .$$

Combining these inequalities there follows, in view of (4.9), by a general theorem on best approximation due to Butzer - Scherer [5] our central result, namely,

THEOREM 2. <u>Let</u> $f \in X$, $r_1, r_2 \in \mathbb{P}$, $0 < \alpha \leq 1$ <u>with</u> $r_2 < \alpha < r_1$. <u>The following six assertions are equivalent to each other:</u>

i) $E_n(f;X) = \mathcal{O}(n^{-2\alpha})$ $\qquad (n \to \infty)$;

ii) $\|D^{r_1} P_n^*\|_X = \mathcal{O}(n^{-2\alpha+2r_1})$ $\qquad (n \to \infty)$;

iii) $f \in W_X^{r_2}$, $\|D^{r_2} f - D^{r_2} P_n^*\|_X = \mathcal{O}(n^{-2\alpha+2r_2})$ $\qquad (n \to \infty)$;

iv) $K(t^{r_1},f;X,W_X^{r_1}) = \mathcal{O}(t^{\alpha}) \qquad (t \to 0+)$;

v) $\omega_{r_1}^{S}(\delta;f;X) = \mathcal{O}(\delta^{\alpha}) \qquad (\delta \to 0+)$;

vi) $f \in W_X^{r_2}\,,\ \omega_{r_1-r_2}^{S}(\delta;D^{r_2}f;X) = \mathcal{O}(\delta^{\alpha-r_2}) \qquad (\delta \to 0+)$.

In view of the equivalence of the assertions i) and vi) one has:

$$E_n(f;X) = \mathcal{O}(n^{-2r-2\alpha}) \qquad (n \to \infty)$$

$$\Longleftrightarrow D^r f \in \begin{cases} \mathrm{Lip}_1^S(\alpha;X) & 0 < \alpha < 1 \\ \mathrm{Lip}_2^S(1;X) & \alpha = 1\,. \end{cases}$$

This is exactly the counterpart of Theorem 0 cited in the introduction.

On the other hand, this theorem gives a solution to a problem posed by Butzer - Johnen [3] in 1971 on the characterization of the rate of best approximation by higher Lipschitz conditions. Indeed, the equivalence i) $\Longleftrightarrow$ v) can for $0 < \alpha < r$ be rewritten as

$$E_n(f;X) = \mathcal{O}(n^{-2\alpha}) \quad (n \to \infty) \Longleftrightarrow f \in \mathrm{Lip}_r^S(\alpha;X)\,.$$

In this respect A.S. Džafarov [8] also gave a result where the modulus of continuity was even defined by means of the r-th difference Δ_h^r; but his proof is wrong, since the "polynomials" used there for deriving the Jackson inequality are actually not polynomials.

For continuous functions the equivalence i) $\Longleftrightarrow$ v) also can be found in Kušnirenko [10] provided $r = 1$.

From the results of Ragozin [14] one can derive (cf. [2]) a characterization of the rate of best approximation of continuous functions by a Lipschitz condition of classical type. In fact for $f \in X$, $0 < \alpha < 1$ one has

$$E_n(f;C(S^k)) = \mathcal{O}(n^{-\alpha}) \qquad (n \to \infty)$$

$$\Longleftrightarrow |f(x) - f(y)| \leqslant L|x-y|^\alpha \qquad (x,y \in S^k) .$$

But if one would wish to derive such a result for $\alpha \geqslant 1$ there arises the difficulty in defining differences of higher order of this classical type one the sphere.

The proofs of the results announced here are to be found in [17] for the case $k = 3$. The methods used there can also be carried over in a modified form to establish corresponding results on best approximation in other function spaces, such as for Fourier - Jacobi expansions of functions defined on (-1,1) (see [7]) or Fourier - Laguerre expansions of functions defined on $(0,\infty)$ (in preperation).

The author was supported by research grant No. II B 4 - FA 8356 of "Der Minister für Wissenschaft und Forschung des Landes Nordrhein-Westfalen" which is gratefully acknowledged. He also would like to thank Professor P.L. Butzer and Dr. R.L. Stens, Aachen, for a critical reading of the manucript as well as for valuable suggestions.

REFERENCES

[1] Bavinck, H., Approximation processes for Fourier - Jacobi expansions. Applicable Anal. 5 (1975/76), 293 - 312.

[2] Berens, H. - Butzer, P.L. - Pawelke S., Limitierungsverfahren mehrdimensionaler Kugelfunktionen und deren Saturationsverhalten. Publ.Res. Inst. Math. Sci., Ser. A 4 (1968/69), 201 - 268.

[3] Butzer, P.L. - Johnen, H., Lipschitz spaces on compact manifolds. J. Functional Analysis 7 (1971), 242 - 266.

[4] Butzer, P.L. - Nessel, R.J., Fourier Analysis and Approximation. (Vol. 1) Birkhäuser Verlag, Basel/Stuttgart, and Academic Press, New York 1971.

[5] Butzer, P.L. - Scherer, K., Über die Fundamentalsätze der klassischen Approximationstheorie in abstrakten Räumen. In: Abstract Spaces and Approximation (Proc. Conf., Oberwolfach, 1968; eds. P.L. Butzer and B. Sz. - Nagy), ISNM Vol. 10, Birkhäuser Verlag, Basel/Stuttgart 1969, pp. 113 - 125.

[6] Butzer, P.L. - Stens, R.L. - Wehrens, M., Approximation of functions by algebraic convolution integrals. In: Approximation Theory and Functional Analysis (Proc. Conf., Campinas, Brazil, 1977; ed. J.B. Prolla), North-Holland Publishing Company 1979, pp. 71 - 120.

[7] Butzer, P.L. - Stens, R.L. - Wehrens, M., Higher order moduli of continuity based on the Jacobi translation operator and best approximation. Math. Rep. Acad. Sci., R. Soc. Can. 2 (1980), No. 2, 83-88.

[8] Džafarov, A.S., On the order of the best approximation of the functions continuous on the unit sphere by means of finite spherical sums (Russian). In: Studies Contemporary Problems Constructive Theory of Functions (Proc. Second All-Union Conf., Baku, 1962; ed. I.I. Ibragimov), Izdat. Akad. Nauk Azerbaidžan. SSR, Baku 1965, pp.46-52.

[9] Fejér, L., A Laplace-féle sorokrol. Mat és Term. Ertésitő 26 (1908), 323 - 373. (see also: Leopold Fejér, Gesammelte Arbeiten I. (ed. P. Turan), Birkhäuser Verlag, Basel/Stuttgart 1970, pp. 361 - 443, containing a translation into German.)

[10] Kušninenko, G.G., The approximation of functions on the unit sphere by finite spherical sums (Russian). Naucn.Dokl. Vysš. Skoly Fiz.-Mat. Nauki No. 4 (1958), 47 - 53.

[11] Lorentz, G.G., Approximation of Functions. Holt, Rinehart and Winston, New York/Chikago/San Francisco/Toronto/London 1966.

[12] Müller, C., Spherical Harmonics. Lecture Notes in Math. 17, Springer Verlag, Berlin/Heidelberg/New York 1966.

[13] Niemeyer, H., Lokale und asymptotische Eigenschaften der Lösungen der Helmholtzschen Schwingungsgleichung. Jber. DMV 65 (1962), 1 - 44.

[14] Ragozin, D.L., Approximation Theory on Compact Manifolds, and Lie Groups, with Applications to Harmonic Analysis. Ph.D. Thesis, Harvard Univ., Cambridge, Mass. 1967.

[15] Stens, R.L. - Wehrens, M., Legendre transform methods and best algebraic approximation. Comment. Math. Prace Mat. 21 (1979), 351 - 380.

[16] Trebels, W., Multipliers for (C,α) - Bounded Fourier Expansions in Banach Spaces and Approximation Theory. Lecture Notes in Math. 329, Springer Verlag, Berlin/Heidelberg/New York 1973.

[17] Wehrens, M., Legendre - Transformationsmethoden und Approximation von Funktionen auf der Einheitskugel im $\mathbb{R}^3$. Doctoral Dissertation, RWTH Aachen 1980.

V Best Approximation

EIN PROBLEM ÜBER DIE BESTE APPROXIMATION IN HILBERTRÄUMEN

Hubert Berens
Mathematisches Institut
Universität Erlangen-Nürnberg
Erlangen

In the beginning sixties V. L. Klee conjectured that there exist nonconvex Chebyshev sets in an infinite dimensional Hilbert space. Up to today no real progress has been made in proving or disproving the conjecture.
The author wants to discuss a modified version of Klees's conjecture which seems to be of some independent interest.

1. Im folgenden sei H ein Hilbertraum über $\mathbb{R}$ mit innerem Produkt $\langle\cdot,\cdot\rangle$ und Norm $|\cdot|$.

Für eine nichtleere Teilmenge K in H bezeichnet $P_K : H \to 2^K$ die metrische Projektion von H auf K und $d_K : H \to \mathbb{R}$ die Distanzfunktion. $\mathcal{D}(P_K)$ bezeichnet den Definitionsbereich von P_K, das ist die Menge $\{x \in H : P_K(x) \neq \emptyset\}$. Es ist gebräuchlich, die metrische Projektion mit ihrem Graphen in $H \times H$ zu identifizieren,

$$(x,k) \in P_K \quad \text{bedeutet somit} \quad k \in P_K(x).$$

Sei K eine nichtleere Teilmenge von H. Als Verallgemeinerung der metrischen Projektion möchten wir die folgende mengenwertige Abbildung auf H in sich betrachten:

$$H \ni x \mapsto \bigcap_{r > d_K(x)} \overline{\text{co}}\{b_r(x) \cap K\},$$

wobei $b_r(x)$ die offene Kugel um das Element x mit Radius r ist und $\overline{\text{co}}\{\ldots\}$ die abgeschlossene konvexe Hülle der Menge $\{\ldots\}$.

Offenkundig ist $\mathcal{D}(\Phi_K) = H$ und $\overline{\text{co}}P_K \subset \Phi_K$. Ist H von endlicher Dimension und ist K eine abgeschlossene Teilmenge in H, dann gilt

(*) $\quad$ für jedes $x \in H \quad \overline{\text{co}}P_K(x) = \Phi_K(x) \quad$ und $\quad \text{ext}\Phi_K(x) = P_K(x)$,

$\text{ext}\{\ldots\}$ ist die Extremalpunktmenge der Menge $\{\ldots\}$.

Wir möchten in den folgenden Abschnitten einige Eigenschaften der verallgemeinerten metrischen Projektion herleiten und ihre Bedeutung für die beste Approximation aufzeigen, siehe hierzu auch [2]. Die Frage, ob die Aussage (*) in Hilberträumen schlechthin gültig ist, führt uns zur Kleeschen Vermutung.

2. Sei K eine nichtleere Teilmenge von H. Als eine erste elementare Eigenschaft von Φ_K zeigen wir

SATZ 1. Φ_K ist monoton, sogar zyklisch monoton. Letzteres besagt: Für $n \in \mathbb{N}$ und $(x_o,\eta_o),\ldots,(x_n,\eta_n) \in \Phi_K$ mit $(x_n,\eta_n) = (x_o,\eta_o)$ gilt

$$0 \leq \sum_{j=1}^{n} < x_j - x_{j-1}, \eta_j >.$$

BEWEIS. Zu jedem $\varepsilon_j \in \mathbb{R}^+$, $j = 0,1,\ldots,n-1$, wählen wir $r_j \in \mathbb{R}^+$ mittels $r_j^2 = d_K^2(x_j) + \varepsilon_j$. Ist $k_j \in b_{r_j}(x_j) \cap K$, dann gilt

$$|x_j - k_j|^2 - \varepsilon_j \leq d_K^2(x_j) \leq |x_j - k_{j+1}|^2$$

oder

$$-\varepsilon_j \leq 2 < x_j, k_j - k_{j+1} > + |k_{j+1}|^2 - |k_j|^2.$$

Wir summieren die n Ungleichungen und erhalten

$$-\frac{1}{2}\sum_{j=0}^{n-1} \varepsilon_j \leq \sum_{j=0}^{n-1} < x_j, k_j - k_{j+1} > = \sum_{j=1}^{n} < x_j - x_{j-1}, k_j >.$$

Die Abschätzung bleibt erhalten, wenn wir die Elemente k_j aus $b_{r_j}(x_j) \cap K$ durch die aus $\overline{\text{co}}\{b_{r_j}(x_j) \cap K\}$ ersetzen, insbesondere durch $\eta_j \in \Phi_K(x_j)$, was zur Ungleichung

$$-\frac{1}{2}\sum_{j=0}^{n-1} \varepsilon_j \leq \sum_{j=1}^{n} < x_j - x_{j-1}, \eta_j >, \quad (x_j,\eta_j) \in \Phi_K,$$

für jedes $\varepsilon_j \in \mathbb{R}^+$, $j = 0,1,\ldots,n-1$, führt.

Tiefliegender als die Aussage von Satz 1 ist die folgende in

SATZ 2. Φ_K ist maximal monoton, d.h., Φ_K besitzt in $H \times H$ keine echte monotone Fortsetzung.

BEWEIS. Für ein $x \in H$ bezeichne

$$v(x,u) = \sup\{< \eta,u > : \eta \in \Phi_K(x)\} \qquad u \in H,\ |u| = 1$$

die Trägerfunktion von $\Phi_K(x)$. Offenkundig wird das Supremum angenommen.

Wir setzen $x_t = x + tu$, $t \in \mathbb{R}$ und betrachten die skalare Funktion

$$\mathbb{R} \ni t \mapsto v(x_t, u).$$

Aus der Monotonie von Φ_K folgt sofort, daß die Funktion monoton wachsend für wachsendes t ist. Wir zeigen, daß sie rechtsseitig stetig ist. Hierzu genügt es

$$v(x,u) = \lim_{t \to 0+} v(x_t, u)$$

zu beweisen. Für ein $t \in \mathbb{R}^+$ sei $\eta_t \in \Phi_K(x_t)$ so gewählt, daß $v(x_t,u) = \langle \eta_t, u \rangle$ gilt. Ist η ein schwacher Häufungspunkt des Netzes $\{\eta_t : t \to 0+\}$, dann ist $\eta \in \Phi_K(x)$. Denn für jedes $\varepsilon \in \mathbb{R}^+$ und jedes $x' \in H$, $|x' - x| < \varepsilon/3$, gilt $b_{r'}(x') \subset$ $\subset b_r(x)$ mit $r' = d_K(x') + \varepsilon/3$ und $r = d_K(x) + \varepsilon$. Daraus folgt aber für jedes x_t, $0 < t < \varepsilon/3$,

$$\eta_t \in \Phi_K(x_t) \subset \overline{co}\{b_{r_t}(x_t) \cap K\} \subset \overline{co}\{b_r(x) \cap K\}, \quad r_t = d_K(x_t) + \varepsilon/3,$$

und somit

$$\text{für jedes } \varepsilon \in \mathbb{R}^+ \quad \eta \in \overline{co}\{b_r(x) \cap K\} \quad \text{oder} \quad \eta \in \Phi_K(x).$$

Aus

$$v(x,u) \leq \lim_{t \to 0+} v(x_t,u) = \lim_{t \to 0+} \langle \eta_t, u \rangle = \langle \eta, u \rangle \leq v(x,u)$$

folgt die rechtsseitige Stetigkeit der Funktion.

Die maximale Monotonie von Φ_K ist nach diesen Vorbereitungen schnell bewiesen. Angenommen es existiere ein Paar $(x,\eta) \in H \times X$, das nicht zu Φ_K gehört, aber für jedes $(x',\eta') \in \Phi_K$

$$0 \leq \langle \eta - \eta', x - x' \rangle$$

erfüllt. Aus den bekannten Trennungssätzen konvexer Mengen in H folgt die Existenz eines Elementes $u \in H$, $|u| = 1$, und die eines Skalars $c \in \mathbb{R}$, so daß

$$v(x,u) < c < \langle \eta, u \rangle$$

gilt. Insbesondere folgt für die Elemente $x_t, t \in \mathbb{R}^+$, aus der oben gemachten Annahme

$$v(x,u) < c < \langle \eta, u \rangle \leq \langle \eta_t, u \rangle \leq v(x_t u),$$

was wegen der rechtsseitigen Stetigkeit von $t \mapsto v(x_t,u)$ zu einem Widerspruch führt. //

Nach einem Resultat von R. T. Rockafellar können wir aus Satz 1 und Satz 2 folgern, daß Φ_K das Subdifferential einer stetigen, konvexen Funktion auf H ist. In der Tat gilt

SATZ 3. Φ_K ist das Subdifferential der Funktion

$$H \ni x \mapsto \varphi_K(x) = \sup\{ \langle x,u \rangle - \frac{|k|^2}{2} : k \in K\}.$$

BEWEIS. Die folgende Form der Funktion $\varphi_K : H \to \mathbb{R}$ ist vom approximationstheoretischen Standpunkt her gesehen um vieles aufschlußreicher:

$$H \ni x \mapsto \varphi_K(x) = \frac{|x|^2}{2} - \frac{d_K^2(x)}{2} .$$

Es genügt $\Phi_K \subset \partial\varphi_K$ zu zeigen, wobei $\partial\varphi_K$ das Subdifferential von φ_K bezeichnet. Sei $x \in H$ vorgegeben. Sei $\varepsilon \in \mathbb{R}^+$, $r^2 = d_K^2(x) + \varepsilon$, $k \in b_r(x) \cap K$ und $x' \in H$. Dann gilt

$$\varphi_K(x) = \frac{|x|^2}{2} - \frac{d_K^2(x)}{2} \leq \frac{|x|^2}{2} - \frac{|x-k|^2 - \varepsilon}{2} =$$

$$= \langle x,k \rangle - \frac{|k|^2}{2} + \frac{\varepsilon}{2} = \langle x',k \rangle - \frac{|k|^2}{2} - \langle x'-x,k \rangle + \frac{\varepsilon}{2} \leq$$

$$\leq \varphi_K(x') - \langle x'-x,k \rangle + \frac{\varepsilon}{2} ,$$

oder

$$\varphi_K(x') \geq \varphi_K(x) + \langle x'-x,k \rangle - \frac{\varepsilon}{2} .$$

Wie im Beweis von Satz 1 bleibt auch hier die Ungleichung erhalten, wenn wir die Elemente k aus $b_r(x) \cap K$ durch die aus $\overline{\text{co}}\{b_r(x) \cap K\}$ ersetzen. Wir erhalten insbesondere für jedes $\varepsilon \in \mathbb{R}^+$, jedes $x' \in H$ und jedes $\eta \in \Phi_K(x)$

$$\varphi_K(x') \geq \varphi_K(x) + \langle x'-x,\eta \rangle - \frac{\varepsilon}{2} ,$$

woraus die Behauptung folgt. //

Als maximal monotoner Operator auf H besitzt Φ_K alle die Eigenschaften, die diesen Operatoren, speziell den Subdifferentialen stetiger konvexer Funktionen, zu eigen sind. Wir möchten an dieser Stelle nur darauf hinweisen, daß $\Phi_K : H \to H_w$ n.o. halbstetig ist. H_w bezeichnet hier den Hilbertraum versehen mit der schwachen Topologie. Im Beweis von Satz 2 haben wir zuerst die n.o. Hemi-Halbstetigkeit von $\Phi_K : H \to H_w$ gezeigt, eine schwächere Stetigkeitsaussage als die oben angegebene, und daraus die maximale Monotonie geschlossen. In diesem Zusammenhang und für das folgende möchten wir auf die Arbeit [9] von E. Zarantonello verweisen, für die Behandlung von monotonen Operatoren auf Hilberträumen schlechthin auf die Monographie [4] von H. Brezis. Schließlich möchten wir noch festhalten, daß für jedes $x \in H$

$$\Phi_K(x) = \bigcap_{\delta>0} \overline{\text{co}}\{\Phi_K(x') : |x'-x| < \delta\}$$

gilt, was von M. B. Suryanarayana [7] gezeigt wurde.

Dem Autor ist die Funktion $\varphi_K : H \to \mathbb{R}$ erstmals in der Arbeit [1] von E. Asplund aus dem Jahre 1969 über Tschebyscheffmengen in Hilberträumen begegnet und die Aussagen der bisherigen Sätze sind nichts als Beschreibungen ihres Subdifferentials. Die folgenden Aussagen gehen direkt auf E. Asplund zurück: Die Eindeutigkeitsmenge $\mathcal{U}_{\Phi_K} = \{x \in H : \#(\Phi_K(x)) = 1\}$ ist eine dichte G_δ-Menge in H, sie beschreibt genau die Elemente in H, in denen φ_K Gateaux-differenzierbar ist. Indirekt geht der erste Teil der Aussage schon auf S. B. Stečkin [6] aus dem Jahre 1963 zurück. Die Funktion φ_K ist darüberhinaus in einer dichten G_δ-Menge Frechét-differenzierbar. Bezeichnen wir diese Punktmenge mit $\mathcal{U}_{P_K}$ dann ist ihr Bild unter Φ_K im Abschluß von K enthalten. Für eine a b g e s c h l o s s e n e Teilmenge K in H - als ein guter Approximationstheoretiker sollte man nur solche Teilmengen betrachten - enthält somit die Tschebyscheffmenge von K eine dichte G_δ-Menge, nämlich die Untermenge, auf der die metrische Projektion stetig ist.

In nichtendlich dimensionalen Hilberträumen sind abgeschlossene Teilmengen im allgemeinen keine Existenzmengen. Dennoch ist für solche Mengen Φ_K durch die metrische Projektion P_K eindeutig bestimmt ist. Es gilt

SATZ 4. Sei K eine abgeschlossene, nichtleere Teilmenge von H. Für jedes $x \in H$ ist

$$\Phi_K(x) = \bigcap_{\delta>0} \overline{co}\{P_K(x') : 0 < |x' - x| < \delta \quad \text{und} \quad x' \in \mathcal{D}(P_K)\}.$$

BEWEIS. Nach obigem ist für jedes $x \in H$ die Punktmenge auf der rechten Seite der Gleichung nicht leer und in $\Phi_K(x)$ enthalten. Die Annahme, daß die Inklusion für ein $x \in H$ echt ist, führt wie im Beweis von Satz 2 zu einem Widerspruch. //

Satz 4 gibt uns eine Möglichkeit, monotone Operatoren monoton fortzusetzen. Für die metrische Projektion P_K auf eine abgeschlossene Teilmenge K in H besagt er, daß P_K eine eindeutig bestimmte maximal monotone Fortsetzung besitzt und daß sie durch

$$H \ni x \mapsto \bigcap_{\delta>0} \overline{co}\{P_K(x') : |x - x'| < \delta \quad \text{und} \quad x' \in \mathcal{D}(P_K)\}.$$

gegeben ist.

3. Es ist wohl bekannt, daß abgeschlossene und konvexe Teilmengen eines Hilbertraumes Tschebyscheffmengen sind. Im $\mathbb{R}^n$ gilt die Umkehrung, was wohl von L.N.H. Bunt im Jahre 1934 erstmals bewiesen wurde, in einem beliebigen Hilbertraum bisher nur unter zusätzlichen Annahmen. Die wohl schwächste Zusatzvoraussetzung wurde von L. P. Vlasov [8] gestellt. Er zeigte: Ist K eine Tschebyscheffmenge in H und gilt für jedes $(x,k) \in P_K$

$$w-\lim_{t \to 1+} P_K(k + t(x-k)) = k,$$

dann ist K abgeschlossen und konvex. In [3] haben U. Westphal und der Autor einen Beweis im Rahmen monotoner Operatoren gegeben. Wir zeigten, daß die Stetigkeitsforderung maximale Monotonie von P_K impliziert, d.h. $P_K = \Phi_K$ auf H, was zur Abgeschlossenheit und Konvexität von K äquivalent ist.

Vlasovs Arbeit erschien 1967. Schon 1965 hatte V. L. Klee [5] auf einer Tagung über Konvexität in Kopenhagen die Vermutung ausgesprochen, daß in nichtendlich dimensionalen, möglicherweise nichtseparablen, Hilberträumen nichtkonvexe Tschebyscheffmengen existieren. Klee stützt seine Vermutung auf Beispiele semi-tschebyscheffscher Mengen, sowie proximinaler Mengen mit in sich zusammenziehbarem Bild eines jeden Elementes im Raum unter der metrischen Projektion, deren Komplement beschränkt, offen und konvex ist. Solche Mengen existieren nicht im $\mathbb{R}^n$.

Das Problem scheint auch heute noch so weit von einer Lösung entfernt zu sein wie vor 15 Jahren. Der wohl schönste Beitrag hierzu geht auf E. Asplund, loc. cit., zurück, der zeigte: Ist die Vermutung richtig, dann existieren Tschebyscheffmengen mit beschränktem, offenem und konvexem Komplement.

Der Autor hat vergeblich versucht, die Vermutung durch die Angabe einer proximinalen Menge K in einem Hilbertraum H zu stützen, für die für ein Element $x \in H$ $\overline{co}P_K(x) \subsetneqq \Phi_K(x)$ gilt. Die Annahme, daß für p r o x i m i n a l e Mengen in Hilberträumen schlechthin $\overline{co}P_K = \Phi_K$ gilt, widerspricht der Kleeschen Vermutung.

Das folgende Beispiel mag die Situation ein wenig erläutern.

BEISPIEL. 5 $\{e_j : j \in \mathbb{N}\}$ sei die natürliche orthonormale Basis von $\ell_2(\mathbb{N})$. Wir wählen sie als Approximationsmenge K. Ist $\sum \alpha_j e_j$ die Orthogonalreihenentwicklung des Elementes x in $\ell_2(\mathbb{N})$, dann gilt

$$\ell_2(\mathbb{N}) \ni x \mapsto \varphi_K(x) = \sup_j \alpha_j - \frac{1}{2} \to -\frac{1}{2} \quad \text{für} \quad j \to \infty.$$

Offenkundig ist $\mathcal{D}(P_K) = \{x \in \ell_2(\mathbb{N}) : \alpha_j \geq 0$ für wenigstens einen Index $j \in \mathbb{N}\}$, und es ist nicht schwer einzusehen, daß

$$\ell_2(\mathbb{N}) \ni x \;\mapsto\; \Phi_K(x) = \begin{cases} \overline{co}P_K(x), & \alpha_j > 0 \text{ für wenigstens einen Index } j \in \mathbb{N}, \\ \overline{co}\{0, e_j : \forall j \in \mathbb{N} \ni \alpha_j = 0\}, & \text{sonst} \end{cases}$$

gilt. Ergänzen wir nun K durch ein Element $y \sim \sum \beta_j e_j$, so daß $K_y = K \cup \{y\}$ eine Existenzmenge wird, dann muß $|y| \leq 1$ und $\beta_j \leq 0$ für jedes $j \in \mathbb{N}$ gelten. Doch welches solche Element y wir auch wählen, es gilt stets

$$\overline{co}P_{K_y} = \Phi_{K_y} .$$

Ist J eine überabzählbare Indexmenge und ist $\{e_j : j \in J\}$ die natürliche orthonormale Basis von $\ell_2(J)$, dann ist $K = \{e_j : j \in J\}$ proximinal. Auch hier gilt: $\overline{co}P_K = \Phi_K$.

Wir möchten mit dem folgenden positiven Ergebnis schließen.

SATZ 6. Jede der folgenden Bedingungen an die Teilmenge K in H ist hinreichend, um $\Phi_K = \overline{co}P_K$ zu garantieren.

(i) K ist approximativ kompakt.

(ii) K ist proximinal und $P_K : H \to H$ n.o. halbstetig.

(iii) K ist proximinal und $\overline{co}P_K : H \to H_w$ n.o. halbstetig.

(iv) K ist proximinal und für jedes $x \in H$ gilt

$$\overline{co}P_K(x) = \bigcap_{\delta>0} \overline{co}\{P_K(x') : |x - x'| < \delta\}.$$

(v) Für jeden abgeschlossenen Halbraum M in H ist $K \cap M$ proximinal.

BEWEIS. Es gelten die Implikationen (i) ⇒ (ii) ⇒ (iii) ⇒ (iv). Bedingung (iv) aber besagt nach Satz 4, daß $\overline{co}P_K$ maximal monoton ist. Es bleibt also die Hinlänglichkeit der Bedingung (v) nachzuweisen. Zuerst halten wir fest, daß K selbst proximinal ist. Angenommen, $\overline{co}P_K(x) \subsetneqq \Phi_K(x)$ für ein $x \in H$. Dann existiert ein $u \in H$, $|u| = 1$, und ein $c \in \mathbb{R}$, so daß für jedes $k \in P_K(x)$

$$< k,u > < c < v(x,u).$$

Sei $M = \{y \in H : c \leq < y,u >\}$. Da $v(x,u) > c$, ist $M \cap K$ nicht leer und $d_K(x) = d_{K\cap M}(x)$. Nach Voraussetzung ist $K \cap M$ proximinal. Wir erhalten also $P_{K\cap M}(x) \subset P_K(x)$ im Widerspruch zu $P_K(x) \subset \complement M$. //

Bedingung (v) wurde erstmals von E. Asplund, loc. cit., formuliert. Er folgerte, daß eine Tschebyscheffmenge mit dieser Eigenschaft abgeschlossen und konvex ist.

LITERATUR

[1] Asplund, E., Chebyshev sets in Hilbert space. Trans. Amer. Math. Soc. 144 (1969), 235-240.

[2] Berens, H., Best approximation in Hilbert space. In "Approximation Theory III", ed. by E. W. Cheney, Academic Press, Inc., New York 1980, 1 - 2o.

[3] Berens, H. - Westphal, U., Kodissipative metrische Projektionen in normierten linearen Räumen. In "Linear Spaces and Approximation", ed. by P. L. Butzer and B. Sz.-Nagy, ISNM Vol. 40, Birkhäuser Verlag, Basel 1978, 119-130.

[4] Brezis, H., Opérateurs Maximaux Monotones. Math. Studies Vol. 5, North-Holland Publ. Co., Amsterdam 1973.

[5] Klee, V., Remarks on nearest points in normal linear spaces. In "Proceedings of the Colloquium on Convexity, Copenhagen 1965". Universität von Kopenhagen 1967, 168-176.

[6] Stečkin, S.B., Approximationseigenschaften von Mengen in normierten linearen Räumen. Rev. Roumaine Math. Pures Appl. 8 (1963), 5-18 (Russ.).

[7] Suryanarayana, M.B., Monotonicity and upper semicontinuity. Bull. Amer. Math. Soc. 82 (1976), 936-938.

[8] Vlasov, L.P., On Čebyšev sets. Dokl. Akad. Nauk SSSR 173 (1967), 491-494 = Soviet Math. Dokl. 8 (1967), 401-404.

[9] Zarantonello, E.H., Dense single-valuedness of monotone operators. Israel J. Math. 15 (1973), 158-166.

ÜBER EXISTENZ- UND EINDEUTIGKEITSMENGEN BEI DER BESTEN KO-APPROXIMATION

Ursula Westphal
Institut für Mathematik
Universität Hannover
Hannover

A condition on a normed linear space is given which implies that every existence and uniqueness set with respect to the best co-approximation is a closed flat. This is applied to L^p spaces.

1. Sei X ein reeller normierter linearer Raum. Für $x \in X$ und $r > 0$ sei $B(x;r)$ die offene Kugel um x mit Radius r und $\overline{B}(x;r)$ ihre Abschließung.

Sei K eine nichtleere Teilmenge von X. Die metrische Projektion P_K von X auf K ist die mengenwertige Abbildung $P_K : X \to 2^K$, die durch

$$P_K(x) = \bigcap_{k' \in K} \overline{B}(x; \|x - k'\|) \cap K \qquad \text{für jedes } x \in X$$

definiert ist.

Neben der metrischen Projektion führen wir folgenden die Approximation durch Elemente aus K betreffenden Operator ein : Es sei $R_K : X \to 2^K$ die mengenwertige Abbildung, die durch

$$R_K(x) = \bigcap_{k' \in K} \overline{B}(k'; \|x - k'\|) \cap K \qquad \text{für jedes } x \in X$$

definiert ist. R_K wird metrische Ko-Projektion von X auf K genannt, und für jedes $x \in X$ heißt $k \in R_K(x)$ ein Element bester Ko-Approximation von x durch Elemente aus K.

Implizit spielt der Operator R_K für eine Teilmenge $K \subset X$ bereits seit langem eine wesentliche Rolle beim Studium der Existenz einer kontraktiven Retraktion auf K. Man siehe bereits die Arbeiten von Kirszbraun [11] und Kakutani [10], sowie Arbeiten jüngeren Datums, z.B. [2],[7]. Explizit wurde R_K 1972 von Franchetti und Furi [8] im Zusammenhang mit der Charakterisierung von Hilberträumen eingeführt. Für lineare Teilräume K haben Papini und Singer

[13] die metrische Ko-Projektion vor allem im Hinblick auf sog. Charakterisierungssätze untersucht; auf sie geht auch die Bezeichnung "beste Ko-Approximation" zurück. Bzgl. der metrischen Ko-Projektion siehe auch [12].

In Analogie zu dem von Efimov und Stečkin für die beste Approximation eingeführten Begriff der Sonne nennen wir $K \subset X$ eine Ko-Sonne, falls für jedes $x \in X$ gilt:

$$(1) \qquad k \in R_K(x) \Rightarrow k \in R_K(k + \lambda(x - k)) \qquad \text{für jedes } \lambda \geq 0;$$

d.h. ist k ein Element bester Ko-Approximation zu x, so ist k auch Element bester Ko-Approximation zu jedem Punkt auf dem Strahl von k durch x. Daher auch nennt man R_K "strahlenförmig" (englisch: "sunny"). Die Bedingung (1) läßt sich äquivalent mit Hilfe des semi-inneren Produktes auf $X \times X$ beschreiben, das für $(x,y) \in X \times X$ wie folgt definiert ist:

$$\langle x,y\rangle_s := \lim_{t \to 0+} \frac{\| y + tx \|^2 - \| y \|^2}{2t} .$$

$K \subset X$ ist genau dann eine Ko-Sonne, wenn K der folgenden Bedingung genügt: Ist $x \in X$ und $k \in R_K(x)$, so gilt $\langle x - k, k - k'\rangle_s \geq 0$ für jedes $k' \in K$. Diese Bedingung ist das Pendant zum verallgemeinerten Kolmogoroff-Kriterium bei der besten Approximation; man siehe dazu die Ausführungen und Literaturangaben in Berens-Westphal [3].

Die Ko-Sonneneigenschaft einer Teilmenge K steht in direktem Zusammenhang mit dem Begriff des Approximationsbereiches, der bereits 1967 von Browder bei der Approximation von Fixpunkten kontraktiver Abbildungen in Hilberträumen betrachtet wurde. In unserer Terminologie ist der Approximationsbereich $A_K(x)$ zwischen $x \in X$ und K durch

$$A_K(x) = \{y \in X; \langle x - y, y - k'\rangle_s \geq 0 \ \forall \ k' \in K\}$$

definiert. Es gilt: K ist eine Ko-Sonne genau dann, wenn $R_K(x) = A_K(x) \cap K$ für jedes $x \in X$. Man siehe hierzu Bruck, Jr. [5].

Ist $K \subset X$ ein affiner Teilraum, so ist leicht einzusehen, daß K eine Ko-Sonne ist. Mehr noch, für jedes $x \in X$ gilt:

$$(2) \qquad k \in R_K(x) \Rightarrow k \in R_K(k + \lambda(x - k)) \qquad \text{für jedes } \lambda \in \mathbb{R} .$$

Verwendet man den Orthogonalitätsbegriff im Sinne von Birkhoff, nämlich für $y,z \in X$ heißt y orthogonal zu z, $y \perp z$, falls $\| y \| \leq \| y + \lambda z \|$ für jedes $\lambda \in \mathbb{R}$, so besagt (2) :

$$k \in R_K(x) \Rightarrow K - k \perp x - k .$$

Analog zu den Begriffsbildungen bei der besten Approximation nennen wir $K \subset X$ eine Existenzmenge bzgl. der besten Ko-Approximation, falls $R_K(x) \neq 0$ für jedes $x \in X$, und eine Eindeutigkeitsmenge bzgl. der besten Ko-Approximation, falls $R_K(x)$ für jedes $x \in X$ einelementig oder leer ist.

Bei der besten Approximation interessiert man sich u.a. für die Charakterisierung der Existenz- und Eindeutigkeitsmengen, der sog. Tschebyscheffmengen. So weiß man in endlich dimensionalen glatten Banachräumen, daß jede Tschebyscheffmenge konvex ist. Ob dies auch in unendlich dimensionalen Räumen gilt, ist nicht einmal im Fall eines Hilbertraumes bekannt. Im Gegensatz dazu ist die Charakterisierung der Existenz- und Eindeutigkeitsmengen bzgl. der besten Ko-Approximation in Hilberträumen kein Problem. Diese Mengen sind genau die abgeschlossenen affinen Teilräume, wie H. Berens und der Autor in einer gemeinsamen Note [4] gezeigt haben.

Die Frage ist, ob auch in allgemeineren normierten linearen Räumen jede Existenz- und Eindeutigkeitsmenge bzgl. der besten Ko-Approximation ein affiner Teilraum sein muß. Für die Räume L^p, $2 < p < \infty$, über einem σ-endlichen Maßraum wird diese Frage im folgenden Abschnitt positiv beantwortet. Daher sind die Existenz- und Eindeutigkeitsmengen bzgl. der besten Ko-Approximation in L^p, $2 < p < \infty$, genau diejenigen abgeschlossenen affinen Teilräume, deren Verschiebung durch den Nullpunkt Wertebereich einer kontraktiven linearen Projektion ist. Im Fall eines endlichen Maßraumes hat Ando [1] solche abgeschlossenen linearen Teilräume von L^p vollständig charakterisiert: Es sind genau diejenigen, die zu einem L^p-Raum über einem geeigneten Maßraum isometrisch isomorph sind. Tzafriri [14] hat dieses Ergebnis auf L^p-Räume über nicht endlichen Maßräumen erweitert.

An dieser Stelle möchte ich H. Berens für seine Anregungen und Hinweise im Zusammenhang mit dieser Arbeit sowie für seine Gesprächsbereitschaft herzlich danken.

2. Sei X ein reeller normierter linearer Raum. Im ersten Satz stellen wir einge unmittelbare Folgerungen der Existenz- und Eindeutigkeitseigenschaft von Teilmengen in X zusammen.

SATZ 1. *Sei $K \subset X$ eine Existenz- und Eindeutigkeitsmenge bzgl. der besten Ko-Approximation mit metrischer Ko-Projektion R_K. Dann gilt:*

(i) *K ist eine abgeschlossene Ko-Sonne.*

(ii) *Ist X strikt konvex, so ist K eine konvexe Menge.*

(iii) *Ist X glatt, so ist R_K eine Kontraktion; daher ist K Wertebereich einer strahlenförmigen kontraktiven Retraktion.*

BEWEIS. Wir zeigen (i) und (iii). (i) Die Abgeschlossenheit folgt unmittelbar aus der Existenzeigenschaft. Sei $x \in X$ und $x_\lambda = R_K(x) + \lambda(x - R_K(x))$ für $\lambda \geq 0$. Für jedes $\lambda \in [0,1]$ und jedes $k' \in K$ gilt $x_\lambda \in \overline{B}(k'; \| x - k' \|)$ und folglich $\overline{B}(k'; \| x_\lambda - k' \|) \subset \overline{B}(k'; \| x - k' \|)$. Daraus folgt $R_K(x_\lambda) \subset R_K(x)$ für jedes $\lambda \in [0,1]$, und da K Existenz- und Eindeutigkeitsmenge ist, $R_K(x_\lambda) = R_K(x)$. Für $\lambda \geq 1$ ergibt eine triviale Abschätzung $R_k(x_\lambda) = R_k(x)$.
(iii) Seien $x, x' \in X$. Da K Ko-Sonne ist, gilt

$$\langle x - R_K(x), R_K(x) - R_K(x') \rangle_s \geq 0 \quad \text{und} \quad \langle x' - R_K(x'), R_K(x') - R_K(x) \rangle_s \geq 0 .$$

Ist X glatt, so ist das semi-innere Produkt $\langle \cdot , \cdot \rangle_s$ in der ersten Koordinate linear, und folglich führt Addieren der beiden Ungleichungen auf die Ungleichung

$$\langle x - x' - (R_K(x) - R_K(x')), R_K(x) - R_K(x') \rangle_s \geq 0,$$

die zu $\quad \| R_K(x) - R_K(x') \|^2 \leq \langle x - x', R_K(x) - R_K(x') \rangle_s$

äquivalent ist. Daraus ergibt sich

$$\| R_K(x) - R_K(x') \| \leq \| x - x' \| . \qquad //$$

Wie bereits erwähnt, ist jeder affine Teilraum K eine Ko-Sonne, die sogar die Bedingung (2) erfüllt. Wir werden nun umgekehrt zeigen, daß für Existenzmengen K in strikt konvexen Räumen diese Bedingung impliziert, daß K ein affiner Teilraum ist.

LEMMA 1. *Sei X strikt konvex und $K \subset X$ eine Existenzmenge bzgl. der besten Ko-Approximation. Folgende Aussagen sind äquivalent:*

(i) *Für jedes $x \in X$ gilt: $k \in R_K(x) \Rightarrow K - k \perp x - k$.*

(ii) *K ist ein affiner Teilraum.*

BEWEIS. Wir zeigen den Schritt (i) $\Rightarrow$ (ii). Wegen der strikten Konvexität von

X ist K eine konvexe Menge. Angenommen, K ist kein affiner Teilraum. Dann existieren $k_1, k_2 \in K$, $k_1 \neq k_2$, so daß $x:=k_1 + t_0(k_2-k_1) \notin K$ für ein $t_0 > 1$. Sei $k \in R_k(x)$ und ohne Beschränkung der Allgemeinheit $k = 0$. Die metrische Projektion P_G von X auf die Gerade $G := \{\lambda x;\ \lambda \in \mathbb{R}\}$ ist einwertig, quasi-additiv und homogen; außerdem besagt Bedingung (i), daß $P_G(k') = 0$ für jedes $k' \in K$. Daher gilt

$$0 = P_G(k_2) = P_G((1-\frac{1}{t_0})k_1 + \frac{1}{t_0}x) = (1-\frac{1}{t_0})P_G(k_1) + \frac{1}{t_0}x = \frac{1}{t_0}x,$$

also $x = 0$, was $x \notin K$ widerspricht. //

Ist also K eine Existenz- und Eindeutigkeitsmenge bzgl. der besten Ko-Approximation in einem strikt konvexen Raum, so ist K ein affiner Teilraum genau dann, wenn sich die nach Satz 1(i) geltende Ko-Sonneneigenschaft zu der Bedingung

$$K - R_K(x) \perp x - R_K(x) \qquad \text{für jedes } x \in X$$

verschärfen läßt.

Für $x \neq 0$ definieren wir

$$H_x := \{y \in X;\ \|y\| \leq \|y - \lambda x\|\ \forall \lambda \in R\} = \{y \in X;\ y \perp x\}.$$

Ist X ein zweidimensionaler Raum oder ein innerer Produktraum, so ist H_x für jedes $x \neq 0$ eine Hyperebene, und diese Eigenschaft charakterisiert gerade alle drei- und höherdimensionalen inneren Produkträume (vgl. [9]).

SATZ 2. X _habe die Eigenschaft_ :

(*) _Zu jedem_ $x \in X \setminus \{0\}$ _existiert ein_ $\eta > 0$, _so daß_
$\lambda x \in \cap \{\bar{B}(y; \|y - x\|);\ y \in H_x\}$ für jedes $\lambda \in [-\eta, 1]$.

Ist $K \subset X$ _Existenz- und Eindeutigkeitsmenge bzgl. der besten Ko-Approximation, so gilt für jedes_ $x \in X$:

$$K - R_K(x) \perp x - R_K(x),$$

_d.h._ $R_K(x) = R_K(x_\lambda)$ _für jedes_ $\lambda \in \mathbb{R}$, _wobei_ $x_\lambda = R_K(x) + \lambda(x - R_K(x))$.

BEWEIS. Sei $x \in X \setminus K$. Da K Ko-Sonne ist, gilt $R_K(x) = R_K(x_\lambda)$ für jedes $\lambda \geq 0$. Um diese Gleichung auch für $\lambda < 0$ zu verifizieren, betrachten wir die Menge

$$L := \{y \in X;\ y - R_K(x) \perp x - R_K(x)\}.$$

Aus der Bedingung (*) folgt, daß zu $x - R_K(x)$ ($\neq 0$) ein $\eta > 0$ existiert, so daß

$$x_\lambda \in \bigcap_{y \in L} \bar{B}(y; \| y - x \|) \qquad \text{für jedes } \lambda \in [-\eta, 1] .$$

Abkürzend setzen wir $Q(x) := x_{-\eta} = R_K(x) - \eta\,(x - R_K(x))$ und wollen zeigen, daß $R_K(Q(x)) = R_K(x)$.

Es ist für jedes $y \in L$ $\| Q(x) - y \| \leqq \| x - y \|$. Um die Gültigkeit dieser Ungleichung auch für jedes $y \in K$ nachzuweisen, zeigen wir, daß zu jedem $k' \in K$ ein $\alpha \in [0,1)$ existiert, so daß $y' := \alpha x + (1-\alpha)k' \in L$.
Ist nämlich $k' \in K$, so existiert ein $\beta \in \mathbb{R}$, für das $\| k' - R_K(x) + \lambda(x - R_K(x))\|$ als Funktion von λ sein Infimum annimmt. Es ist dann

$$k' - R_K(x) + \beta(x - R_K(x)) \perp x - R_K(x) .$$

Ist $\beta > 0$, so folgt daraus

$$\frac{\beta}{\beta + 1}\, x + \frac{1}{\beta + 1}\, k' - R_K(x) \perp x - R_K(x) .$$

D.h. $\alpha x + (1-\alpha)k' \in L$, wobei $\alpha := \frac{\beta}{\beta + 1} \in (0,1)$. Ist $\beta \leqq 0$, so ergibt sich, da K eine Ko-Sonne ist:

$$\| R_K(x) - k' \| \leqq \| R_K(x) - \beta(x - R_K(x)) - k' \| \leqq \| R_K(x) + \lambda(x - R_K(x)) - k' \|$$

für jedes $\lambda \in \mathbb{R}$. Also ist $k' - R_K(x) \perp x - R_K(x)$, d.h. $k' \in L$.
Für jedes $k' \in K$ gilt nun:

$$\| Q(x) - k' \| \leqq \| Q(x) - y' \| + \| y' - k' \| \leqq \| x - y' \| + \| y' - k' \| = \|x - k'\|.$$

Da K Existenz- und Eindeutigkeitsmenge ist, folgt damit

$$\| R_K(Q(x)) - k' \| \leqq \| Q(x) - k' \| \leqq \| x - k' \| \quad \text{für jedes } k' \in K$$

und $R_K(Q(x)) = R_K(x)$. Wegen der Ko-Sonneneigenschaft ist dann sogar $R_K(x_\lambda) = R_K(x)$ für jedes $\lambda \leqq 0$. //

Es ist klar, daß jeder innere Produktraum die Eigenschaft (*) für $\eta = 1$ besitzt. Wir wollen nun zeigen, daß sie in den Räumen L^p, $2 \leqq p < \infty$, auch für $p \neq 2$ erfüllt ist. Zuvor die folgende

BEMERKUNG. _Ist_ X _gleichmäßig konvex, so gibt es zu jedem_ $x \in X \backslash \{0\}$ _und jeder Kugel_ $\bar{B}(0;r)$ _ein_ $\eta > 0$, _so daß_

$$\lambda x \in \bigcap \{\bar{B}(y; \| y - x \|);\ y \in H_x \cap \bar{B}(0;r)\} \quad \text{\underline{für jedes} } \lambda \in [-\eta, 1].$$

BEWEIS. Sei $x \neq 0$ und setze $H = H_x$. Für jedes $y \in H$ ist $\| y - \lambda x \|$ als Funktion von λ stetig und streng monoton wachsend für $\lambda \geqq 0$ sowie streng monoton fallend für $\lambda \leqq 0$. Daher existiert eine Funktion $h : H \to \mathbb{R}^+$, so daß

$$(3) \qquad \| y - x \| = \| y + h(y)x \| \qquad \text{für jedes } y \in H .$$

Sei $r > 0$. Dann ist $\inf \{h(y); y \in H \cap \bar{B}(0;r)\} > 0$. Angenommen, dies ist nicht der Fall. Dann gibt es eine Folge $(y_k)_{k\in \mathbb{N}}$, $y_k \in H$, so daß $\lim_{k\to\infty} h(y_k) = 0$ und $\lim_{k\to\infty} \| y_k - \alpha x \| = : d > 0$ für jedes $\alpha \in [0,1]$. Nun gilt für $u_k : = \| y_k - x \|^{-1} y_k$ und $v_k : = \| y_k - x \|^{-1} (y_k - x)$, daß $\| u_k \| \leqq 1, \| v_k \| \leqq 1$ und $\lim_{k\to\infty} \| \frac{1}{2}(u_k + v_k) \| = 1$. Da X gleichmäßig konvex, folgt daraus, daß $0 = \lim_{k\to\infty} \|u_k - v_k \| = \| x \| d^{-1}$, was im Widerspruch zu $x \neq 0$ steht. Sei $\eta : = \inf \{h(y); y \in H \cap \bar{B}(0;r)\}$. Dann ist für jedes $\lambda \in [-\eta,1]$

$$\lambda x \in \cap \{\bar{B}(y; \| y - x \|); y \in H \cap \bar{B}(0;r)\} . \qquad //$$

Sei nun X einer der gleichmäßig konvexen, glatten Banachräume $L^p = L^p(\Omega, \mathfrak{A}, \mu)$, wobei $p > 2$ und $(\Omega, \mathfrak{A}, \mu)$ ein σ-endlicher Maßraum ist. Für $x,y \in L^p$ ist das semi-innere Produkt durch

$$< x, y >_s = \frac{1}{\| y \|^{p-2}} \int x \mid y \mid^{p-1} \operatorname{sgn} y \, d\mu$$

gegeben. Da in glatten Räumen $y \perp x \Leftrightarrow < x, y >_s = 0$ ist, gilt für $x \neq 0$

$$H = H_x = \{y \in L^p; \int x \mid y \mid^{p-1} \operatorname{sgn} y \, d\mu = 0\} .$$

Der Konvexitätsmodul $\delta_p(\varepsilon)$, $0 < \varepsilon \leqq 2$, von L^p, definiert durch

$$\delta_p(\varepsilon) = \inf \{1 - \frac{\| u + v \|}{2}; u, v \in L^p, \| u \| \leqq 1, \| v \| \leqq 1, \| u - v \| \geqq \varepsilon\} ,$$

wurde im Fall $p > 2$ bereits von Clarkson mit

$$\delta_p(\varepsilon) = 1 - (1 - (\tfrac{\varepsilon}{2})^p)^{1/p}$$

angegeben; vgl. z.B. [6]. Sind daher $u, v \in L^p$, so daß $\| u \| \leqq 1$, $\| v \| \leqq 1$ und $\| u - v \| \geqq \varepsilon$, so gilt

$$(4) \qquad (\tfrac{\varepsilon}{2})^p \leqq 1 - \frac{\| u + v \|^p}{2^p} .$$

LEMMA 2. Die Räume L^p, $2 < p < \infty$, genügen der Bedingung (*).

BEWEIS. Sei $x \neq 0$ in L^p. Sei $h : H \to \mathbb{R}^+$ die durch (3) definierte Funktion. Nach obiger Bemerkung ist auf jeder beschränkten Teilmenge von H $\inf h > 0$. Wir werden nun zeigen, daß auch für $\| y \| \to \infty$ $h(y)$ durch eine positive Konstante nach unten beschränkt ist. Angenommen, es ist $\inf \{ h(y);\ y \in H \} = 0$. Dann gibt es eine Folge $(y_k)_{k \in \mathbb{N}}$, $y_k \in H$, so daß $\lim_{k\to\infty} \| y_k \| = \infty$ und $\lim_{k\to\infty} h(y_k) = 0$. Wir zeigen zunächst, daß $\lim_{k\to\infty} \| y_k - x \|^p - \| y_k \|^p = 0$ ist. Dazu betrachten wir im Falle $m < p \leqq m+1$, $m = 2, 3, \ldots$, das asymptotische Verhalten der Differenzen

$$\| y_k - \tfrac{1}{\ell} x \|^p - \| y_k \|^p = \| y_k + \tfrac{1}{\ell} h(\ell y_k)\, x \|^p - \| y_k \|^p$$

für $\ell = 1, \ldots, m-1$. Wegen $h(\ell y_k) \leqq \ell\, h(y_k)$ für $\ell \geqq 1$ ist auch $\lim_{k\to\infty} h(\ell y_k) = 0$. Setzt man

$$a_{k,i} := (-1)^i \binom{p}{i} \int x^i \, | y_k |^{p-i} (\operatorname{sgn} y_k)^i \, d\mu$$

für $i = 2, \ldots, m$, so gilt für $\ell = 1, \ldots, m-1$

$$(5) \quad \| y_k - \tfrac{1}{\ell} x \|^p - \| y_k \|^p - \sum_{i=2}^{m} \frac{1}{\ell^i} a_{k,i} = o(\| x \|^p)$$

$$(6) \quad \| y_k - \tfrac{1}{\ell} x \|^p - \| y_k \|^p - \sum_{i=2}^{m} \frac{1}{\ell^i} (-1)^i h^i(\ell y_k) a_{k,i} = o(h^p(\ell y_k) \| x \|^p) .$$

Aus (5) und (6) ergeben sich (m-1) beschränkte Linearkombinationen der (m-1) Ausdrücke $a_{k,i}$ $(i = 2, \ldots, m)$, nämlich für $\ell = 1, \ldots, m-1$

$$\sum_{i=2}^{m} \frac{1}{\ell^i} (1 - (-1)^i h^i(\ell y_k))\, a_{k,i} = o(\| x \|^p) .$$

Der Betrag der Koeffizientendeterminante dieses Systems von Linearkombinationen ist durch eine positive Konstante nach unten beschränkt, falls k groß genug ist. Daher ist jedes $a_{k,i}$ $(i = 2, \ldots, m)$ für $k \to \infty$ beschränkt. Aus (6) folgt dann

$$(7) \qquad \lim_{k\to\infty} \| y_k - x \|^p - \| y_k \|^p = 0 .$$

Sei nun für $k \in \mathbb{N}$ $u_k = \| y_k - x \|^{-1} y_k$, $v_k = \| y_k - x \|^{-1} (y_k - x)$ und $\varepsilon_k = \| y_k - x \|^{-1} \| x \|$. Dann folgt aus (4), wenn man dort u_k, v_k, ε_k für u, v, ε setzt,

$$\left(\frac{\|x\|}{2}\right)^p \leq \|y_k - x\|^p - \|y_k - \frac{x}{2}\|^p \leq \|y_k - x\|^p - \|y_k\|^p .$$

Für $k \to \infty$ führt diese Ungleichung wegen (7) zu $x = 0$, was der Voraussetzung $x \neq 0$ widerspricht. Also ist $\eta := \inf\{h(y);\ y \in H\}$ positiv und

$$\lambda x \in \cap\{\overline{B}(y;\|y - x\|);\ y \in H\} \qquad \text{für jedes } \lambda \in [-\eta, 1] . \qquad //$$

Für $1 < p < 2$ ist der Konvexitätsmodul $\delta_p(\varepsilon)$ vom Potenztyp 2 und nicht p :

$$\delta_p(\varepsilon) = \frac{p-1}{8}\varepsilon^2 + o(\varepsilon^2) ;$$

daher ist die Argumentation von Lemma 2 auf diesen Fall nicht anwendbar. Für $p \geq 2$ fassen wir zusammen:

SATZ 3. In L^p, $2 \leq p < \infty$, über einem σ-endlichen Maßraum sind die Existenz- und Eindeutigkeitsmengen bzgl. der besten Ko-Approximation diejenigen abgeschlossenen affinen Teilräume, deren Verschiebungen durch den Nullpunkt Wertebereiche kontraktiver linearer Projektionen sind.

LITERATUR

[1] Ando, T., Contractive projections in L_p spaces. Pacific J. Math. 17 (1966), 391 - 405.

[2] Beauzamy, B., Projections contractantes dans les espaces de Banach. Bull. Sci. Math. (2) 102 (1978), 43 - 47.

[3] Berens, H. - Westphal, U., Kodissipative metrische Projektionen in normierten linearen Räumen. In "Linear Spaces and Approximation", ed. by P.L. Butzer and B. Sz.-Nagy, ISNM vol 40, Birkhäuser Verlag, Basel 1978, 119 - 130.

[4] Berens, H. - Westphal, U., On the best co-approximation in a Hilbert space. In "Quantitative Approximation", ed. by R.A. DeVore and K. Scherer, Academic Press, New York 1980, 7 - 10.

[5] Bruck, Jr., R.E., Nonexpansive projections on subsets of Banach spaces. Pacific J. Math. 47 (1973), 341 - 355.

[6] Day, M.M., Normed Linear Spaces. Springer Verlag, Berlin / Heidelberg / New York (1973).

[7] DeFigueiredo, D.G. - Karlovitz, L.A., The extension of contractions and the intersection of balls in Banach spaces. J. Funct. Anal. 11 (1972), 168 - 178.

[8] Franchetti, C. - Furi, M., Some characteristic properties of real Hilbert spaces. Rev. Roumaine Math. Pures Appl. 17 (1972), 1045 - 1048.

[9] James, R.C., Inner products in normed linear spaces. Bull. Amer. Math. Soc. 53 (1947), 559 - 566.

[10] Kakutani, S., Some characterizations of Euclidean space. Japan. J. Math. 16 (1940), 93 - 97.

[11] Kirszbraun, M.D., Über die zusammenziehenden und Lipschitzschen Transformationen. Fund. Math. 22 (1934), 77 - 108.

[12] Papini, P.L., Approximation and strong approximation in normed spaces via tangent functionals. J. Appr. Theory 22 (1978), 111 - 118.

[13] Papini, P.L. - Singer, I., Best coapproximation in normed linear spaces. Monatsh. Math. 88 (1979), 27 - 44.

[14] Tzafriri, L., Remarks on contractive projections in L_p-spaces. Israel J. Math. 7 (1969), 9 - 15.

INVERSE APPROXIMATION THEOREMS OF LEBEDEV AND TAMRAZOV

L. Bijvoets, W. Hogeveen and J. Korevaar
Mathematisch Instituut
Universiteit van Amsterdam

In the period 1966-73, Lebedev and Tamrazov obtained very general inverse approximation theorems for polynomial approximation on plane compacta K. Their work extends the inverse theorems of Dzjadyk for the interval $[-1,1]$ and other well-behaved continua; because of its generality it is rather complicated. The present paper, based on the joint "Master's thesis" of the first two authors, deals with a simpler situation. It is assumed that K is a continuum with connected complement and that for an f in A(K), the approximation by polynomials p_n of degree $\leqslant n$ on $L = \partial K$ is at most of order $d(x,L^{1/n})^s$. Here L^u is the level curve $|\phi| = e^u$ of the exterior mapping function ϕ and $s = k+\alpha$, k a nonnegative integer, $0 < \alpha \leqslant 1$. The conclusion is that f is of class Λ^s on L and also on K, that is, f is in C^k and $f^{(k)}$ is in Lip α (if $\alpha < 1$) or the Zygmund class (if $\alpha = 1$). Except for integral s this theorem is a very special case of the results of Lebedev and Tamrazov [5], [7].

1. Model Theorems for Degree of Approximation

For degree of approximation problems generally, the model theorem is the classical Jackson-Bernstein-Zygmund result (1911-12-45) for the circle or $C_{2\pi}$:

$$f \in \Lambda^s \ \substack{\leftarrow \\ \rightarrow}\ E_n^{trig}(f) = O(n^{-s}), \quad s > 0.$$

Here Λ^s = Lip s if $0 < s < 1$, Λ^1 is the Zygmund class and $f \in \Lambda^{1+t}$, $t > 0$ means $f \in C^1$ and $f' \in \Lambda^t$. We will not consider more general smoothness classes.

1.1 The Interval [-1, 1]. For results on polynomial approximation in the complex plane, a better model is provided by the Timan-Dzjadyk characterization

(1951-56) for the interval [-1, 1]:

$$f \in \Lambda^s \rightleftarrows \begin{cases} \text{there exist polynomials } p_n \text{ and a constant } M \text{ such that} \\ |f(x) - p_n(x)| \le M\Delta_n(x)^s, \ -1 \le x \le 1, \quad n \ge 1, \end{cases}$$

where

$$\Delta_n(x) = \max\,(\sqrt{1 - x^2}/n, \quad 1/n^2).$$

The characteristic rate of approximation is better at the ends of the interval! ([6]. The term $1/n^2$ in the definition of $\Delta_n(x)$ may be omitted, Teljakovskiǐ [10], but such improvement would get in the way of our story).It is remarkable that this theorem was discovered, and is usually proved, by real variable methods. Complex methods would seem to be very natural here, especially for the "inverse theorem"!

What is the geometric meaning of $\Delta_n(x)$? What quantity should one use in the case of more general continua K in the plane with connected complement? The function $\Delta_n(x)$ is comparable to the distance between $x \in [-1, 1]$ and the ellipse with foci ±1 and major axis $e^{1/n} + e^{-1/n}$. For general K one may use the distance $d(x, L^{1/n})$ between x on $L = \partial K$ and the curve $L^{1/n}$ (Dzjadyk 1958).Here L^u is the level curve $\{g = u\}$ of the Green function $g(z, \infty)$ for the complement $\Omega_0 = \mathbb{C}^* \setminus K$ with pole at ∞. Equivalently, it is the level curve $\{|\phi| = e^u\}$ for the 1 - 1 conformal mapping $w = \phi(z)$ of Ω_0 onto $\{|w| > 1\}$ such that $\phi(\infty) = \infty$, $\phi'(\infty) > 0$:

$$g(z, \infty) = \log |\phi(z)|.$$

In the case of the closed unit disc, $K = \bar{B}(0, 1)$, one has $\phi(z) = z$, $g(z, \infty) = \log |z|$,

$$d(z, L^{1/n}) = e^{1/n} - 1 \sim 1/n.$$

In the case $K = [-1, 1]$ the conformal map is given by $z = \frac{1}{2}(w + 1/w)$; one can use it to prove the characterization theorem, in particular the inverse theorem.

2. Known Results for the Complex Plane

Let K be a compact set with connected complement. We consider only functions in A(K), that is, continuous functions on K which are holomorphic on the interior K^0. The N i k o l ' s k i ǐ p r o b l e m (1956) is to characterize the classes Λ^s, $s > 0$ by polynomial approximation properties. In the period 1959-65, Dzjadyk considered continua K with nice boundary L: piecewise rather smooth, no cusps. For such K, he obtained the result

$$f \in \Lambda^s \begin{array}{c}\rightarrow\\ \leftarrow\end{array} \begin{cases} \text{there exist polynomials } p_n \text{ and a constant M such that} \\ |f(x) - p_n(x)| \leq Md(x, L^{1/n})^s, \quad x \in L, \quad n \geq 1. \end{cases}$$

The proof depended on properties of the mapping function ϕ and a generalized Jackson formula.

2.1 Recent Results. Recent work by members of the Dzjadyk school and others has been directed to the problem of weakening the conditions on K.

Belyĭ [1] has obtained a very powerful direct theorem. He proved " $\rightarrow$ " under the sole condition that $L = \partial K$ be a quasi-circle or quasiconformal curve, that is, the image of a circle under a quasiconformal mapping of the plane onto itself. (Equivalently, a quasi-circle is the image of a quasi-line under a fractional linear transformation; a quasi-line is a Jordan curve Γ through ∞ in $\mathbb{C}^*$ for which there is a constant B, such that for any three finite points z_1, z_2, z_3 on Γ, with z_2 "between" z_1 and z_3, one has $|z_1 - z_2| / |z_1 - z_3| \leq B$.) The proof depends on very sophisticated use of extremal lengths to obtain the necessary properties of the mapping function ϕ and on further generalization of the Jackson formula. Belyĭ's work continues; the restriction that Γ be a quasi-circle can be relaxed; some cusps are permissible.

It is quite surprising that for the inverse Nikol'skiĭ problem practically no conditions on K are necessary. Continuing Dzjadyk's work, Lebedev-Tamrazov [5] and Tamrazov [7] (cf. also [8]) have obtained extremely general and definitive inverse theorems. The results involve very general moduli of continuity and very general compacta with connected complement. Because of this generality, the statements and proofs are rather complicated. The present paper is intended as an introduction to this work. In order to bring out the basic ideas, we will discuss only the simple case involving the classes Λ^s and continua K with connected complement.

Before we start on this discussion, we briefly remark that there is a parallel problem to that of Nikol'skiĭ, namely, to characterize the functions f in A(K) for which $E_n(f) = O(n^{-s})$. Also on that problem, Dzjadyk and other

Soviet mathematicians have done a good deal of work. Recent contributors have been Kövari, Andersson and Dyn'kin [2]. Here one is led to consider closed Jordan domains K of bounded boundary rotation. For such (and somewhat more general) K, the condition $f \circ \psi_C \in \Lambda^s$ (on the unit circle C), where $\psi = \phi^{-1}$, implies that $E_n(f) = O(n^{-s})$. A limited converse holds.

Additional references for this section are Dzjadyk [3], Korevaar [4] and Tamrazov [9].

3. The Principal Theorem

In the following $\omega_E(f, \delta)$ will denote the usual modulus of continuity of f on the compact set E. An equivalent modulus may be obtained with the aid of local best approximation by constants. To define a *Zygmund class* Λ^1 for E one introduces a second order modulus of continuity. Such a modulus may be based on local best approximation by general or special linear functions. We will use

$$\omega_E^*(f, \delta) = \max_{z_0 \in E} \min_{c \in \mathbb{C}} \max_{|z-z_0| \leq \delta,\ z \in E} |f(z) - f(z_0) - c(z - z_0)|$$

and say that f is in Λ^1 on E when $\omega_E^*(f, \delta) = O(\delta)$.

Except for integral s, the following theorem is a very special case of the results in [5], [7].

THEOREM. *Let K be a continuum with connected complement* Ω_0, *set* $L = \partial K$ *and let* L^u, $u > 0$ *be the level curve* $\{g = u\}$ *of the Green function* $g(z, \infty)$ *for* Ω_0 *with pole at* ∞. *Suppose that for some* f *in* A(K) *and* $s > 0$ *there exist polynomials* p_n *of degree* $\leq n$, *a constant* M *and a positive integer* m *such that*

$$|f(x) - p_n(x)| \leq Md(x, L^{1/n})^s, \quad x \in L, \quad n \geq m. \tag{3.1}$$

Then if $0 < s < 1$

$$\omega_L(f - p_m, \delta) \leq C_1(s)M\delta^s, \quad \delta > 0, \tag{3.2}$$

$$\omega_K(f - p_m, \delta) \leq C_2(s)M\delta^s, \quad \delta > 0. \tag{3.3}$$

If $s = 1$ *then* (3.2) *and* (3.3) *hold with* ω *replaced by* ω^*.

If $s > 1$ *then* f *is differentiable on* K, f' *is in* $A(K)$ *and*

$$|f'(x) - p_n'(x)| \le C_3(s)Md(x, L^{1/n})^{s-1}, \quad x \in L, \quad n \ge m. \tag{3.4}$$

Constants $C_i(s)$ *independent of* K *are easily determined.*

COROLLARY. *Under the hypotheses of the theorem* f *will be of class* Λ^s *on* L *and* K: *if* $0 < s < 1$ *then* f *is in* Lip s, *if* $s = 1$ *then* f *is in the Zygmund class and if* $s > 1$ *then* f *is in* C^1 *and* f' *in* Λ^{s-1}.

3.1 Introduction to the Proof (cf. [5]). Dividing by M we may assume $M = 1$, subtracting p_m from f and the p_n we may assume $p_m = 0$. As in the Bernstein case of $C_{2\pi}$ one takes $m = m_0 < m_1 < \ldots$ and writes

$$f = \lim p_n = (f - p_{m_k}) + \Sigma_1^k q_{m_j}, \quad q_{m_j} = p_{m_j} - p_{m_{j-1}}. \tag{3.5}$$

If $0 < s < 1$ one can proceed with the following direct estimate:

$$|f(x_0) - f(x)| \le |f(x_0) - p_{m_k}(x_0)| + |f(x) - p_{m_k}(x)| + \Sigma_1^k |q_{m_j}(x_0) - q_{m_j}(x)|. \tag{3.6}$$

In the case of trigonometric approximation and $E_n^{trig}(f) = O(n^{-s})$ it is convenient to take $m_j = 2^j$, $j \ge j_0$. To deal with the last term in (3.6), Bernstein invented his inequality for the derivative of a trigonometric polynomial of order $\le n$, namely, $||T_n'|| \le cn\, ||T_n||$ (where one may take $c = 1$). He finally took k such that $m_k \sim 1/\delta$ and readily concluded (when $0 < s < 1$) that $\omega(f, \delta) = O(\delta^s)$.

We proceed with the problem of proving (3.2). For x_0 and x on $L = \partial K$ and $|x - x_0| \le \delta$, (3.6) and (3.1) with $M = 1$ give

$$|f(x_0) - f(x)| \le d_k^s + (\delta + d_k)^s + \Sigma_1^k |q_{m_j}(x_0) - q_{m_j}(x)|. \tag{3.7}$$

Here we have introduced the notation

$$d_j = d(x_0, L^{1/m_j}) \tag{3.8}$$

and we have used the fact that $d(x, L^u) \le |x - x_0| + d(x_0, L^u)$. There is an immediate complication. For reasons to be explained below, one would like the sequence $\{d_j\}$ to decrease exponentially but not faster and one would also like to keep the sequence of ratios m_{j+1}/m_j bounded. The price will be that the sequence $\{m_j\}$ and the corresponding numbers d_j in general have to depend on the point x_0.

Let us look more closely at the estimation of $|q(x_0) - q(x)|$ where

$$q = \tfrac{1}{2}q_{m_{j+1}} = \tfrac{1}{2}(f - p_{m_j}) - \tfrac{1}{2}(f - p_{m_{j+1}}). \tag{3.9}$$

This is a polynomial of degree $\le m_{j+1}$ such that on L, by (3.1),

$$|q(x)| \le \tfrac{1}{2}d(x, L^{1/m_j})^s + \tfrac{1}{2}d(x, L^{1/m_{j+1}})^s < d(x, L^u)^s, \quad u = 1/m_j.$$

One may first estimate $|q(z)|$ near x_0. On every bounded component Ω_i of K^0 function $\log |q(z)|$ is subharmonic and has boundary values $\le s \log d(x, L^u)$. Let $H_i\{z, \log d(x, L^u)\}$ be the harmonic function on the simply connected domain Ω_i which solves the Dirichlet problem with the continuous boundary values $\log d(x, L^u)$. Then the function $\log |q|$ will be majorized by sH_i on Ω_i. On $\Omega_0 = \mathbb{C}^* \setminus K$ there is a similar upper bound, although not for $\log |q|$ (which in general tends to $+\infty$ as $z \to \infty$), but for $\log |q| - m_{j+1}g(\cdot, \infty)$.

We thus require a good local estimate for the continuous function H on C* which is equal to H_i on Ω_i, $i = 0, 1, \ldots$. Such a result exists (cf. [5] and section 4):

$$H\{z, \log d(x, L^u)\} \le \log d(x_0, L^u) + \log 23, \quad |z - x_0| \le d(x_0, L^u). \tag{3.10}$$

Setting $g_0 = g$ on Ω_0 and $g_0 = 0$ elsewhere, (3.10) will give the inequality

$$\log |q(z)| \le m_{j+1}g_0(z, \infty) + s \log 23\, d(x_0, L^{1/m_j}), \quad z \in \bar{B}(x_0, d_j). \tag{3.11}$$

Thus for z on the circle $C(x_0, \rho)$ and whatever value s has in (3.1),

$$\tfrac{1}{2}|q_{m_{j+1}}(z)| = |q(z)| \le \begin{cases} e^{m_{j+1}/m_j}\, 23^s\, d_j^s & \text{if } \rho = d_j, \\ e \cdot 23^s\, d_j^s & \text{if } \rho = d_{j+1}. \end{cases} \tag{3.12}$$

One may next use Cauchy's formula for z on $\bar{B}(x_0, \frac{1}{2}\rho)$ to obtain

$$(3.13)\quad \begin{aligned} |q(x_0) - q(z)| &= \Big| \frac{1}{2\pi i} \int_C q(w)\Big(\frac{1}{w-x_0} - \frac{1}{w-z}\Big)dw\Big| \\ &\le \max_C |q(w)|\cdot|x_0 - z| \frac{1}{2\pi}\int_0^{2\pi} \frac{dt}{\rho|e^{it} - \frac{1}{2}|} \approx 1.073 \ \max_C|q|\cdot|x_0 - z| \ / \ \rho. \end{aligned}$$

If $m_{j+1}/m_j \le 2$, say, one could take $\rho = d_j$, using the first line (3.12). For large m_{j+1}/m_j one could take $\rho = d_{j+1}$ and use the second line. However, the bound for $|q(x_0) - q(z)|$ in the second case would be larger than the first bound by roughly a factor d_j/d_{j+1}. The second bound will be useful only if the ratio d_j/d_{j+1} is not too large. A satisfactory balanced construction of the sequences $\{m_j\}$ and $\{d_j\}$ is possible (see [5] and section 5.1). Putting everything together and choosing k in a suitable manner, one will obtain a bound for $|f(x_0) - f(x)|$ which does not involve the sequences $\{m_j\}$ and $\{d_j\}$; inequality (3.2) will follow (see section 5.2).

For (3.3) more work is required. One first shows that the value $\omega_K(f, \delta)$ is attained when one of the variable points in its definition lies on the boundary $L = \partial K$. Denoting it by x_0 we have to estimate $|f(x_0) - f(z)|$. In (3.6) with z instead of x the above estimate for $|q(x_0) - q(z)|$ can be used again. However, this time we also have to estimate $|f(z) - p_{m_k}(z)|$ at points z of K^0. Here too the solutions H_i, $i \ge 1$ of the earlier Dirichlet problems can be used as majorants. We finally need a more flexible estimate than the one in (3.10) and proceed to prove one.

4. The Key Lemma

Let K and L^u, $u > 0$ be as in the theorem (section 3) and let $H\{z, \log d(x, L^u)\}$ be the combined solution of the simultaneous Dirichlet problems, for $\Omega_0 = \mathbb{C}^* \setminus K$ and the bounded components Ω_i of K^0, with boundary values $\log d(x, L^u)$ when $x \in L = \partial K$.

KEY LEMMA. <u>For</u> $x_0 \in L$ <u>and</u> $z \in \bar{B}(x_0, r)$ <u>one has</u>

$$H\{z, \log d(x, L^u)\} \le \begin{cases} \log r + \log 23 & \text{if } r \ge d(x_0, L^u), \\ \log r + \log 65 & \text{if } r \ge d(x_0, L^u)/8. \end{cases}$$

More generally, for $z \in \bar{B}(x_0, r)$ *with* $r \geq d(x_0, L^u)/a \quad (a > 0)$,

$$H\{z, \log d(x, L^u)\} \leq \log r + \log \min_{\lambda>0} \max \{a + 2 + 2\lambda, (a + 2)\sigma(1/\lambda)\},$$

where

$$\sigma(t) = 3 + 8t + 2\sqrt{2(1 + 2t)(1 + 4t)}.$$

The special cases $a = 1$ and $a = 8$ are obtained by taking $\lambda = 1/t = 10$ and $= 27$, respectively.

4.1 Auxiliary Lemma. For the proof of the key lemma we need an auxiliary result from conformal mapping.

LEMMA [5]. *Let* D *be a simply connected domain,* z_0 *in* D *such that* $0 < d(z_0, \partial D) = r < \infty$ *and* $\operatorname{diam} \partial D \geq \lambda r$ *where* $0 < \lambda < \infty$. *Then for* $|z - z_0| = r$ *and with* $\sigma(t)$ *as above, the Green function* $g_D(z, z_0)$ *satisfies the inequality*

$$g_D(z, z_0) \leq \log \sigma(1/\lambda).$$

PROOF. As is well-known, the Green function is given by

$$g_D(z, z_0) = - \log |F(z)|$$

where F is any $1 - 1$ conformal map of D onto the unit disc $B(0, 1)$ such that $F(z_0) = 0$. We will obtain such an F with the aid of an auxiliary transformation. Choose points z_1 and z_2 on ∂D such that $d(z_0, z_1) = r$ and $d(z_1, z_2) = \frac{1}{2}\lambda r$. Let T be the fractional linear transformation which takes z_0 to 0, z_1 to $\frac{1}{4}$ and z_2 to ∞:

$$T(z) = \frac{1}{4} \frac{(z_1 - z_2)(z - z_0)}{(z_1 - z_0)(z - z_2)} .$$

We denote the image TD by G and let Φ be the $1 - 1$ conformal map of G onto $B(0, 1)$ such that $\Phi(0) = 0$, $\Phi'(0) > 0$. Then the composition $\Phi \circ T$ gives an appropriate F.

The inverse Ψ of Φ is a $1 - 1$ conformal map of $B(0, 1)$ onto G such that $\Psi(0) = 0$ and the boundary of G contains the point $\frac{1}{4}$. It follows that $c = \Psi'(0) \leq 1$. Indeed, if $c = 1$ then by Koebe's $\frac{1}{4}$ theorem, $\Psi B(0, 1)$ must contain the disc $B(0, \frac{1}{4})$. Thus our $G = \Psi B(0, 1)$ must contain the disc $B(0, \frac{1}{4}c)$ and since the point $\frac{1}{4}$ is not in G, c must be ≤ 1.

The map $\Psi_0 = \Psi/c$ has $\Psi_0(0) = 0$, $\Psi_0'(0) = 1$, hence it belongs to the class S of normalized univalent functions. For this class one has Koebe's distortion theorem which shows that

$$|\Psi_0(w)| \le \frac{|w|}{(1 - |w|)^2} .$$

We now observe that

$$\frac{1}{c} T = \frac{1}{c} \Psi \circ F = \Psi_0 \circ F.$$

Taking $|z - z_0| = r$ we have on the one hand

$$\frac{1}{c}|T(z)| \ge 1 \cdot \frac{1}{4} \frac{\frac{1}{2}\lambda r \cdot r}{r(2r + \frac{1}{2}\lambda r)} = \frac{1}{4 + 16t}, \quad t = 1/\lambda.$$

On the other hand, by the distortion theorem,

$$\frac{1}{c}|T| \le |F|/(1 - |F|)^2 = v/(v - 1)^2, \quad v = 1/|F|.$$

Solving the resulting quadratic inequality for $v > 1$ we obtain

$$1/|F| = v \le 3 + 8t + \sqrt{(3 + 8t)^2 - 1} = \sigma(t) = \sigma(1/\lambda).$$

4.2 Proof of the Key Lemma (cf. [5]). We fix a and $u > 0$ and take $r \ge d(x_0, L^u)/a$. We next select z_0 on $B(x_0, r)$; let z_0 lie in $\bar{\Omega}_i$ where $i \ge 0$. We finally introduce a parameter $\lambda > 0$.

(i) Suppose $\operatorname{diam} L \le (2 + 2\lambda)r$. Then for any $z \in \mathbb{C}^*$

$$\begin{aligned} H\{z, \log d(x, L^u)\} &\le \max_{x \in L} \log d(x, L^u) \\ &\le \log \{d(x_0, L^u) + \operatorname{diam} L\} \le \log (a + 2 + 2\lambda)r. \end{aligned} \tag{4.1}$$

(ii) Suppose now that $\operatorname{diam} L \ge (2 + 2\lambda)r$. Then there is a continuum $L^* \subset L$ with $d(z_0, L^*) = r$ and $\operatorname{diam} L^* \ge \lambda r$. [There must be a point with $d(z_0, x_1) \ge (1 + \lambda)r$ or else L would belong to the disc $B(z_0, (1+\lambda)r)$.]

We choose such an L^* and define D as the component of $\mathbb{C}^* \setminus L^*$ which contains $B = B(z_0, r)$ and Ω_i. Then D is a simply connected domain which contains x_0 and hence Ω_0. The boundary ∂D will contain L^* so that diam $\partial D \geq \lambda r$.

Clearly

$$d(x, L^u) \leq |x - z_0| + |z_0 - x_0| + d(x_0, L^u)$$

$$\leq \begin{cases} (a + 2)r & \text{if} \quad x \in \bar{B} \cap \partial\Omega_i, \\ (a + 2)|x - z_0| & \text{if} \quad x \in B^c \cap \partial\Omega_i. \end{cases}$$

It will follow that $\log d(x, L^u)$ is majorized by $g_D(x, z_0) + \log (a+2)|x - z_0|$ everywhere on $\partial\Omega_i$. This is clear for $x \in B^c \cap \partial\Omega_i$ since $g_D \geq 0$. For $z \in \partial B$ likewise $\log (a + 2)r \leq g_D(z, z_0) + \log (a + 2)|z - z_0|$ and since the right-hand member of the inequality is harmonic on B, the inequality will hold throughout $\bar{B}$ and in particular on $\bar{B} \cap \partial\Omega_i$.

The conclusion is that the solution H_i of the Dirichlet problem for Ω_i with boundary values $\log d(x, L^u)$ satisfies the inequality

$$H_i\{z, \log d(x, L^u)\} \leq g_D(z, z_0) + \log (a + 2)|z - z_0| \tag{4.2}$$

throughout $\bar{\Omega}_i$.

We now apply the auxiliary lemma to the present domain D. It shows that for $z \in \partial B$, the right-hand side of (4.2) is majorized by $\log \sigma(1/\lambda) + \log (a + 2)r$. The same constant will majorize that harmonic right-hand side throughout B; it will in particular majorize the (limit) value of the right-hand side at z_0. It follows that

$$H_i\{z_0, \log d(x, L^u)\} \leq \log \sigma(1/\lambda) + \log (a + 2)r. \tag{4.3}$$

This inequality will hold for each z_0 in $\bar{\Omega}_i \cap B(x_0, r)$ and each $i \geq 0$, hence it will hold for H throughout $B(x_0, r)$. Combining (4.1) and (4.3) we conclude that for every $\lambda > 0$,

$$H\{z, \log d(x, L^u)\} \leq \log r + \log \max \{a + 2 + 2\lambda, (a + 2)\sigma(1/\lambda)\}$$

throughout $B(x_0, r)$ and hence $\bar{B}(x_0, r)$.

5. Proof of the Theorem

5.1 The Sequences $\{m_j\}$ and $\{d_j\}$(cf. [5]). Let x_0 on L be fixed. We remark that $d(x_0, L^u)$, $u > 0$ is strictly increasing as a function of u and tends to 0 as $u \downarrow 0$ (think of the exterior conformal mapping of Ω_0 in connection with the level curves L^u).

To start the construction one defines $m_0 = m \geq 1$, where m is given by the theorem. Suppose now that $m_0 < \ldots < m_j$ have been defined; we write

$$(5.1) \qquad d_j = d(x_0, L^{1/m_j}).$$

Let n_j be the smallest positive integer such that $d(x_0, L^{1/n_j}) \leq \frac{1}{2}d_j$. We then take $m_{j+1} = n_j$ if

(a) either $d(x_0, L^{1/n_j}) \geq \frac{1}{4}d_j$ or $n_j/m_j \leq 2$ (or both).

In this case $\frac{1}{4}d_j \leq d_{j+1} \leq \frac{1}{2}d_j$ or $m_{j+1}/m_j \leq 2$. However, if

(b) $d(x_0, L^{1/n_j}) < \frac{1}{4}d_j$ and $n_j/m_j > 2$

we define $m_{j+1} = n_j - 1$ ($> m_j$). Then $\frac{1}{2}d_j < d_{j+1} < d_j$.

In case (b), $d(x_0, L^{1/n_j}) < \frac{1}{2}d_{j+1}$ so that $n_{j+1} = n_j$. Also, $n_{j+1}/m_{j+1} = 1 + 1/m_{j+1} \leq 2$. Hence for j+1 we will be in case (a): $m_{j+2} = n_{j+1} = n_j$, $d_{j+2} < \frac{1}{2}d_{j+1}$ and $m_{j+2}/m_{j+1} \leq 2$.

Consequences of the construction. The sequence $\{m_j\}$ is strictly increasing, the sequence $\{d_j\}$ strictly decreasing. For each j, at least one of the following is true:

(i) $d_{j+1} \geq \frac{1}{4}d_j$.

(ii) $m_{j+1}/m_j \leq 2$ and $d_{j+1} \leq \frac{1}{2}d_j$.

Furthermore, for each j,

(iii) $d_{j+2} < \frac{1}{2}d_j$.

5.2 Proof of (3.2). We complete the proof begun in section 3.1. We have $0 < s < 1$ and take $M = 1$, $p_m = 0$. Fixing $\delta > 0$, we choose $x_0 \in L$ and $x \in L \cap \bar{B}(x_0, \delta)$; we let $\{m_j\}$ and $\{d_j\}$ be the sequences associated with x_0 as in section 5.1.

For any $k \geq 0$, (3.1), (3.5) and (3.6) give (3.7). Via (3.9) - (3.11), appealing to the key lemma, we will arrive at (3.12). Continuing, we distinguish two cases

(i) The case $d_{j+1} \geq \frac{1}{4}d_j$. We use the second line in (3.12). For $z \in \bar{B}(x_0, d_j/8) \subset \bar{B}(x_0, \frac{1}{2}d_{j+1})$ formula (3.13) then shows that

$$|q_{m_{j+1}}(x_0) - q_{m_{j+1}}(z)| \leq (1.08)2e\cdot 23^s d_j^s|x_0 - z|/d_{j+1} \leq 24\cdot 23^s d_j^{s-1}|x_0 - z|.$$

(5.2)

(ii) The case $d_{j+1} < \frac{1}{4}d_j$. Now $m_{j+1}/m_j \leq 2$ (section 5.1) and we use the first line in (3.12). Via a different middle step we easily obtain the estimate (5.2) even on $\bar{B}(x_0, \frac{1}{2}d_j)$.

Summing the results (5.2) for $j = 0,\dots, k-1$ we find that for $z \in \bar{B}(x_0, d_{k-1}/8)$, using the fact that $d_{j+2} < \frac{1}{2}d_j$,

$$(5.3) \qquad \Sigma_1^k|q_{m_j}(x_0) - q_{m_j}(z)| \leq 24\cdot 23^s|x_0 - z|\,\Sigma_0^{k-1}d_j^{s-1}$$

$$\leq 24\cdot 23^s|x_0 - z|2d_{k-1}^{s-1}(1 + 2^{s-1} + 4^{s-1} + \dots) \leq 48\,\frac{23^s}{1 - 2^{s-1}}|x_0 - z|d_{k-1}^{s-1}.$$

We finally determine $k \geq 0$ such that $d_k \leq 8\delta < d_{k-1}$ (setting $d_{-1} = \infty$). Applying (5.3) to $z = x \in L \cap \bar{B}(x_0, \delta)$ and substituting in (3.7), we obtain the following result (a fortiori valid when $k = 0$)

$$(5.4) \qquad |f(x_0) - f(x)| \leq \left(8^s + 9^s + 12\,\frac{92^s}{2^{1-s} - 1}\right)\delta^s, \quad |x - x_0| \leq \delta.$$

The right-hand side of (5.4) is independent of the sequence $\{m_j\}$; it majorizes $|f(x_0 - f(x)|$ whenever x_0, $x \in L$ and $|x - x_0| \leq \delta$, hence it majorizes $\omega_L(f, \delta)$.

5.3 Proof of (3.3). Again $0 < s < 1$, $M = 1$, $p_m = 0$. We fix $\delta > 0$ and now have to estimate

$$\omega_K(f, \delta) = \max|f(z_1) - f(z_2)|, \quad z_1, z_2 \in K, \quad |z_1 - z_2| \leq \delta.$$

We claim that the maximum is attained when one of the points z_1, z_2 lies on L (but perhaps not only in that case). Indeed, suppose the maximum

$\omega_K(f, \delta)$ is attained for points z_1 and $z_2 = z_1 + h$ in K^0 (with $|h| \le \delta$). Write $\min\{d(z_1, L), d(z_2, L)\} = \rho > 0$; we may assume $d(z_1, L) = \rho$. The function $\phi(z) = f(z) - f(z + h)$ is continuous on $\bar{B}(z_1, \rho)$ and holomorphic on $B(z_1, \rho)$, and $|\phi|$ attains its maximum at z_1. Thus by the maximum principle ϕ is constant on $\bar{B}(z_1, \rho)$, hence $|\phi| = \omega_K(f, \delta)$; the circle $C(z_1, \rho)$ contains a point of L. The conclusion is that there is a point $x_0 \in L$ such that

$$\omega_K(f, \delta) = \max |f(x_0) - f(z)|, \quad z \in K \cap \bar{B}(x_0, \delta).$$

For such a point (or any point) $x_0 \in L$ we again let $\{m_j\}$ and $\{d_j\}$ be as in section 5.1. For $z \in K \cap \bar{B}(x_0, \delta)$ we may write, cf. (3.6),

$$(5.5) \qquad |f(x_0) - f(z)| \le d_k^s + |f(z) - p_{m_k}(z)| + \Sigma_1^k |q_{m_j}(x_0) - q_{m_j}(z)|.$$

For the last term we have the estimate from (5.3) provided $z \in \bar{B}(x_0, d_{k-1}/8)$.

It only remains to estimate the next to last term in (5.5). We know from (3.1) that

$$\log |f(x) - p_{m_k}(x)| \le s \log d(x, L^{1/m_k}), \quad x \in L.$$

Since we are dealing with a subharmonic function, it follows that

$$\log |f(z) - p_{m_k}(z)| \le s\, H\{z, \log d(x, L^{1/m_k})\}, \quad z \in K.$$

This time we will use the second inequality in the key lemma (section 4). We again determine k such that $d_k \le 8\delta < d_{k-1}$. Now taking $r = \delta$ so that $r \ge d_k/8$, the key lemma shows that our H is bounded by $\log 65\delta$ on $\bar{B}(x_0, \delta)$, hence

$$(5.6) \qquad |f(z) - p_{m_k}(z)| \le 65^s \delta^s, \quad z \in K \cap \bar{B}(x_0, \delta).$$

Combining (5.5), (5.6) and (5.3) we conclude that

$$|f(x_0) - f(z)| \le \left(8^s + 65^s + 12\, \frac{92^s}{2^{1-s} - 1}\right) \delta^s, \quad z \in K \cap \bar{B}(x_0, \delta).$$

5.4 Proof for the Case $s = 1$. We again take $M = 1$, $p_m = 0$ and fix $\delta > 0$. In order to obtain bounds for the second order moduli $\omega^*(f, \delta)$ (section 3) we will estimate $|f(x_0) - f(z) - c(x_0 - z)|$ for $x_0 \in L$, $z \in \bar{B}(x_0, \delta)$ and suitable c depending on x_0 and δ. This will be good enough also in the case of $\omega_K^*(f, \delta)$: one can show as in section 5.3 that the modulus is attained when one of the points used in the definition lies on L.

We again use the sequences $\{m_j\}$ and $\{d_j\}$ associated with x_0. By (3.5),

$$(5.7)\quad \begin{aligned} |f(x_0) - f(z) - p'_{m_k}(x_0)(x_0 - z)| &\le |f(x_0) - p_{m_k}(x_0)| + |f(z) - p_{m_k}(z)| \\ &+ \Sigma_1^k\, |q_{m_j}(x_0) - q_{m_j}(z) - q'_{m_j}(x_0)(x_0 - z)|. \end{aligned}$$

For z on $C(x_0, \rho)$ with $\rho = d_j$ or d_{j+1} we have the estimates (3.12) for $q = \frac{1}{2}q_{m_{j+1}}$. Distinguishing two cases as in section 5.2, we now use the Cauchy formulas to obtain, cf. (3.13),

$$(5.8)\quad \begin{aligned} 2|q(x_0) - q(z) - q'(x_0)(x_0 - z)| &= \frac{1}{\pi}\left|\int_C q(w)\left\{\frac{1}{w - x_0} - \frac{1}{w - z} - \frac{x_0 - z}{(w - x_0)^2}\right\}dw\right| \\ &\le 94\cdot 23^s |x_0 - z|^2 d_j^{s-2}, \quad z \in \bar{B}(x_0, d_j/8). \end{aligned}$$

Taking s = 1 and summing over $0 \le j \le k-1$ we find that for $z \in \bar{B}(x_0, d_{k-1}/8)$, cf. (5.3),

$$(5.9)\quad \Sigma_1^k\, |q_{m_j}(x_0) - q_{m_j}(z) - q'_{m_j}(x_0)(x_0 - z)| \le 94\cdot 92\, |x_0 - z|^2/d_{k-1}.$$

As before we determine k such that $d_k \le 8\delta < d_{k-1}$. The first two terms on the right-hand side of (5.7) are bounded by 8δ and 9δ when $z = x \in L \cap \bar{B}(x_0, \delta)$; inequality (5.6) gives the upper bound 65δ for the second term when $z \in K \cap \bar{B}(x_0, \delta)$. Combining (5.7) and (5.9) we thus obtain for $z = x \in L \cap \bar{B}(x_0, \delta)$

$$|f(x_0) - f(x) - p'_{m_k}(x_0)(x_0 - x)| \le 1100\,\delta,$$

hence $\omega_L^*(f, \delta) \le 1100\,\delta$. Similarly $\omega_K^*(f, \delta) \le 1155\,\delta$.

<u>5.5 Proof for s > 1.</u> We mostly write $s = 1 + t$ and again take $M = 1$ and $p_m = 0$. The proof goes in two steps. We show first that (3.1) implies uniform conver-

gence of the sequence $\{p'_n\}$ on K to a function g in A(K). The function g will be approximated by the polynomials p'_n to order $d(x, L^{1/n})^t$ on L. In the second step we show that f is differentiable on K and that f' = g.

(i) We choose x_0 on L and let $\{m_j\}$ and $\{d_j\}$ be as in section 5.1. Taking positive integers $n \geq m$ an $v > n$ we want to estimate $|p'_v(x_0) - p'_n(x_0)|$. Let j be such that $m_j \leq n < m_{j+1}$.

Supposing first that $v \leq m_{j+1}$ we write $\frac{1}{2}(p_v - p_n) = q$, comparable to the last member of (3.9) but with m_j replaced by n and m_{j+1} by v. The analysis of section 3 now leads to the inequalities (3.12) with m_j and d_j replaced by n and $d(x_0, L^{1/n})$ etc. Distinguishing the same two cases as in section 5.2 we use these inequalities and Cauchy's formula to estimate $q'(x_0)$. The result will be

$$(5.10)\qquad |p'_v(x_0) - p'_n(x_0)| \leq 8e\cdot 23^s\, d(x_0, L^{1/n})^{s-1}, \quad m_j \leq n < v \leq m_{j+1}.$$

When $v > m_{j+1}$ we write

$$p'_v - p'_n = (p'_{m_{j+1}} - p'_n) + (p'_{m_{j+2}} - p'_{m_{j+1}}) + \dots + (p'_v - p'_{m_k})$$

with suitable k. Repeated application of (5.10) gives the general formula

$$(5.11)\qquad \begin{aligned} |p'_v(x_0) - p'_n(x_0)| &\leq 8e\cdot 23^{1+t}\{d(x_0, L^{1/n})^t + d^t_{j+1} + d^t_{j+2} + \dots\} \\ &\leq 24e\,\frac{23^{1+t}}{1 - 2^{-t}}\, d(x_0, L^{1/n})^t, \quad v > n \geq m. \end{aligned}$$

As $n \to \infty$ the distance $d(x, L^{1/n})$ tends to 0 uniformly for $x \in L$ (think of the exterior conformal mapping in connection with the level curves). Thus by (5.11) the polynomials p'_n converge uniformly on L and hence on K. The limit function g will be continuous on K and holomorphic at interior points.

Letting v tend to ∞ in (5.11) we see that g is approximated by the polynomials p'_n on L to order $d(x, L^{1/n})^t$.

(ii) Since $p_n \to f$ uniformly on K, complex analysis shows that $p'_n \to f'$ at interior points, hence f' = g on K^0. It remains to prove the corresponding result for boundary points,

$$(5.12)\qquad \lim \frac{f(x_0) - f(z)}{x_0 - z} = g(x_0), \quad z \to x_0 \in L, \quad z \in K.$$

Taking the sequences $\{m_j\}$ and $\{d_j\}$ corresponding to x_0 we start with (5.7)

$$|f(x_0) - f(z) - g(x_0)(x_0 - z)| \le |f(x_0) - p_{m_k}(x_0)| + |f(z) - p_{m_k}(z)|$$
$$+ |g(x_0) - p'_{m_k}(x_0)|\cdot|x_0 - z| + \Sigma_1^k |q_{m_j}(x_0) - q_{m_j}(z) - q'_{m_j}(x_0)(x_0 - z)| =$$
$$T_1 + T_2 + T_3 + T_4, \tag{5.13}$$

say. For $z \in K$, $z \neq x_0$ we write $|x_0 - z| = \delta$ and determine k such that $d_k \leqslant 8\delta < d_{k-1}$. Then each of the terms T_i will be $o(\delta)$ as $\delta \to 0$! For T_1 this follows from (3.1), for T_2 from (5.6), for T_3 from the definition of g or from (5.11) as $v \to \infty$ and for T_4 from (5.8) (we have to sum over $0 \le j \le k-1$; note the special case s = 2). Thus (5.13) establishes (5.12).

REFERENCES

[1] Belyǐ, V.I., Conformal mappings and the approximation of analytic functions in domains with a quasiconformal boundary. Math. USSR Sbornik 31 (1977), 289-317.

[2] Dyn'kin, E.M., Uniform approximation of functions in Jordan domains. Siberian Math. J. 18 (1977), 548-557.

[3] Dzjadyk, V.K., Introduction to the Theory of Uniform Approximation of Functions by Polynomials (Russian). Nauka, Moscow 1977.

[4] Korevaar, J., Polynomial and rational approximation in the complex domain. In Aspects of Contemporary Complex Analysis (J.G. Clunie, ed.). Acad. Press, New York/London 1980, pp. 251-291.

[5] Lebedev, N.A. - Tamrazov, P.M., Inverse approximation theorems on regular compacta of the complex plane. Math USSR Izvestija 4 (1970), 1355-1405.

[6] Lorentz, G.G., Approximation of Functions. Holt, Rinehart and Winston, New York 1966.

[7] Tamrazov, P.M., The solid inverse problem of polynomial approximation of functions on a regular compactum. Math. USSR Izvestija 7 (1973), 145-162.

[8] Tamrazov, P.M., Smoothness and Polynomial Approximation (Russian). Izdat. Naukova Dumka, Kiev 1975.

[9] Tamrazov, P.M., Structural and approximation properties of functions in the complex domain. In Linear Spaces and Approximation (P.L. Butzer-B. Sz.-Nagy, eds.).(ISNM, vol. 40) Birkhäuser Verlag, Basel/Stuttgart 1978, pp. 503-514.

[10] Teljakovskiĭ, S.A., Two theorems on approximation of functions by algebraic polynomials. Mat. Sbornik 70 (1966), 252-265; Amer. Math. Soc. Transl. 77 (1968), 163-178.

BEST MULTIPOINT LOCAL APPROXIMATION

Rick Beatson and Charles K. Chui[1]
Department of Mathematics, University of Texas, Austin, Texas 78712
Department of Mathematics, Texas A&M University, College Station, Texas 77843

The problem of best multipoint local approximation is posed and discussed. In special cases, these approximants are solutions of certain minimax problems.

1. Introduction

Let f be a sufficiently smooth function and m_ε be a best approximant of f, from a class of functions M, on a disjoint union I_ε of k nondegenerate closed intervals, with the L_p norm. In this paper we investigate the behaviour of the net of best approximants $\{m_\varepsilon\}$ as I_ε shrinks to a union X of k points when $\varepsilon \to 0^+$.

Suppose that for each $\varepsilon > 0$, f has a unique best approximant m_ε from M on I_ε. Then it is natural to ask if the net $\{m_\varepsilon\}$ has a cluster point, m_0, as $\varepsilon \to 0^+$, with respect to some topology on M. If m_0 exists, it will be called a best k-point local approximant of f (on the set X with respect to the topology on M). If f has a best k-point local approximant m_0, then the following questions are of interest. Is m_0 unique, and if so how is it characterized? When I_ε is "symmetric" and M is a "nice" $d = Nk$ parameter family, then instinct tells us that the net $\{m_\varepsilon\}$ should have a unique limit m_0 characterized by the interpolation conditions: $(m_0 - f)^{(j)}(x) = 0$, $j = 0, \ldots, N-1$, for each $x \in X$.

[1] Supported in part by the U. S. Army Research Office under Grant No. DAAG 29-78-G-0097.

We will prove a result of this type for algebraic polynomials, where the best k-point local approximants are Hermite interpolatory polynomials. If d is not divisible by k, then the problem is much more complicated. We will prove that at least in special cases, m_0 is characterized as the solution of a minimax problem.

To be more precise, we let $X = \{x_1, \ldots, x_k\}$ where $x_1 < \ldots < x_k$ and $x_{i+1} - x_i > 2\delta > 0$ for $i = 1, \ldots, k-1$. For $0 < \varepsilon \le \delta$, we will consider the case where I_ε is the disjoint union of k nondegenerated closed intervals each of length ε so that $X \subset I_\varepsilon$. Let π_ℓ denote, as usual, the linear space of all polynomials with degrees not exceeding ℓ. Our first result is the intuitively obvious

THEOREM 1. Let $1 \le p \le \infty$ and $f \in C^{N-1}(I_\delta)$ where $N \ge 1$. For each ε, $0 < \varepsilon \le \delta$, let P_ε be the best $L_p(I_\varepsilon)$ approximant to f from π_{kN-1}. Then the net $\{P_\varepsilon\}$ converges coefficientwise as $\varepsilon \to 0^+$ to some $P_0 \in \pi_{kN-1}$. Furthermore P_0 is the unique polynomial in π_{kN-1} which satisfies the interpolation conditions

$$P_0^{(j)}(x_i) = f^{(j)}(x_i), \quad i = 1, \ldots, k, \quad j = 0, \ldots, N-1 . \tag{1.1}$$

In Theorem 1 the restriction that the number of parameters is a multiple of k is most undesirable. In special cases we will show that this restriction is indeed not needed, and the best k-point local approximant turns out to be the solution of a certain minimax problem.

Let $J_\varepsilon = [-1, -1+\varepsilon] \cup [1-\varepsilon, 1]$ where $0 < \varepsilon \le \delta < 1$. For each $f \in C^N(I_\delta)$, let

$$\mathcal{F}_f = \{p \in \pi_{2N} : (p - f)^{(j)}(\pm 1) = 0, \quad j = 0, \ldots, N-1\}. \tag{1.2}$$

We have the following result:

THEOREM 2. Let $f \in C^N(J_\delta)$, $N \ge 0$. For each ε, $0 < \varepsilon \le \delta$, let $p_\varepsilon(f)$ be the best uniform approximant of f on J_ε from π_{2N}. Then the net $\{p_\varepsilon(f)\}$ converges coefficientwise as $\varepsilon \to 0^+$, to some $p_0 \in \pi_{2N}$. Furthermore p_0 is the unique polynomial in $\mathcal{F}_f$ which minimizes

(1.3) $$\max\{|(p-f)^{(N)}(-1)|,\ |(p-f)^{(N)}(1)|\}$$

over all $p \in F_f$.

In Theorem 2, $M = \pi_{2N}$ is a $2N + 1$ parameter family. Of course approximation from the $2N$ parameter family π_{2N-1} is a special case of Theorem 1. The L_2 analogue of Theorem 2 was established by Su [6]. However, the linear methods in [6] are not applicable to the uniform norm setting.

Let X be a set of k distinct points as before. For $0 < \varepsilon \le \delta$, let $K_\varepsilon = \bigcup_{i=1}^{k} [x_i - \varepsilon, x_i + \varepsilon]$. For each $f \in C^1(K_\delta)$, let

(1.4) $$G_f = \{p \in \pi_{2k-2} : (p-f)(x_i) = 0,\ i = 1, \ldots, k\}.$$

We have the following result:

THEOREM 3. *Let* $f \in C^1(K_\delta)$. *For each* ε, $0 < \varepsilon \le \delta$, *let* $p_\varepsilon(f)$ *be the best uniform approximant of* f *on* K_ε *from* π_{2k-2}. *Then the net* $\{p_\varepsilon(f)\}$ *converges coefficientwise as* $\varepsilon \to 0^+$ *to some* $p_0 \in \pi_{2k-2}$. *Furthermore* p_0 *is the unique polynomial in* G_f *which minimizes* $\max\{|(p-f)'(x_i)| : i = 1, \ldots, k\}$ *over all* $p \in G_f$.

Results related to best 1-point local approximation are contained in Walsh [7,8,9]; Chui, Shisha and Smith [2,3]; Chui, Smith and Ward [4]; Chui [1]; and Wolfe [10].

2. Best k-Point Local Approximation from π_{kN-1}

Given a bounded measurable set $E \subset \mathbb{R}$ and $1 \le p < \infty$, we define

$$\|f\|_{L_p(E)} = \left(\int_E |f|^p\right)^{1/p},$$

$$\|f\|_{L_p^*(E)} = \left(\int_E |f|^p \Big/ \int_E 1\right)^{1/p}, \quad \text{and}$$

$$\|f\|_{L_\infty(E)} = \|f\|_{L_\infty^*(E)} = \operatorname{ess\,sup}\{|f(x)| : x \in E\}.$$

Then for every non-negative integer d, there exists a constant $C > 0$, depending on d alone, so that

$$(2.1) \qquad C\|h\|_{L_\infty[a,b]} \le \|h\|_{L_p^*[a,b]} \le \|h\|_{L_\infty[a,b]},$$

for all $h \in \pi_d$, $1 \le p \le \infty$, and $-\infty < a < b < \infty$. When $a = 0$ and $b = 1$ (2.1) is trivial since $\|\cdot\|_{L_p^*[0,1]} = \|\cdot\|_{L_p[0,1]}$. Hence, since the normalized L_p norm, $\|\ \|_{L_p^*}$, is unchanged by a linear change of variable, (2.1) follows for arbitrary values of a and b.

LEMMA 2.1. <u>Suppose that</u> $1 \le p \le \infty$ <u>and</u> $\{Q_\varepsilon\} \subset \pi_{kN-1}$ <u>is a net with</u>

$$\|Q_\varepsilon\|_{L_p^*(I_\varepsilon)} = o(\varepsilon^{N-1}),$$

<u>as</u> $\varepsilon \to 0^+$. <u>Then</u> $Q_\varepsilon \to \theta$, <u>the zero polynomial, coefficientwise as</u> $\varepsilon \to 0^+$.

PROOF. Since I_ε is the union of k disjoint intervals each of length ε we find from (2.1) above that

$$D\|Q\|_{L_\infty^*(I_\varepsilon)} \le \|Q\|_{L_p^*(I_\varepsilon)} \le \|Q\|_{L_\infty^*(I_\varepsilon)} \quad .$$

for all $Q \in \pi_{kN-1}$, $1 \le p \le \infty$, and $0 < \varepsilon \le \delta$, where $D > 0$ is a constant depending only on k and N. Hence it suffices to prove the lemma when $p = \infty$.

Let h_{ij} be the unique polynomial in π_{kN-1} with

$$h_{ij}^{(r)}(x_e) = \delta_{ie}\delta_{jr}, \quad i, e = 1, \ldots, k; \quad j, r = 0, \ldots, N-1,$$

where δ_{ie}, δ_{jr} are the Kronecker deltas. Then $Q_\varepsilon = \sum_{i=1}^{k} \sum_{j=0}^{N-1} a_{i,j,\varepsilon} h_{ij}$ for some $\{a_{i,j,\varepsilon}\}$. Applying the Markov inequality on each of the k intervals of length ε comprising I_ε, we find

$$\|Q_\varepsilon^{(j)}\|_{L_\infty(I_\varepsilon)} = O\big(\varepsilon^{-j}\|Q_\varepsilon\|_{L_\infty(I_\varepsilon)}\big) = o(1), \quad j = 0, 1, \ldots, N-1 .$$

This implies that

$$|a_{i,j,\varepsilon}| = |Q_\varepsilon^{(j)}(x_i)| \le \|Q_\varepsilon^{(j)}\|_{L_\infty(I_\varepsilon)} = o(1),$$

as $\varepsilon \to 0^+$, $i = 1, \ldots, k$; $j = 0, 1, \ldots, N-1$. Hence, $Q_\varepsilon \to 0$, the zero polynomial, coefficientwise as $\varepsilon \to 0^+$. This completes the proof.

We are now ready to prove Theorem 1 which is an easy corollary of Lemma 2.1. Let $f \in C^{N-1}(I_\delta)$ and for each δ, $0 < \varepsilon \le \delta$, let $P_\varepsilon(f) \in \pi_{kN-1}$ be the best $L_p(I_\varepsilon)$ approximant of f from π_{kN-1}. Let P_0 be the unique polynomial in π_{kN-1} such that $(f - P_0)^{(j)}(x_i) = 0$, $i = 1, \ldots, k$; $j = 0, \ldots, N-1$. Since $P_\varepsilon(f)$ is the best approximant of f in $L_p(I_\varepsilon)$, we have

$$\|f - P_\varepsilon\|_{L_p^*(I_\varepsilon)} \le \|f - P_0\|_{L_p^*(I_\varepsilon)} = o(\varepsilon^{N-1})$$

implying

$$\|P_0 - P_\varepsilon\|_{L_p^*(I_\varepsilon)} = o(\varepsilon^{N-1}) .$$

It follows from Lemma 2.1 that $P_\varepsilon \to P_0$ coefficientwise as $\varepsilon \to 0^+$ as required.

3. Best 2-Point Local Approximation from π_{2N}

This section will be devoted to the proof of Theorem 2 stated in the introduction. To facilitate our proof we need a sequence of six lemmas.

As in Section 1, we set $J_\varepsilon = [-1, -1+\varepsilon] \cup [1-\varepsilon, 1]$, where $0 < \varepsilon \le \delta$ and $0 < \delta < 1$. Throughout this section we denote by $\|\cdot\|_{J_\varepsilon}$ the uniform norm on J_ε. Let $P(x) = x^{2N+1}$, $P_0 \in \pi_{2N-1}$ be determined uniquely by the interpolation conditions $(P_0 - P)^{(j)}(\pm 1) = 0$ $j = 0, \ldots, N-1$, and $P_\varepsilon \in \pi_{2N}$ be the best uniform approximant of P on J_ε from π_{2N}. Since P is an odd function, P_ε must be a polynomial with odd degree, so that it is also the best approximant of P on J_ε from π_{2N-1}. Hence, as a consequence of Theorem 1, we have the following

LEMMA 3.1. _The net $\{P_\varepsilon\}$ converges coefficientwise to P_0 as $\varepsilon \to 0^+$._

We next derive an error bound on the convergence of $\{P_\varepsilon\}$ to P.

LEMMA 3.2. *The following estimates hold:*

$$\|P - P_\varepsilon\|_{J_\varepsilon} = O(\varepsilon^N) \tag{3.1}$$

as $\varepsilon \to 0^+$, *but*

$$\liminf_{\varepsilon \to 0^+} \varepsilon^{-N}\|P - P_\varepsilon\|_{J_\varepsilon} > 0 . \tag{3.2}$$

To obtain (3.1), we simply note that $\|P - P_\varepsilon\|_{J_\varepsilon} \le \|P - P_0\|_{J_\varepsilon} = O(\varepsilon^N)$. Assume, on the contrary, that (3.2) does not hold. Then there is a sequence $\varepsilon_k \to 0^+$ such that $\|P - P_{\varepsilon_k}\|_{J_{\varepsilon_k}} = o(\varepsilon_k^N)$. Hence, by Lemma 2.1 we may conclude that $P - P_{\varepsilon_k} \to \theta$ coefficientwise. This is a contradiction since $P - P_{\varepsilon_k}$ is a polynomial in π_{2N+1} with leading coefficient 1.

We next consider the following minimax problem. Let $F_P = \{Q \in \pi_{2N} : (Q - P)^{(j)}(\pm 1) = 0,\ j = 0, \ldots, N-1\}$. We will study the extremal problem

$$\min_{Q \in F_P} \max_{x=\pm 1} |(Q - P)^{(N)}(x)| . \tag{3.3}$$

LEMMA 3.3. *The extremal problem* (3.3) *has a unique solution given by* $Q = P_0$.

To prove this result, we note that every $Q \in F_P$ can be written as $Q(x) = P_0(x) + a(x^2 - 1)^N$ for some constant a. Hence, we have $(Q - P)(x) = (P_0 - P)(x) + a(x^2 - 1)^N$, where $P_0 - P$ is an odd function and $a(x^2 - 1)^N$ is an even function. It follows immediately that $|(Q - P)^{(N)}(1)| = |(P_0 - P)^{(N)}(1) + a2^N N!|$, $|(Q - P)^{(N)}(-1)| = |(P_0 - P)^{(N)}(1) - a2^N N!|$, and $|(P_0 - P)^{(N)}(-1)| = |(P_0 - P)^{(N)}(1)|$. Hence we have $\max_{x=\pm 1} |(Q - P)^{(N)}(x)| \ge \max_{x=\pm 1} |(P_0 - P)^{(N)}(x)|$, where equality holds if and only if $a = 0$. This completes the proof of the lemma.

In order to apply the above results to an arbitrary function $f \in C^N(J_\delta)$, we consider the polynomial

$$h_f(x) = a_0 + a_1 x + \ldots + a_{2N+1} x^{2N+1} \tag{3.4}$$

which is uniquely determined by the interpolation conditions $(f-h_f)^{(j)}(\pm 1)=0$, $j=0,\ldots,N$. Let

$$p_f(x):=a_0+a_1x+\cdots+a_{2N}x^{2N}+a_{2N+1}P_0(x) \tag{3.5}$$

$$=h_f(x)+a_{2N+1}(P_0(x)-P(x)).$$

Hence, $p_f\in\pi_{2N}$ and satisfies the interpolation conditions $(f-p_f)^{(j)}(\pm 1)$ $=0$ for $j=0,\ldots,N-1$. That is, $p_f\in F_f$. We also note that

$$(p_f-f)^{(N)}(\pm 1)=a_{2N+1}(P_0-P)^{(N)}(\pm 1), \tag{3.6}$$

and that if $a_{2N+1}\neq 0$ a polynomial q is in F_f if and only if

$$q=h_f+a_{2N+1}(Q-P) \tag{3.7}$$

where Q is in F_p. Hence, Lemma 3.3 yields the following

LEMMA 3.4. The extremal problem $\min_{q\in F_f}\max_{x=\pm 1}|(q-f)^{(N)}(x)|$ has a unique solution given by $q=p_f$.

It is now intuitively clear that h_f is a "polynomial representer" of the given function f useful in discussing best 2-point local approximation of f. We will therefore study the approximation properties of h_f. Let $p_\varepsilon(h_f)$ be the best approximant of h_f in $C(J_\varepsilon)$ from π_{2N}, and set $E_{2N,\varepsilon}(h_f)=\|h_f-p_\varepsilon(h_f)\|_{J_\varepsilon}$. It is clear that $E_{2N,\varepsilon}(h_f)=0$ if and only if $a_{2N+1}=0$. Since π_{2N} is a Chebyshev system we can find $x_i=x_i(\varepsilon)\in J_\varepsilon$, $i=0,\ldots,2N+1$, with $x_0<\cdots<x_{2N+1}$, such that $(h_f-p_\varepsilon(h_f))(x_i)=\sigma(-1)^iE_{2N,\varepsilon}(h_f)$ for $i=0,\ldots,2N+1$, where $\sigma=-1$ or 1.

Let the integer $\ell\geq 0$ and the $\ell+2$ points $z_0<z_1<\cdots<z_{\ell+1}$ be given. For each j, let $e_j\in\pi_\ell$ be uniquely determined by the interpolation conditions:

$$(-1)^ie_j(z_i)=-1,\quad \text{for } i\neq j,\ 0\leq i\leq \ell+1. \tag{3.8}$$

Then we have the following result of Maehly and Witzgall which is contained in the proof of the first lemma in [5].

LEMMA 3.5. Let a be a positive number and q be a polynomial in π_ℓ satisfying the inequalities $(-1)^i q(z_i) \geq -a$, $i = 0,\ldots \ell+1$. Then for each $x \in \mathbb{R}$, q also satisfies the inequalities

$$a \min_{0\leq j\leq \ell+1} e_j(x) \leq q(x) \leq a \max_{0\leq j\leq \ell+1} e_j(x) .$$

When $\ell = 2N$ and $z_i = x_i(\varepsilon)$, $i = 0, \ldots, 2N+1$, we shall denote the polynomials e_j by $e_{\varepsilon,j}$. We then have

LEMMA 3.6. Suppose $a_{2N+1} \neq 0$. Then for each j, $0 \leq j \leq 2N+1$, $\|e_{\varepsilon,j}\|_{J_\varepsilon} = O(1)$ as $\varepsilon \to 0^+$.

To prove Lemma 3.6, we note that the alternant $\{x_0, \ldots, x_{2N+1}\}$ of $h_f - p_\varepsilon(h_f)$ is also the alternant of $(P - P_\varepsilon)(x) := x^{2N+1} - P_\varepsilon(x)$, where P_ε is the best approximant of P in $C(J_\varepsilon)$ from π_{2N}. Set $g_\varepsilon = P - P_\varepsilon$. Applying Markov's inequality to (3.1), we have $\|g_\varepsilon'\|_{J_\varepsilon} = O(\varepsilon^{N-1})$. For each i, $0 \leq i \leq 2N$, we also have

$$\begin{aligned} 2\|g_\varepsilon\|_{J_\varepsilon} &= |g_\varepsilon(x_{i+1}) - g(x_i)| \leq \|g_\varepsilon'\|_{[x_i,x_{i+1}]}(x_{i+1} - x_i) \\ &\leq \|g_\varepsilon'\|_{J_\varepsilon}(x_{i+1} - x_i) . \end{aligned}$$

Hence, by (3.2) in Lemma 3.2, there exists an $\eta > 0$ such that

$$|x_{i+1} - x_i| \geq \eta\varepsilon \tag{3.9}$$

for all sufficiently small $\varepsilon > 0$. Since g_ε is a nontrivial polynomial in π_{2N+1}, a simple zero counting argument shows that the alternant $\{x_0, \ldots, x_{2N+1}\}$ is unique. Also, since $P(x) = x^{2N+1}$ is odd, we have $-1 \leq x_0 < \ldots < x_N \leq -1+\varepsilon < 1-\varepsilon \leq x_{N+1} < \ldots < x_{2N+1} \leq 1$.

To estimate $\|e_{\varepsilon,j}\|_{J_\varepsilon}$, we rename the points in $A_j = \{x_0,\ldots,x_{j-1}, x_{j+1},\ldots,x_{2N+1}\}$ as $B = \{y_0,\ldots,y_{2N}\}$ so that all even indexed points in B lie in the half of J_ε containing $N + 1$ points of A_j, and all the odd indexed points in B lie in the other half of J_ε, containing the remaining N points of A_j. By Newton's formula, we may write

$$(3.10) \qquad e_{\varepsilon,j}(x) = e_{\varepsilon,j}(y_0) + (x - y_0)[y_0, y_1]e_{\varepsilon,j} + \ldots + (x - y_0) \ldots (x - y_{2N-1})[y_0,\ldots,y_{2N}]e_{\varepsilon,j}.$$

Let $0 < \varepsilon \le 1/2$. By (3.9) and the separation of the consecutive y_i's it is clear that the following estimates can be obtained:

$$|e_{\varepsilon,j}(y_0)| = 1, \quad |[y_0,\ldots,y_i]e_{\varepsilon,j}| \le 2^i(\eta\varepsilon)^{-[i/2]},$$

$$\|(x - y_0) \ldots (x - y_i)\|_{J_\varepsilon} \le 2^{[i/2]+1}\, \varepsilon^{[(i+1)/2]}$$

$i = 0, \ldots, 2N$, where for any nonnegative number α, $[\alpha]$ denotes, as usual, its integer part. Hence, by using (3.10), we have $\|e_{\varepsilon,j}\|_{J_\varepsilon} = O(1)$ as $\varepsilon \to 0^+$, for each $j = 0, \ldots, 2N+1$. This completes the proof of Lemma 3.6.

We are now ready to prove Theorem 2. Let $f \in C^N(J_\delta)$, and for each ε, $0 < \varepsilon \le \delta$, let $p_\varepsilon(f) \in \pi_{2N}$ be the best approximant of f in $C(J_\varepsilon)$ from π_{2N}. In view of Lemma 3.4, it is sufficient to prove that the net $\{p_\varepsilon(f)\}$ converges coefficientwise to $p_0 := p_f$ as $\varepsilon \to 0^+$, where p_f is defined in (3.5).

Let us first consider the trivial case when $a_{2N+1} = 0$. In this case, $p_f \equiv h_f$. Since $(f - h_f)^{(j)}(\pm 1) = 0$ for $0 \le j \le N$, we have $\|f - p_f\|_{J_\varepsilon} = o(\varepsilon^N)$. Hence, it follows that $\|p_f - p_\varepsilon(f)\|_{J_\varepsilon} \le \|p_f - f\|_{J_\varepsilon} + \|f - p_\varepsilon(f)\|_{J_\varepsilon} \le 2\|p_f - f\|_{J_\varepsilon} = o(\varepsilon^N)$ so that, by Lemma 2.1, $p_\varepsilon(f) \to h_f \equiv p_f$ coefficientwise as $\varepsilon \to 0^+$.

Now assume that $a_{2N+1} \ne 0$. Then since $(f - h_f)^{(j)}(\pm 1) = 0$ for $0 \le j \le N$, we have

$$(3.11) \qquad \|f - h_f\|_{J_\varepsilon} = o(\varepsilon^N) .$$

Let $E_{2N,\varepsilon}(f) = \|f - p_\varepsilon(f)\|_{J_\varepsilon}$. Then by (3.11), we have

$$(3.12) \qquad E_{2N,\varepsilon}(f) - E_{2N,\varepsilon}(h_f) = o(\varepsilon^N) .$$

Since $a_{2N+1} \ne 0$, Lemma 3.2 yields

$$(3.13)\qquad 0 < \liminf_{\varepsilon\to 0^+} \varepsilon^{-N} E_{2N,\varepsilon}(h_f) \le \limsup_{\varepsilon\to 0^+} \varepsilon^{-N} E_{2N,\varepsilon}(h_f) < \infty.$$

From (3.11) and (3.12), we obtain

$$(3.14)\qquad \begin{aligned}(f - p_\varepsilon(h_f))(x_i) &= \sigma(-1)^i E_{2N,\varepsilon}(h_f) + o(\varepsilon^N)\\ &= \sigma(-1)^i E_{2N,\varepsilon}(f) + o(\varepsilon^N).\end{aligned}$$

Hence, by the first inequality in (3.13), the signs of $(f - p_\varepsilon(h_f))(x_i)$ alternate for $i = 0, \ldots, 2N+1$ for all sufficiently small $\varepsilon > 0$. Set $q_\varepsilon(f) := p_\varepsilon(f) - p_\varepsilon(h_f) = (p_\varepsilon(f) - f) + (f - p_\varepsilon(h_f))$. We therefore have $-E_{2N,\varepsilon}(f) + (f - p_\varepsilon(h_f))(x_i) \le q_\varepsilon(x_i) \le E_{2N,\varepsilon}(f) + (f - p_\varepsilon(h_f))(x_i)$, and by (3.14), it follows that $q_\varepsilon(x_i) \ge -a_\varepsilon$ if $\operatorname{sgn}(f - p_\varepsilon(h_f))(x_i) = \sigma(-1)^i > 0$, $q_\varepsilon(x_i) \le a_\varepsilon$ if $\operatorname{sgn}(f - p_\varepsilon(h_f))(x_i) = \sigma(-1)^i < 0$, where $0 < a_\varepsilon = o(\varepsilon^N)$. Hence, by Lemmas 3.5 and 3.6, we have $\|q_\varepsilon\|_{J_\varepsilon} \le a_\varepsilon \max_j \|e_{\varepsilon,j}\|_{J_\varepsilon} = o(\varepsilon^N)$. Again, Lemma 2.1 implies that $q_\varepsilon = p_\varepsilon(f) - p_\varepsilon(h_f) \to \theta$ coefficientwise as $\varepsilon \to 0^+$. However, we also have $p_\varepsilon(h_f) \to a_{2N+1}P_0 + (a_0 + \ldots + a_{2N}x^{2N})$ $:= p_f$ by Lemma 3.1. That is, the net $\{p_\varepsilon(f)\}$ converges to p_f coefficientwise as $\varepsilon \to 0^+$. This completes the proof of Theorem 2.

4. Best k-Point Local Approximation from π_{2k-2}

This section will be devoted to the proof of Theorem 3 stated in the introduction.

As in Section 1, we let $X = \{x_1,\ldots,x_k\}$ be a set of k distinct points spaced at least 2δ apart and set $K_\varepsilon = \bigcup_{i=1}^{k} [x_i-\varepsilon, x_i+\varepsilon]$ where $0 < \varepsilon \le \delta$. Throughout this section we denote by $\|\cdot\|_{K_\varepsilon}$ the uniform norm on K_ε. Let $P(x) = x^{2k-1}$ and $G_P = \{Q \in \pi_{2k-2} : (Q - P)(x_i) = 0,$ $i = 1,\ldots,k\}$. We first study the extremal problem

$$(4.1)\qquad \min_{Q\in G_P}\ \max_{i=1,\ldots,k} |(Q - P)'(x_i)| .$$

LEMMA 4.1. There is a unique polynomial $P_0 \in G_P$ such that $(P_0-P)'(x_{i+1}) = (P_0-P)'(x_i)$, for all $1 \le i \le k-1$. Furthermore, $(P_0-P)'(x_i) \ne 0$.

Let $r(x) = a_0 + a_1k + \ldots + a_{2k-2}\, x^{2k-2} - cx^{2k-1}$. Now given data

f_i, f_i', $i = 1,\ldots,k$, it is well known that there is a unique solution $(a_0, a_1, \ldots, a_{2k-2}, c)^T$ to the system of equations

$$(4.2) \qquad r(x_i) = f_i \quad \text{and} \quad r'(x_i) = f_i', \quad i = 1,\ldots,k .$$

Let $(a_0^*, \ldots, a_{2k-2}^*, c^*)^T$ be the solution of the system (4.2) with $f_i = 0$ and $f_i' = 1$, $i = 1,\ldots,k$. Then it is easy to see that $r(x)$ has at least $2k-1$ sign changes, one at each x_i, and one in each interval (x_i, x_{i+1}). It follows that $r(x)$ is of exact degree $2k-1$. Hence, $c^* \neq 0$.

Next, let d be a real number. From above, the solution of (4.2) for the data $f_i = 0$, $i = 1,\ldots,k$, $f_i' = d$, $i = 1,\ldots,k$, is unique and equals $(da_0^*, \ldots, da_{2k-2}^*, dc^*)^T$. There is one and only one d^* for which $d^*c^* = 1$. Hence, there is one and only one $Q \in G_P$ such that $(Q - P)'(x_{i+1}) = (Q - P)'(x_i)$ for all $i = 1,\ldots,k$. We note that with this choice of Q, $(Q - P)'(x_i) = d^* \neq 0$. This completes the proof of the lemma. We also have the following

LEMMA 4.2. <u>The extremal problem</u> (4.1) <u>has a unique solution given by</u> $Q = P_0$.

To prove this result, we note that every $Q \in G_P$ can be written as $Q(x) = P_0(x) + w(x)v(x)$ where $w(x) = (x-x_1)(x-x_2)\ldots(x-x_k)$ and $v \in \pi_{k-2}$. Hence,

$$(4.3) \qquad (Q - P)'(x_i) = (P_0 - P)'(x_i) + w'(x_i)v(x_i), \quad i = 1,\ldots,k.$$

Suppose now Q is a solution of the extremal problem (4.1). Then, in particular,

$$(4.4) \qquad |d^* + w'(x_i)v(x_i)| \le |d^*|, \quad i = 1,\ldots,k,$$

where d^* is the non-zero constant value of $(P_0-P)'(x_i)$. But $\operatorname{sgn} w'(x_i) = (-1)^{k-i}$ for all $i = 1,\ldots,k$. Thus (4.4) implies that $\sigma(-1)^{k-i}v(x_i) \le 0$, $i = 1,\ldots,k$, where $\sigma = \operatorname{sgn} d^*$. Hence, $v \in \pi_{k-2}$ has at least $k-1$ zeros and is identically zero. Thus, P_0 is the unique solution of the minimax problem, yielding Lemma 4.2.

We let $h_f \in \pi_{2k-1}$ be the unique Hermite interpolating polynomial satisfying $(h_f-f)^{(j)}(x_i) = 0$ for $j = 0,1$, and $i = 1,\ldots,k$. If

(4.5) $$h_f(x) = a_0 + a_1x + \dots + a_{2k-2}x^{2k-2} + a_{2k-1}x^{2k-1},$$

we let

(4.6) $$p_f(x) = a_0 + a_1x + \cdots + a_{2k-2}x^{2k-2} + a_{2k-1}P_0(x)$$
$$= h_f(x) + a_{2k-1}(P_0(x) - P(x)).$$

Hence, $p_f \in \pi_{2k-2}$ and satisfies the interpolation conditions $(p_f-f)(x_i) = 0$ for $i = 1,\dots,k$. That is, $p_f \in G_f$. Then since if $a_{2k-1} \neq 0$, $q \in G_f$ if and only if $q = h_f + a_{2k-1}(Q-P)$ for some $Q \in G_p$, Lemmas 4.1 and 4.2 yield

LEMMA 4.3. p_f _is the unique solution to the minimax problem_

(4.7) $$\min_{q \in G_f} \max_{i=1,\dots,k} |(q-f)'(x_i)|.$$

Furthermore p_f _also satisfies_

(4.8) $$(p_f-f)'(x_{i+1}) = (p_f-f)'(x_i), \quad \text{for} \quad i = 1,\dots,k-1.$$

We now define $\{z_0,\dots,z_{2k-1}\}$ by $z_{2i} = x_{i+1} - \varepsilon$, $z_{2i+1} = x_{i+1} + \varepsilon$ for $0 \leq i \leq k-1$, and let $e_{\varepsilon,j} \in \pi_{2k-2}$ be the polynomial specified by the interpolation conditions (3.8).

LEMMA 4.4. _For each_ j, $0 \leq j \leq 2k-1$, $\|e_{\varepsilon,j}\|_{K_\varepsilon} = O(1)$ _as_ $\varepsilon \to 0^+$.

To prove this lemma, we rename the points in $A_j = \{z_0,\dots,z_{j-1}, z_{j+1},\dots,z_{2k-1}\}$ as $B = \{y_0,\dots,y_{2k-2}\}$ so that $y_0,\dots,y_{k-1}$ all lie in distinct subintervals of K_ε and y_{k+i} lies in the same subinterval as y_i for $0 \leq i \leq k-2$. Then for distinct i and j $|y_i - y_j| \geq 2\delta$ unless $|i-j| = k$ in which case $|y_i - y_j| = 2\varepsilon$. By Newton's formula

(4.9) $$e_{\varepsilon,j}(x) = e_{\varepsilon,j}(y_0) + (x-y_0)[y_0,y_1]e_{\varepsilon,j} + \cdots$$
$$+ \cdots + (x-y_0)\cdots(x-y_{2k-3})[y_0,\dots,y_{2k-2}]e_{\varepsilon,j}.$$

Now from the spacing of the y_i's it is clear that for $i = 0,\dots,2k-2$

(4.10) $$|[y_0,\dots,y_i]e_{\varepsilon,j}| = O(\varepsilon^{-[i/k]})$$

and $\|(x-y_0)\ \dots\ (x-y_i)\|_{K_\varepsilon} = O(\varepsilon^{[(i+1)/k]})$. Hence, by using (4.9) we have $\|e_{\varepsilon,j}\|_{K_\varepsilon} = O(1)$ as $\varepsilon \to 0+$ for each $j = 0,1,\dots,2k-1$. This completes the proof of the lemma.

We are now ready to prove Theorem 3. Let $f \in C^1(K_\delta)$ and for each ε, $0 < \varepsilon \le \delta$, let $p_\varepsilon(f) \in \pi_{2k-2}$ be the best approximant of f in $C(K_\varepsilon)$ from π_{2k-2}. In view of Lemma 4.3, it is sufficient to prove that the net $p_\varepsilon(f)$ converges coefficientwise to $p_0 := p_f$ as $\varepsilon \to 0^+$ where p_f is defined in (4.6).

In the trivial case when $a_{2k-1} = 0$ we have $p_f = h_f$ and $\|f - p_f\|_{K_\varepsilon} = o(\varepsilon)$. The result follows after an application of Lemma 2.1 to $(p_\varepsilon(f) - p_f)$.

Now assume that $a_{2k-1} \neq 0$. Then by Lemmas 4.1 and 4.3, we have $(p_f - f)'(x_{i+1}) = (p_f - f)'(x_i) \neq 0$, for $1 \le i \le k-1$. Thus by the continuity of $(f - p_f)'$ on K_δ $(f - p_f)(x_i + (-1)^j\varepsilon) = (-1)^j\sigma\|f - p_f\|_{K_\varepsilon} + o(\varepsilon)$, $j = 0,1$, $1 \le i \le k$, where $\sigma = -1$ or $\sigma = 1$. Setting $q_\varepsilon(f) = p_\varepsilon(f) - p(f)$ and noting that $\|f - p_\varepsilon(f)\|_{K_\varepsilon} \le \|f - p_f\|_{K_\varepsilon}$ we see that $(-1)^j\sigma q_\varepsilon(x_i+(-1)^j\varepsilon) \ge -a_\varepsilon$, for $1 \le i \le k, j = 0,1$, where $0 < a_\varepsilon = o(\varepsilon)$. Hence by Lemmas 3.5 and 4.4, we have $\|q_\varepsilon\|_{K_\varepsilon} = o(\varepsilon)$. Then Lemma 2.1 implies that $q_\varepsilon = p_\varepsilon(f) - p_f \to \theta$ coefficientwise as $\varepsilon \to 0^+$. That is, the net $\{p_\varepsilon(f)\}$ converges coefficientwise to p_f as $\varepsilon \to 0^+$. This concludes the proof of Theorem 3.

REFERENCES

[1] Chui, C.K., Recent results on Padé approximants and related problems, in Approximation Theory II. G.G. Lorentz, C.K. Chui and L.L. Schumaker eds., Academic Press, Inc., New York 1976.

[2] Chui, C.K. - Shisha, O. - Smith, P.W., Padé approximants as limits of Chebyshev rational approximants. J. Approx. Th. 12 (1974), 201-204.

[3] Chui, C.K. - Shisha, O. - Smith, P.W., Best local approximation. J. Approx. Th. 15 (1975), 371-381.

[4] Chui, C.K. - Smith, P.W. - Ward, J.D., Best L_2 local approximation. J. Approx. Th. 22 (1978), 254-261.

[5] Maehly, H. - Witzgall, C., Tschebyscheff - Approximationen in kleinen Intervallen I. Approximation durch Polynome. Numer Math. 2 (1960), 142-150.

[6] Su, L.Y., Best Local Approximation. Ph.D. Thesis, Texas A§M University, College Station, 1979.

[7] Walsh, J.L., On approximation to an analytic function by rational functions of best approximation. Math. Z. 38 (1934), 163-176.

[8] Walsh, J.L., Padé approximants as limits of rational functions of best approximation. J. Math. Mech. 13 (1964), 305-312.

[9] Walsh, J.L., Padé approximants as limits of rational functions of best approximation, real domain. J. Approx. Th. 11 (1974), 225-230.

[10] Wolfe, J., Interpolation and best L_p local approximation. J. Approximation Theory, (to appear).

VI Approximation by Linear Operators

ON THE LEBESGUE FUNCTION OF INTERPOLATION

Pål Erdős and Péter Vértesi
Mathematical Institute
of the Hungarian Academy of Sciences
Budapest

Solving an old problem of P.Erdős, we prove the best possible in order estimation for the Lebesgue function of Lagrange interpolation.

1. Introduction

Let $Z=\{x_{kn}\}$, $n=1,2,\dots;\ 1\le k\le n$, be a triangular matrix where

$$(1.1)\qquad -1\le x_{nn}<x_{n-1,n}<\dots<x_{1n}\le 1 \qquad (n=1,2,\dots)$$

are n arbitrary points in $[-1,1]$ (shortly $x_k=x_{kn}$).

Putting

$$(1.2)\qquad \omega(x)=\omega_n(Z,x)=\prod_{k=1}^{n}(x-x_k) \qquad (n=1,2,\dots),$$

$$(1.3)\qquad \ell_k(x)=\ell_{kn}(Z,x)=\frac{\omega(x)}{\omega'(x_k)(x-x_k)} \qquad (k=1,2,\dots,n)$$

are the corresponding fundamental polynomials of the Lagrange interpolation. It is well known that the so called Lebesgue function and Lebesgue constant

$$\lambda_n(x)=\lambda_n(Z,x)=\sum_{k=1}^{n}|\ell_k(x)|\ ,\qquad \lambda_n=\lambda_n(Z)=\max_{-1\le x\le 1}\lambda_n(x)$$

play a decisive role in the convergence and divergence properties of Lagrange interpolation.

G.Faber [1] proved that

$$\lambda_n > \frac{1}{12} \ln n$$

for arbitrary matrix Z. Later S.Bernstein [1] obtained that for any system of nodes (1.1)

$$\overline{\lim_{n\to\infty}}\, \lambda_n(x_o) = \infty \tag{1.4}$$

for a certain $x_o \in (-1,1)$.

In 1961, P.Erdős [5] improved an earlier result of P.Erdős and P.Turán [6] proving

$$\lambda_n > \frac{2}{\pi} \ln n - c \qquad (n \geq n_o)$$

for all system (1.1) again. (Here and later $c, c_1, c_2, \ldots,$ will denote positive absolute constants.)

Finally we quote the result of P.Erdős [4] which says as follows.

THEOREM 1.1. Let ε and A be any given positive numbers. Then, considering arbitrary matrix Z, the measure of the set in x $(-\infty<x<\infty)$ for which

$$\lambda_n(x) \leq A \quad \text{if} \quad n \geq n_o(A,\varepsilon), \tag{1.5}$$

is less than ε.

2. Results

Here we prove the following improvement of Theorem 1.1.

THEOREM 2.1. Let $\varepsilon > 0$ be any given number. Then for arbitrary matrix Z there exist sets H_n with $|H_n| \leq \varepsilon$ and $\eta(\varepsilon) > 0$

such that

(2.1) $\lambda_n(x)>\eta(\varepsilon)\ln n$ whenever $x\in[-1,1]\setminus H_n$ and $n\ge n_o(\varepsilon)$.

The case of Chebyshev nodes shows that the order of (2.1) is best possible.

By this theorems it is easy to obtain the following

COROLLARY 2.2. Let $\varepsilon>0$ and $\eta(\varepsilon)>0$ be as above. If $S_n\subseteq[-1,1]$ are arbitrary measurable sets then for any matrix Z

(2.2) $\int_{S_n}\lambda_n(x)dx>(|S_n|-\varepsilon)\eta(\varepsilon)\ln n$ whenever $n\ge n_o(\varepsilon)$.

The case $S_n\equiv S=[a,b]$ was treated by P.Erdős and J.Szabados [7].

2.1. The relation (2.1) is obviously valid if $|x|\ge 1+\varepsilon$ because of $x^{n-1}\equiv\sum_{k=1}^{n}x_k^{n-1}\ \ell_k(x)$ which means $|x|^{n-1}\le\sum_{k=1}^{n}|\ell_k(x)|$. So we have (2.1) on the whole real line apart from a set of measure $\le 3\varepsilon$ $(n\ge n_o(\varepsilon))$.

2.2. Nearly 50 years ago S.Bernstein [1] conjectured that

$$\min_Z \lambda_n(Z)$$

is assumed if all the $n+1$ maxima in $(-1,1)$ of $\lambda_n(x)$ are the same. P.Erdős conjectured that the smallest of these $n+1$ maxima is largest again if all these $n+1$ maxima are the same. Erdős further conjectured that if the z_i are on the unit circle then the corresponding extremal problems are solved if the z_i are the n-th roots of unity.

All these conjectures were recently proved in a series of remarkable papers by T.A.Kilgore [10], C.de Boor and A.Pinkus [2] and L.Bratman [3].

3. Proof

3.1. In what follows, sometimes omitting the superfluous notations, let $x_{on}\equiv 1$, $x_{n+1,n}\equiv -1$ and

$$(3.1)\qquad J_{kn}=[x_{k+1,n},x_{kn}] \qquad (k=0,1,\dots,n;\ n=1,2,\dots) .$$

Let us define the index-sets K_{1n} and K_{2n}, further the sets D_{1n} and D_{2n} by

$$(3.2)\qquad |J_{kn}| \begin{cases} \le n^{-1/6} \overset{\text{def}}{=} \delta_n & \text{iff} \quad k\in K_{1n} , \\ > \delta_n & \text{iff} \quad k\in K_{2n} , \end{cases}$$

$$D_{1n}=\bigcup_{k\in K_{1n}} J_{kn} , \qquad D_{2n}=[-1,1]\setminus D_{1n} .$$

If $|J_k|\le\delta_n$ (which means $k\in K_{1n}$ and $J_k\subset D_{1n}$) we say that the interval is short; the others are the long ones.

3.2. In our common paper [8] we proved

LEMMA 3.1. Let $|J_{kn}|>\delta_n$ (k is fixed, $0\le k\le n$). Then for any fixed $0<\bar{q}<1/4$ we can define the index $t=t(k,n)$ and the set $h_{kn}\subset J_{kn}$ so that $|h_{kn}|\le 4\bar{q}|J_{kn}|$, moreover

$$(3.3)\qquad |\ell_{tn}(x)|\ge 3^{n\delta_n^5} \quad \text{if} \quad x\in J_{kn}\setminus h_{kn} \quad \text{and} \quad n\ge n_1(\bar{q}) .$$

(See [8], Lemma 4.4. In [8] $\delta_n=1/\ln n$ but this does not make any difference in the proof.)

Now, if $\bar{q}=\varepsilon/32$, for the long intervals we obtain (2.1) (see (3.3)) if $x\in D_{2n}\setminus H_{1n}$.

Here $H_{1n} \overset{\text{def}}{=} \bigcup_{k\in K_{2n}} h_{kn}$, which means $|H_{1n}|\le 4\bar{q} \sum_k |J_k| \le \varepsilon/4$ $(n\ge n_2(\varepsilon))$.

3.3. To settle the short intervals we introduce the following notations

$$J_k(q)=J_{kn}(q)=[x_{k+1}+q|J_k|,\ x_k-q|J_k|] \qquad (0\le k\le n)$$

where $0\le q\le 1/2$. Let $z_k=z_{kn}(q)$ be defined by

(3.4) $|\omega_n(z_k)| = \min_{x\in J_k(q)} |\omega_n(x)|$, $k=0,1,\dots,n$,

finally let

$$|J_i,J_k|=\max(|x_{i+1}-x_k|,|x_{k+1}-x_i|) \qquad (0\le i,\ k\le n)\ .$$

In [8],Lemma 4.2 we proved

LEMMA 3.2. If $1\le k,\ r<n$ then for arbitrary $0<q\le 1/2$

$$|\ell_k(x)|+|\ell_{k+1}(x)|\ge q^2 \frac{|\omega_n(z_r)|}{|\omega_n(z_k)|}\frac{|J_k|}{|J_r,J_k|} \quad \text{if} \quad x\in J_r(q)\ . \tag{3.5}$$

3.4. Later we shall also use the

LEMMA 3.3. Let $I_k=[a_k,b_k]$, $1\le k\le t$, $t\ge 2$, be any t intervals in $[-1,1]$ with $|I_k\cap I_j|=0$ $(k\ne j)$, $|I_k|\le\rho$ $(1\le k\le t)$, $\sum_{k=1}^{t}|I_k|=\mu$. Supposing that for certain integer $R\ge 2$ we have $\mu\ge 2^R\rho$, there exists the index s, $1\le s\le t$, such that

$$S=\sum_{k=1}^{t}\frac{|I_k|}{|I_s,I_k|}\ge \frac{R}{8}\mu\ . \tag{3.6}$$

I_s will be called accumulation interval of $\{I_k\}_{k=1}^{t}$.

(Here and later mutatis mutandis we apply the notations of 3.3. for arbitrary intervals.)

Note that we do not require $b_k\le a_{k+1}$.

The lemma and its proof correspond to [8], 4.1.3. Indeed, dropping the interval I_j containing the middle point of $[-1,1]$ and bisecting the same interval $[-1,1]$, we have (say) in $[0,1]$ a set of measure $\geq(\mu-|I_j|)/2 \geq (\mu-\rho)/2$ consisting of certain I_k. Doing the same, after the ℓ-th bisection we obtain that interval of length $2^{1-\ell}$ which contains certain I_k's of aggregate measure $>2^{-\ell}\mu-\rho\geq 2^{-\ell-1}\mu\geq\rho$ for $1\leq\ell\leq p \overset{\text{def}}{=} R-1$.

Consider these intervals $L_1^*,L_2^*,\dots,L_p^*$ (Fig. 1).

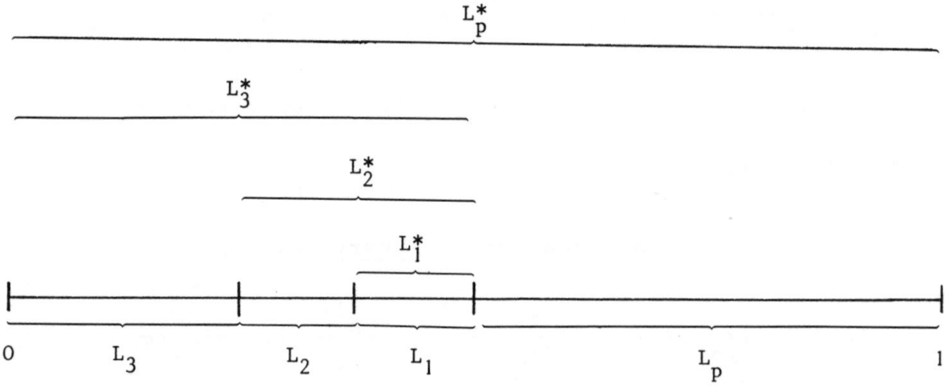

Figure 1.

Obviously $|L_\ell^*|=2^{\ell-p}$. Further each L_ℓ^* contains at least $2^{\ell-1}$ intervals I_k because

$$(3.7)\qquad \sum_{\substack{k\\ I_k\subset L_\ell^*}} |I_k| \geq 2^{\ell-p-2}\mu \qquad (1\leq\ell\leq p)\,.$$

Let $L_1=L_1^*$, further $L_\ell=L_\ell^*\setminus L_{\ell-1}^*$ $(2\leq\ell\leq p)$ (see Figure 1). If s is an index, for which $I_s\subseteq L_1$, we can write

$$(3.8)\qquad S \geq \sum_{\ell=1}^{p} \sum_{\substack{k\\ I_k\subset L_\ell}} \frac{|I_k|}{|I_s,I_k|} \overset{\text{def}}{=} B\,.$$

To estimate B, let

$$(3.9)\qquad \sum_{\substack{k\\ I_k\subset L_\ell}} |I_k| \overset{\text{def}}{=} \alpha_\ell\mu \qquad (1\leq\ell\leq p)\,.$$

By (3.7) and construction we can write

$$(3.10)\qquad \mu\sum_{\ell=1}^{i}\alpha_\ell \ge 2^{i-p-2}\mu \qquad (1\le i\le p)\ ,$$

$$(3.11)\qquad |I_s, I_i| \le 2^{\ell-p} \quad \text{if} \quad I_i\subset L_\ell \qquad (1\le \ell\le p)\ .$$

It is worth to remark that

$$(3.12)\qquad \alpha_\ell \le 2^{\ell-2}\alpha_1 \qquad (2\le\ell\le p)\ .$$

(Indeed, by construction $\alpha_2\le\alpha_1$, $\alpha_\ell \le \sum_{i=1}^{\ell-1}\alpha_i \le 2\sum_{i=1}^{\ell-2}\alpha_i$, $3\le\ell\le p$, from where we get (3.12).)

Now by (3.11), (3.9), (3.10), finally by the Abel transformation we obtain as follows

$$B\ge\mu 2^p\sum_{\ell=1}^{p}2^{-\ell}\alpha_\ell = \mu 2^p\Big[\sum_{\ell=1}^{p-1}2^{-\ell-1}\Big(\sum_{i=1}^{\ell}\alpha_i\Big) + 2^{-p}\sum_{i=1}^{p}\alpha_i\Big] \ge$$

$$\ge \mu 2^p\Big(\sum_{\ell=1}^{p-1}2^{\ell-p-2-\ell-1}+2^{-p-2}\Big) = [2^{-3}(p-1)+2^{-2}]\mu = \frac{p+1}{8}\mu\ ,$$

which was to be proven.

<u>3.5.</u> Suppose $x\in J_{kn}(q)\subset D_{1n}$ $(1\le k\le n-1)$; whenever $\lambda_n(x)\le$ $\le\eta(\varepsilon)\ln n$ (η will be determined later), the point x, the intervals J_{kn} and $J_{kn}(q)$, finally the index k will be called e x c e p t i o n a l . Let $q=\varepsilon/12$.

W e s h a l l p r o v e

$$(3.13)\qquad \sum_k{}' |J_{kn}| \overset{\text{def}}{=} \mu_n \le \frac{\varepsilon}{6} \qquad (n\ge n_o=n_o(\varepsilon))\ .$$

Here and later the dash indicates that the summation is extended only over the exceptional indices k . To prove (3.13) it is enough to consider those indices $\{n_i\}_{i=1}^{\infty} \overset{\text{def}}{=} N$ for which $\mu_{n_i} \geq \varepsilon/10$.

We can apply Lemma 3.3 for the exceptional J_{kn}'s with $\mu=\mu_n$, $\rho=\delta_n$ and $R=[\log^2 n^{1/7}]+1$ if $n\in N$ and $n\geq n_0(\varepsilon)$ (shortly $n\in N_1$).

Denote by $M_1=M_{1n}$ the accumulation interval. Dropping M_1, we apply Lemma 3.3. again for the remaining exceptional intervals with $\mu=\mu_n-|M_1|>\mu_n/2$ and the above ρ and R, supposing $\mu_n\geq\rho 2^{R+1}$ whenever $n\in N_1$. We denote the accumulation interval by M_2. At the i-th step $(2\leq i\leq\psi_n)$ we drop $M_1, M_2, \dots M_{i-1}$ and apply Lemma 3.3. for the remaining exceptional intervals with $\mu=\mu_n-\sum_{j=1}^{i-1}|M_i|$ using the same ρ and R.

Here ψ_n is the first index for which

$$(3.14)\qquad \sum_{i=1}^{\psi_n-1}|M_i|\leq\frac{\mu_n}{2}\quad\text{but}\quad\sum_{i=1}^{\psi_n}|M_i|>\frac{\mu_n}{2},\qquad n\in N_1 .$$

If we denote by $M_{\psi_n+1}, M_{\psi_n+2} \dots M_{\varphi_n}$ the remaining (i.e. not accumulation) exceptional intervals (by $|M_i|\leq\delta_n$, $(\varepsilon/20)n^{1/6}<\psi_n<\varphi_n)$, by (3.6) we can write

$$(3.15)\qquad \sum_{k=r}^{\varphi_n}\frac{|M_k|}{|M_r,M_k|}\geq\frac{\mu_n \ln n}{112}\qquad\text{if } 1\leq r\leq\psi_n \quad (n\in N_1) .$$

<u>3.6.</u> To go further in proving (3.13) let $\eta=c_1\varepsilon^3/6$, $u_{in}\in M_{in}(q)$ $(1\leq i\leq\varphi_n,\ n\in N_1)$ be exceptional points, where c_1 will be determined later.

If for a fixed $n\in N_1$ there exists t, $1\leq t\leq\varphi_n$, such that

$$(3.16)\qquad \lambda_n(u_{tn})\geq c_1\varepsilon^2\mu_n\ln n ,$$

by $\eta\ln n\geq\lambda_n(u_{tn})$ we obtain (3.13) for this n. We shall prove (3.16) for arbitrary $n\in N_1$. Indeed, let us suppose that for a certain $m \in N_1$

(3.17) $\lambda_m(u_{rm}) < c_1 \varepsilon^2 \mu_m \ln m$ where $u_{rm} \in M_{rm}(q)$, $1 \le r \le \varphi_m$.

By (3.17) we obtain

$$(3.18) \quad \sum_{r=1}^{\varphi_m} |M_{rm}| \lambda_m(u_{rm}) < c_1 \varepsilon^2 \mu_m^2 \ln m \quad \text{where} \quad m \in N_1 .$$

On the other hand, by (3.5), for a r b i t r a r y $n \in N_1$

$$|M_r| \sum_{k=1}^{n} |\ell_k(u_r)| \ge \frac{1}{2} |M_r| \sum_k{}' [|\ell_k(u_r)| + |\ell_{k+1}(u_r)|] \ge$$

$$\ge \frac{q^2}{2} \sum_{k=1}^{\varphi_n} \left|\frac{\omega(\bar{z}_r)}{\omega(\bar{z}_k)}\right| \frac{|M_r||M_k|}{|M_r, M_k|}, \quad (1 \le r \le \varphi_n),$$

so, by (3.14) and (3.15) we have

$$\sum_{r=1}^{\varphi_n} |M_r| \lambda_n(u_r) = \sum_{r=1}^{\varphi_n} |M_r| \sum_{k=1}^{n} |\ell_k(u_r)| \ge \frac{q^2}{2} \sum_{r=1}^{\varphi_n} \sum_{k=1}^{\varphi_n} \left|\frac{\omega(\bar{z}_r)}{\omega(\bar{z}_k)}\right| \frac{|M_r||M_k|}{|M_r, M_k|} \ge$$

$$\ge \frac{1}{2} \frac{q^2}{2} \sum_{r=1}^{\varphi_n} \sum_{k=r}^{\varphi_n} \left[\frac{\omega(\bar{z}_r)}{\omega(\bar{z}_k)} + \frac{\omega(\bar{z}_k)}{\omega(\bar{z}_r)}\right] \frac{|M_r||M_k|}{|M_r, M_k|} \ge$$

$$\ge \frac{q^2}{4} \sum_{r=1}^{\psi_n} |M_r| \sum_{k=r}^{\varphi_n} \frac{|M_k|}{|M_r, M_k|} > \frac{q^2}{4} \frac{\mu_n}{2} \frac{\mu_n \ln n}{112} = c_1 \varepsilon^2 \mu_n^2 \ln n$$

if $c_1 = 8.144.112$. This contradicts to (3.18), i.e. (3.16) is valid for arbitrary $n \in N_1$, which proves (3.13).

<u>3.7.</u> By definition, if the short J_{kn} is not exceptional, then for any $x \in J_{kn}(q)$ (2.1) valid, supposing that $k \ne 0, n$. If J_{0n} is short it should belong to H_n. The same should be done with

J_{nn}. Moreover, the sets $J_{kn}\setminus J_{kn}(q)$ of aggregate measure c_2 should belong to H_n, too. Obviously $c_2 \le 2q \sum_{k=0}^{n} |J_{kn}| = 4q = \varepsilon/3$. So using these, 3.2 and (3.13), we obtain

$$|H_n| \le |H_{1n}| + \mu_n + 2\delta_n + c_2 \le \varepsilon/4 + \varepsilon/6 + \varepsilon/4 + \varepsilon/3 = \varepsilon ,$$

which completes the proof.

- . -

The authors are indebted to G.Halász for his valuable remarks and suggestions.

REFERENCES

[1] Bernstein, S., Sur la limitation des valeurs d'un polynome. Bull. Acad. Sci. de l'URSS. 8 (1931), 1025-1050.

[2] de Boor, C. - Pinkus, A., Proof of the conjectures of Bernstein and Erdös concerning the optimal nodes for polynomial interpolation. J. Approximation Theory. 24 (1978), 289—303.

[3] Bratman, L., On the polynomial and rational projections in the complex plane. SIAM J. Numer. Anal. (to appear).

[4] Erdös, P., Problems and results on the theory of interpolation I. Acta Math. Acad. Sci. Hungar. 9 (1958), 381-388.

[5] Erdös, P., Problems and results on the theory of interpolation II. Acta Math. Acad. Sci. Hungar. 12 (1961), 235-244.

[6] Erdös, P. - Turán, P., An extremal problem in the theory of interpolation. Acta Math. Acad. Sci. Hungar. 12 (1961), 221-234.

[7] Erdös, P. - Szabados, J., On the integral of the Legesgue function of interpolation. Acta Math. Acad. Sci. Hungar. 32 (1978), 191-195.

[8] Erdös, P. - Vértesi, P., On the almost everywhere divergence of Lagrange interpolatory polynomials for arbritrary system of nodes. Acta Math. Acad. Sci. Hungar. (to appear).

[9] Faber, G., Über die interpolatorische Darstellung stetiger Funktionen. Jahresber. der Deutschen Math. Ver. 23 (1914), 191-210.

[10] Kilgore, T.A., A characterization of the Lagrange interpolating projection with minimal Tchebycheff norm. J. Approx. Theory 24 (1978), 273 - 288.

A UNIFORM BOUNDEDNESS PRINCIPLE WITH RATES AND AN APPLICATION TO LINEAR PROCESSES

Werner Dickmeis*) and Rolf Joachim Nessel
Lehrstuhl A für Mathematik
Rheinisch-Westfälische Technische Hochschule
Aachen

It is shown that in the classical uniform boundedness principle the condition of strong (pure) boundedness of a sequence of bounded linear operators on a Banach space X may indeed be replaced by boundedness with rates on corresponding subsets of X. The method of proof employed is the gliding hump method but now equipped with rates. Some applications are given to linear polynomial convolution operators, regaining and extending relevant work of Dahmen - Görlich 1974 and Baskakov 1977.

1. Introduction

With X a Banach space (with norm $\|\cdot\|_X$), Y a normed linear space, and $[X,Y]$ the space of bounded linear operators of X into Y, the classical uniform boundedness principle (UBP) reads ($\mathbb{N}$:= set of natural numbers):

UBP. <u>If for $\{T_n\}_{n\in\mathbb{N}} \subset [X,Y]$ one has (pointwise) strong boundedness</u>

$$\|T_n f\|_Y = \mathcal{O}_f(1) \qquad (f \in X,\ n \to \infty), \tag{1.1}$$

<u>then the operators are also uniformly bounded, i.e.,</u>

$$\|T_n\|_{[X,Y]} = \mathcal{O}(1) \qquad (n \to \infty). \tag{1.2}$$

The aim of this paper is to develop a version of a UBP with rates in the sense that, if in (1.1) one replaces the whole space X by a certain subset, but correspondingly the strong (pure) boundedness on X by boundedness with an appropriate rate, then this nevertheless implies the uniform estimate (1.2). For details see Sec. 2. Some applications to linear approximation pro-

*) The contribution of this author was supported by Grant No. II B4 FA 7888 awarded by the Minister für Wissenschaft und Forschung des Landes NRW.

cesses are given in Sec.3: In Sec. 3.1 we regain and extend results of W. Dahmen and E. Görlich concerning a conjecture of M. Golomb as well as work of V.A. Baskakov concerning a problem of P.P. Korovkin. Whereas these contributions are settled in the frame of one-dimensional trigonometric expansions, Sec. 3.2 outlines extensions to regular biorthogonal systems in Banach spaces, now possible in view of the general treatment of Sec. 2.

The authors thank Professor E. Görlich for many valuable suggestions in connection with Sec. 3.1 and for a critical reading of the manuscript.

2. A Uniform Boundedness Principle with Rates

Let $U \subset X$ be a seminormed linear subset of X (with seminorm $|\cdot|_U$). Then for each $f \in X$, $t \geq 0$ the K-functional is defined by

$$K(t,f) := K(t,f;X,U) := \inf\{\|f-g\|_X + t|g|_U;\ g \in U\} \tag{2.1}$$

and serves as an abstract measure of smoothness (cf. (3.1)). Let ω be a modulus of continuity, thus a continuous, increasing function on $[0,\infty)$ satisfying

$$\omega(0) = 0, \quad \omega(t) > 0 \quad \text{for } t > 0, \tag{2.2}$$

$$\omega(s+t) \leq \omega(s) + \omega(t) \quad \text{for } s,t \geq 0.$$

Employing the additional assumption

$$\sup\{\omega(t)/t;\ t > 0\} = \infty, \tag{2.3}$$

we consider the intermediate spaces $U \subset X_{\omega,o} \subset X_\omega \subset X$,

$$X_\omega := \{f \in X;\ K(t,f) = \mathcal{O}_f(\omega(t)),\ t \to 0+\},$$

$$X_{\omega,o} := \{f \in X;\ K(t,f) = o_f(\omega(t)),\ t \to 0+\},$$

endowed with the seminorm

$$|f|_\omega := \sup\{K(t,f)/\omega(t);\ t > 0\}. \tag{2.4}$$

Let $\{\varphi_n\}_{n \in \mathbb{N}}$ denote a sequence of positive numbers with

$$\lim_{n\to\infty} \varphi_n = 0. \tag{2.5}$$

Then one has the following version of a UBP with rates.

THEOREM 1. *Let* T_n, $T \in [X,Y]$, *and let* $\{\varphi_n\}$ *satisfy* (2.5). *Suppose that for each* $n \in \mathbb{N}$ *there exists* $h_n \in U$ *such that*

$$\|h_n\|_X \leq C_1, \tag{2.6}$$

$$|h_n|_U \leq C_2/\varphi_n, \tag{2.7}$$

$$\|T_n\|_{[X,Y]} \leq C_3 \|T_n h_n\|_Y. \tag{2.8}$$

Let ω *be a modulus of continuity satisfying* (2.2/3). *Then the (pointwise) strong boundedness condition (with rates on* $X_{\omega,o}$)

$$\|T_n f - Tf\|_Y = \mathcal{O}_f(\omega(\varphi_n)) \qquad (f \in X_{\omega,o},\ n \to \infty) \tag{2.9}$$

implies that the operators T_n *are uniformly bounded, i.e.,*

$$\|T_n\|_{[X,Y]} = \mathcal{O}(1) \qquad (n \to \infty). \tag{2.10}$$

PROOF. First note that (2.3) is equivalent to

$$\lim_{t \to 0} \omega(t)/t = \infty. \tag{2.11}$$

Moreover, each modulus ω satisfies

$$\omega(t)/t \leq 2\omega(s)/s \quad \text{for } t \geq s > 0. \tag{2.12}$$

Assume that (2.10) does not hold, i.e.,

$$\sup_{n \in \mathbb{N}} \|T_n\|_{[X,Y]} = \infty. \tag{2.13}$$

Then one may successively construct a subsequence $\{n_k\}_{k \in \mathbb{N}}$ satisfying $(k \geq 2)$:

$$n_{k-1} < n_k, \quad \varphi_{n_{k-1}} > \varphi_{n_k}, \tag{2.14}$$

$$\omega(\varphi_{n_k}) \leq (1/2)\omega(\varphi_{n_{k-1}}), \tag{2.15}$$

$$\sum_{j=1}^{k-1} \omega(\varphi_{n_j})/j\varphi_{n_j} \leq \omega(\varphi_{n_k})/k\varphi_{n_k}, \tag{2.16}$$

(2.17) $$\|T_{n_{k-1}} - T\|_{[X,Y]} \leq \frac{1}{8C_1C_3}\frac{k}{k-1} M_{k-2}^2 \omega(\varphi_{n_{k-1}})/\omega(\varphi_{n_k}),$$

(2.18) $$\|T_{n_k}\|_{[X,Y]} \geq \max\{M_{k-1}^2;\ 2C_1C_3\|T\|_{[X,Y]}\},$$

where the constants M_k are defined via

$$g_k := \sum_{j=1}^{k} (\omega(\varphi_{n_j})/j)h_{n_j} \in U \subset X_{\omega,o}$$

by $M_o = 1$ and for $k \geq 1$ by (cf. (2.9))

(2.19) $$M_k := \max\{8C_3(k+1);\ \sup_{n\in\mathbb{N}} \|T_n g_k - Tg_k\|_Y/\omega(\varphi_n)\}.$$

Since X is complete and (cf. (2.6/15))

(2.20) $$\sum_{j=k}^{\infty} (\omega(\varphi_{n_j})/j)\|h_{n_j}\|_X \leq (C_1/k)\sum_{j=k}^{\infty} \omega(\varphi_{n_j}) \leq 2C_1\omega(\varphi_{n_k})/k,$$

the case k=1 implies that $g_\omega := \sum_{j=1}^{\infty} (\omega(\varphi_{n_j})/j)h_{n_j}$ is well-defined as an element in X. Moreover, $g_\omega \in X_{\omega,o}$. Indeed, for each $t \in (0,\varphi_{n_1})$ there exists $k \in \mathbb{N}$ such that $\varphi_{n_{k+1}} \leq t < \varphi_{n_k}$ (cf. (2.5/14)). Using the corresponding $g_k \in U$ and conditions (2.6/7), (2.16/20), and finally (2.12) one obtains

$$\begin{aligned} K(t,g_\omega) &\leq \|g_\omega - g_k\|_X + t|g_k|_U \\ &\leq \|\sum_{j=k+1}^{\infty} (\omega(\varphi_{n_j})/j)h_{n_j}\|_X + t|\sum_{j=1}^{k} (\omega(\varphi_{n_j})/j)h_{n_j}|_U \\ &\leq 2C_1\omega(\varphi_{n_{k+1}})/(k+1) + t2C_2\omega(\varphi_{n_k})/k\varphi_{n_k} \\ &\leq (2C_1 + 4C_2)\omega(t)/k = o(\omega(t)) \qquad (t \to 0+). \end{aligned}$$

This proves $g_\omega \in X_{\omega,o}$. Applying $T_{n_k} - T$ to

$$g_\omega = (\omega(\varphi_{n_k})/k)h_{n_k} + g_{k-1} + (g_\omega - g_k)$$

yields by (2.8/17-20) that

$$\begin{aligned} \|T_{n_k}g_\omega - Tg_\omega\|_Y \geq{}& \|T_{n_k}(\omega(\varphi_{n_k}/k)h_{n_k})\|_Y - \|T\|_{[X,Y]}C_1\omega(\varphi_{n_k})/k \\ &- \|T_{n_k}g_{k-1} - Tg_{k-1}\|_Y - \|T_{n_k} - T\|_{[X,Y]}\|g_\omega - g_k\|_X \end{aligned}$$

$$\geqslant M_{k-1}\omega(\varphi_{n_k})\{\frac{M_{k-1}}{2C_3k} - 1 - \frac{M_{k-1}}{4C_3k}\} + O(\omega(\varphi_{n_k})),$$

which is a contradiction to (2.9) proving the theorem.

Apart from obvious modifications, the preceding proof also establishes the following version, where $\{\psi_n\}_{n\in\mathbb{N}}$ is such that

$$\psi_n \geqslant 1 \qquad (n \in \mathbb{N}). \tag{2.21}$$

COROLLARY 1. _Let_ $X,Y,U,\omega,\{\varphi_n\},\{T_n\},T,$ _and_ $\{\psi_n\}$ _satisfy the hypotheses of Thm._ 1 _and_ (2.21), _respectively. Then_

$$\|T_nf - Tf\|_Y = O_f(\psi_n\omega(\varphi_n)) \qquad (f \in X_{\omega,o},\ n\to\infty)$$

implies the growth condition

$$\|T_n\|_{[X,Y]} = O(\psi_n) \qquad (n\to\infty). \tag{2.22}$$

Let us give a few remarks concerning the limiting cases $\omega_1(t) = 1$, $\omega_2(t) = t$, exluded by (2.2/3). Since $X_{\omega_1} = X$, the first case is covered by the classical UBP, even without (2.6-8). Concerning ω_2 we may mention

COROLLARY 2. _Let_ $X,Y,U,\{\varphi_n\},\{T_n\},T,$ _and_ $\{\psi_n\}$ _satisfy the hypotheses of Cor._ 1. _Furthermore, let_ U _be complete in the sense that_

(2.23) _for every sequence_ $\{f_n\}_{n\in\mathbb{N}} \subset U$ _with_ $\lim_{n,m\to\infty}|f_n - f_m|_U = 0$ _there exists_ $f_o \in U$ _such that_ $\lim_{n\to\infty}|f_n - f_o|_U = 0$,

and let $T_n - T$ _be bounded operators of_ U _into_ Y _in the sense that they satisfy the Jackson-type inequality_

$$\|T_nf - Tf\|_Y \leqslant C_n|f|_U \qquad (f \in U). \tag{2.24}$$

Then $\|T_nf - Tf\|_Y = O_f(\psi_n\varphi_n)$ _for all_ $f \in U$ _implies_ (2.22).

PROOF. In view of (2.23/24) an application of the classical UBP on U gives

$$\|T_nf - Tf\|_Y \leqslant C\psi_n\varphi_n|f|_U \qquad (f \in U).$$

Hence conditions (2.6-8/21) deliver

$$\|T_n\|_{[X,Y]} \leq C_3\|T_nh_n\|_Y \leq C_3\{\|T_nh_n - Th_n\|_Y + \|Th_n\|_Y\}$$
$$\leq C_3\{(C_2/\varphi_n)C\psi_n\varphi_n + \|T\|_{[X,Y]}C_1\} = O(\psi_n).$$

Of course, a corresponding treatment would also be possible for the intermediate spaces $X_{\omega,o}$, X_ω, provided one makes additional (but unnecessary, see Thm.1) assumptions which ensure completeness of $X_{\omega,o}, X_\omega$ relative to (2.4) and boundedness of $T_n - T$ on $X_{\omega,o}, X_\omega$ (cf. (2.23/24)).

Analogously to Thm.1 there holds a "oh"-version in the sense that

$$\|T_nf\|_Y = o_f(\omega(\varphi_n)) \qquad (f \in X_\omega,\ n \to \infty)$$

even implies

$$\|T_n\|_{[X,Y]} = O(1) \qquad (n \to \infty).$$

Again the proof proceeds via a gliding hump method with rates. In fact, this method of proof was inspired by recent work of Teljakovskii [13], Pochuev [12], and Mertens-Nessel [10] concerning multipliers of strong convergence. Of course, the latter results can now easily be deduced from the present functional analytical principles. For further details, however, see [7,8].

3. Applications to Linear Polynomial Processes

3.1 Trigonometric Convolution Operators. Let $X_{2\pi}$ be one of the spaces $L^p_{2\pi}$, $1 \leq p < \infty$, or $C_{2\pi}$ of 2π-periodic functions, p-integrable or continuous with

$$\|f\|_p := \{\frac{1}{2\pi}\int_{-\pi}^{\pi}|f(u)|^p du\}^{1/p}, \quad \|f\|_C := \sup_{-\pi\leq u\leq\pi}|f(u)|,$$

respectively. For $r \in \mathbb{N}$ let

$$X^{(r)}_{2\pi} := \{f \in X_{2\pi};\ f^{(j)} \in X_{2\pi},\ 0 \leq j \leq r\}, \quad |f|_{X^{(r)}_{2\pi}} := \|f^{(r)}\|_{X_{2\pi}}.$$

Then the corresponding K-functional is equivalent to the *rth* modulus of continuity of the function $f \in X_{2\pi}$,

$$\omega_r(t,f;X_{2\pi}) := \sup_{|h|\leq t}\|\sum_{k=0}^{r}\binom{r}{k}f(\cdot + kh)\|_{X_{2\pi}},$$

in the sense that there exist constants c_1, $c_2 > 0$ independent of $f \in X_{2\pi}$ and $t \geq 0$ such that

(3.1) $$c_1\omega_r(t,f;X_{2\pi}) \leqslant K(t^r,f;X_{2\pi},X_{2\pi}^{(r)}) \leqslant c_2\omega_r(t,f;X_{2\pi}).$$

Hence the intermediate spaces $(X_{2\pi})_\omega$ are Lipschitz spaces (cf. [2, p. 191ff]). We consider polynomial operators $T_n \in [X_{2\pi}]$ $(:=[X_{2\pi},X_{2\pi}])$ of convolution type, i.e.,

(3.2) $$(T_nf)(x) := (1/2\pi)\int_{-\pi}^{\pi} f(x-u)t_n(u)du \qquad (f \in X_{2\pi}),$$

where $t_n \in \Pi_n$ (:= set of trigonometric polynomials of degree n). It follows that conditions (2.6-8) are always satisfied for the spaces and operators under consideration.

LEMMA 1. <u>For sequences of operators of type</u> (3.2) <u>there exist elements</u> $h_n \in X_{2\pi}^{(r)}$ <u>such that conditions</u> (2.6-8) <u>hold true with</u> $\varphi_n = n^{-r}$.

PROOF. By definition of an operator norm there exists $f_n \in X_{2\pi}$ with

(3.3) $$\|f_n\|_{X_{2\pi}} \leqslant 1, \quad \|T_n\|_{[X_{2\pi}]} \leqslant 2\|T_nf_n\|_{X_{2\pi}}.$$

Applying the delayed means of de La Vallée Poussin

(3.4) $$(V_{n,2n}f)(x) := \sum_{k=-2n}^{2n} r(|k|/n)f^\wedge(k)e^{ikx} \qquad (f \in X_{2\pi}),$$

$$f^\wedge(k) := (1/2\pi)\int_{-\pi}^{\pi} f(u)e^{-iku}du, \quad r(t) := \begin{cases} 1, & 0 \leqslant t \leqslant 1, \\ 2-t, & 1 \leqslant t \leqslant 2, \\ 0, & t \geqslant 2, \end{cases}$$

then already furnishes appropriate candidates via

(3.5) $$h_n := V_{n,2n}f_n.$$

Indeed, since the operators $V_{n,2n}$ are uniformly bounded in $[X_{2\pi}]$, one has (2.6), and since they are polynomial of degree 2n, also (2.7) with $\varphi_n = n^{-r}$ via the classical Bernstein inequality. Furthermore, $T_nf_n = T_nh_n$ which establishes (2.8), too.

In the present setting Dahmen - Görlich [5,6] proved the following result, in fact one step in their verification of a conjecture of M. Golomb concerning asymptotically optimal linear approximation processes:

PROPOSITION 1. <u>Let</u> $\{T_n\}_{n\in\mathbb{N}}$ <u>be given via</u> (3.2). <u>Then</u>

(3.6) $$\sup\{\|T_n f - f\|_C;\ \|f^{(r)}\|_C \leq 1\} \leq cn^{-r} \qquad (n \in \mathbb{N})$$

<u>implies the uniform bound</u>

$$\|T_n\|_{[C_{2\pi}]} = \|t_n\|_1 = \mathcal{O}(1) \qquad (n \to \infty).$$

A similar result concerning the Lipschitz spaces

$$H_\omega := \{f \in C_{2\pi};\ \omega_1(t,f;C_{2\pi}) \leq \omega(t),\ t \geq 0\}$$

was obtained by Baskakov [1] as a contribution to a problem of P.P. Korovkin.

PROPOSITION 2. <u>Let</u> ω, $\{\psi_n\}$ <u>satisfy</u> (2.2/21), <u>respectively</u>. <u>Then</u>

(3.7) $$\sup\{\|T_n f - f\|_C;\ f \in H_\omega\} \leq c\psi_n \omega(1/n) \qquad (n \in \mathbb{N})$$

<u>implies the uniform growth condition</u>

$$\|T_n\|_{[C_{2\pi}]} = \|t_n\|_1 = \mathcal{O}(\psi_n) \qquad (n \to \infty).$$

PROOFS of Prop. 1-2 (in $X_{2\pi}$). With h_n as given by (3.5) first note that for $g_{n,\omega} := c_o\omega(\varphi_n)h_n$, $c_o > 0$, one has by La.1 (cf. 2.2/5-7/12))

$$K(t^r, g_{n,\omega}; X_{2\pi}, X_{2\pi}^{(r)}) \leq c_o\omega(\varphi_n)\cdot \begin{cases} \|h_n\|_{X_{2\pi}} \leq c_o C_1 \omega(t), & t \geq \varphi_n, \\ t|h_n|_{X_{2\pi}^{(r)}} \leq 2c_o C_2 \omega(t), & t \leq \varphi_n. \end{cases}$$

In view of (3.1) this implies $g_{n,\omega} \in H_\omega$ for c_o sufficiently small. Thus Prop.2 follows since by (2.8/21) and (3.7)(with $r = 1$, $\varphi_n = n^{-1}$)

$$\begin{aligned} \|T_n\|_{X_{2\pi}} &\leq C_3\|T_n h_n\|_{X_{2\pi}} \leq C_3\{\|T_n h_n - h_n\|_{X_{2\pi}} + \|h_n\|_{X_{2\pi}}\} \\ &\leq C_3\{(c_o\omega(\varphi_n))^{-1}\|T_n g_{n,\omega} - g_{n,\omega}\|_{X_{2\pi}} + C_1\} = \mathcal{O}(\psi_n). \end{aligned}$$

Using $g_n := c_o\varphi_n h_n$, Prop.1 is established quite analogously.

Let us mention that Dahmen and Görlich [5,6] apparently were the first who pointed out that the uniform boundedness of a sequence of convolution operators can be concluded from the rate of approximation on a subspace. In fact, the original proof of Prop.1 in [6] has nearly the same structure as that given here: One uses a set which determines the operator norms in the sense of (2.8) and smoothes via delayed means so that the Bernstein inequality is applicable. The result of Prop.1 has been extended by Dahmen [4] to exponential rates of approximation, even including an asymptotic expansion of the (then necessarily not uniformly bounded) operator norms in terms of n (and $\omega(\varphi_n)$). Concerning the original proof of Prop.2, however, Baskakov [1] employed rather specific arguments, only available in $C_{2\pi}$.

On the other hand, Thm.1 now admits further extensions. To this end, note that the left-hand sides of (3.6/7) in fact represent operator norms. The following result shows that these assumptions in a uniform operator topology may indeed be replaced by corresponding (pointwise) strong ones.

COROLLARY 3. Let $\{T_n\}$ be given via (3.2), and let $\omega,\{\psi_n\}$ satisfy (2.2/3), (2.21), respectively. Then either of the conditions

$$\|T_nf-f\|_{X_{2\pi}} = \mathcal{O}_f(\psi_n/n^r) \qquad (f\in X_{2\pi}^{(r)},\ n\to\infty), \tag{3.8}$$

$$\|T_nf-f\|_{X_{2\pi}} = \mathcal{O}_f(\psi_n\omega(1/n^r)) \qquad (f\in (X_{2\pi})_{\omega,o},\ n\to\infty), \tag{3.9}$$

implies that the operator norms satisfy

$$\|T_n\|_{[X_{2\pi}]} = \mathcal{O}(\psi_n) \qquad (n\to\infty). \tag{3.10}$$

Indeed, (3.8) ⇒ (3.10) by Cor.2, and (3.9) ⇒ (3.10) by Cor.1, La.1. Note that assumption (3.9) is only needed on $(X_{2\pi})_{\omega,o}$ (instead of $(X_{2\pi})_\omega$) and that the result is stated for any $X_{2\pi}$-space.

3.2 Summation Processes of Fourier Expansions in Banach Spaces. With X^* the dual of X let $\{f_k,f_k^*\}_{k=0}^{\infty}\subset X\times X^*$ be a biorthogonal system on X. The system is assumed to be total, i.e., $f_k^*(f)=0$ for all k implies $f=0$, and to be regular, i.e., there exists some $\alpha\geqslant 0$ such that the (C,α)-Cesàro means

$$(C,\alpha)_nf := \sum_{k=0}^{n} (A_{n-k}^{\alpha}/A_n^{\alpha})f_k^*(f)f_k, \quad A_n^{\alpha} := \binom{n+\alpha}{n},$$

are uniformly bounded:

$$\|(C,\alpha)_nf\|_X \leqslant C_\alpha\|f\|_X \qquad (f\in X).$$

With $\mathbb{C}$ the set of complex numbers let

$$\Pi := \bigcup_{n=o}^{\infty} \Pi_n, \quad \Pi_n := \{f \in X;\ f = \sum_{k=o}^{n} a_k f_k,\ a_k \in \mathbb{C}\}$$

be the set of polynomials and of those of degree n, respectively.

A basic feature of regular systems is that the powerful tool of de La Vallée Poussin (or delayed) means is still available. Indeed, with an arbitrarily often differentiable function λ satisfying

$$0 \leqslant \lambda(t) \leqslant 1, \quad \lambda(t) = \begin{cases} 1, & 0 \leqslant t \leqslant 1, \\ 0, & t \geqslant 2, \end{cases}$$

(generalized) de La Vallée Poussin means are defined by

$$V_n f := \sum_{k=o}^{\infty} \lambda(k/n) f_k^*(f) f_k \qquad (f \in X). \tag{3.11}$$

As for one-dimensional trigonometric expansions one has

LEMMA 2. <u>Let</u> X <u>be</u> <u>a</u> <u>Banach</u> <u>space</u> <u>with</u> <u>total</u> <u>biorthogonal</u> <u>system</u> $\{f_k, f_k^*\} \subset X \times X^*$ <u>which</u> <u>is</u> <u>regular</u> (<u>for</u> <u>some</u> $\alpha \geqslant 0$). <u>Then</u> <u>the</u> <u>means</u> (3.11) <u>possess</u> <u>the</u> <u>properties</u>

(i) $V_n f \in \Pi_{2n}$ <u>for</u> <u>each</u> $f \in X$, (ii) $V_n p = p$ <u>for</u> <u>each</u> $p \in \Pi_n$,

(iii) $\|V_n\|_{[X]} \leqslant D_\alpha \int_0^2 t^{[[\alpha]]+1} |\lambda^{([[\alpha]]+2)}(t)| dt$,

<u>where</u> $[[\alpha]]$ <u>denotes</u> <u>the</u> <u>greatest</u> <u>integer</u> <u>less</u> <u>than</u> <u>or</u> <u>equal</u> <u>to</u> α.

Indeed, (i), (ii) are obvious, and (iii) follows by general multiplier criteria for regular systems (see [3,11] and the literature cited there).

LEMMA 3. <u>Let</u> $U \subset X$ <u>be</u> <u>a</u> <u>seminormed</u> <u>linear</u> <u>subspace</u> <u>for</u> <u>which</u> <u>one</u> <u>has</u> <u>the</u> <u>Bernstein-type</u> <u>inequality</u> ($\{\varphi_n\}$ satisfying (2.5))

$$|p_n|_U \leqslant (C/\varphi_n) \|p_n\|_X \qquad (p_n \in \Pi_n). \tag{3.12}$$

<u>Then</u> <u>for</u> <u>any</u> <u>linear</u> <u>polynomial</u> <u>summation</u> <u>process</u> <u>given</u> <u>via</u> ($a_{kn} \in \mathbb{C}$)

$$T_n f := \sum_{k=o}^{n} a_{kn} f_k^*(f) f_k \tag{3.13}$$

<u>there</u> <u>exist</u> <u>elements</u> $h_n \in U$ <u>so</u> <u>that</u> <u>conditions</u> (2.6-8) <u>hold</u> <u>true</u>.

Indeed, in view of La.2 one may proceed as for La.1. Concerning a treatment of Bernstein - type inequalities in the setting of regular systems in Banach spaces one may consult [3,9].

COROLLARY 4. Let X be a Banach space with regular total biorthogonal system $\{f_k, f_k^*\}$, and let a sequence of polynomial operators $\{T_n\} \subset [X]$ be given via (3.13). Let $U \subset X$ be a seminormed linear subspace with Bernstein - type inequality (3.12). If $\omega, \{\varphi_n\}, \{\psi_n\}$ are subject to (2.2/3/5/21), respectively, then

$$\|T_n f - f\|_X = \mathcal{O}_f(\psi_n \omega(\varphi_n)) \qquad (f \in X_{\omega,o})$$

necessarily implies the uniform growth condition

$$\|T_n\|_{[X]} = \mathcal{O}(\psi_n) \qquad (n \to \infty).$$

In view of La.3 the proof follows by Cor.1. Obviously, one may also formulate a counterpart for the limiting case $\omega_2(t) = t$, using Cor.2.

REFERENCES

[1] Baskakov, V.A., Über eine Hypothese von P.P. Korovkin. In: Linear Spaces and Approximation, Proc. Conf. Oberwolfach 1977 (P.L. Butzer - B.Sz.-Nagy, Eds.), ISNM 40, Birkhäuser, Basel 1978, 389-393.

[2] Butzer, P.L. - Berens, H., Semi - Groups of Operators and Approximation. Springer, Berlin 1967.

[3] Butzer, P.L. - Nessel, R.J. - Trebels, W., Multipliers with respect to spectral measures in Banach spaces and approximation, I: Radial multipliers in connection with Riesz - bounded spectral measures. J. Approximation Theory 8 (1973), 335-356.

[4] Dahmen, W., Trigonometric approximation with exponential error orders, I: Construction of asymptotically optimal processes, generalized de la Vallée Poussin sums; II: Properties of asymptotically optimal processes, impossibility of arbitrarily good error estimates. Math. Ann. 230 (1977), 57-74; J. Math. Anal. Appl. 68 (1979), 118-129.

[5] Dahmen, W. - Görlich, E., A conjecture of M. Golomb on optimal and nearly-optimal linear approximation. Bull. Amer. Math. Soc. 80 (1974), 1199-1202.

[6] Dahmen, W. - Görlich, E., Asymptotically optimal linear approximation processes and a conjecture of Golomb. In: Linear Operators and Approximation II, Proc. Conf. Oberwolfach 1974 (P.L. Butzer - B.Sz.-Nagy, Eds.), ISNM 25, Birkhäuser, Basel 1974, 327-335.

[7] Dickmeis, W. - Nessel, R.J., A unified approach to certain counterexamples in approximation theory in connection with a uniform boundedness principle with rates. J. Approximation Theory (in print).

[8] Dickmeis, W. - Nessel, R.J., On uniform boundedness principles and Banach-Steinhaus theorems with rates. (to appear).

[9] Görlich, E. - Nessel, R.J. - Trebels, W., Bernstein-type inequalities for families of multiplier operators in Banach spaces with Cesàro decompositions, I:General theory; II: Applications. Acta Sci. Math. (Szeged) 34 (1973), 121-130; 36 (1974), 39-48.

[10] Mertens, H.J. - Nessel, R.J., An equivalence theorem concerning multipliers of strong convergence. J. Approximation Theory (in print).

[11] Mertens, H.J. - Nessel, R.J. - Wilmes, G., Über Multiplikatoren zwischen verschiedenen Banach-Räumen im Zusammenhang mit diskreten Orthogonalentwicklungen. Forschungsberichte des Landes NRW 2599, Westdeutscher Verlag, Opladen 1976.

[12] Pochuev, V.R., On multpliers of uniform convergence and multipliers of uniform boundedness of partial sums of Fourier series (Russ.). Izv. Vyss. Ucebn. Zaved. Matematika 21 (1977), 74-81 = Soviet Math. 21 (1977), 60-66.

[13] Teljakovskii, S.A., Uniform convergence factors for Fourier series of functions with a given modulus of continuity (Russ.). Mat. Zametki 10 (1971), 33-40 = Math. Notes 10 (1971), 444-448.

SLOW APPROXIMATION WITH CONVOLUTION OPERATORS

P.C. Sikkema
Department of Mathematics
University of Technology
Delft, Netherlands

In this paper it is proved that the class B of functions β with which the convolution operators U_ρ are constructed contains elements such that if f belongs to a certain class M and if $f''(x)$ exists the rate of approximation of f(x) by $(U_\rho f)(x)$ $(\rho \to \infty)$ is very small.

1. Introduction

The class M consists of all real functions f(t), defined, bounded and Lebesgue measurable on the real axis R. The class B consists of all real functions $\beta(t)$ defined on R and possessing the following four properties:

1. $\beta(t) \geq 0$ on R,
2. $\beta(t)$ is continuous at $t = 0$ and $\beta(0) = 1$,
3. for all $\delta > 0$ is $\sup_{|t| \geq \delta} \beta(t) < 1$,
4. $\beta(t)$ belongs to the Lebesgue class L_1.

In [2] the author studied approximation properties of operators U_ρ of convolution type defined on the class M by

$$(U_\rho f)(x) = I_\rho^{-1} \int_{-\infty}^{\infty} f(x-t)\beta^\rho(t)dt \qquad (\rho \geq 1) \tag{1}$$

with

$$I_\rho = \int_{-\infty}^{\infty} \beta^\rho(t)dt \qquad (\rho \geq 1). \tag{2}$$

It was proved that the approximation property holds, i.e. that

$$(U_\rho f)(x) - f(x) \to 0 \qquad (\rho \to \infty)$$

for each $\beta \in B$ and each $f \in M$ at a point x where f is continuous.
Moreover, the speed with which the approximation takes place was studied under an additional assumption concerning the behaviour of $\beta(t)$ as $t \to 0$. In fact, it was assumed that there exist four positive constants α, α', c, c' such that

$$\beta(t) = 1 - ct^{\alpha} + o(t^{\alpha}) \qquad \text{if } t \downarrow 0,$$
$$\beta(t) = 1 - c'|t|^{\alpha'} + o(|t|^{\alpha'}) \qquad \text{if } t \uparrow 0.$$

It was shown e.g. that if $f''(x)$ exists and if $\alpha > \alpha'$, the order of approximation is $\rho^{1/\alpha}$; to be more precise

$$\rho^{1/\alpha}\{(U_\rho f)(x) - f(x)\} = -c^{-1/\alpha}\Gamma(2/\alpha)\,\{\Gamma(1/\alpha)\}^{-1}\; f'(x) + o(1) \qquad (\rho \to \infty).$$

In [3] the author considered a class of elements $\beta \in B$ which yield a very high order of approximation. As an example a β was given to which the order of approximation is $\exp(d_1\rho^{1/2} + d_2\rho^{1/4})$, d_1 and d_2 being constants.
In the present paper it is shown that if β is properly chosen in B, if $f \in M$ and if $f''(x)$ exists at a point x, the order of approximation is very low, viz. $\log_2\rho$. (For shortness log logρ is denoted by $\log_2\rho$, log log logρ by $\log_3\rho$ etc.). This result is formulated in Theorem 2.

More results on the convolution operators U_ρ and related literature can be found in [1], [2].

Whenever in this paper the order symbols O and o are used they are always related to ρ tending to infinity. Hence "$\rho \to \infty$" will be omitted almost everywhere.

2. Choice of $\beta(t)$

Let $\alpha > 0$ and

$$\delta = 2^{-1/\alpha}. \tag{3}$$

The function $\beta(t)$ is defined by

$$\beta(t) = \begin{cases} \exp(-t^2) & (t \le 0) \\ 1 - \exp(-\exp(t^{-\alpha})) & (0 < t \le \delta) \\ \exp(-t^2) & (t > \delta). \end{cases} \tag{4}$$

Clearly $\beta \in B$. Hence if $f \in M$ and f is continuous at x the approximation theorem holds at x.

If $f''(x)$ exists Taylor's expansion gives

$$f(x-t) - f(x) = -tf'(x) + \tfrac{1}{2}t^2 f''(x) + t^2\gamma_x(t),$$

where $\gamma_x(t)$ is bounded on R and by setting $\gamma_x(0) = 0$, $\gamma_x(t)$ is continuous at $t = 0$. Multiplication of both sides with $I_\rho^{-1}\beta^\rho(t)$, $\rho \geq 1$, $\beta(t)$ being given by (4) and I_ρ by (2), and integration from $-\infty$ to ∞ gives because of (1)

$$(U_\rho f)(x) - f(x) = I_\rho^{-1} \int_{-\infty}^{\infty} \{f(x-t) - f(x)\}\, \beta^\rho(t)dt$$

$$= I_\rho^{-1} \int_{-\delta}^{\delta} \{-tf'(x) + \tfrac{1}{2}t^2 f''(x) + t^2\gamma_x(t)\}\, \beta^\rho(t)dt$$

$$+ I_\rho^{-1} \int_{|t|\geq\delta} \{f(x-t) - f(x)\}\, \beta^\rho(t)dt$$

(5) $$= -f'(x)\, I_\rho^{-1}\, I_{1\rho}(\delta) + \tfrac{1}{2}f''(x)\, I_\rho^{-1}\, I_{2\rho}(\delta) + I_\rho^{-1}\, J_\rho(\delta) + I_\rho^{-1}\, K_\rho(\delta),$$

with

(6) $$I_{\nu\rho}(\delta) = \int_{-\delta}^{\delta} t^\nu \beta^\rho(t)dt$$ ($\nu = 1, 2$, later on also considered with $\nu = 0$)

(7) $$J_\rho(\delta) = \int_{-\delta}^{\delta} t^2\gamma_x(t)\, \beta^\rho(t)dt,$$

and

(8) $$K_\rho(\delta) = \int_{|t|\geq\delta} \{f(x-t) - f(x)\}\, \beta^\rho(t)dt.$$

The asymptotic behaviour of (6), (7) and (8) for $\rho \to \infty$ will be determined in section 4.

3. Investigation of a Function $\lambda(\rho,u)$

In section 4 the function

(9) $$\lambda(\rho,u) = \gamma^{-1}\log u - \rho\log(1 - \exp(-e^u)) \quad , \quad \gamma = \alpha/(\nu+\alpha+1)$$

$\nu = 1,2$, plays an important rôle. In the present section $\lambda(\rho,u)$ will be investigated.

THEOREM 1. *For all sufficiently large* ρ $\lambda(\rho,u)$ *possesses one minimum on the interval* $[2,\infty)$, *say at* $u = u_0$, *with* $(\rho \to \infty)$

$$(10) \qquad u_0 = \log_2\rho + \frac{\log_2\sigma}{\log\sigma} + \frac{\log_3\sigma}{\log\sigma} + O\left(\frac{\log_2^2\sigma}{\log^2\sigma}\right) \qquad (\sigma = \gamma\rho).$$

PROOF. Setting $\sigma = \gamma\rho$ it follows from (9) that

$$(11) \qquad \frac{d\lambda}{du} = (\gamma u)^{-1} - \rho e^u \{\exp(e^u - 1)\}^{-1}$$

$$= \{\exp(e^u) - 1 - \sigma u e^u\} \ (\gamma u)^{-1} \ \{\exp(e^u) - 1\}^{-1}.$$

Then $\frac{d\lambda}{du} = 0$ leads to the equation

$$(12) \qquad \exp(e^u) - 1 = \sigma u e^u.$$

If (12) has a solution that tends to infinity if $\rho \to \infty$ a first approximation to this solution has to satisfy the equation $\exp(e^u) = \sigma u e^u$ and hence the equation

$$(13) \qquad e^u = \log\sigma + \log u + u.$$

Consequently the first approximation is $u = \log_2\sigma$. In order to obtain a second approximation the substitution $u = \log_2\sigma + \phi(u)$ $(\phi(u) = o(\log_2\sigma))$ in (13) is made. It gives

$$\log\sigma \exp(\phi(\rho)) = \log\sigma + \log(\log_2\sigma + \phi(\rho)) + \log_2\sigma + \phi(\rho)$$

and thus

$$\exp(\phi(\rho)) = 1 + \frac{1}{\log\sigma}\left\{ \log_2\sigma + \log_3\sigma + \log\left(1 + \frac{\phi(\rho)}{\log_2\sigma}\right) + \phi(\rho)\right\}$$

from which it follows that

$$\phi(\rho) = \frac{\log_2\sigma}{\log\sigma} + \frac{\log_3\sigma}{\log\sigma} + O\left(\frac{\log_2^2\sigma}{\log^2\sigma}\right).$$

Hence a second approximation to a solution of (12) tending to infinity if $\rho \to \infty$ is

$$(14) \qquad u = u_o = \log_2\sigma + \frac{\log_2\sigma}{\log\sigma} + \frac{\log_3\sigma}{\log\sigma} + O\left(\frac{\log_2^2\sigma}{\log^2\sigma}\right).$$

Calcultation of $\frac{d^2\lambda}{du^2}$ at $u = u_o$ yields $\frac{d^2\lambda}{du_0^2} = \log\sigma.(\gamma\log_2\sigma)^{-1}\cdot(1 + o(1))$.

Thus $\lambda(\rho,u)$ has a minimum at $u = u_o$.

If (12) has a solution that remains bounded if $\rho \to \infty$ (12) shows that it tends to zero if $\rho \to \infty$. Therefore (12) is written as

$$\exp(1 + u + O(u^2)) - 1 = \sigma(u + u^2 + O(u^3))$$

which leads to

$$\sigma^{-1}e(1 + u + O(u^2)) - \sigma^{-1} = u + u^2 + O(u^3).$$

From this equation it follows that

$$u = u_o^* = (e-1)(\sigma^{-1} + \sigma^{-2}) + o(\sigma^{-2}).$$

Recalling that $\sigma = \gamma\rho$ it appears that $\lambda(\rho,u)$ has a maximum at $u = u_o^*$ and also that u_o satisfies the inequality $u_o^* < 2$ for all sufficiently large values of ρ. Consequently, for all sufficiently large values of ρ $\lambda(\rho,u)$ has **one (interior) extremum** on $[2,\infty)$, it lies at $u = u_o$ given by (14) and it is a minimum. This proves the theorem.

LEMMA 1. <u>If</u> $\rho \to \infty$ <u>the following two relations hold with the notation</u> $\sigma = \gamma\rho$ ((10)):

$$(16) \qquad (1 - \exp(-e^{u_o}))^\rho = 1 - (\gamma\log\sigma\,\log_2\sigma)^{-1}(1 + o(1)),$$

$$(17) \qquad \exp(-\lambda(\rho,u_o)) = (\log_2\sigma)^{-1/\gamma}(1 + o(1)).$$

PROOF. From (14) it follows that if $\rho \to \infty$

$$e^{u_o} = \log\sigma\,\exp\left(\frac{\log_2\sigma}{\log\sigma}\right)\exp\left(\frac{\log_3\sigma}{\log\sigma}\right)\exp\left\{O\left(\frac{\log_2^2\sigma}{\log^2\sigma}\right)\right\}$$

$$= \log\sigma\left\{1 + \frac{\log_2\sigma}{\log\sigma} + O\left(\frac{\log_2^2\sigma}{\log^2\sigma}\right)\right\}\left\{1 + \frac{\log_3\sigma}{\log\sigma} + O\left(\frac{\log_3^2\sigma}{\log^2\sigma}\right)\right\}.$$

$$(18)\qquad \cdot \left\{1 + O\left(\frac{\log_2^2\sigma}{\log^2\sigma}\right)\right\} = \log\sigma + \log_2\sigma + \log_3\sigma + O\left(\frac{\log_2^2\sigma}{\log\sigma}\right).$$

Consequently,

$$(19)\qquad \exp(e^{u_o}) = \sigma\log\sigma\,\log_2\sigma\left(1 + O\left(\frac{\log_2^2\sigma}{\log\sigma}\right)\right).$$

This leads to

$$(1 - \exp(-e^{u_o}))^{\rho} = \exp\{\rho\log(1 - \exp(-e^{u_o}))\}$$

$$= \exp\left\{\rho\log\left\{1 - (\sigma\log\sigma\log_2\sigma)^{-1}\left(1 + O\left(\frac{\log_2^2\sigma}{\log\sigma}\right)\right)\right\}\right\}$$

$$= \exp\left\{-(\gamma\log\sigma\log_2\sigma)^{-1}\left(1 + O\left(\frac{\log_2^2\sigma}{\log\sigma}\right)\right)\right\}$$

$$(20)\qquad = 1 - (\gamma\log\sigma\log_2\sigma)^{-1}(1 + O(1))$$

which proves (16).

Again, substitution of (14) and (20) in (9) with $u = u_o$ yields

$$\lambda(\rho,u_o) = \gamma^{-1}\log_3\sigma(1 + O(\log^{-1}\sigma)) + (\gamma\log\sigma\log_2\sigma)^{-1}(1 + o(1))$$

$$= \gamma^{-1}\log_3\sigma\,(1 + O(\log^{-1}\sigma)).$$

Thus (17) is established.

LEMMA 2. <u>For all sufficiently large values of ρ is</u>

$$(21)\qquad \exp(-\lambda(\rho,u_o - \log_3\sigma)) < (\log_2\sigma)^{-1/\gamma}\exp(-\sigma^{1/2}).$$

PROOF. Since by (14)

$$u_o - \log_3\sigma = \log_2\sigma - \log_3\sigma + \frac{\log_2\sigma}{\log\sigma} + \frac{\log_3\sigma}{\log\sigma} + O\left(\frac{\log_2^2\sigma}{\log\sigma}\right)$$

we have in view of (18)

$$\exp(u_o - \log_3\sigma) = \log_2^{-1}\sigma\exp u_o = \frac{\log\sigma}{\log_2\sigma} + 1 + \frac{\log_3\sigma}{\log_2\sigma} + O\left(\frac{\log_2\sigma}{\log\sigma}\right)$$

and thus

$$\exp(-\exp(u_0 - \log_3\sigma)) = \left\{e \exp\left(\frac{\log\sigma}{\log_2\sigma}\right)(1 + o(1)\right\} = e^{-1}\,\sigma^{-1/\log_2\sigma}\,(1 + o(1)).$$

Consequently, by (9)

$$\lambda(\rho, u_0 - \log_3\sigma) = \gamma^{-1} \log(u_0 - \log_3\sigma) - \rho\log\{1 - \exp(-\exp(u_0 - \log_3\sigma))\}$$

$$= \gamma^{-1} \log_3\sigma\,(1 + o(1)) - \rho\log\{1 - e^{-1}\,\sigma^{-1/\log_2\sigma}(1+ o(1))\}$$

$$= \gamma^{-1} \log_3\sigma\,(1 + o(1)) + (e\gamma)^{-1}\,\sigma^{1-(1/\log_2\sigma)}\,(1+ o(1)).$$

From this it follows that

$$\exp(-\lambda(\rho, u_0 - \log_3\sigma)) = (\log_2\sigma)^{-1/\gamma}\exp\{-(e\gamma)^{-1}\,\sigma^{1-(1/\log_2\sigma)}\}\,(1 + o(1))$$

which proves the lemma.

4. Asymptotic Behaviour of $I_{\nu\rho}(\delta)$, $J_\rho(\delta)$, $K_\rho(\delta)$

Setting for $\nu = 0, 1, 2$

$$(22) \qquad A_{\nu\rho}(\delta) = \int_0^\delta t^\nu \beta^\rho(t)dt \quad , \qquad B_{\nu\rho}(\delta) = \int_0^\delta t^\nu \beta^\rho(-t)dt \quad ,$$

it follows from (6) that

$$(23) \qquad I_{\nu\rho}(\delta) = A_{\nu\rho}(\delta) + (-1)^\nu\, B_{\nu\rho}(\delta) \qquad\qquad (\nu = 0, 1, 2).$$

In order to derive the asymptotic behaviour for $\rho \to \infty$ of $A_{\nu\rho}(\delta)$ the substitution $t^{-\alpha} = u$ in the first part of (22) is made. In view of (22), (4) and (3) this gives

$$A_{\nu\rho}(\delta) = \int_0^\delta t^\nu \exp(\rho\,\log(1 - \exp(-\exp t^\alpha)))dt$$

$$(24) \qquad = \alpha^{-1} \int_2^\infty u^{-(\nu+\alpha+1)/\alpha}\,\exp(\rho\,\log(1 - \exp(-e^u)))du\,.$$

Setting

$$(25) \qquad \lambda(\rho,u) = \gamma^{-1}\log u - \rho\log(1 - \exp(-e^u)) \quad , \quad \gamma = \alpha/(\nu+\alpha+1)$$

it follows from (24) that

$$(26)\qquad \alpha A_{\nu\rho}(\delta) = \int_2^\infty \exp(-\lambda(\rho,u))du.$$

$\lambda(\rho,u)$ was investigated in section 3.

It is supposed from now on that ρ is so large that Theorem 1 holds. That means that $\lambda(\rho,u)$ has one minimum on $[2,\infty)$, at $u = u_o$. Then the integral at the right-hand side of (26) is written as the sum of three integrals:

$$(27)\quad \alpha A_{\nu\rho}(\delta) = \left(\int_2^{u_o-\log_3\sigma} + \int_{u_o-\log_3\sigma}^{u_o} + \int_{u_o}^{\infty} \right) \exp(-\lambda(\rho,u))du = I_1 + I_2 + I_3.$$

The behaviour of I_3 if $\rho \to \infty$ will be investigated first. The function $\log(1 - \exp(-e^u))$ is monotonically increasing on the interval $[u_o,\infty)$. Hence using the definition (9) of $\lambda(\rho,u)$

$$(28)\quad \exp(\rho \log(1 - \exp(e^{-u_o}))) \int_{u_o}^{\infty} u^{-1/\gamma}du \le I_3 \le \int_{u_o}^{\infty} u^{-1/\gamma}du \quad .$$

By **Lemma** 1, (16), the factor before the integral in the left-hand side of (28) tends to 1. Therefore and because $0 < \gamma < 1$

$$I_3 = \int_{u_o}^{\infty} u^{-1/\gamma}du\ (1 + o(1)) = \gamma/(1-\gamma)\ u_o^{1-(1/\gamma)}\ (1 + o(1)).$$

From this, from the definition of γ in (9) and from (14) it follows that

$$(29)\qquad I_3 = \alpha/(\nu + 1)(\log_2\sigma)^{-(\nu + 1)/\alpha}(1 + o(1)).$$

Secondly, the asymptotic behaviour of I_1 and I_2 if $\rho \to \infty$ will be derived. $\lambda(\rho,u)$ is monotonically decreasing on $[2,u_o]$ and thus $\exp(-\lambda(\rho,u))$ is monotonically increasing there. Hence

$$I_1 \le (u_o - \log_3\sigma - 2)\exp(-\lambda(\rho,u_o - \log_3\sigma)) < u_o \exp(-\lambda(\rho,u_o - \log_3\sigma))$$

from which it follows by (14) and lemma 2, (21), that if ρ is large enough,

$$I_1 < (\log_2\sigma)^{1-(1/\gamma)} \exp(-\sigma^{1/2}) = (\log_2\sigma)^{-(\nu+1)/\alpha}\exp(-\sigma^{1/2}).$$

According to (29) this means that

(30) $$I_1 = o(I_3).$$

Again,

$$I_2 \le (\log_3\sigma)\exp(-\lambda(\rho,u_0))$$

and by Lemma 1,(17), this leads to

(31) $$I_2 \le \log_3\sigma\,(\log_2\sigma)^{-1/\gamma}(1 + o(1)) = o(I_3).$$

Combining (27), (29), (30) and (31) and using the definitions of σ in (10) and γ in (9) $A_{\nu\rho}(\delta)$ is getting the form

(32) $$A_{\nu\rho}(\delta) = (\nu + 1)^{-1}\{\log_2(\alpha\rho/(\nu+\alpha+1))\}^{-(\nu+1)/\alpha}(1 + o(1)).$$

Returning to (23) it follows from the definition of $B_{\nu\rho}(\delta)$ in (22) and from (4) that

$$B_{\nu\rho}(\delta) = \int_0^\infty t^\nu\exp(-\rho t^2)dt = \int_0^\infty t^\nu\exp(-\rho t^2)dt(1 + o(1))$$
$$= O(\rho^{-(\nu+1)/2}) = o(A_{\nu\rho}(\delta))$$

by (32). Consequently,

(33) $$I_{\nu\rho}(\delta) = A_{\nu\rho}(\delta)\ (1 + o(1)).$$

Further, (2), (4) and (33) yield

(34) $$I_\rho = I_{o\rho}(\delta)\ (1 + o(1)) = A_{o\rho}(\delta)\ (1 + o(1)).$$

Combination of (33), (32) and (34) gives the asymptotic relations

(35) $$I_\rho^{-1}\ I_{1\rho}(\delta) = (A_{o\rho}(\delta))^{-1} A_{1\rho}(\delta)\ (1 + o(1)) = 2^{-1}(\log_2\rho)^{-1/\alpha}(1 + o(1))$$

and

$$(36)\qquad I_\rho^{-1}\, I_{2\rho}(\delta) = (A_{o\rho}(\delta))^{-1}\, A_{2\rho}(\delta)\ (1 + o(1)) = 3^{-1}\ (\log_2\rho)^{-2/\alpha}(1+ o(1)).$$

Again, concerning (7) it is obvious from the definition of $\beta(t)$ in (4) that

$$J_\rho(\delta) = \int_0^\delta t^2\gamma_x(t)\ \beta^\rho(t)dt\ (1 + o(1)).$$

As was noticed in section 2 $\gamma_x(0)$ is put equal to zero making $\gamma_x(t)$ continuous at $t = 0$. Hence to each $\varepsilon > 0$ there exists an η, $0 < \eta < \delta$, such that $|\gamma_x(t)| < \varepsilon$ if $0 \le t \le \eta$. Moreover $\gamma_x(t)$ is bounded on $[0,\delta]$ which means that there exists a constant $M > 0$ such that $|\gamma_x(t)| \le M$ on $[\eta,\delta]$.
Then

$$|J_\rho(\delta)| \le \left(\int_0^\eta + \int_\eta^\delta\right) t^2\ |\gamma_x(t)|\ \beta^\rho(t)dt < \varepsilon \int_0^\eta t^2\ \beta^\rho(t)dt + M\int_\eta^\delta t^2\ \beta^\rho(t)dt$$

$$< \varepsilon\ I_{2\rho}(\delta) + M\delta^3\beta^\rho(\eta)\ = \varepsilon\ I_{2\rho}(\eta) + M\delta^3\ \{1 - \exp(-\exp(\eta^{-\alpha}))\}^\rho$$

$$= \varepsilon\ I_{2\rho}(\delta) + o(I_{2\rho}(\delta))\quad \text{because of (33) and (32).}$$

Since $\varepsilon > 0$ is arbitrary it follows that $J_\rho(\delta) = o(I_{2\rho}(\delta))$ and hence

$$(37)\qquad I_\rho^{-1}\ J_\rho(\delta) = o(I_\rho^{-1}\ I_{2\rho}(\delta)).$$

Finally, concerning (8) it is noticed that f is bounded on R. Hence there exists a constant $P > 0$ such that $|f| \le P$ on R. In view of (4) this means that

$$|K_\rho(\delta)| \le 2P\int_{|t|\ge\delta} \beta^\rho(t)dt\ = 4P\int_\delta^\infty e^{-\rho t^2}dt$$

$$(38)\qquad < 4P\exp(-2^{-1}\ \rho\delta^2)\int_0^\infty e^{-\frac{1}{2}\rho t^2}dt = o(I_{2\rho}(\delta))$$

(because of (33) and (32)). Combination of (5), (35), (36), (37) and (38) leads to

$$(U_\rho f)(x) - f(x) = -2^{-1}(\log_2\rho)^{-1/\alpha}(1 + o(1))f'(x) + o(\log_2\rho)^{-1/\alpha}\ .$$

This proves the following theorem:

THEOREM 2. <u>If</u> $\beta(t)$ <u>is given by</u> (4), <u>if</u> $f(t)$ <u>is defined, bounded and Lebesgue measurable on</u> R, <u>if</u> $f''(x)$ <u>exists at a point</u> $x\in R$ <u>and if</u> $f'(x) \neq 0$ <u>then the operators</u> U_ρ $(\rho \geq 1)$ <u>defined by</u> (1), (2) <u>have the property that</u>

$$\lim_{\rho\to\infty} (\log_2\rho)^{1/\alpha}\{(U_\rho f)(x) - f(x)\} = -2^{-1}f'(x).$$

Thus under the conditions of Theorem 2 the order of approximation at x is $(\log\log\rho)^{1/\alpha}$.

REMARKS

1. If p is a positive integer, $p \geq 2$, if $f\in M$, $f^{(p)}(x)$ exists at a point $x\in R$ and $f'(x) = \ldots = f^{(p-1)}(x) = 0$, $f^{(p)}(x) \neq 0$ then for the operators U_ρ $(\rho \geq 1)$ used in theorem 2 it can be proved that

$$\lim_{\rho\to\infty} (\log_2\rho)^{1/\alpha}\{(U_\rho f)(x) - f(x)\} = (-1)^p(p+1)!^{-1} f^{(p)}(x).$$

Hence the order of approximation at x is then $(\log\log\rho)^{p/\alpha}$.

2. $\beta(t)$ as defined by (4) is not the only element in B for which **Theorem** 2 holds. In fact, in order to maintain the result of Theorem 2 $\beta(t)$ as defined by (4) can be altered on $(-\infty,0)$ and on (δ,∞) in such a way that properties 1.-4. are retained and $1-\beta(t)$ increases more quickly if t leaves the origin towards the left than if it leaves the origin towards the right; $\beta(t)$ as defined on $(0,\delta]$ by (4) remains unchanged.

3. A still more slow approximation than is given by theorem 2 can be achieved by altering $\beta(t)$ in its definition (4) on the interval $0 < t \leq \delta$ in the following way: on $0 < t \leq \delta$ $\beta(t)$ is defined by

$$(39) \qquad \beta(t) = 1 -\exp(-\phi(t)) \ , \qquad \phi(t) = \exp(\exp(\ldots(\exp t^{-\alpha})\ldots)),$$

where $\phi(t)$ contains exp $(n-1)$ times, n being a positive integer $(n \geq 2)$. On $(-\infty,0]$ and on (δ,∞) $\beta(t)$ remains as it is given by (4). Then the order of approximation which results from this new function $\beta(t)$ is equal to $(\log_n\rho)^{1/\alpha}$. For sake of completeness it should be added here that if in (39) $\phi(t)$ contains no exp, so $\phi(t) = t^{-\alpha}$, then the order of approximation is $(\log\rho)^{1/\alpha}$, which yields a faster approximation than Theorem 2 gives.

REFERENCES

[1] Sikkema, P.C., Estimations involving a modulus of continuity for a generalization of Korovkin's operators. Linear Spaces and Approximation, eds. P.L. Butzer and B.Sz.-Nagy. (ISNM, vol. 40) Birkhäuser Verlag, Basel/Stuttgart 1978, 289-303.

[2] Sikkema, P.C., Approximation formulae of Voronovskaya - type for certain convolution operators. J. Approximation Theory 26 (1979), 26-45.

[3] Sikkema, P.C., Voronovskaya type formulae for convolution operators approximating with great speed. Approximation Theory III. Proc. Conf. on Approximation Theory, Austin (Tex.) U.S.A., 8-12 Jan. 1980.

A REMARK ON ASYMPTOTICALLY OPTIMAL APPROXIMATION BY FABER SERIES

E. Görlich
Lehrstuhl A für Mathematik
Rheinisch-Westfälische Technische Hochschule
Aachen

In the real domain, the generalized de La Vallée Poussin means of Fourier series yield asymptotically optimal approximation. The purpose of this remark is to transfer this result to Faber expansions on a complex domain.

Let G be a **Jordan** domain in $\mathbb{C}$ with rectifiable boundary Γ, Ψ the mapping of $\{w;\ |w|>1\}$ onto the exterior of G such that $\Psi'(\infty)>0$, and $\{F_k(z)\}_{k=0}^{\infty}$ the associated sequence of Faber polynomials. Denoting by $A(\overline{G})$ the space of continuous functions on $\overline{G}$ which are regular in G with maximum norm, the Faber coefficients of $f\in A(\overline{G})$ are defined by

$$(1)\qquad a_k(f)=\frac{1}{2\pi i}\int_{|w|=1} f(\Psi(w))\, w^{-k-1}\,dw \qquad (k\in\mathbb{P}=\{0,1,2,\dots\}).$$

The de La Vallée Poussin means (delayed means) of the Faber series and their rate of approximation have been considered by Kövari [5]; see also Gaier [4, p. 56] and the literature cited there. The purpose of this remark is to indicate one (certainly not the best possible) way to extend these results to generalized de **La** Vallée Poussin means. This will cover exponential rates of approximation as treated by Dahmen [2], [3] in the real domain. Concerning the required degree of smoothness of Γ we content ourselves here with $\Gamma\in C(4,\varepsilon)$ for some $\varepsilon>0$. For $r\in\mathbb{P}$, $\varepsilon>0$, $\Gamma\in C(r,\varepsilon)$ means that the representation $z=z(s)$ of Γ via are length s has an r th derivative in Lip ε.

Given an increasing sequence $\{m(n)\}_{n\in\mathbb{P}}$ of integers with $0\leqslant m(n)<n$ $\forall\, n\in\mathbb{P}$, the generalized de la Vallée Poussin means $V_{n,m(n)}(f;z)$ of $f\in A(\overline{G})$ are defined by

$$V_{n,m(n)}(f;z)=\sum_{k=0}^{n} v_{n,m}(k)\,a_k(f)\,F_k(z) \qquad (z\in\overline{G}),$$

where

$$v_{n,m}(k) = \begin{cases} 1 & ; \quad 0 \leqslant k \leqslant n - m(n), \\ \dfrac{n-k+1}{m(n)+1} & ; \quad n - m(n) < k \leqslant n, \\ 0 & ; \quad k > n. \end{cases}$$

If $m(n) = [n/2]$ or $m(n) = 0$, where $[n/2]$ denotes the integral part of $n/2$, the $V_{n,m(n)}$ reduce to the delayed means or to the partial sum operators, respectively. Concerning their operator norms on $A(\overline{G})$ one has the following

PROPOSITION 1. <u>Let</u> $\Gamma \in C(4,\varepsilon)$ <u>for some</u> $\varepsilon > 0$. <u>There is a constant</u> M <u>such that</u>

$$\|V_{n,m(n)}\|_{[A(\overline{G})]} = M \log \frac{n}{m(n)} + \mathcal{O}(1), \qquad n \to \infty.$$

PROOF. The Faber polynomials are related to the trigonometric functions via

$$F_k(\Psi(e^{i\theta})) = e^{ik\theta} + \sum_{j=1}^{\infty} \alpha_{kj}\, e^{-ij\theta}, \tag{2}$$

where the coefficients satisfy the Grunsky law of symmetry

$$\alpha_{jk}/j = \alpha_{kj}/k. \tag{3}$$

Under the hypothesis $\Gamma \in C(r+2,\varepsilon)$, $r \in \mathbb{P}$, $0 < \alpha < 1$, they satisfy the following estimates

$$\Big|\sum_{j=1}^{\infty} \alpha_{kj}\, e^{-ij\theta}\Big| \leqslant M\, k^{-r-\alpha}, \qquad k \in \mathbb{P}, \tag{4}$$

$$|\alpha_{k,j}| \leqslant M\, k^{-r-\alpha}, \tag{5}$$

$$|\alpha_{k,j}| \leqslant M\, k\, j^{-r-1-\alpha}, \tag{6}$$

uniformly in k and j, for some constant M. Indeed, (4) follows by combining (2) with Suetin's [6; p. 128] result

$$F_k(\Psi(w)) = w^k + \mathcal{O}(k^{-r-\alpha}), \qquad k \to \infty \tag{7}$$

for $|w| \geqslant 1$, and (4) implies (5). Inequality (6) follows by (3) and (5).

Let $f \in A(\overline{G})$. Using (5), (6) with $r = 2$ and setting $\gamma = (4+\varepsilon)/(6+2\varepsilon)$ one has

$$(8)\quad \begin{aligned} \Big|\sum_{j=1}^{\infty}\{\sum_{k=1}^{\infty} a_k(f)\,\alpha_{kj}\}e^{-ij\theta}\Big| &\leqslant \|f\|_{A(\overline{G})}\sum_{j=1}^{\infty}\sum_{k=1}^{\infty}|\alpha_{kj}|^{\gamma}|\alpha_{kj}|^{1-\gamma} \\ &\leqslant M^2\,\|f\|_{A(\overline{G})}\sum_{j=1}^{\infty} j^{-1-\varepsilon/2}\sum_{k=1}^{\infty} k^{-1-\varepsilon/2} < \infty. \end{aligned}$$

Thus the function

$$(9)\quad K(\theta) = f(\Psi(e^{i\theta})) - \sum_{j=1}^{\infty}\{\sum_{k=1}^{\infty} a_k(f)\,\alpha_{kj}\}\,e^{-ij\theta}$$

belongs to $C_{2\pi}$, the space of continuous, 2π periodic functions on $\mathbb{R}$, and has Fourier coefficients

$$(10)\quad K^{\wedge}(k) = \frac{1}{2\pi}\int_{-\pi}^{\pi} K(\theta)\,e^{-ik\theta}\,d\theta = \begin{cases} a_k(f)\,; & k\in\mathbb{P},\\ 0\,; & -k\in\mathbb{N},\end{cases}$$

and, as in Curtiss [1, p. 592/593] it follows that

$$(11)\quad \begin{aligned} \|V_{n,m(n)}\,f\|_{A(\overline{G})} &\leqslant \|\sum_{k=0}^{n} v_{n,m}(k)\,K^{\wedge}(k)\,e^{ik\theta}\|_{C_{2\pi}} + \\ &+ \|\sum_{k=0}^{n} v_{n,m}(k)\,a_k(f)\sum_{j=1}^{\infty}\alpha_{kj}\,e^{-ij\theta}\|_{C_{2\pi}} = I_1 + I_2, \end{aligned}$$

say. Here $\|\cdot\|_{C_{2\pi}}$ is the maximum norm with respect to θ. From the corresponding real variable result (cf. [3], Remark 4.1) one has $I_1 \leqslant \|K\|_{C_{2\pi}}\{\frac{4}{\pi^2}\log\frac{n}{m(n)} + \mathcal{O}(1)\}$, $n\to\infty$, where $\|K\|_{C_{2\pi}} \leqslant M\,\|f\|_{A(\overline{G})}$, by (8),(9). By (4), it follows that $I_2 \leqslant M\,\|f\|_{A(\overline{G})}$ for some constant M, which proves the upper estimate of $\|V_{n,m(n)}\|_{[A(\overline{G})]}$.

To show that this is best possible with respect to the rate of increase it suffices to consider $V_{n,m(n)}(f_n;z)$, where f_n are polynomials defined by

$$a_k(f_n) = \begin{cases} 0\,; & k=0,n+1,\quad k\geqslant 2n+2,\\ (n+1-k)^{-1}\,; & 1\leqslant k\leqslant n,\quad n+2\leqslant k\leqslant 2n+1,\end{cases}$$

which satisfy $\|f_n\|_{A(\overline{G})} = \mathcal{O}(1)$, $n\to\infty$. The lower estimate then follows as above.

We want to study the rate of approximation furnished by the generalized de La Vallée Poussin means, in comparison with the rate of best approximation. It is customary to express this in terms of asymptotically optimal approximation. Let $E_n[f] = \inf_{p_n \in P_n} \|f - p_n\|_{A(\overline{G})}$ denote the error of best approximation to $f \in A(\overline{G})$ by polynomials p_n of degree $\leqslant n$. Given a subset $W \subset A(\overline{G})$ with the property that $\sup_{f \in W} E_n[f]$ exists for each $n \in \mathbb{P}$, a sequence of bounded linear operators U_n from $A(\overline{G})$ into P_n is said to yield *asymptotically optimal approximation on* W if there is a constant M such that

$$\sup_{f \in W} \|f - U_n f\|_{A(\overline{G})} \leqslant M \sup_{f \in W} E_n[f] \qquad (n \in \mathbb{P}).$$

It is known, for example, that the delayed means $V_{n,[n/2]}$ yield asymptotically optimal approximation on $W = \{f \in A(\overline{G});\ f(\Psi(e^{i\theta})) \in \mathrm{Lip}^* \alpha\}$, $0 < \alpha < 2$, if Γ is of bounded rotation, see Kövari [5], cf. Gaier [4], p. 55/56. The following proposition contains an extension of this result to certain general classes $W = B_\varphi$, under the more restrictive hypothesis $\Gamma \in C(4,\varepsilon)$, $\varepsilon > 0$.

In order to define the classes B_φ we suppose that φ is an element of the following set Ω of "orders of approximation". Let $\mathbb{R}^+ = [0,\infty)$ and

$$\Omega_0 = \{\varphi(x);\ \varphi: \mathbb{R}^+ \to \mathbb{R}^+,\ \varphi(0) = 1,\ \varphi'(x)\ \exists,\ \text{continuous and} > 0 \text{ on } (0,\infty),$$
$$\lim_{x \to \infty} \varphi(x) = +\infty\},$$

$$\Omega = \{\varphi \in \Omega_0;\ \varphi(x) = e^{g(x)},\ g'''(x)\ \exists,\ \text{continuous on } (0,\infty),\ \exists\, x_0 > 0 \text{ with}$$
$$g''(x) \leqslant 0,\ g'''(x) \geqslant 0\ \forall\, x > x_0,\ \limsup_{x \to \infty} \frac{|g''(x)|}{(g'(x))^2} < 1\}.$$

Cf. e.g. [2] for a discussion of these assumptions. For each $\varphi \in \Omega$ we set

$$B_\varphi = \{f \in A(\overline{G});\ \|f\|_{A(\overline{G})} \leqslant 1,\ E_n[f] \leqslant \frac{1}{\varphi(n)}\ \forall\, n \in \mathbb{P}\}.$$

We can now state

PROPOSITION 2. <u>Let</u> $\Gamma \in C(4,\varepsilon)$ <u>for some</u> $\varepsilon > 0$. <u>Given</u> $\varphi \in \Omega$, <u>the generalized de La Vallée Poussin means</u> $V_{n,m(n)}$ <u>with</u> $m(n) = [1/(\log \varphi)'(n)]$ <u>yield asymptotically optimal approximation on</u> B_φ.

PROOF. As in (11) one has, with an obvious notation,

$$\|f - V_{n,m(n)}\, f\|_{A(\overline{G})} \leq \|K(\theta) - V_{n,m(n)}\,(K;\theta)\|_{C_{2\pi}} +$$

$$+\Big\| \sum_{k=1}^{n} v_{n,m}(k)\, a_k(f) \sum_{j=1}^{\infty} \alpha_{kj}\, e^{-ij\theta} - \sum_{k=1}^{\infty} a_k(f) \sum_{j=1}^{\infty} \alpha_{kj}\, e^{-ij\theta}\Big\|_{C_{2\pi}}$$

$$= I_3 + I_4,$$

say. If it can be shown that the trigonometric best approximation of order n to $K(\theta)$ in $C_{2\pi}$ behaves like $\mathcal{O}(1/\varphi(n))$ as $n \to \infty$, it follows by [3], Thm. 5.2, that $I_3 = \mathcal{O}(1/\varphi(n))$, $n \to \infty$. To this end $K(\theta)$ will be approximated by trigonometric polynomials $t_n(\theta)$ defined as follows. Let $p_n^*(z)$ denote the polynomial of best approximation to f on $\overline{G}$ and let

$$(12) \qquad \lambda_j = \sum_{k=1}^{\infty} a_k(f)\, \alpha_{kj}, \qquad \lambda_j' = \sum_{k=1}^{\infty} a_k(p_n^*)\, \alpha_{kj}$$

with $a_k(p_n^*) = 0$ for $k > n$. The first series in (12) converges since

$$(13) \qquad |a_k(f)| \leq E_{k-1}[f]$$

and $\sum_{k=1}^{\infty} 1/\varphi(k) < \infty$ for each $\varphi \in \Omega$, cf. [2], La. 2.3. Setting

$$t_n(\theta) = \sum_{j=0}^{n} a_j(p_n^*)\, e^{ij\theta} + \sum_{j=-n}^{-1} (\lambda'_{-j} - \lambda_{-j})\, e^{ij\theta}$$

it follows in view of (8), (9) and the hypothesis that

$$\|K(\theta) - t_n(\theta)\|_{C_{2\pi}} \leq E_n[f] + \Big\| \sum_{j=n+1}^{\infty} (\lambda_j' - \lambda_j)\, e^{-ij\theta}\Big\|_{C_{2\pi}}$$

$$\leq E_n[f]\, \{1 + \sum_{j=n+1}^{\infty} \sum_{k=1}^{\infty} |\alpha_{kj}|\} = \mathcal{O}(1/\varphi(n)), \qquad n \to \infty.$$

Rewriting the sum I_4 as

$$I_4 = \Big\| \sum_{k=1}^{n} (v_{n,m}(k) - 1)\, a_k(f) \sum_{j=1}^{\infty} \alpha_{kj}\, e^{-ij\theta} - \sum_{k=n+1}^{\infty} a_k(f) \sum_{j=1}^{\infty} \alpha_{kj}\, e^{-ij\theta}\Big\|_{C_{2\pi}}$$

and using that $|v_{n,m}(k) - 1| \leq C\, \varphi(k)/\varphi(n)$ uniformly in $n \in \mathbb{N}$, $k \in \mathbb{Z}$ for some constant C it follows by (13) and (4) that

$$I_4 \leqslant M \sum_{k=1}^{n} \frac{\varphi(k)}{\varphi(n)} \frac{k^{-2-\varepsilon}}{\varphi(k-1)} + M \sum_{k=n+1}^{\infty} \frac{k^{-2-\varepsilon}}{\varphi(k-1)} = \mathcal{O}\left(\frac{1}{\varphi(n)}\right), \qquad n \to \infty,$$

and the proof is complete.

REFERENCES

[1] Curtiss, J.H., Faber polynomials and the Faber series. Amer. Math. Monthly 78 (1971), 577-596.

[2] Dahmen, W., Trigonometric approximation with exponential error orders. I. Construction of asymptotically optimal processes; generalized de la Vallée Poussin sums. Math. Ann. 230 (1977), 57-74.

[3] Dahmen, W., Trigonometric approximation with exponential error orders. II. Properties of asymptotically optimal processes; impossibility of arbitrarily good error estimates. J. Math. Anal. Appl. 68 (1979), 118-129.

[4] Gaier, D., Vorlesungen über Approximation im Komplexen. Birkhäuser Verlag, Basel 1980.

[5] Kövari, T., On the order of polynomial approximation for closed Jordan domains. J. Approximation Theory 5 (1972), 362-373.

[6] Suetin, P.K., The basic properties of Faber polynomials (Russian). Uspehi Mat. Nauk 19 (1964), 125-154. Translation: Russian Math. Surveys 19 (1964), 121-149.

VII Strong and Müntz Approximation

STRONG APPROXIMATION AND GENERALIZED LIPSCHITZ CLASSES

László Leindler
Bolyai Institute
Attila József University
Szeged

1. Introduction

Recently several papers (see e.g. [2], [3], [5], [7], [10]) deal with problems of imbedding of classes of functions connected with strong approximation of Fourier series. At such problems the main question is to find conditions implying that a certain class of functions should be imbedded into another one, and one of the classes in question is determined by certain properties of the strong approximation of Fourier series.

The aim of this note is to present some new relations of this type introducing the concept of the enlarged Lipschitz class.

2. Definitions and Theorem

In a previous paper [8] we investigated a certain class of functions showing great similarity to the classical Lipschitz class; consequently we shall call this class of functions the *enlarged Lipschitz class* and denote it by $\mathrm{Lip}^{(e)}\alpha$. More precisely we shall say that a modulus of continuity $\omega(\delta)=\omega_\alpha(\delta)$ belongs to the class $\mathrm{Lip}^{(e)}\alpha$ if for any $\alpha'>\alpha$ there exists a natural number $\mu=\mu(\alpha')$ such that

$$(2.1)\qquad 2^{\mu\alpha'}\omega_\alpha(2^{-n-\mu}) > 2\omega_\alpha(2^{-n})$$

holds for all $n\geqslant 1$; and simultaneously for any natural number ν there exists another natural number $N(\nu)$ such that if $n>N(\nu)$ then

$$(2.2) \qquad 2^{\nu\alpha}\omega_\alpha(2^{-n-\nu}) \leq 2\omega_\alpha(2^{-n})$$

holds.

It is clear that the classical class $\mathrm{Lip}\alpha$ is imbedded into $\mathrm{Lip}^{(e)}\alpha$ strictly, i.e.,

$$\mathrm{Lip}\alpha \subset \mathrm{Lip}^{(e)}\alpha.$$

Before formulating the theorem we give some known definitions and notations.

Let $f(x)$ be a continuous and 2π-periodic function and let

$$(2.3) \qquad f(x) \sim \frac{a_o}{2} + \sum_{n=1}^{\infty} (a_n \cos nx + b_n \sin nx)$$

be its Fourier series. Denote by $s_n = s_n(x) = s_n(f;x)$ the n-th partial sum of (2.3) and let $f^{(r)}$ denote the r-th derivative of f. For any positive β and p we define the following strong mean

$$h_n(f,\beta,p) := \left\| \left\{ \frac{1}{(n+1)^\beta} \sum_{k=o}^{n} (k+1)^{\beta-1} |s_k - f|^p \right\}^{1/p} \right\|,$$

where $\| \ \|$ denotes the usual maximum norm.

Let $\omega(\delta)$ be a modulus of continuity, i.e. a nondecreasing continuous function on the interval $[0,2\pi]$ having the properties:
$\omega(0)=0$, $\omega(\delta_1+\delta_2) \leq \omega(\delta_1)+\omega(\delta_2)$ for any $0 \leq \delta_1 \leq \delta_2 \leq \delta_1+\delta_2 \leq 2\pi$.

Let $E_n(f)$ denote the best approximation of f by trigonometric polynomials of order at most n.

We define the following classes of functions:

$$(2.4) \qquad \begin{aligned} H(\beta,p,r,\omega) &:= \{f : h_n(f,\beta,p) = \mathcal{O}(n^{-r}\omega(1/n))\}, \\ W^rH^\omega &:= \{f : \omega(f^{(r)};\delta) = \mathcal{O}(\omega(\delta))\}, \\ W^rH^\omega \ln H &:= \{f : \omega(f^{(r)};\delta) = \mathcal{O}(\omega(\delta)\ln(1/\delta))\}, \\ W^rH^* &:= \{f : f^{(r)} \subset \Lambda_*\} \end{aligned}$$

where Λ_* denotes the class of Zygmund (see [11], p.43), and $\omega(f,\delta)$ is the modulus of continuity of f. In the case $\omega(\delta)=\delta^\alpha$ we write W^rH^α and

$H(\beta,p,r,\alpha)$ instead of $W^rH^{\delta^\alpha}$ and $H(\beta,p,r,\delta^\alpha)$, respectively; and if $r=0$, H^ω stands for W^oH^ω.

Generalizing a result of Alexits and Králik [1] we([5]) proved the following equivalence and imbedding relations:

Let β,p and α be positive numbers and r be a nonnegative integer, and additionally

if $\beta > (r+\alpha)p$ then

$$H(\beta,p,r,\alpha) \equiv W^rH^\alpha \qquad (\alpha < 1),$$

$$W^rH^1 \subset H(\beta,p,r,1) \equiv W^rH^* \qquad (\alpha = 1);$$

and if $\beta = (r+\alpha)p$ then

$$H(\beta,p,r,\alpha) \subset W^rH^\alpha \qquad (\alpha < 1),$$

$$H(\beta,p,r,1) \subset W^rH^* \qquad (\alpha = 1).$$

The aim of this note is to extend these relations to the classes defined under (2.4) assuming that the modulus of continuity $\omega(\delta) = \omega_\alpha(\delta)$ in question belongs to the class $\mathrm{Lip}^{(e)}\alpha$.

THEOREM. Let β,p and α be positive numbers, r be a nonnegative integer and let $\omega_\alpha = \omega_\alpha(\delta)$ belong to the class $\mathrm{Lip}^{(e)}\alpha$.

Additionally if $\beta > (r+\alpha)p$ then

$$H(\beta,p,r,\omega_\alpha) \equiv W^rH^{\omega_\alpha} \qquad (\alpha < 1), \tag{2.5}$$

$$W^rH^{\omega_1} \subset H(\beta,p,r,\omega_1) \qquad (\alpha = 1); \tag{2.6}$$

and if $\beta = (r+\alpha)p$ then

$$H(\beta,p,r,\omega_\alpha) \subset W^rH^{\omega_\alpha} \qquad (\alpha < 1), \tag{2.7}$$

$$H(\beta,p,r,\omega_1) \subset W^rH^{\omega_1}\ln H \qquad (\alpha = 1). \tag{2.8}$$

3. Required Propositions and Lemmas

PROPOSITION 1. For any positive β and p we have

$$(3.1)\qquad h_n(f,\beta,p) \leq K\{\frac{1}{n^\beta}\sum_{k=0}^{n}(k+1)^{\beta-1}E_k^p(f)\}^{1/p}$$

This is a trivial consequence of Theorem 1 in [4].

PROPOSITION 2. (Corollary 2 in [6]). For any β and p

$$(3.2)\qquad E_n(f) \leq K\, h_n(f,\beta,p).$$

PROPOSITION 3. ([9, pp. 59 and 61]). We have for any $r \geq 0$

$$(3.3)\qquad \omega(f^{(r)};\frac{1}{n}) \leq K\left\{\frac{1}{n}\sum_{k=1}^{n} k^r E_k(f) + \sum_{k=n+1}^{\infty} k^{r-1}E_k(f)\right\}.$$

LEMMA 1. (Lemma 3 in [8]). For any nonnegative sequence $\{a_n\}$ the inequality

$$(3.4)\qquad \sum_{k=1}^{m} a_n \leq Ka_m \qquad (m=1,2,\ldots;\ K>0)$$

holds if and only if there exists a positive number c and a natural number μ such that for any n

$$(3.5)\qquad a_{n+1} > ca_n$$

and

$$(3.6)\qquad a_{n+\mu} > 2a_n$$

are valid.

LEMMA 2. Condition (3.4) implies that for any positive p

$$\sum_{n=1}^{m} a_n^p \leq K_1 a_m^p$$

also holds.

This is an obvious consequence of Lemma 1.

4. Proof of Theorem

First we prove that

$$(4.1)\qquad W^rH^{\omega_\alpha} \subset H(\beta,p,r,\omega_\alpha)$$

holds if $\beta > (r+\alpha)p$ and $\alpha < 1$.

Assuming that f belongs to $W^rH^{\omega_\alpha}$ we have the inequality

$$\omega(f^{(r)};\tfrac{1}{n}) = O(\omega_\alpha(\tfrac{1}{n})),$$

which by the following well-known inequalities

$$(4.2)\qquad E_n(f) \leqslant K\omega(f;\tfrac{1}{n}) \quad\text{and}\quad E_n(f) \leqslant Kn^{-r}E_n(f^{(r)})$$

implies that

$$(4.3)\qquad E_n(f) \leqslant Kn^{-r}\omega_\alpha(\tfrac{1}{n}).$$

Hence, by (3.1), we get that

$$(4.4)\qquad \begin{aligned} h_n(f,\beta,p) &\leqslant K_1 \left\{ \frac{1}{n^\beta} \sum_{k=1}^{n} k^{\beta-1}(k^{-r}\omega_\alpha(\tfrac{1}{k}))^p \right\}^{1/p} \\ &\leqslant K_2 \left\{ \frac{1}{n^\beta} \sum_{m=1}^{\log n} 2^{m(\beta-rp)}\omega_\alpha^p(\tfrac{1}{2^m}) \right\}^{1/p}. \end{aligned}$$

Using Lemma 1 and 2, by $\beta > (r+\alpha)p$ and (2.1), (4.4) gives that

$$(4.5)\qquad h_n(f,\beta,p) \leqslant K_3 n^{-r}\omega_\alpha(\tfrac{1}{n}),$$

i.e., $f \in H(\beta,p,r,\omega_\alpha)$, and this proves (4.1).

So, in order to prove (2.5), it is enough to show

$$(4.6)\qquad H(\beta,p,r,\omega_\alpha) \subset W^rH^{\omega_\alpha}.$$

If $f \in H(\beta,p,r,\omega_\alpha)$ then (4.5) holds and this, by (3.2), implies (4.3). Then, by (3.3) and (4.3), we obtain that

$$\omega(f^{(r)};\tfrac{1}{n}) \leqslant K \left\{ \frac{1}{n} \sum_{k=1}^{n} \omega_\alpha(\tfrac{1}{k}) + \sum_{k=n+1}^{\infty} k^{-1}\omega_\alpha(\tfrac{1}{k}) \right\}.$$

The first term in the brackets, using Lemma 1 and the conditions $\alpha < 1$ and (2.1), can be estimated by $O(\omega_\alpha(1/n))$.

Next we show that the second term has the same order and this will verify (4.6). It is clear that

$$\sum_{k=n+1}^{\infty} k^{-1}\omega_\alpha(\frac{1}{k}) \leqslant K \sum_{m=\log n}^{\infty} \omega_\alpha(\frac{1}{2^m}),$$

so if we choose ν such that $\nu\alpha > 1$ then (2.2) implies that

$$\sum_{m=\log n}^{\infty} \omega_\alpha(\frac{1}{2^m}) \leqslant K_1\omega_\alpha(\frac{1}{n}),$$

which completes the proof of (4.6).

(4.1) and (4.6) jointly prove (2.5).

The proof of (2.6) is shorter. Using the same consideration as before we obtain that $f \in W^r H^{\omega_1}$ implies

$$h_n(f,\beta,p) \leqslant K \left\{ \frac{1}{n^\beta} \sum_{m=1}^{\log n} 2^{m(\beta-rp)} \omega_1^p(\frac{1}{2^m}) \right\}^{1/p},$$

and hence, by the arguments used in the proof of (4.5), we get that

$$h_n(f,\beta,p) \leqslant K_1 n^{-r}\omega_1(\frac{1}{n})$$

holds, which verifies that $f \in H(\beta,p,r,\omega_1)$ and so we concluded the proof of (2.6).

An examination of the proof of (4.6) shows that we did not use the condition $\beta > (r+\alpha)p$ in its proof, so (4.6) holds for any β, r and p; the only important condition is $\alpha < 1$. In view of this the relation (2.7) does not require a new proof.

Finally we prove (2.8). If $f \in H(\beta,p,r,\omega_1)$ then

$$h_n(f,\beta,p) \leqslant K\, n^{-r}\omega_1(\frac{1}{n}),$$

whence by (3.2)

$$E_n(f) \leqslant K\, n^{-r}\omega_1(\tfrac{1}{n}).$$

Putting these estimates into (3.3) we get that

$$\omega(f^{(r)};\tfrac{1}{n}) \leqslant K\{\frac{1}{n}\sum_{k=1}^{n}\omega_1(\tfrac{1}{k}) + \sum_{k=n+1}^{\infty} k^{-1}\omega_1(\tfrac{1}{k})\}. \tag{4.7}$$

Since

$$\sum_{k=1}^{n}\omega_1(\tfrac{1}{k}) \leqslant K \sum_{m=1}^{\log n} 2^m\omega_1(\tfrac{1}{2^m}),$$

so by (2.2) it is clear that

$$\frac{1}{n}\sum_{k=1}^{n}\omega_1(\tfrac{1}{k}) \leqslant K_1\,\frac{1}{n}\ln n \leqslant K_2\omega_1(\tfrac{1}{n})\ \ln n. \tag{4.8}$$

On the other hand, by (2.2) (e.g. choosing $\nu=2$),

$$\sum_{k=n+1}^{\infty} k^{-1}\omega_1(\tfrac{1}{k}) \leqslant K \sum_{m=\log n}^{\infty}\omega_1(2^{-m}) \leqslant K_1\omega_1(\tfrac{1}{n}). \tag{4.9}$$

Summing up, (4.7), (4.8) and (4.9) give that

$$\omega(f^{(r)};\tfrac{1}{n}) \leqslant K\omega_1(\tfrac{1}{n})\ \ln n$$

and this verifies that f belongs to the class $W^rH^{\omega_1}\ \ln H$ in accordance with statement (2.8).

Thus the proof is complete.

REFERENCES

[1] Alexits, G. - Králik, D., Über den Annährungsgrad der Approximation im starken Sinne von stetigen Funktionen. Magyar Tud. Akad. Mat. Kutató Int. Közl. 8 (1963), 317-327.

[2] Krotov, V.G., Strong approximation by Fourier series and differentiability properties of functions. Analysis Math. 4 (1978), 199-214.

[3] Krotov, V.G. - Leindler, L., On the strong summability of Fourier series and the classes H^{ω}. Acta Sci. Math. (Szeged) 40 (1978), 93-98.

[4] Leindler, L., Über die Approximation im starken Sinne. Acta Math. Acad. Sci. Hungar. 16 (1965), 255-262.

[5] Leindler, L., Strong approximation and classes of functions. Mitt. Math. Sem. Giessen 132 (1978), 29-38.

[6] Leindler, L., Strong and best approximation of Fourier series and the Lipschitz classes. Analysis Math. 4 (1978), 101-116.

[7] Leindler, L., Imbedding theorems and strong approximation. Acta Sci. Math. (Szeged) (to appear).

[8] Leindler, L., Generalizations of Prössdorf's theorems. Studia Sci. Math. Hungar. (to appear).

[9] Lorentz, G.G., Approximation of functions. Holt, Rinehart and Winston, New York / Chicago / Toronto, 1966.

[10] Totik, V., On structural properties of functions arising from strong approximation of Fourier series. Acta Sci. Math. (Szeged) 41 (1979), 227-251.

[11] Zygmund, A., Trigonometric Series. Cambridge, 1968.

STRONG APPROXIMATION AND THE BEHAVIOUR OF FOURIER SERIES

Vilmos Totik
Bolyai Institute
József Attila University
Szeged

Since 1963, a work of G. Alexits and D. Králik [1], the so called strong approximation of Fourier series has developed very rapidly. The difference between ordinary and strong approximation is that the latter examines means of type

$$\{\sum_k t_{nk}|s_k - f|^p\}^{1/p} \qquad (t_{nk} \geq 0,\ p > 0)$$

where $s_k(x) = s_k(f;x)$ is the k-th partial sum of the Fourier series of the 2π periodic function f, or of even more general types. In this work we apply mostly known strong approximation results for proving theorems concerning the behaviour of Fourier series. For our purposes the case $p = 1$ will be sufficient, therefore, the cited results are presented only in this particular case. In the first two paragraphs we prove two approximation theorems, and in the last two ones we estimate $\sigma_n^\alpha(f) - f$, $\alpha > -1/2$, for "almost all n".

1. A Nikol'skii Type Approximation Result.

In this point we give an example how strong approximation results can be directly applied to ordinary approximation. We do not strive for too much generality, only indicate the possible results.

Let $\{\lambda_k\}_{k=0}^{\infty}$ be a nonnegative sequence, $\{\lambda_k^*\}_{k=2^m+1}^{2^{m+1}}$ the monotone increasing rearrangement of the finite subsequence $\{\lambda_k\}_{k=2^m+1}^{2^{m+1}}$ and

$$\Lambda_m = \sum_{k=1}^{2^m} \lambda^*_{2^m+k} \log(2^m/(2^m+1-k)) \qquad (m=1,2,\ldots),$$

$\Lambda_0 = \lambda_0 + \lambda_1$. In [12] we proved that

$$\sum_k \lambda_k |s_k(f;x) - f(x)| \leqslant K \sum_m \Lambda_m E_{2^m}(f) \quad ^{1)}$$

where $E_n(f)$ is the best uniform approximation of f by trigonometric polynomials of order at most n and K is an absolute constant. This implies:

THEOREM 1. *If $\sum_k \lambda_k = 1$ then we have*

$$|f(x) - \sum_k \lambda_k s_k(f;x)| \leqslant K \sum_m \Lambda_m E_{2^m}(f). \tag{1.1}$$

(1.1) cannot be sharpened in general, e.g.,if $0 < \alpha < 1$ and $r \geqslant 0$ is an integer,then there is a function $f_{r,\alpha}$ with $f_{r,\alpha}^{(r)} \in \mathrm{Lip}\,\alpha$ such that

$$|f_{r,\alpha}(0) - \sum_k \lambda_k s_k(f_{r,\alpha};0)| \geqslant \sum_m \Lambda_m E_{2^m}(f) \tag{1.2}$$

for every sequence $\{\lambda_k\}$ the finite subsequences $\{\lambda_k\}_{k=2^m+1}^{2^{m+1}}$ of which are all monotone (see the proof of [12, Theorem 2]).

(1.1) and (1.2) can be applied to almost all of the known summation methods: Abel, Euler, Borel, de La Vallée Poussin, Lindelöf methods, many of Riesz and Norlund type methods etc. (see [2,3,12]). We mention only one corollary:

COROLLARY. *If $(\lambda_k^{(n)})_{k \leqslant n}$ is a triangular matrix with $\lambda_k^{(n)} \leqslant \lambda_{k+1}^{(n)}$, $0 \leqslant k < n$, $\sum_{k=0}^n \lambda_k^{(n)} = 1$ and $f^{(r)} \in \mathrm{Lip}\,\alpha$, $0 < \alpha < 1$, then*

$$f(x) - \sum_{k=0}^{n} \lambda_k^{(n)} s_k(f;x) = O\Big(\sum_{k=1}^{n} \lambda_k^{(n)} \frac{1}{k^{r+\alpha}} + \frac{1}{n^{r+\alpha}} \sum_{k=1}^{n} \lambda_k^{(n)} \log \frac{n}{n+1-k}\Big),$$

furthermore, this is already the best possible estimate.

Results of similar character are contained in [6, Chapter 8].

---

1) K,A,c denote (mostly absolute) constants, besides K,c are not necessarily the same at each occurence.

2. Generalized de La Vallée Poussin Means.

It it well-known that

$$\frac{1}{n}\sum_{k=n+1}^{2n} s_k(f;x) - f(x) = \mathcal{O}(\omega(f;\tfrac{1}{n}))$$

where $\omega(f;\delta)$ is the modulus of continuity of f. In this connection a very interesting question arises: what can we say about the approximation properties of the means $\frac{1}{n}\sum_{i=n+1}^{2n} s_{k_i}(f;x)$ where $\{k_i\}$ is an arbitrary subsequence of the natural numbers. Concerning strong approximation we proved a result of this type ([7]):

THEOREM A. *If* $E_n(f) \leq K\rho_n$ *and* $i\rho_{2^i n} \leq K\rho_n$, *then for every sequence* $\{k_i\}$ *we have*

$$\frac{1}{n}\sum_{i=n+1}^{2n} |s_{k_i}(f;x) - f(x)| \leq A\rho_n$$

and here A *depends only on* K.

The answer to the above problem is given by the next two theorems.

In the following by ω we denote always an arbitrary modulus of continuity. Let H^ω and H_o^ω be the class of functions f for which $\omega(f;\delta) \leq K_f\omega(\delta)$, $\delta \in [0,2\pi]$, is satisfied with a constant K_f and with $K_f = 1$, respectively.

THEOREM 2. *Let* $\omega_*(\delta) = \sup_{\varepsilon \geq 1}\omega(\varepsilon\delta)\log(1/\varepsilon)$. *If* $f \in H_o^\omega$, *then, whatever the sequence* $\{k_i\}$ *be, we have*

$$|\frac{1}{n}\sum_{i=n+1}^{2n} s_{k_i}(f;x) - f(x)| \leq K\omega_*(\tfrac{1}{n}) \tag{2.1}$$

with an absolute constant K.

THEOREM 3. *For every modulus of continuity* ω *with* $\omega_*(1) < \infty$ *there is an* $f \in H^\omega$ *and a sequence* $\{k_i\}$ *with the property*

$$\overline{\lim_{n\to\infty}} |\frac{1}{n}\sum_{i=n+1}^{2n} s_{k_i}(f;0) - f(0)|/\omega_*(\tfrac{1}{n}) > 0.$$

We mention that if $\omega_*(\delta)\to 0$, $\delta\to 0$, then $\omega_*(\delta)$ is again a modulus of continuity. Our proof shows that beyond (2.1)

$$(2.2)\qquad \frac{1}{n}\sum_{i=n+1}^{2n} |s_{k_i}-f| \leq K\omega_*(\frac{1}{n})$$

is true as well for any set $\{k_i\}_{i=n}^{2n}$ of numbers k_i greater than n, and so, if we take into account the estimate ([2,10]):

$$\frac{1}{n+1}\sum_{k=0}^{n} |s_k-f| \leq K\frac{1}{n}\int_{1/n}^{2\pi}\frac{\omega(t)}{t^2}\,dt,$$

we obtain

COROLLARY. <u>If</u> $f\in H_o^\omega$ <u>and</u> $k_1,\dots,k_n$ <u>are arbitrary distinct natural numbers then</u>

$$\left|\frac{1}{n}\sum_{i=0}^{n} s_{k_i}(x)-f(x)\right| \leq K(\frac{1}{n}\int_{1/n}^{\pi}\frac{\omega(t)}{t^2}\,dt+\omega_*(\frac{1}{n}))$$

<u>with an absolute constant</u> K.

Thus, e.g., if $f\in \mathrm{Lip}\,\alpha$, then

$$\frac{1}{n}\sum_{i=0}^{n} s_{k_i}(f)-f = \begin{cases} \mathcal{O}(n^{-\alpha}), & \text{if } 0<\alpha<1 \\ \mathcal{O}(\frac{\log n}{n}), & \text{if } \alpha=1, \end{cases}$$

uniformly in $\{k_i\}$ and x.

Of course, the latter corollary is not the only consequence of our results, we could consider means of type $\sum t_{nk}s_k$ with quite general (t_{nk}) (see e.g. [2,10]), but we do not go further in this direction.

Finally, we remark that from the proof of (2.2) it will follow that if f is r-times differentiable, $r\geq 1$, then we have for every $\{k_i\}$

$$\left|\frac{1}{n}\sum_{i=n+1}^{2n} s_{k_i}(x)-f(x)\right| \leq A\,\frac{1}{n^r}\,\omega(f^{(r)};\frac{1}{n}).$$

PROOF of Theorem 2. We prove (2.2). The following inequality will be used (see [10]):

LEMMA 1. <u>If</u> $1\leq k_1<\dots<k_r\leq n$ <u>are arbitrary, then</u>

$$(2.3)\qquad \frac{1}{r}\sum_{i=1}^{r} |s_{k_i}(f;x) - f(x)| \leq A\, E_{k_1}(f)\log\frac{2n}{r}$$

<u>holds with an absolute constant</u> A.

Let ν_i be the number of those k_t for which

$$2^i n < k_t \leq 2^{i+1} n \qquad (n < t \leq 2n,\ i = 0,1,\dots).$$

By (2.3) and by Jackson's theorem

$$\sum_{\substack{2^i n < k_t \leq 2^{i+1} n \\ n < t \leq 2n}} |s_{k_i}(x) - f(x)| \leq K\,\nu_i\,\omega\left(\frac{1}{2^i n}\right)\log\frac{2^{i+1} n}{\nu_i}$$

and thus, it is enough to prove that

$$(2.4)\qquad S = \frac{1}{n}\sum_{\nu_i > 0}\nu_i\omega\left(\frac{1}{2^i n}\right)\log\frac{2^{i+1} n}{\nu_i} \leq K\omega_*(\tfrac{1}{n}).$$

But

$$S \leq K\Big(\frac{1}{n}\sum_{\nu_i > 0}\nu_i\omega\left(\frac{1}{2^i n}\right)(i+1) + \frac{1}{n}\sum_{\nu_i > 0}\omega\left(\frac{1}{2^i n}\right)\nu_i\log\frac{n}{\nu_i}\Big) = S_1 + S_2,$$

and clearly

$$S_1 \leq K\omega_*(\tfrac{1}{n})\frac{1}{n}\sum_{\nu_i > 0}\nu_i \leq K\omega_*(\tfrac{1}{n}).$$

Using that $\omega\left(\frac{1}{2^i n}\right) \leq K\frac{1}{i+1}\omega_*(\frac{1}{n})$, we obtain $\omega(1/(2^i n)) \leq K\omega^*(1/n)/(i+1)$

$$S_2 \leq K\omega_*(\tfrac{1}{n})\sum_{\nu_i > 0}\frac{1}{i+1}\nu_i\log\frac{n}{\nu_i} = S_{21} + S_{22}$$

where the summation in S_{21} is extended to the i's satisfying the condition $\log(n/\nu_i)/(i+1) \leq 1$. Thus we have

$$S_{21} \leq K\omega_*(\tfrac{1}{n})\frac{1}{n}\sum_{\nu_i > 0}\nu_i \leq K\omega_*(\tfrac{1}{n}),$$

and since $\log(n/\nu_i) > (i+1)$ implies $\log(n/\nu_i)/(n/\nu_i) \leq (i+1)/e^{i+1}$, we obtain

$$S_{22} \leqslant K\omega_*(\frac{1}{n}) \sum_{i=o}^{\infty} \frac{1}{i+1} \frac{i+1}{e^{i+1}} \leqslant K\omega_*(\frac{1}{n})$$

which, together with the previous estimates, prove (2.4).

We have proved Theorem 2.

PROOF of Theorem 3. For every n we can choose an $m_n > e^{100}$ with $\omega(1/(m_n n)) \log m_n \geqslant c\omega_*(1/n)$, $c > 0$. We use the following lemma (see the proof of [10, Lemma 5]):

LEMMA 2. *There is an* $f \in H^{\omega}$ *such that*

$$s_{m_n n+\lambda}(f;0) - f(0) > \omega(\frac{1}{m_n n}) \log \frac{m_n n}{\lambda} \qquad (0 < \alpha < \frac{m_n n}{e^{100}}) \tag{2.5}$$

is satisfied for infinitely many n.

Let $\{k_i\}$ be a sequence such that for infinitely many n satisfying (2.5) we should have $k_{n+1} = m_n n + 1$, $k_{n+2} = m_n m + 2, \ldots, k_{2n} = m_n n + n$. For this $\{k_n\}$ and the above f we have

$$\frac{1}{n} \sum_{i=n+1}^{2n} s_{k_i}(f;0) - f(0) \geqslant \omega(\frac{1}{m_n n}) \sum_{\lambda=1}^{n} \log \frac{m_n n}{\lambda}$$

$$\geqslant \omega(\frac{1}{m_n n}) \log m_n > c\omega_*(\frac{1}{n})$$

for infinitely many n and this was to be proved.

3. Approximation by the Partial Sums of the Fourier Series for "almost all n"

The first strong approximation result is due to G. Alexits and D. Krâlik [1] who proved that in the case $f \in \text{Lip}\,\alpha$, $0 < \alpha < 1$,

$$\frac{1}{n} \sum_{k=o}^{n} |s_k - f| = O(n^{-\alpha})$$

holds. They remarked that this implies:

THEOREM B. *If* $\lambda_n \nearrow \infty$ *arbitrarily, then for every fixed* x

$$|s_n(f;x) - f(x)| \leq \lambda_n n^{-\alpha}$$

holds for all n *but a sequence* n_k *with zero density, i.e., with* $k/n_k = o(1)$.

The important inequality

$$\frac{1}{n} \sum_{k=n+1}^{2n} |s_k - f| = O(E_n(f))$$

of L. Leindler [2] shows that a similar statement holds for $f^{(r)} \in H^{\omega}$. We now prove another result of similar kind:

THEOREM 4. *There exists an absolute constant* B *such that if* $f^{(r)} \in H_o^{\omega}$ *and* $x \in [0,2\pi)$ *then the sequence of the natural numbers can be decomposed into two subsequences* $\{n_k\}$ *and* $\{n_k'\}$ *in such a way that*

$$|s_{n_k}(f;x) - f(x)| \leq B\, 2^r \frac{1}{n_k} \omega(\frac{1}{n_k}) \log \log n_k \qquad (k = 1,2,\dots) \tag{3.1}$$

and

$$\sum_{k=1}^{\infty} \frac{1}{n_k'} < \infty$$

are satisfied.

This theorem cannot be strenghtened in general, as is shown by

THEOREM 5. *If* $0 < \alpha < 1$ *and* $r \geq 0$ *integer, then there exist a function* f *with* $f^{(r)} \in \text{Lip}\,\alpha$, *a constant* $c > 0$ *and a sequence* $\{n_k\}$ *with* $\sum_{k=1}^{\infty} (1/n_k) = \infty$ *for which*

$$s_{n_k}(f;0) - f(0) \geq c \frac{1}{n_k^{r+\alpha}} \log \log n_k \qquad (k = 1,2,\dots).$$

PROOF of Theorem 4. Let us start from the inequality (2.3). By this, if

$$H_n(x) = \{n < k \leq 2n \mid |s_k(f;x) - f(x)| > 6A\, 2^{r+2} \frac{1}{k^r} \omega(\frac{1}{k}) \log \log k\},$$

then

$$|H_n(x)|\, 6A \frac{1}{n^r} \omega(\frac{1}{n}) \log \log n \leq 6A\, 2^{r+2} \sum_{k \in H_n(x)} \frac{1}{k^r} \omega(\frac{1}{k}) \log \log k \leq$$

$$\leq \sum_{k\in H_n(x)} |s_k(f;x)-f(x)| \leq A|H_n(x)|E_n(f)\log\frac{2n}{|H_n|}$$

$$\leq |H_n(x)|\, 3A\,\frac{1}{n^r}\,\omega(\tfrac{1}{n})\log\frac{2n}{|H_n(x)|}$$

(in the last step we used that $E_n(f)\leq (3/n^r)\omega(f^{(r)};1/n)$; see [6, p. 293]), by which

$$\frac{|H_n(x)|}{n} \leq \frac{2}{(\log n)^2}\,. \tag{3.2}$$

Thus, if $B = 24A$ then those n's, for which

$$|s_n(f;x)-f(x)| \leq \frac{B2^r}{n^r}\,\omega(\tfrac{1}{n})\log\log n$$

is not satisfied, will belong to $\bigcup_n H_{2^n}(x)$, and we have to remark only that, according to (3.2),

$$\sum_{k\in \bigcup_n H_{2^n}(x)} \frac{1}{k} = O\left(\sum_n \frac{|H_{2^n}(x)|}{2^n}\right) = O\left(\sum_n \frac{1}{n^2}\right) = O(1).$$

PROOF of Theorem 5. Let us put

$$f(x) = \sum_{n=2}^{\infty} 2^{-n(r+\alpha)} Q_{2^n,2^{n-2}}(x),$$

where

$$Q_{n,m}(x) = \sum_{i=1}^{m}\left(\frac{\cos(n-i)x}{i} - \frac{\cos(n+i)x}{i}\right)$$

is the well-known Fejér-polynomial. As $|Q_{n,m}|\leq 4$, $f^{(r)}\in \mathrm{Lip}\,\alpha$ is clear. For $0<\lambda<2^{n-2}/n-1$ we have

$$f(0)-s_{2^n+\lambda}(f;0) = -s_{2^n+\lambda}(Q_{2^n,2^{n-2}};0) = \frac{1}{2^{n(r+\alpha)}}\sum_{i=\lambda+1}^{2^{n-2}}\frac{1}{i} > \frac{1}{2^{n(r+\alpha)}}\log\frac{2^{n-2}}{\lambda+1}$$

$$> \frac{1}{2^{n(r+\alpha)}}\log n > c\,\frac{1}{(2^n+\lambda)^{r+\alpha}}\log\log(2^n+\lambda).$$

Now, the proof can be completed easily, since

$$\sum_{n:1\leqslant\lambda<2^{n-2}/n-1} \frac{1}{2^n+\lambda} \geqslant c \sum_n \frac{1}{n} = \infty.$$

4. Approximation by the (C,α)-Means, α ≠ 0, of the Fourier Series for "almost all n".

In this section we examine the behaviour of the Cesaro means $(\sigma_k^\alpha(x))$ of Fourier series. If $\alpha>0$ and $f\in H^\omega$ then (see [4,8])

$$(4.1) \qquad \sigma_n^\alpha(f;x)-f(x) = O(\omega^*(\tfrac{1}{n})); \quad \tilde{\sigma}_n^\alpha(f;x)-\tilde{f}(x) = O(\omega^{**}(\tfrac{1}{n}))$$

where

$$\omega^*(\delta) = \int_\delta^{2\pi} \frac{\omega(t)}{t^2}\,dt \text{ and } \omega^{**}(\delta) = \int_0^\delta \frac{\omega(t)}{t}\,dt$$

are two moduli of continuity associated to ω. For negative α the best possible estimate is (see e.g. [9, Theorem 2]):

$$\sigma_n^\alpha(x)-f(x) = O(n^{-\alpha}\omega(\tfrac{1}{n})); \quad \tilde{\sigma}_n^\alpha(x)-\tilde{f}(x) = O(\omega^{**}(\tfrac{1}{n})+n^{-\alpha}\omega(\tfrac{1}{n})).$$

But, if we require an estimate only for most of the indices then we nearly obtain (4.1):

THEOREM 6. _If_ $0>\alpha>-1/2$, $f\in H^\omega$, $x\in(0,2\pi]$, _and_ $\lambda_n\to\infty$, _then_

$$|\sigma_n^\alpha(x)-f(x)| \leqslant \lambda_n\omega^*(\tfrac{1}{n}); \quad |\tilde{\sigma}_n^\alpha(x)-\tilde{f}(x)| \leqslant \lambda_n\omega^{**}(\tfrac{1}{n})$$

for every n _but a sequence_ $\{n_k\}$ _with density_ 0.

This follows immediately from the estimates (see [8, (3.4), (3.5)])

$$\frac{1}{n}\sum_{k=n+1}^{2n} |\sigma_k^\alpha(x)-f(x)| \leqslant K\,\omega^*(\tfrac{1}{n}),$$

$$\frac{1}{n}\sum_{k=n+1}^{2n} |\tilde{\sigma}_k^\alpha(x)-\tilde{f}(x)| \leqslant K\,\omega^{**}(\tfrac{1}{n}).$$

A "more dense" "good" approximation is given by

THEOREM 7. _If_ $0>\alpha>-1/2$, $f\in H^\omega$, $x\in(0;2\pi]$, _and_ $\varepsilon>0$, _then we have_

$$|\sigma_n^\alpha(x) - f(x)| \leqslant K\ \omega^*(\tfrac{1}{n}) + \omega(\tfrac{1}{n})\ \log^{-\alpha+\varepsilon} n,$$

$$|\tilde{\sigma}^\alpha(x) - \tilde{f}(x)| \leqslant K\ \omega^{**}(\tfrac{1}{n}) + \omega(\tfrac{1}{n})\ \log^{-\alpha+\varepsilon} n$$

for every n but a sequence $\{n_k\}$ with $\sum_k 1/n_k < \infty$.

For example, in the case $f \in \mathrm{Lip}\ \beta$, $0 < \beta < 1$, this theorem gives:

$$|\sigma_n^\alpha(x) - f(x)| \leqslant n^{-\beta} \log^{-\alpha+\varepsilon} n$$

for every n but a sequence $\{n_k\}$ with $\sum_k 1/n_k < \infty$.

Now the following theorem shows that, in general, Theorem 7 cannot be strenghtened: $\varepsilon > 0$ is necessary in it.

THEOREM 8. If $0 > \alpha > -1$, $1 > \beta > 0$, $1 > -\alpha+\beta$, then there is an $f \in \mathrm{Lip}\ \beta$ and a sequence $\{n_k\}$ with $\sum_k 1/n_k = \infty$ such that

$$\lim_{k\to\infty} |\sigma_{n_k}^\alpha(0) - f(0)| \ /\ n_k^{-\beta} \log^{-\alpha} n_k = \infty. \tag{4.2}$$

We mention that both Theorem 7 and Theorem 8 hold in a sharper form:

COROLLARY. Let $\{\varphi(n)\}$ be an increasing sequence with $\varphi(2n) = \mathcal{O}(\varphi(n))$. If $\sum_k k^{-1}(\varphi(k))^{1/\alpha} < \infty$, then

$$|\sigma_n^\alpha(x) - f(x)| \leqslant K\ \omega^*(\tfrac{1}{n}) + \omega(\tfrac{1}{n})\varphi(n)$$

holds for every n but a sequence $\{n_k\}$ with $\sum_k 1/n_k < \infty$. On the other hand, if $\sum_k k^{-1}(\varphi(k))^{1/\alpha} = \infty$, then there exist an $f \in \mathrm{Lip}\ \beta$, $\beta-\alpha < 1$, and a sequence $\{n_k\}$ with $\sum_k 1/n_k = \infty$ for which

$$|\sigma_{n_k}^\alpha(0) - f(0)| \ /\ n_k^{-\beta} \varphi(n_k) \to \infty \qquad (k \to \infty).$$

The proofs of these statements are similar to those of Theorem 7 and Theorem 8.

Before we prove Theorem 7 we state an analogous result for integrable functions. In [5] G. Sunouchi proved:

THEOREM C. If $0 > \alpha > -1/2$ and $f \in L^{1/(1+\alpha)}$, then

$$\frac{1}{n}\sum_{k=o}^{n}|\sigma_k^\alpha(x)-f(x)| = o_x(1) \qquad \text{(a.e.)}.$$

This implies immediately

THEOREM 9. If $0>\alpha>-1/2$ and $f\in L^{1/(1+\alpha)}$, then for almost all $x\in(0;2\pi]$ the relation

$$\sigma_n^\alpha(x)-f(x) = o_x(1)$$

holds not counting an index-sequence $\{n_k\}$ of density 0.

In [11] we proved that there is an f such that $f\in L^\beta$ for every $\beta<1/(1+\alpha)$ but

$$\sup_n \frac{1}{n}\sum_{k=o}^{n}|\sigma_k^\alpha(x)-f(x)| = \infty \qquad \text{(a.e.)}.$$

A closer examination of the proof shows that the following statement holds, as well

THEOREM 10. There exists a function f such that $f\in L^\beta$ for every $\beta<1/(1+\alpha)$ but for almost all x there is a sequence $\{n_k^{(x)}\}$ not of density zero and a positive number $\varepsilon^{(x)}$ such that

$$|\sigma^\alpha_{n_k^{(x)}}(x)-f(x)| > \varepsilon^{(x)}.$$

PROOF of Theorem 7. We consider only the first estimate. Using the inequality (see [9, Lemma 1])

$$\frac{1}{\lambda}\sum_{i=1}^{\lambda}|\sigma_{k_i}^\alpha(x)-f(x)| \leqslant A(\omega^*(\tfrac{1}{n})+(\tfrac{\lambda}{n})^\alpha\omega(\tfrac{1}{n})),$$

which is valid for arbitrary n and $n<k_1<\ldots<k_\lambda\leqslant 2n$, the proof is similar to that of Theorem 4: if

$$H_n(x) = \{n<k\leqslant 2n \,|\, |\sigma_k^\alpha(x)-f(x)| > A\omega^*(\tfrac{1}{n})+\omega(\tfrac{1}{n})(\log n)^{-\alpha+\varepsilon}\},$$

then

$$A\omega^*(\tfrac{1}{n})+\omega(\tfrac{1}{n})(\log n)^{-\alpha+\varepsilon} \leqslant \frac{1}{|H_n(x)|}\sum_{k\in H_n(x)}|\sigma_k^\alpha(x)-f(x)| \leqslant$$

$$\leq A\left(\omega^*(\frac{1}{n}) + \left(\frac{|H_n(x)|}{n}\right)^{\alpha}\omega(\frac{1}{n})\right)$$

by which

$$\frac{|H_n(x)|}{n} \leq A^{-\frac{1}{\lambda}} \frac{1}{(\log n)^{1+\varepsilon/(-\alpha)}}$$

from which the statement follows easily (see section 3).

PROOF of Theorem 8. Let

$$\varphi_n(t) = \begin{cases} \sin\left[(2^n+\frac{1}{2}+\frac{\alpha}{2})t - \frac{\pi\alpha}{2}\right], \text{if} & \frac{\pi(1+\frac{\alpha}{2})}{2^n+\frac{1}{2}+\frac{\alpha}{2}} \leq t \leq \pi \\ 0 & , \text{elsewhere in } (-\pi,\pi] \end{cases}$$

and

$$f(x) = \sum_{n=2}^{\infty} 2^{-n\beta}\varphi_n(x).$$

The relation $f \in \mathrm{Lip}\ \beta$ can be proved easily.

We use that

$$\sigma_n^{\alpha}(x) - f(x) = \frac{2}{\pi}\int_0^{\pi} \varphi_x(t)K_n^{\alpha}(t)dt,$$

where $\varphi_x(t) = (f(x+t)+f(x-t)-2f(x))/2$ and $K_n^{\alpha}(t)$ is the k-th (C,α)-kernel. It is well-known (see [13 pp. 94-94]) that

$$|K_n^{\alpha}(t)| \leq 2n; \qquad K_n^{\alpha}(t) = \frac{1}{A_n^{\alpha}}\frac{\sin\{(n+\frac{\alpha}{2}+\frac{1}{2})t - \frac{\pi\alpha}{2}\}}{(2\sin t/2)^{1+\alpha}} + \frac{2\theta(t)\alpha}{n(2\sin t/2)^2}$$

$$(|\theta| \leq 1,\ t \in (\tfrac{1}{n};\pi],\ A_n^{\alpha} = \tbinom{n+\alpha}{n},\ n = 1,2,\ldots),$$

by which

$$|\sigma_{2^n+\lambda}^{\alpha}(0) - f(0)| = \frac{1}{\pi}\frac{1}{A_{2^n+\lambda}^{\alpha}}\int_{2^{-n}}^{\pi}\frac{f(t)\sin\{(2^n+\lambda+\frac{\alpha}{2}+\frac{1}{2})t - \frac{\pi\alpha}{2}\}}{(2\sin t/2)^{1+\alpha}}dt + O(2^{-n\beta}).$$

Here the absolute value of the integral for $\lambda \leq 2^n/(n\log n)$ can be estimated as $\geq I_n - \sum_{k\neq n} I_k$, where

$$I_k = \left|\frac{1}{2^{k\beta}} \int_{2^{-n}}^{\pi} \frac{\varphi_k(t)\sin\{(2^n+\lambda+\frac{\alpha}{2}+\frac{1}{2})t-\frac{\pi\alpha}{2}\}}{(2\sin t/2)^{1+\alpha}}\,dt\right|.$$

By the second mean value theorem we have:

a) for $k > n$

$$I_k = O(2^{-k\beta}2^{n(1+\alpha)}2^{-k}) = O(2^{-k\beta}2^{n\alpha}),$$

b) for $k < n$ (with $a_k = \pi(1+\alpha/2)/(2^k+\alpha/2+1/2)$)

$$I_k = 2^{-k\beta}\left|\int_{a_k}^{\pi}\right| \leqslant K\,2^{-k\beta}2^{k(1+\alpha)}2^{-n} = K\,2^{k(1+\alpha-\beta)}2^{-n},$$

c)

$$I_n \geqslant \frac{1}{2}2^{-n\beta}\int_{2^{-n}}^{\pi}\frac{\cos\lambda t}{(2\sin t/2)^{1+\alpha}}\,dt - \left|\frac{1}{2}2^{-n\beta}\int_{2^{-n}}^{\pi}\frac{\cos[(2\cdot 2^n+\lambda+1+\alpha)t-\pi\alpha]}{(2\sin t/2)^{1+\alpha}}\,dt\right|$$

$$\geqslant I_n^* - K\,2^{-n\beta}2^{n(1+\alpha)}2^{-n}.$$

Now, an easy calculation shows that

$$\left|I_n^* - 2^{-n\beta}\int_{2^{-n}}^{\pi}\frac{\cos\lambda t}{t^{1+\alpha}}\,dt\right| \leqslant K\,2^{-n\beta}\frac{1}{\lambda}\mathrm{Var}\left(\frac{1}{(2\sin t/2)^{1+\alpha}} - \frac{1}{t^{1+\alpha}}\right) \leqslant K\,2^{-n\beta}\frac{1}{\lambda}$$

and thus, with a suitable λ_o, we get for $\lambda \geqslant \lambda_o$

$$I_n^* \geqslant \frac{\lambda^\alpha}{2^{n\beta}}\int_{\lambda/2^n}^{\lambda\pi}\frac{\cos t}{t^{1+\alpha}}\,dt - K\,2^{-n\beta}\frac{1}{\lambda} \geqslant c\lambda^\alpha 2^{-n\beta}$$

with c independent of n and $\lambda_o \leqslant \lambda \leqslant 2^n/(n\log n)$, since

$$\int_o^\infty \frac{\cos t}{t^{1+\alpha}}\,dt = \mathrm{Re}\left(\int_o^\infty t^{-1-\alpha}e^{-it}dt\right) = \mathrm{Re}(e^{i\pi\alpha/2}\Gamma(-\alpha)) = \cos\frac{\pi\alpha}{2}\Gamma(-\alpha) > 0.$$

Collecting the above estimates and taking into account that $A_n^\alpha \sim n^\alpha$ and that $\alpha-\beta > -1$, we get for $\lambda_o \leqslant \lambda \leqslant 2^n/(n\log n)$

$$|\sigma^\alpha_{2^n+\lambda}(0) - f(0)| \geqslant c\lambda^\alpha 2^{-n\alpha}2^{-n\beta} - K\cdot 2^{-n\beta} - K\sum_{k=n+1}^{\infty} 2^{-k\beta} - K\,2^{-n\alpha}\sum_{k=1}^{n} 2^{-n}2^{k(1+\alpha-\beta)}$$

$$\geqslant c\left(\frac{\lambda}{2^n}\right)^\alpha 2^{-n\beta} - K\,2^{-n\beta} \geqslant cn^{-\alpha}(\log n)^{-\alpha}2^{-n\beta} \geqslant c(\log n)^{-\alpha}(\log(2^n+\lambda))^{-\alpha}(2^n+\lambda)^{-\beta}.$$

Thus, if we arrange the numbers $2^n+\lambda$, $n=1,\ldots,\infty$, $\lambda_o<\lambda\leqslant 2^n/(n\log n)$, into a sequence $\{n_k\}$, we get that (4.2) is satisfied and

$$\sum_k \frac{1}{n_k} \geqslant \sum_n \left(\frac{2^n}{n\log n}-\lambda_o\right)\frac{1}{2^n}=\infty$$

by which we proved our theorem.

REFERENCES

[1] Alexits, G. - Králik, D., Über den Annäherungsgrad im starken Sinne von stetigen Funktionen. Magyar Tud. Akad. Mat. Kut. Int. Közl. 8 (1963), 317-327.

[2] Leindler, L., Über die Approximation im starken Sinne. Acta Math. Acad. Sci. Hungar. 16 (1965), 255-262.

[3] Leindler, L., Bemerkung zur Approximation im starken Sinne. Acta Math. Acad Sci. Hungar. 18 (1967), 273-277.

[4] Leindler, L., On strong approximation of Fourier series. Periodica Math. Hungar. 1 (1971), 157-162.

[5] Sunouchi, G., On the strong summability of power series and Fourier series. Tohoku Math. J. (2) 6 (1954), 220-225.

[6] Timan, A.F., Theory of Approximation of a Real Variable. Hindustan Publishing Company, Delhi, 1966.

[7] Totik, V., On the very strong and mixed approximations. Acta Sci. Math. 41 (1979), 419-428.

[8] Totik, V., On the strong approximation by the (C,α)-means of Fourier series I. Analysis, Math. 6 (1980), 57-85.

[9] Totik, V., On the strong approximation by the (C,α)-means of Fourier series II. Analysis Math. 6 (1980), 165-184.

[10] Totik, V., On the strong approximation of Fourier series. Acta Math. Acad. Sci. Hungar. 35 (1980), 151-172.

[11] Totik, V., On the strong summability by the (C,α)-means of Fourier series. Periodica Math. Hungar. (to appear)

[12] Totik, V., A general theorem on strong means. Magyar Tud. Akad. Math. Kut. Int. Közl. (to appear)

[13] Zygmund, A., Trigonometric Series. I. Oxford University Press, Cambridge, 1959.

ON THE RATE OF APPROXIMATION BY MÜNTZ POLYNOMIALS SATISFYING CONSTRAINTS

Dany Leviatan
Department of Mathematics
California Institute of Technology
Pasadena

The rate of approximation to functions in $C_0[0,1]$ by means of Müntz polynomials the coefficients of which satisfy some growth restrictions is discussed. Relations between the size of the restricting constants and the speed of the approximation process are derived in various cases.

1. Introduction

Given a sequence of non negative constants $\mathcal{A} = \{A_k\}(k \geq 1)$ denote by $P_{\mathcal{A}}$ the set of polynomials

$$\{p(x) : p(x) = \sum_{k=1}^{n} a_k x^k, \text{ n arbitrary and } |a_k| \leq A_k^k\}.$$

It has been shown by v. Golitschek [3] and Roulier [5] that for $P_{\mathcal{A}}$ to be dense in $C_0[0,1]$ it is necessary and sufficient that there should exist a subsequence $\{k_j\}(j \geq 1)$ such that $\sum_{j=1}^{\infty} 1/k_j = \infty$ for which $\lim_{j\to\infty} A_{k_j} = \infty$. Since we require nothing from the other A_k's we see that the denseness property of $P_{\mathcal{A}}$ depends on a Müntz sequence of coefficients i.e. we may impose arbitrary conditions on the other A_k's in particular that all of them be zero. Namely, we may allow non zero coefficients only for a Müntz subsequence $\{k_j\}(j \geq 1)$ and obtain Müntz polynomials. Indeed one can extend the above result to Müntz polynomials with non integral exponents and have a similar characterization of those $P_{\mathcal{A}}$ that are dense in $C_0[0,1]$.

Given a Müntz subsequence of the integers that is $\{k_j\}(j \geq 1)$ satisfying $\sum_{j=1}^{\infty} 1/k_j = \infty$ the rate of approximation to functions in $C_0[0,1]$ by means of Müntz polynomials is given by the well known Müntz-Jackson theorem (see [2, Thm 1]).

THEOREM A. *There exists an absolute constant* $C > 0$ *independent of the sequence* $\{k_j\}$ *such that for any* $f \in C_0[0,1]$ *and each* $n \geq 1$ *there is a Müntz polynomial* $p_n(x) = \sum_{j=1}^{n} a_j x^{k_j}$ *with*

$$\|f - p_n\| \leq C\ \omega(f, \epsilon_n), \tag{1}$$

where

$$\epsilon_n = \max_{\mathrm{Re}\,z=1} \frac{1}{z}\left|\prod_{j=1}^{n} \frac{z-k_j}{z+k_j}\right| . \tag{2}$$

Here $\omega(f,\cdot)$ *is the modulus of continuity of* f. *Moreover this is best possible in the sense that there are functions the approximation to which is not better than the rate in* (1).

If the sequence $\{k_j\}$ satisfies $k_j \geq 2j\ \forall j$, then

$$\epsilon_n \sim \exp\left[-2 \sum_{j=1}^{n} 1/k_j\right]. \tag{3}$$

2. Rate of Approximation by $P_{\mathcal{A}}$

One may ask what if anything does one lose in the rate of approximation by restricting the coefficients, alternatively, what are the restrictions that would still guarantee the Müntz-Jackson rate of approximation (1).

For ordinary polynomials Bak, v. Golitschek and the author [1] have recently shown the following

THEOREM B. *If* $A_k \geq \delta k^2$ *for some* $\delta > 0$ *and all* $k \geq k_0$, *then the rate of approximation to functions in* $C_0[0,1]$ *by means of polynomials in* $P_{\mathcal{A}}$ *is at least that guaranteed by Jackson's theorem, namely, there exists a constant* $C > 0$ *such that for any* $f \in C_0[0,1]$ *and all sufficiently large* n, *there is a* $p_n \in P_{\mathcal{A}}$ *such that*

$$\|f - p_n\| \leq C\ \omega(f, \tfrac{1}{n}). \tag{4}$$

Moreover, for a sequence $\mathcal{A} = \{A_k\}$, *if the Jackson rate of approximation to functions in* $C_0[0,1]$ *by polynomials in* $P_{\mathcal{A}}$ *is guaranteed, then for every* $\epsilon > 0$ *and any sequence* $\{S_k\}$ *with* $S_k = O(k^{2-\epsilon})$ *there is a subsequence*

$\{k_i\}$ such that $A_{k_i} \geq S_{k_i}$.

This means that for ordinary polynomials if we would like to guarantee the Jackson rate of approximation our restricting sequence $\mathcal{A} = \{A_k\}$ should behave like $\{k^2\}$. So that although there is a gap between the necessary and sufficient parts of Theorem B it gives a good idea of the "best possible" restriction. But Theorem B does not take advantage of the fact that we may a-priori have only a Müntz subsequence of non zero A_k's, thus instead of trying to obtain (4) we should try to get the rate (1).

We prove the following

THEOREM 1. *Let* $\{k_j\}$ $(j \geq 1)$ *be a subsequence of the integers satisfying* $k_j \geq 2j$ $\forall j$ *and let* $e_n = \exp\left[-2 \sum_{j=1}^{n} 1/k_j\right]$. *Then there exists a constant* $C > 0$ *such that for any* $f \in C_0[0,1]$ *and any* $\epsilon > 0$ *there are Müntz polynomials* $p_n(x) = \sum_{j=1}^{n} a_{jn} x^{k_j}$ *with the properties that for all sufficiently large* n

$$(5) \qquad \|f - p_n\| \leq C\, \omega(f, e_n^{1-\epsilon})$$

and

$$(6) \qquad |a_{jn}| \leq 2\|f\|\, e^{2k_j} e_j^{-k_j(1-\epsilon)/\epsilon}, \quad 1 \leq j \leq n.$$

Comparing (5) and (1) we see by virtue of (3) that we may get arbitrarily close to the Müntz-Jackson rate and the payment is in the right hand side of inequality (6) the exponent in which grows to infinity as ϵ approaches zero. As we shall see this phenomenon occurs in other results in this theory.

COROLLARY. *Let* $\{k_j\}$ $(j \geq 1)$ *be a subsequence of the integers satisfying* $k_j \geq 2j$ $\forall j$ *and assume* $\mathcal{A} = \{A_k\}$ *is such that* $A_{k_j} \geq e^2 e_j^{-(1-\epsilon)/\epsilon}$ *for some* $\epsilon > 0$. *Then there exists a* $C > 0$ *such that for any* $f \in C_0[0,1]$ *and all sufficiently large* n *there is a polynomial* $p_n \in P_{\mathcal{A}}$ *of degree* $\leq k_n$ *for which* (5) *holds.*

PROOF of Theorem 1. Fix $0 < \epsilon < 1$ and let $m(\epsilon,n)$ be chosen so that

$$(7) \qquad \exp\left[(1-\epsilon)\sum_{j=1}^{m-1} 1/k_j - \epsilon \sum_{j=m}^{n} 1/k_j\right] \leq 1$$

and

$$\text{(8)} \qquad \exp\left[(1-\epsilon)\sum_{j=1}^{m} 1/k_j - \epsilon \sum_{j=m+1}^{n} 1/k_j\right] > 1.$$

Now apply Theorem A to the set of exponents $\{k_m, \ldots, k_n\}$. We conclude that there exists a constant $C > 0$ independent of the sequence $\{k_j\}$ and of m,n and ϵ such that for any $f \in C_0[0,1]$ there is a Müntz polynomial $p_n(x) = \sum_{j=m}^{n} a_{jn} x^{k_j}$ such that

$$\text{(9)} \qquad \|f - p_n\| \le C\ \omega(f, \eta_n)$$

where

$$\eta_n = \max_{\text{Re}z=1} \left| \frac{1}{z} \prod_{j=m}^{n} \frac{z-k_j}{z+k_j} \right| .$$

Assume that the maximum is achieved at $|z| = r$. Then since $k_j \ge 2j\ \forall j$ we have

$$\eta_n \le \frac{1}{r} \exp\left[-2 \sum_{j=m}^{n} \frac{k_j}{r^2+k_j^2}\right]$$

$$= \frac{1}{r} \exp\left[-2 \sum_{j=m}^{n} 1/k_j\right] \exp\left[2 \sum_{j=m}^{n} \frac{r^2}{k_j(r^2+k_j^2)}\right]$$

$$\le \frac{1}{r} \exp\left[-2 \sum_{j=m}^{n} 1/k_j\right] \exp \frac{1}{2} \log(1+r^2)$$

$$\le 2 \exp\left[-2 \sum_{j=m}^{n} 1/k_j\right] .$$

By virtue of (7) we have

$$\eta_n \le 2\ e_n^{1-\epsilon} .$$

Hence by (9) we have established (5).

In order to prove (6) note that by (5) our polynomials p_n tend to f in the uniform norm and so for sufficiently large n $\|p_n\| \le 2\|f\|$. Applying

[4, Lemma 2] we have

$$a_{jn} \leq 2\|f\| (1+k_j)^{1/2} e^{3k_j/2} \varphi(m,n)^{1+2k_j}, \quad m \leq j \leq n,$$

where

$$\varphi(n,m) = \exp\left[\sum_{j=m+1}^{n} 1/k_j\right].$$

By (8)

(10) $$\varphi(m,n) \leq e_m^{(1-\epsilon)/2\epsilon}$$

and for $m \leq j \leq n$

(11) $$\varphi(m,n)^{1/k_j} \leq \exp\left[\frac{1-\epsilon}{\epsilon m} \sum_{j=1}^{m} 1/j\right]$$

$$\leq \exp\left[\frac{1-\epsilon}{\epsilon m} \log m\right]$$

$$\leq e^{1/4}$$

for sufficiently large n since (8) implies that $m(\epsilon,n) \to \infty$ as $n \to \infty$. Now the proof of (6) is complete by combining (9) (10) and (11).

Theorem 1 can be extended to differentiable functions.

THEOREM 2. _Let $\{k_j\}$ be a subsequence of the integers satisfying $k_j \geq 2j$ $\forall j$ and let $r \geq 0$. There exists a constant $C_r > 0$ independent of the sequence $\{k_j\}$ such that for any $\epsilon > 0$ and any $f \in C^{(r)}[0,1]$ which satisfies $f^{(i)}(0) = 0$ for $i = 0, 1, \ldots, r$, there are Müntz polynomials $p_n(x) = \sum_{j=1}^{n} a_{jn} x^{k_j}$ such that_

$$\|f - p_n\| < C_r e_n^{(1-\epsilon)r} \omega(f^{(r)}, e_n^{1-\epsilon})$$

and the coefficients a_{jn} satisfy (6).

Let us return now to the phenomenon of having to "pay" in the rate of growth of the coefficients in order to obtain better rate of approximation by means of the Müntz polynomials. One case of this type is apparent in the second part of Theorem B. Another example of a little different nature is the following. It is easily seen that the function $f(x) = x^{1/2}$ is approximable by means of ordinary polynomials at the rate of n^{-1} although Jackson's theorem guarantees the rate $n^{-1/2}$. Recently Bak, v. Golitschek and the author [1] have shown that for any $\epsilon > 0$ $f(x) = x^{1/2}$ can be approximated at the rate of $n^{-(1-\epsilon)}$ by polynomials $p_n(x) - \sum_{k=1}^{n} a_{kn}x^k$ such that

$$|a_{kn}| \leq k^{2(1-\epsilon)k/\epsilon} . \tag{12}$$

However the rate n^{-1} cannot be achieved if (12) is satisfied no matter how small ϵ is.

REFERENCES

[1] Bak, J.-Golitschek, M.v. - Leviatan, D., The rate of approximation by means of polynomials with restricted coefficients. Israel J. Math. 26 (1977), 265-275.

[2] Bak, J.- Leviatan, D. - Newman, D.J. - Tzimbalario, J., Generalized polynomial approximation. Israel J. Math. 15 (1973), 337-349.

[3] Golitschek, M. v., Permissible bounds on the coefficients of generalized polynomials. Approx. Theory, (Proc. Conf. on Approx. Theory, Austin, Texas 1973) G. G. Lorentz Ed. Academic Press N.Y. 1973, 353 - 357.

[4] Golitschek, M.v. - Leviatan, D., Permissible bounds on the coefficients of approximating polynomials with real or complex exponents. J. Math. Anal. and Appl. 60 (1977), 123-138.

[5] Roulier, J., Restrictions on the coefficients of approximating polynomials. J. Approx. Theory 6 (1972), 276-282.

VIII Number Theory and Probability

THE ASYMPTOTIC DISTRIBUTION OF LATTICE POINTS IN EUCLIDEAN AND NON-EUCLIDEAN SPACES

Peter D. Lax and Ralph S. Phillips

Courant Institute, New York University, New York, New York

Department of Mathematics, Stanford University, Stanford, California

The counting numbers for discrete subgroups of motions in Euclidean and non-Euclidean spaces are obtained using the wave equation as the principal tool. In dimensions 2 and 3 the error estimates are close to the best known.

1. Introduction

Counting the number of lattice points in a circle is a classical number theoretic problem. One can describe the lattice points in the plane as the orbit of the origin when acted on by the group generated by unit translations in the horizontal and vertical directions. This suggests that we take as the non-Euclidean analogue of this problem the counting of orbit points, in a non-Euclidean circle, created by a discrete subgroup Γ of the motions of the hyperbolic plane:

(1.1) $$N(s;x,z) = \#[\gamma \in \Gamma;\ \text{dist}(x,\gamma z) \leqslant s]\,.$$

The estimation of $N(s;x,z)$ was first studied by H. Huber [1,2] who considered Fuchsian subgroups for which the fundamental domain was compact. Somewhat later S.J. Patterson [3] was able to handle all discrete Fuchsian subgroups with fundamental domains of finite area. A. Selberg (Stanford University lectures in 1980) treated the same problem in real hyperbolic spaces of arbitrary dimension, again for fundamental domains of finite volume. Selberg's error estimates are markedly better than those of Huber and Patterson.

In the present work we extend the previous results for 2 and 3 spatial dimensions, to discrete subgroups with the *finite geometric property*. This property requires that the polygonal representation of the fundamental domain have a finite number of sides; the volume may be finite or infinite. Our error estimates are essentially the same as Selberg's.

The main tool in our approach is the wave equation. The fact that signals travel with finite speed makes the wave equation especially well suited for this task. To understand why this is so, consider a spherically symmetric solution $u_o(z,t)$ with initial support in the ball $\{|z-x| < \Delta\}$ emanating from a point x in 3-space. Because of Huygens' principle, the solution will be different from zero at time t only in an annulus about x of inner and outer radii $t-\Delta$ and $t+\Delta$, respectively. To obtain a weighted count of the number of orbital points $\{\gamma z\}$ in this annulus we need only sum the solution over the group:

$$u(z,t) = \sum_{\gamma \in \Gamma} u_o(\gamma z,t). \tag{1.2}$$

Note that u(z,t) is the "automorphic solution" of the wave equation for the subgroup Γ and can be estimated directly. Summing over disjoint annulii of radius $\leq T$ and letting Δ tend to zero we obtain a progressively more accurate count. In Section 2 we carry out the details for the lattice problem in Euclidean 3 space.

Our results are summarized in the following two theorems:

THEOREM 1.1. *The number of lattice points* N(s) *in a sphere of radius* r *in* $\mathbb{R}_n$, $n=2$ *or* 3, *about a point* x *is*

$$N(s) = A_n(s) + \mathcal{O}\left(s^{\alpha_n+\varepsilon}\right) \tag{1.3}$$

where $A_2 = \pi s^2$, $A_3 = \frac{4}{3}\pi s^3$, $\alpha_2 = 2/3$ *and* $\alpha_3 = 3/2$.

Next let Γ be a discrete subgroup of the motions in a hyperbolic n-space. Γ having the finite geometric property. In this case the Laplace-Beltrami operator on the fundamental domain F has a finite number of eigenvalues $\lambda_1 \geq \lambda_2 \geq \ldots \geq \lambda_k$ lying above the continuous spectrum; that is $\lambda_i > -(\frac{n-1}{2})^2$. Denote the corresponding normalized eigenfunctions by $\{\varphi_i, i=1,\ldots,k\}$; $\varphi_1(x) \geq 0$ and is constant if and only if $\mathrm{vol}(F) < \infty$.

THEOREM 1.2. *Set*

$$(1.4) \qquad \mu_i = \frac{n-1}{2} + \sqrt{\lambda_i + (\frac{n-1}{2})^2} \; .$$

Then for $n = 2$

$$(1.5)_2 \quad N(s;x,z) = \sum_{\mu_i > \beta_2} e^{\mu_i s} \varphi_i(x)\varphi_i(z)\pi^{1/2} \frac{(\mu_i - 3/2)!}{\mu_i!} + \mathcal{O}(\exp(\beta_2 + \varepsilon)s)$$

where $\beta_2 = (\mu_1 + 1)/3$; and for $n = 3$

$$(1.5)_3 \quad N(s;x,z) = \sum_{\mu_i > \beta_3} e^{\mu_i s} \varphi_i(x)\varphi_i(z) \frac{\pi}{\mu_i(\mu_i - 1)} + \mathcal{O}\left(e^{(\beta_3 + \varepsilon)s}\right)$$

where $\beta_3 = (\mu_1 + 1)/2$.

It should be noted that when $\mathrm{vol}(F)=\infty$ send hence φ_1 is not constant, then the leading term in $N(s;x,z)$ depends on x and z. This makes it very unlikely that a purely geometric argument for (1.5) will be forthcoming.

2. The Number of Lattice Points in a Ball.

Denote by $N(s)$ the number of integer lattice points in a ball of radius s about a point x in 3-dimensional Euclidean space. It is well known that as $s \to \infty$, $N(s)$ is asymptotically equal to the volume of the sphere:

$$(2.1) \qquad N(s) \simeq \frac{4\pi}{3} s^3 .$$

It is further well known that the deviation of $N(s)$ from the asymptotic value above does not exceed the area of the sphere bounding the ball:

$$(2.2) \qquad |N(s) - \frac{4\pi}{3} s^3| \leqslant \mathcal{O}(s^2).$$

In this section we show that

THEOREM 2.1.

$$(2.3) \qquad |N(s) - \frac{4\pi}{3} s^3| \leqslant \mathcal{O}(s^{1.5} \log^{1/4} s).$$

Our proof is based on the behavior for large t of solutions of the wave equation

$$(2.4) \qquad u_{tt} - \Delta u = 0$$

in $\mathbb{R}^3$. A spherically symmetric solution of this equation is of the form

$$(2.5)\qquad \frac{h(r-t)}{r}, \quad r = |x|, \quad t \geq 0,$$

h some C_0^2 function supported on the positive axis. In what follows h shall be normalized by

$$(2.6)\qquad \int h(r)dr = 1;$$

in addition, h shall depend on a small parameter α in the following fashion:

$$(2.7)\qquad h(r) = \frac{1}{\alpha} h_1(\frac{r}{\alpha})$$

where h_1 is some C^2 function supported on (0,1), satisfying (2.6). We take h_1 to be ≥ 0.

We are interested in those solutions of the wave equation which are periodic in x. To construct such a solution out of the outgoing spherical wave (2.5) we sum over the group of all integer translations n:

$$(2.8)\qquad u(x,t) = \sum_n \frac{h(r_n - t)}{r_n}$$

where $r_n = |x_n| = |x-n|$. Note that for t bounded only a finite number of terms in (2.8) are $\neq 0$. Using the counting function N(s) defined above we can rewrite (2.8) as a Stieltjes integral:

$$(2.8)'\qquad u(x,t) = \int \frac{h(s,t)}{s} dN(s).$$

We form the integral

$$(2.9)\qquad I = \int_o^T u(x,t)t\, dt.$$

I is a function of T and of the parameter α which enters the function h via (2.7).

Setting (2.8)' into (2.9) we get after interchanging the order of integrations

$$(2.10)\qquad I = \int_o^T \int \frac{h(s-t)}{s} dN(s)t\, dt = \int_o^\infty g(s,T)\, dN(s),$$

where

$$(2.11)\qquad g(s,t) = \int_o^T \frac{h(s-t)}{s} tdt.$$

Integrating by parts in (2.10) gives

$$I = -\int g_s(s,T)N(s)\,ds. \tag{2.12}$$

Differentiating (2.11) gives

$$g_s = \int_0^T \frac{h'(s-t)}{s}\,t\,dt - \int_0^T \frac{h(s-t)}{s^2}\,t\,dt. \tag{2.13}$$

Integrating the first term by parts leads to

$$g_s = -h(s-T)\frac{T}{s} + s^{-2}\int_0^T h(s-t)(s-t)\,dt = -h(s-T)\frac{T}{s} + f(s,T). \tag{2.14}$$

LEMMA 2.2.

$$f(s,T) = \begin{cases} \mathcal{O}(\alpha s^{-2}) & \underline{\text{for}}\ s < T+\alpha \\ 0 & \underline{\text{for}}\ s > T+\alpha\ . \end{cases} \tag{2.15}$$

PROOF. It follows from (2.7) that the range of t integration in the second term in (2.14) can be restricted to $0 < s-t < \alpha$. For $s > T+\alpha$ this interval has empty intersection with $(0,T)$; this proves the second part of Lemma 2.2. For $s < T+\alpha$

$$f(s,T) \leqslant s^{-2}\int_{s-\alpha}^{s} h(s-t)(s-t)\,dt \leqslant s^{-2}\int_0^{\alpha} h(p)p\,dp \leqslant \alpha s^{-2}$$

by (2.7). This proves the lemma.

Setting (2.14) into (2.12) we get

$$I = \int h(s-T)\,\frac{T}{s}\,N(s)ds + \int f(s,T)N(s)\,ds.$$

Using (2.15) and the trivial estimate $N(s) = \mathcal{O}(s^3)$ we obtain

$$I = \int h(s-T)\,\frac{T}{s}\,N(s)ds + \mathcal{O}(T^2\alpha). \tag{2.16}$$

It follows from (2.7) that the range of the s integration in (2.16) is $T \leqslant s \leqslant T+\alpha$; since N(s) is an increasing function, we get using (2.6) the following upper and lower bounds for $I(T,\alpha)$:

$$\frac{T}{T+\alpha}N(T) + \mathcal{O}(T^2\alpha) \leqslant I(T,\alpha) \leqslant N(T+\alpha) + \mathcal{O}(T^2\alpha).$$

From this we deduce that

$$(2.17)\qquad I(T-\alpha,\alpha) + O(T^2\alpha) \leq N(T) \leq \frac{T+\alpha}{T} I(T,\alpha) + O(T^2\alpha).$$

We shall give now an independent asymptotic evaluation of I, based on splitting off the mean value of u. Define

$$(2.18)\qquad m(t) = \int_F u(x,t)\,dx,$$

where F is the unit cube in x - space. Integrating (2.4) over F shows that

$$\frac{d^2}{dt^2} m(t) = 0,$$

i.e. that m is a linear function:

$$(2.19)\qquad m(t) = at + b.$$

From (2.18),

$$(2.19)'\qquad a = \int_F u_t(x,0)\,dx, \quad b = \int_F u(x,0)\,dx.$$

For t small and α small, the sum (2.8) has only a single nonzero term:

$$(2.20)\qquad u_t(x,0) = -\frac{h'(r)}{r}, \quad u(x,0) = \frac{h(r)}{r}.$$

Setting this into (2.19) and using polar coordinates, $dx = 4\pi r^2\,dr$, we get, using (2.6), that

$$(2.21)\qquad \begin{aligned} a &= -4\pi \int h'(r)r\,dr = 4\pi, \\ b &= 4\pi \int h(r)r\,dr = O(\alpha). \end{aligned}$$

We now decompose $u = m(t) + v$; clearly v is a solution of the wave equation

$$v_{tt} - \Delta v = 0$$

and v has mean value zero;

$$(2.22)\qquad \int_F v(x,t)dx \equiv 0.$$

Using (2.19) and (2.21) we can write this decomposition as

$$u(x,t) = 4\pi t + O(\alpha) + v(x,t).$$

Setting this into the definition (2.9) of I gives

$$(2.23)\qquad I(T,\alpha) = \int_0^T u(x,t)t\,dt = \frac{4\pi}{3}T^3 + O(\alpha T^2) + V,$$

where

$$(2.24)\qquad V(x) = \int_0^T v(x,t)t\,dt.$$

We shall now estimate V with the aid of a Sobolev type inequality:

LEMMA 2.3. *Let* V(x) *be a periodic function in the unit cube* F *with mean value* = 0. *Let* q *be any number* $>3/4$; *then*

$$(2.25)\qquad |V(x)| \leq c(q)\|\Delta^q V\|,$$

where $\|\;\|$ *denotes the* L_2 *norm over* F, *and* c *is a constant whose dependence on* q *is*

$$(2.26)\qquad c(q) \leq C(q-\tfrac{3}{4})^{-1/2}.$$

For the sake of completeness we give a proof: Expand V in a Fourier series:

$$(2.27)\qquad V(x) = \sum a_n e^{2\pi i n\cdot x},\quad n \text{ integer vector.}$$

$a_o = \int V(x)dx$ which by assumption $=0$. Since the exponentials are eigenfunctions of Δ,

$$\Delta^q V = \sum a_n |2\pi n|^{2q} e^{2\pi i n\cdot x}.$$

According to the Parseval relation

$$(2.28)\qquad \|\Delta^q V\|^2 = (2\pi)^{4q}\sum |a_n|^2|n|^{4q}.$$

Using (2.27), the Schwarz inequality, and (2.28) we get

$$|V(x)| \leq \sum |a_n| = \sum |a_n|\;|n|^{2q}\frac{1}{|n|^{2q}} \leq \left\{\sum |a_n|^2|n|^{4q}\right\}^{1/2}\left\{\sum_{n\neq o}\frac{1}{|n|^{4q}}\right\}^{1/2}$$

$$= \|\Delta^q V\|\,c,$$

where

$$c = \frac{1}{(2\pi)^q} \left\{ \sum_{n \neq o} \frac{1}{|n|^{4q}} \right\}^{1/2}.$$

Replacing the sum by the integral $\int_{|x| \geq 1} r^{-4q} dx$ shows that c satisfies the inequality (2.26). □

Since v(x,t) has mean value 0 for every t, it follows from (2.24) that so does V. To apply Lemma 2.3 we need an estimate for $\|\Delta^q V\|$ for q close to but greater than 3/4. As we show below it is technically easier to start with ΔV and replace $\Delta^q V$ by $(1-\Delta)^{-p}\Delta V$. Because of this, we shall use the following variant of Lemma 2.3, whose proof is identical with that of Lemma 2.3:

LEMMA 2.3'. *Let* V(x) *be a periodic function in the unit cube with mean value* = 0. *Let* p *be any number* < 1/4; *then*

$$|V(x)| \leq c(1-p)\|(1-\Delta)^{-p}\Delta V\|, \tag{2.29}$$

where c *is the function defined before and satisfying* (2.26).

Using the definition (2.24) of V and the fact that v satisfies the wave equation we get, after an integration by parts:

$$\Delta V = \int_0^T \Delta v\, t\, dt = \int_0^T v_{tt}\, t\, dt = T v_t(T) - \int_0^T v_t\, dt. \tag{2.30}$$

Next we introduce the function w:

$$w = (1-\Delta)^{-p} v. \tag{2.31}$$

Applying $(1-\Delta)^{-p}$ to (2.30) we get

$$(1-\Delta)^{-p}\Delta V = T w_t(T) - \int_0^T w_t\, dt. \tag{2.32}$$

From this we deduce that

$$\|(1-\Delta)^{-p}\Delta V\| \leq 2T \max_t \|w_t\|. \tag{2.33}$$

Since v satisfies the wave equation, so does w. To estimate $\|w_t\|$ we apply the law of conservation of energy, which asserts that for solutions w of the wave equation the quantity

(2.34) $$E(w) = \int(|\partial_x w|^2 + |w_t|^2)\,dx$$

is independent of t. From this we conclude that

(2.35) $$\max_t \|w_t\| \leqslant E^{1/2}(w).$$

Conbining this with (2.29) and (2.33) gives

(2.36) $$|V(x)| \leqslant c(1-p)\ T\ E^{1/2}(w).$$

The last estimate that we need is obtained from

LEMMA 2.4. For w defined by (2.31) and for $p < 1/4$

(2.37) $$E(w) = \mathcal{O}(\alpha^{4p-3}).$$

Before proving this estimate we show how it can be used to prove Theorem 2.1. Setting it into (2.36) and using (2.26) gives

(2.38) $$|V(x)| \leqslant \mathcal{O}((1-4p)^{-1/2}\ T\ \alpha^{2p-3/2}).$$

We choose p optimally by minimizing the right side of (2.38); we find that

$$1 - 4p = \frac{1}{|\log\ \alpha|}\ .$$

Inserting this in (2.38) gives

(2.38)' $$|V(x)| \leqslant \mathcal{O}(T\ \alpha^{-1}\ |\log\ \alpha|^{1/2}).$$

Setting this into (2.23) yields

$$I(T,\alpha) = \frac{4\pi}{3}\,T^3 + \mathcal{O}(\alpha T^2) + \mathcal{O}(T\alpha^{-1}\ |\log\ \alpha|^{1/2}),$$

and combining this with (2.17), we get

$$|N(T) - \frac{4\pi}{3}\,T^3| \leqslant \mathcal{O}(\alpha T^2) + \mathcal{O}(T\ \alpha^{-1}\ |\log\ \alpha|^{1/2}).$$

The optimal choice for α is $\alpha = T^{-1/2}|\log\ \alpha|^{1/4}$, which yields the inequality (2.3) of Theorem 2.1.

We turn now to the proof of Lemma 2.4. Recall that v is obtained from u by removing the zero component. Therefore it follows that $(1-\Delta)^{-p}v = w$ is obtained in the same way from $(1-\Delta)^{-p}u = z$. In particular,

$$E(w) \leqslant E(z). \tag{2.39}$$

We estimate $E(z)$ by evaluating it at $t=0$. Setting (2.7) into formulas (2.20) for $u(0)$ and $u_t(0)$ we have

$$u(x,0) = \frac{1}{\alpha r} h_1(\frac{r}{\alpha}) \quad \text{and} \quad u_t(x,0) = \frac{-1}{\alpha^2 r} h_1'(\frac{r}{\alpha}). \tag{2.40}$$

We proceed to expand these functions in their Fourier series:

$$u(x,0) = \sum a_n e^{2\pi i\, n\cdot x} \quad \text{and} \quad u_t(x,0) = \sum b_n e^{2\pi i\, n\cdot x} \tag{2.41}$$

and estimate their Fourier coefficients a_n and b_n.

$$(2.42)_a \quad a_n = 2\pi \int \int_{-\pi/2}^{\pi/2} \exp(-2\pi i\, r\,|n|\cos\theta)\, \frac{1}{\alpha r} h_1(\frac{r}{\alpha}) r^2 \sin\theta \, d\theta \, dr$$

$$= 2\int \frac{\sin 2\pi r\,|n|}{|n|} \alpha^{-1} h_1(\frac{r}{\alpha})\, dr = O(\frac{1}{|n|})$$

$$= -\frac{2}{|2\pi n|^2} \int \frac{\sin 2\pi r\,|n|}{|n|} \alpha^{-3} h_1''(\frac{r}{\alpha})\, dr = O(\frac{1}{\alpha^2 |n|^3});$$

$$(2.42)_b \quad b_n = 2\int \frac{\sin 2\pi r\,|n|}{|n|} \alpha^{-2} h_1'(\frac{r}{\alpha})\, dr$$

$$= 4\pi \int (\cos 2\pi r\,|n|)\, \alpha^{-1} h_1(\frac{r}{\alpha})\, dr = O(1)$$

$$= \frac{2}{|2\pi n|} \int \frac{\cos 2\pi r\,|n|}{|n|} \alpha^{-3} h_1''(\frac{r}{\alpha})\, dr = (\frac{1}{\alpha^2 |n|^2}).$$

In terms of the coefficients a_n and b_n we can write

$$E(z) = \sum \frac{|n|^2 |a_n|^2}{(1+4\pi^2|n|^2)^{2p}} + \sum \frac{|b_n|^2}{(1+4\pi^2|n|^2)^{2p}} \tag{2.43}$$

We break up these sums into two parts

$$\sum{}' = \sum_{|n| \leqslant \ell} \quad \text{and} \quad \sum{}'' = \sum_{|n| > \ell}.$$

Using the estimates (2.42) we see that for p close to 1/4

$$E(z) \leqslant C \left\{ \sum{}' \frac{1}{(1 + |n|^2)^{2p}} + \sum{}'' \frac{\alpha^{-4}}{|n|^4 (1 + |n|^2)^{2p}} \right\}$$

$$\leqslant C [\ell^{3-4p} + \alpha^{-4} \ell^{-1-4p}] .$$

Choosing $\ell = 1/\alpha$, this becomes

$$E(z) = O(\alpha^{4p-3})$$

as asserted in Lemma 2.4. This completes the proof of Theorem 2.1.

REFERENCES

[1] Huber, H., Über eine neue Klasse automorpher Funktionen und ein Gitterpunktproblem in der hyperbolischen Ebene. Comm. Math. Helv. 30 (1956), 20-62.

[2] Huber, H., Zur analytischen Theorie hyperbolischer Raumformen und Bewegungsgruppen I. Math. Ann. 138 (1959), 1-26; II, Math. Ann. 142 (1961), 385-398 and 143 (1961), 463-464.

[3] Patterson, S.J., A lattice point problem in hyperbolic space. Mathematika 22 (1975), 81-88.

ON THE APPROXIMATION OF INDICATOR FUNCTIONS BY SMOOTH FUNCTIONS IN BANACH SPACES

Vygantas Paulauskas
Department of Mathematics
Vilnius V. Kapsukas University
Vilnius

In this note a problem of the approximation of indicator functions of some sets in Banach spaces by smooth functions is considered. The problem had arisen in limit theorems of probability theory in Banach spaces. One method of constructing such an approximation in Banach spaces with sufficiently smooth norm and sets with smooth boundary are given; some examples of such sets are considered.

The main purpose of this short note is to draw attention to mathematicians working on functional analysis to a problem which had arisen when considering limit theorems of probability theory in Banach spaces, but which itself can be formulated purely in terms of functional analysis. Some results connected with the problem are given.

We shall start with the formulation of the problem on limit theorems in Banach spaces, and this is done only for the purpose to show the role which the problem we shall speak about, takes in limit theorems. Let ξ_i, $i \geq 1$, be independent random variables, defined on some probability space $(\Omega, \mathcal{F}, P)$ with values in a Banach space B. Let $S_n = \sum_{i=1}^n \xi_i$ and $F_n(A) = P\{\omega : S_n(\omega) \in A\}$, A being a Borel set in B. Let η_i, $i \geq 1$, be another sequence of independent B-valued random variables, $Z_n = \sum_{i=1}^n \eta_i$, $G_n(A) = P\{\omega : Z_n(\omega) \in A\}$. The sums Z_n are chosen as approximations for S_n, so usually G_n (and of course distributions of η_i, $i \geq 1$) are Gaussian or stable measures. Then one wants to know how good the approximation is, that is to estimate the quantity

$$(1) \qquad \sup_{A \in \mathscr{E}} |F_n(A) - G_n(A)| = \sup_{A \in \mathscr{E}} \left| \int_B \chi_A(x)(F_n - G_n)(dx) \right|,$$

where $\mathscr{E}$ is some class of Borel sets and χ_A denotes as usual the indicator function of the set A. There are some methods for estimating (1) but it is necessary to mention that at present the infinite-dimensional case is very far from the completeness which is achieved in finite-dimensional case (see, for example, [5], [7], [8] for the infinite-dimensional case and [1] for the finite-dimensional case).

The so-called Trotter method for estimating (1) is based on the following idea: the indicator function $\chi_A(x)$ is approximated by a sufficiently smooth function, say $g_{A,\varepsilon}(x)$, which coincides with χ_A everywhere except for the set $A_\varepsilon \setminus A$, where $A_\varepsilon = \{x : \|x-y\| < \varepsilon, \; y \in A\}$. The error of such an approximation is of the order $G_n(A_\varepsilon \setminus A)$, which is usually of order ε, and the integral

$$\int_B g_{A,\varepsilon}(x)(F_n - G_n)(dx)$$

is estimated by expanding $g_{A,\varepsilon}$ in a Taylor series. If we consider the approximation with Gaussian measure G_n, then it is natural to consider random variables ξ_i with $E\|\xi_i\|^3 < \infty$ (E denotes the expectation with respect to measure P), therefore we need the expansion of $g_{A,\varepsilon}$ with three terms. The derivatives of $g_{A,\varepsilon}$ (if B_1 and B_2 are Banach spaces and $f : B_1 \to B_2$, then $D^{(i)}f(x)$, $D^{(i)}f(x)(h_1,h_2,\ldots h_i)$, $\|D^{(i)}f(x)\|$ denote the i-th derivative in the sense of Fréchet, the differential and norm of the derivative, respectively; for differentiation in normed spaces we refer to [4]) must satisfy the relation

$$\|D^{(i)}g_{A,\varepsilon}(x)\| \leqslant C\varepsilon^{-1}\chi_{A_\varepsilon \setminus A}(x),$$

where the constant C may depend on A and the space B.

Thus we are faced with the following problem of pure functional analysis. Let B be a Banach space and A be a connected Borel set in B. For which spaces B and sets A for any $\varepsilon > 0$ is it possible to construct the family of functions $g_{A,\varepsilon} : B \to [0,1]$, having the properties:

(2) $$g_{A,\varepsilon}(x) = \begin{cases} 1, & x \in A, \\ 0, & x \notin A_\varepsilon, \end{cases}$$

(3) $$\|D^{(i)}g_{A,\varepsilon}(x)\| \leqslant C\varepsilon^{-i}\chi_{A_\varepsilon \setminus A}(x), \qquad i = 1,2,3,$$

where $C = C(B,A)$ is a constant depending only on B and A. Since the problem is rather general, it is clear that it is difficult to find an exhaustive answer to it. Moreover, from the beginning we can exclude some Banach spaces, such as $C(0,1)$, ℓ_1, since it is known that in these spaces there does not exist any non-trivial real-valued differentiable function having bounded support (i.e., vanishing outside some bounded set). More precisely, let C^3 (C^∞) denote the class of functions $f : B \to R$ which are three (infinitely many) times differentiable. The Banach space B is said to be C^3-smooth, if the class C^3 contains non-trivial functions with bounded support. We refer the reader for this and more general definitions to [2]. Thus at once we can restrict ourselves to the class of C^3-smooth Banach spaces. It is known [2] that if the B norm as a map $B \setminus \{0\} \to R$ is three times differentiable, then B is C^3-smooth. (In [2] it was asked if the converse statement is true - if B is C^3-smooth, then there exists an equivalent norm which is three times differentiable - but it seems that till now the answer is not known). But even in Hilbert space, the norm of which is in the class C^∞, the above formulated problem is not trivial. Due to the partition of unity in H (see, for example [6]), for any closed connected set A and for any $\varepsilon > 0$ one can construct a function $g_{A,\varepsilon} \in C^\infty$ satisfying (2), but the troubles come with the estimation (3). If one looks carefully through the proof of the partition of unity, presented in [6], then it is easy th see that due to the non-constructivity of Lindelöf's lemma it is impossible to get a bound of the type (3) for the derivatives of $g_{A,\varepsilon}$, constructed in this way, and it seems unlikely that for any closed connected set A the solution will be affirmative.

In this note we propose one rather simple method for constructing $g_{A,\varepsilon}$ for some class of sets with sufficiently smooth boundary in Hilbert space and some Banach spaces, having norm in C^3. The idea, roughly speaking, is to transform the set A into the ball by means of a three times differentiable transformation and for the ball there is no difficulty to construct the required function.

We shall say that the Banach space B belongs to the class $\mathcal{D}$ if the norm $\phi(x) = \|x\|$, considered as a map $\phi : B \setminus \{0\} \to R$, is of class C^3 and satisfies the inequalities

$$(4) \qquad \|D^{(i)}\phi(x)\| < C\|x\|^{1-i}, \qquad i = 1,2,3.$$

It is easy to verify that ℓ_p, $p = 2$ or $3 < p < \infty$ are in $\mathcal{D}$.

Let $\mathcal{A} = \mathcal{A}(M_1, \dots M_5)$ stand for the class of closed sets A, satisfying the following conditions:

($\mathcal{A}$1) A is connected, $0 \in A$ and every ray tx, $t > 0$, $\|x\| = 1$ intersects the boundary ∂A of A at one point;

($\mathscr{A}$2) the functional $d_A(x) = \sup\{t>0 : tx\|x\|^{-1} \in A\}$ is three times differentiable for all $x \neq 0$ and

$$\|D^{(i)} d_A(x)\| \leqslant M_i \|x\|^{-i}, \qquad i=1,2,3, \tag{5}$$

$$\inf_{\|x\|=1} d_A(x) = M_4 > 0, \qquad \sup_{\|x\|=1} d_A(x) = M_5 < \infty. \tag{6}$$

Now we can formulate the following result.

THEOREM 1. *Let* $B \in \mathscr{D}$ *and* $A \in \mathscr{A}(M_1,\ldots M_5)$. *Then for any* $\varepsilon > 0$ *it is possible to construct a function* $g_{A,\varepsilon}$, *satisfying* (2) *and* (3), *and the constant in* (3) *will be dependent on* M_i, $i=1,\ldots 5$ *and the constant from* (4).

REMARKS. 1. It is easy to see that the requirement $0 \in A$ is not essential, since $g_{A-a,\varepsilon}(x) = g_{A,\varepsilon}(x+a)$.
2. In limit theorems, as a rule, one considers the family of sets of the form $A(r) = A\cdot r = \{x \in B : x = r\cdot y,\ y \in A\}$, $r>0$. It is possible to show that for all $r>0$ one can construct functions $g_{A(r),\varepsilon}$, and estimates of derivatives will be uniform with respect to r.

The main step in the proof of Theorem 1 is the following

LEMMA 2. *Let the set* A *satisfy conditions* ($\mathscr{A}$1) *and* ($\mathscr{A}$2), *and let the operator* $K : B \to B$ *be defined by equality*

$$K(x) = M_4 x (d_A(x))^{-1}. \tag{7}$$

Then: (i) $K(A) \equiv \{y \in B : y = K(x),\ x \in A\} = V_{M_4} \equiv \{x \in B : \|x\| < M_4\}$;

(ii) *there exists a set* $\bar{A}$, $A \subset \bar{A} \subset A_\varepsilon$ *such that*

$$K(\bar{A}) = V_{M_4(1+\varepsilon M_5^{-1})};$$

$$\text{(iii)}\quad K \in C^3,\quad \|D^{(i)}K(x)\| < L_i \|x\|^{-i+1}, \qquad i=1,2,3,\ x \neq 0, \tag{8}$$

where L_i *depends on* M_j, $j=1,\ldots,5$.

PROOF. (i) follows from the definition of K, since if $x \in A$ then $\|x\| \leqslant d_A(x)$, and if $x \in \partial A$ then $\|x\| = d_A(x)$. Now let $x \in (A_\varepsilon)^c$, $(A^c = B \setminus A)$, $e_x = x\|x\|^{-1}$, $x_1 \in \partial A$, $x_2 \in \partial A$, $x_i = t_i e_x$ for some $t_i > 0$, $i = 1,2$. Then

$$\begin{aligned}\|K(x)\| &= M_4\|x\|(d_A(x))^{-1} > M_4 d_{A_\varepsilon}(x)(d_A(x))^{-1} \\ &= M_4(d_A(x) + \|x_2 - x_1\|)(d_A(x))^{-1} \\ &> M_4(1 + \varepsilon M_5^{-1})\end{aligned}$$

This means that $K((A_\varepsilon)^c) \subset (V_{M_4(1+\varepsilon M_5^{-1})})^c$, or equivalently $K(A_\varepsilon) \supset V_{M_4(1+\varepsilon M_5^{-1})}$. Since $K(A) = V_{M_4}$ there exists a set $\overline{A}$, $A \subset \overline{A} \subset A_\varepsilon$, such that $K(A) = V_{M_4(1+\varepsilon M_5^{-1})}$, and (ii) is proved. To prove (iii) one needs to calculate three first derivatives of K. We give here the expressions of differentials of K

$$DK(x)(h_1) = M_4 h_1 (d_A(x))^{-1} - M_4 x (d_A(x))^{-2} Dd_A(x)(h_1)$$

$$\begin{aligned}D^{(2)}K(x)(h_1,h_2) &= 2(d_A(x))^{-3} M_4 x Dd_A(x)(h_1) Dd_A(x)(h_2) \\ &- (d_A(x))^{-2} M_4 [xD^{(2)}d_A(x)(h_1,h_2) + h_2 Dd_A(x)(h_1) + h_1 Dd_A(x)(h_2)]\end{aligned}$$

$$\begin{aligned}D^{(3)}K(x)(h_1,h_2,h_3) &= -(d_A(x))^{-2} M_4 [xD^{(3)}d_A(x)(h_1,h_2,h_3) + h_3 D^{(2)}d_A(x)(h_1,h_2) \\ &+ h_2 D^{(2)}d_A(x)(h_1,h_3) + h_1 D^{(2)}d_A(x)(h_2,h_3)] \\ &+ 2(d_A(x)^{-3} M_4 \{h_3 Dd_A(x)(h_1) Dd_A(x) h_2 + h_2 Dd_A(x)(h_1) Dd_A(x)(h_3) \\ &+ h_1 Dd_A(x)(h_2) Dd_A(x)(h_3) + x[D^{(2)}d_A(x)(h_1,h_2) Dd_A(x)(h_3) \\ &+ D^{(2)}d_A(x)(h_1,h_3) Dd_A(x)(h_2) + D^{(2)}d_A(x)(h_2,h_3) Dd_A(x)(h_1)]\} \\ &- 6(d_A(x)^{-4} M_4 x Dd_A(x)(h_1) Dd_A(x)(h_2) Dd_A(x)(h_3),\end{aligned}$$

$$h_1, h_2, h_3 \in B$$

Having these formulae and the estimates (5) one easily derives (8), and the lemma is proved.

PROOF OF THEOREM 1. Let us define the family of functions $f_\varepsilon : R_1 \to [0,1]$, $\varepsilon > 0$, having the properties:

$$f_\varepsilon(u) = \begin{cases} 1, & u < M_4 \\ 0, & u > M_4 + M_6\varepsilon, \qquad M_6 = M_4 M_5^{-1} \end{cases}$$

$$|f_\varepsilon^{(3)}(u)| \leq C(M_6\varepsilon)^{-3}\chi_{(M_4, M_4 + M_6\varepsilon)}(u),$$

$$f_\varepsilon^{(i)}(M_4) = 0, \qquad i = 1,2,3.$$

Now we put $g_{A,\varepsilon} = f_\varepsilon(\|K(x)\|)$, where $K : B \to B$ is defined in (7). From Lemma 2, (i) and (ii), it follows that (2) is fulfilled. In order to prove (3) we need expressions of $D^{(i)} g_{A,\varepsilon}(x)$. Now $g_{A,\varepsilon}$ is the composition of three functions and we must apply the formula of differentiation of compound functions twice (see [4]). For example, if we denote $\phi(x) = \|x\|$, $k(x) = (\phi o K)(x) = \|K(x)\|$, then

$$(9) \qquad D^{(3)} g_{A,\varepsilon}(x)(y,y,y) = f_\varepsilon'(k(x))D^{(3)}k(x)(y,y,y) + 3f_\varepsilon''(k(x))D^{(2)}k(x)(y,y)Dk(x)(y) + f_\varepsilon'''(k(x))(Dk(x)(y))^3.$$

By means of the same formula we get derivatives of k, for example

$$D^{(2)}k(x)(y,y) = D\phi(K(x))(D^{(2)}K(x)(y,y)) + D^{(2)}\phi(K(x))(DK(x)(y), DK(x)(y)).$$

Putting $D^{(i)}k(x)$, $i = 1,2,3$ into (9), and using the estimates (4), (8) we get the desired estimate (3). Theorem 1 is proved.

Now we shall consider the case when a set A is defined by means of some functional, e.g. $A = A_{r,f} = \{x \in B : f(x) < r\}$, and we shall formulate the conditions which f must satisfy in order that $A_{r,f}$ satisfies ($\mathscr{A}$1) and ($\mathscr{A}$2).

PROPOSITION 3. _Let_ $B \in \mathscr{D}$, _and_ $f : B \to R_1$ _satisfy the conditions_:

(F1) $f(x) \geq 0$ _for all_ $x \in B$, $f(tx) = t^\alpha f(x)$, $t > 0$, _for some_ $0 < \alpha < \infty$,

(F2) $\inf_{\|x\| = 1} f(x) = n_1 > 0$, $\sup_{\|x\| = 1} f(x) = n_2 < \infty$,

(F3) $f \in C^3$, $\|D^{(i)} f(x)\| < n_3 \|x\|^{-i+\alpha}$, $\qquad i = 1,2,3, \quad x \neq 0$.

Then $A = A_{r_o,f}$ *satisfies* ($\mathscr{A}1$) *and* ($\mathscr{A}2$) *with*

$$M_i = C(\alpha)(r_o n_1^{-1})^{1/\alpha}\max[(n_3 n_1^{-1})^i, 1], \; i = 1,2,3;$$

$$M_4 = (r_o n_2^{-1})^{1/\alpha}, \quad M_5 = (r_o n_1^{-1})^{1/\alpha}.$$

PROOF. It is easy to see that

$$d_A(x) = f^{-1/\alpha}(x) r_o^{1/\alpha}\|x\|,$$

and $M_4 = \inf_{\|x\|=1} d_A(x) = (r_o n_2^{-1})^{1/\alpha}, \qquad M_5 = \sup_{\|x\|=1} d_A(x) = (r_o n_1^{-1})^{1/\alpha}.$

The calculations of the derivatives of $d_A(x)$ are rather tedious, thus they are omitted.

The rest of this note is devoted to examples of functionals satisfying (F1) - (F3). For this purpose we need some further definitions, used in non-linear functional analysis (for details see [3]). Let B_1 and B_2 be Banach spaces.

DEFINITION 4. *An operator* $Q : B_1 \to B_2$ *is called an homogeneous polynomial operator of order* k, *if there exists a* k-*linear operator* $\overline{Q} : B_1 \times B_1 \times \ldots \times B_1 \to B_2$, *symmetric with respect to all its arguments (this means that the value of* $\overline{Q}$ *is the same for any rearrangement of arguments) such that* $Q(x) = \overline{Q}(x,\ldots,x)$ *for all* $x \in B_1$. *The operator* $\overline{Q}$ *is called polar operator for* Q.

DEFINITION 5. *An homogenous polynomial operator* $Q : B \to B^*$ *of order* k *is positive (positive definite) on the set* $M \subset B$, *if*

$$<\overline{Q}(x,x,\ldots,x,h),h> \; > 0, \; (<\overline{Q}(x,\ldots,x,h),h> \; > C\|x\|^{k-1}\|h\|^2, \; C > 0)$$

for all $x \in M$ *and* $h \in B$, $h \neq 0$, $x \neq 0$.

LEMMA 6[3]. *Let* $Q : B \to B^*$ *be a symmetric homogeneous polynomial operator of order* $k > 1$, *positive on the convex open set* M. *Then* $f(x) = <Q(x), x>$ *is a convex functional on* M, *and* $Df(x) = (k+1)Q(x)$. *(Symmetricity of* Q *means that* $<\overline{Q}(x_1,\ldots,x_k),x_{k+1}>$ *is symmetric with respect to all arguments).*

If we assume that $M = B$ and Q is positive definite on B, then it follows

$$\inf_{\|x\|=1} f(x) = \inf_{\|x\|=1} \langle Q(x),x\rangle > C.$$

Further, if we assume that $\overline{Q}$ is bounded, then Q is bounded, too, namely $\|Q\| < \|\overline{Q}\|$ (see [3]), and we get

$$\sup_{\|x\|=1} f(x) < \|\overline{Q}\|.$$

There are no difficulties to obtain the estimates

$$\|D^{(i)}f(x)\| \leqslant C\|x\|^{k+1-i}, \qquad i < k+1.$$

Thus we have the following

PROPOSITION 7. _Let_ $B \in \mathscr{D}$, _and_ $Q : B \to B^*$ _be a symmetric, homogeneous polynomial bounded operator of order_ $k \geqslant 3$, _positive definite on_ B. _Then_ $f(x) = \langle Q(x),x\rangle$ _satisfies_ (F1) - (F3).

As an example (see [3]) we can give the following operator of order 3, satisfying Proposition 7 in the case $B = L_2(0,1)$:

$$Q(x) \equiv y(s) = x(s) \int_0^1 K(s,t)x^2(t)dt,$$

where $K(s,t)$ is continuous and $0 < C_1 \leqslant K(s,t) \leqslant C_2 < \infty$. Then

$$f(x) = \int_0^1\int_0^1 K(s,t)x^2(s)x^2(t)\, ds\, dt.$$

A similar example can be given in the spaces L_p, for example, if $p = 2m$, $m \geqslant 1$ being an integer, $K(s,t)$ is the same as above, then

$$Q : L_p \to L_q, \quad Q(x) = x^{p-1}(s) \int_0^1 K(s,t)x^p(t)dt, \quad q = p(p-1)^{-1},$$

$$f(x) = \int_0^1\int_0^1 K(s,t)x^p(s)x^p(t)\, ds\, dt.$$

Another class of differentiable functionals one can get by means of operators of Hammerstein and Nemyckii. We recall (for details see [9]) that the operator of Nemyckii, acting from one space of functions to another, is of the form $N(u) = q(u(y),y)$, and the operator of Hammerstein - $\Gamma(u) = \int_0^1 K(x,y)q(u(x),x)\,dx$ (there for simplicity we took the interval $[0,1]$). In [9] there are given conditions in order for the operator Γ to have the first Fréchet derivative. Using the same ideas we can prove the following result. Let

$$(10)\qquad f(x) = \int_0^1 h(t) \int_0^1 K(t,y)q(x(y),y)\,dy\,dt, \qquad x \in L_p \equiv L_p(0,1), \quad p > 4.$$

PROPOSITION 8. <u>Let the following conditions hold</u>:

1) $h \in L_q$, $q = p(p-1)^{-1}$, $h(t) > 0$ <u>for all</u> $t \in [0,1]$,

2) $q(u,s)$ <u>is measurable with respect to</u> s <u>for fixed</u> u;

$$q(ku,s) = k^{p-1}q(u,s) \geqslant 0 \text{ for all } u \in R_1, \quad s \in [0,1];$$

$$|q(u,s)| \leqslant C|u|^{p-1},$$

3) $q_u^{(3)}(u,s)$ <u>is continuous with respect to</u> u <u>and</u>

$$|q_u^{(i)}(u,s)| \leqslant C|u|^{p-1-i}, \qquad i = 1,2,3,$$

4) $\int_0^1\int_0^1 (K(x,y))^p\,dx\,dy < \infty$, $K(x,y) > 0$ <u>for all</u> $x,y \in [0,1]$.

<u>Then</u> f <u>from</u> (10) <u>satisfies</u> (F1), (F3) <u>and</u> $\sup_{\|x\|=1} f(x) = n_2$.

REMARK. The boundedness from below is not achieved by means of conditions 1)-4). This can be done under stronger conditions on h, K and q.

The proof of proposition 8 consists of calculating the derivatives of f, and since f is the composition of a linear functional and operator Γ, thus we need to calculate $D^{(i)}\Gamma(x)$. We omit all calculations and give the final result:

$$D\Gamma(u)(v) = \int_0^1 K(x,y)q_u'(u(y),y)v(y)\,dy,$$

$$D^{(2)}\Gamma(u)(v,z) = \int_0^1 K(x,y)q_u''(u(y),y)v(y)z(y)\,dy,$$

$$D^{(3)}\Gamma(u)(v,z,w) = \int_0^1 K(x,y)q_u'''(u(y),y)v(y)z(y)w(y)\,dy.$$

From here it is easy to get the estimates for $\|D^{(i)}f(x)\|$.

ACKNOWLEDGMENT. It was a great pleasure to receive a proposition from Prof. P.L. Butzer to present a paper for the proceedings of the conference, inspite of the fact that I was not able to participate in the conference.

REFERENCES

[1] Bhattacharya, R.N. - Rao,R.R., Normal Approximation and Asymptotic Expansions. John Willey and Sons, N.Y. 1976.

[2] Bonic, R. - Frampton, J., Smooth functions on Banach manifolds. J. Math. Mech. 15 (1966) 877-893.

[3] Burysek,S., Eigenvalue problem, bifurcations and equations with analytic operators in Banach spaces. Theory of Non-linear Operators, Schriftenreihe des Zentralinstituts für Math. und Mech., DDR, H 20 (1975) 1-15.

[4] Cartan, A., Differential Calculus, Differential Forms. Moscow, 1971 (in Russian).

[5] Kuelbs, J., - Kurtz, T., Berry-Esseen estimates in Hilbert space and an application to the law of the iterated logarithm. Ann. Probab. 2,3 (1974) 387-407.

[6] Lang,S., Introduction to the Theory of Differentiable Manifolds. Moscow 1967 (in Russian).

[7] Paulauskas,V., On the rate of convergence in the central limit theorem in some Banach spaces. Teor. verojat. i primen. 21, 4 (1976) 775-791 (in Russian).

[8] Paulauskas, V., The rates of convergence in the central limit theorem in Banach spaces. Probab. Theory on Vector Spaces II, Lecture Notes in Math, 828 (1980) 234-243.

[9] Vainberg, M.M., Variational Methods for Analysis of Non-Linear Operators. Moscow, 1956 (in Russian).

ON THE o-CLOSENESS OF THE DISTRIBUTION OF TWO WEIGHTED SUMS OF BANACH SPACE VALUED MARTINGALES WITH APPLICATIONS

Marie-Theres Roeckerath
Lehrstuhl A für Mathematik
Aachen University of Technology
Aachen, W.-Germany

For two Banach space valued martingale difference sequences (MDS) $(X_i, F_i)_{i\in\mathbb{P}}$, $(Z_i, G_i)_{i\in\mathbb{P}}$, and a normalizing function $\varphi : \mathbb{N} \to (0,\infty)$, the closeness of the distributions of the weighted sums $\varphi(n)S_n := \varphi(n)\sum_{i=1}^n X_i$ and $\varphi(n)T_n := \varphi(n)\sum_{i=1}^n Z_i$ will be examined. For this purpose, a general theorem concerning weak convergence, equipped with little-o estimates will be established. By applying this theorem to a sequence of independent, mean-zero Gaussian random variables (r.vs.), this yields the central limit theorem (CLT) for martingales in Banach spaces.

1. Introduction

Let B be a real Banach space with a normalized, countable basis $(e_k)_{k\in\mathbb{N}}$, $\mathbb{N} = \{1,2,\ldots\}$, and norm $\|\circ\|_B$, $(X_i)_{i\in\mathbb{N}}$ a sequence of B-valued integrable random variables (r.v.) defined on a common probability space (Ω, A, P), and let $(F_i)_{i\in\mathbb{P}}$, $\mathbb{P} := \mathbb{N}\cup\{0\}$, be an increasing sequence of sub-σ algebras of A such that X_i is F_i-measurable for each $i\in\mathbb{N}$. Then $(X_i, F_i)_{i\in\mathbb{P}}$, $X_0 := 0$, is called a martingale difference sequence (MDS) if

$$E(X_i | F_{i-1}) = 0 \text{ a.s.} \qquad (i\in\mathbb{N}) . \tag{1.1}$$

This is equivalent to $(S_n, F_n)_{n\in\mathbb{P}}$ being a martingale, i.e., that

$$E(S_n | F_{n-1}) = S_{n-1} \text{ a.s.} \qquad (n\in\mathbb{N}) . \tag{1.2}$$

The results of this paper will include the case of two sequences of independent mean-zero B-valued r.vs. $(X_i)_{i\in\mathbb{N}}$ and $(Z_i)_{i\in\mathbb{N}}$, since $(X_i, F_i)_{i\in\mathbb{P}}$, $X_0 \equiv 0$ forms a MDS by choosing $F_0 := \{\emptyset, \Omega\}$ and $F_i := A(X_1,\ldots,X_i)$ (the generated

σ - algebra). It will be seen that our results for B - valued MDS yield the same order of approximation as is already known from the independent Hilbert-space case; compare [3,4]. Indeed, the material of this paper may be regarded as a generalization of that of [3] to martingales in the context of Banach spaces.

Of course our original aim was to try to extend the matter to the case of dependent r.v. in B-spaces. However in this respect, the use of conditional expectations in the proof is rather natural and seems to lead to a type of martingales (see also [8]). This paper will deal with little - o estimates, in contrast to that of [6], which is concerned with the large - O case. Here it will turn out that a Lindeberg-type condition will be of basic importance in the case of MDS, just as in the independent situation [3] (Note that pure convergence assertions are also of little - o type, namely $o(1)$; in this case a conditional Lindeberg condition has to be assumed (comp. [7] and [9]).) In 1975 Paulauskas [10], following up work initiated by V.M. Zolotarev [13] and H. Bergström [2], considered the closeness of the distribution of two weighted sums of independent r.v. in Hilbert spaces, and obtained large O - rates of convergence; these are of the same order as ours.

Stimulated by Basu's paper [1] of 1976 in the real case, the proof of our main theorem will be modelled upon Lévy's version (1925) of Lindeberg's method (1922) as developed into an operator method by Trotter [12] in (1959); the latter, however, tailored and applicable only for independent r.vs. has to be generalized in order to cover the case for MDS. The aim is to deduce estimates for the difference

$$(1.3) \qquad E[f(\varphi(n)S_n] - E[f(\varphi(n)T_n)] \qquad (f \in K),$$

where K is a function class essentially characterized by differentiability conditions upon $f : B \to \mathbb{R}$ (Theorem 3.1). As one possible application, the central limit theorem will be deduced by choosing the limiting sequence of r.v. Z_i to be Gaussian distributed in a suitable fashion (Theorem 4.1), see also [11].

2. Notations and Preliminaries

If B is a real Banach space with normalized basis $(e_k)_{k \in \mathbb{N}}$, then for each $x \in B$ there exists an unique sequence of reals $(x^{(k)})_{k \in \mathbb{N}}$ such that

$$(2.1) \qquad x = \sum_{k=1}^{\infty} x^{(k)} e_k ,$$

or, more precisely, $\lim_{n\to\infty} \| x - \sum_{k=1}^{n} x^{(k)} e_k \|_B = 0$. Such spaces are of course separable. Examples are any separable Hilbert spaces, as well as the spaces $L^p[0,1]$, l^p, $1 \leq p < \infty$, $C[0,1]$ and c_o. Let B^j denote the j - fold product space $B x \dots x B$ endowed with the max - norm $\|\mathfrak{x}\|_{B^j} := \max_{1 \leq k \leq j} \|x_k\|_B$, where $\mathfrak{x} := (x_1, \dots, x_j) \in B^j$. Then the space $L_j \equiv L_j(B^j, \mathbb{R})$ of all real valued multilinear continuous functions $g : B^j \to \mathbb{R}$ is a Banach space under the norm

$$\|g\|_{L_j} := \sup_{\|\mathfrak{x}\|_{B^j} = 1} |g(\mathfrak{x})| = \sup_{\substack{\mathfrak{x} \in B^j \\ x_k \neq 0}} \frac{|g(\mathfrak{x})|}{\|x_1\|_B \dots \|x_j\|_B}$$

Let f be a real valued function defined on B with (sup-norm) $\|f\|_\infty := \sup_{x \in B} |f(x)|$ (may be infinite), whose Fréchet derivatives $f^{(j)} : B \to L_j$ exist and are continuous for $1 \leq j \leq r$, $r \in \mathbb{N}$. Then one has Taylor's formula

$$(2.2) \qquad \begin{aligned} f(x+y) &= f(y) + \sum_{j=1}^{r} \frac{f^{(j)}(y)[x]^j}{j!} \\ &+ \frac{1}{(r-1)!} \int_0^1 (1-t)^{r-1} \{f^{(r)}(y+tx)[x]^r - f^{(r)}(y)[x]^r\} dt , \end{aligned}$$

where $x, y \in B$ and $[x]^j := (x, \dots, x) \in B^j$. Furthermore, one has for a j - times continuously differentiable function f

$$(2.3) \qquad f^{(j)}(y)(x, \dots, x) = \sum_{v_1=1, \dots, v_j=1}^{\infty} x^{(v_1)} \dots x^{(v_j)} \cdot f^{(j)}(y)(e_{v_1}, \dots, e_{v_j}),$$

where $v_k \in \mathbb{N}$, and $x^{(v_k)}$ are the unique components of x, $1 \leq k \leq j$, $y \in B$. Indeed, (2.3) follows immediately from (2.1) and the fact that the $f^{(j)}(y)$ belong to L_j, i.e., are multilinear and continuous. To abbreviate (2.3), we choose the following notations for $v = (v_1, \dots, v_j) \in \mathbb{N}^j$:

$$(2.4) \qquad |v| := j , \qquad x^v := \prod_{k=1}^{j} x^{(v_k)}$$

(2.4) $$f^{[v]}(\bullet) := f^{(j)}(\bullet)(e_{v_1}, \ldots, e_{v_j}) : B \to \mathbb{R}.$$

Then (2.3) takes on the form

(2.5) $$f^{(j)}(y)[x]^j = \sum_{|v|=j} x^v f^{[v]}(y) \qquad (x,y \in B).$$

The following further function classes are needed, $r \in \mathbb{N}$:

$$C_B^o \equiv C_B := \{f : B \to \mathbb{R};\ f \text{ uniformly continuous and bounded on } \mathbb{R}\}$$

$$C_B(L_r) := \{g : B \to L_r;\ g \text{ uniformly continuous and bounded on } L_r\}$$

$$C_B^r := \{f \in C_B; f^{(j)} \in C_B(L_j),\ 1 \leq j \leq r\}.$$

On C_B^r a seminorm is defined by

$$|f|_{C_B^r} := \sup_{x \in B} \|f^{(r)}(x)\|_{L_r} = \| \, \|f^{(r)}(\circ)\|_{L_r} \|_\infty .$$

Given an arbitrary probability space (Ω, A, P), let us now consider a B-valued r.v. $Z : \Omega \to B$, B endowed with the Borel σ-algebra $\mathcal{B}_B$, with distribution P_Z on $\mathcal{B}_B$ defined by $P_Z(B) := P(\{\omega \in \Omega | Z(\omega) \in B\})$ for all $B \in \mathcal{B}_B$. The expectation of Z is defined as $E(Z) := \int_\Omega Z(\omega)P(d\omega)$ and understood in the sense of Bochner. With (2.1) one has the representation $Z(\omega) = \sum_{k=1}^\infty (Z(\omega))^{(k)} e_k$, so that one can define the real-valued component r.vs. $Z^{(k)}$ by $Z^{(k)}(\omega) = (Z(\omega))^{(k)}$. For a j-tuple $v = (v_1, \ldots, v_j) \in \mathbb{N}^j$ define the r.v. $Z^v := \prod_{k=1}^j Z^{(v_k)}$.

3. General Martingale Convergence Theorem with Rates

The reason for assuming that the Banach space B has a countable normalized basis is that instead of posing all conditions upon the B-valued r.v. X_i, $i \in \mathbb{N}$, it allows one to pose them upon the associated real components $X_i^{(k)}$, $k \in \mathbb{N}$. For instance, it is not hard to verify (see [5]) that one version for $E(X_i \mid F_{i-1})$ can be written as $\sum_{k=1}^\infty E(X_i^{(k)} \mid F_{i-1}) e_k$. So $(X_i, F_i)_{i \in \mathbb{P}}$ is a MDS, iff the real components of X_i satisfy

(3.1) $$E(X_i^{(k)} | F_{i-1}) = 0 \quad \text{a.s.} \qquad (k, i \in \mathbb{N}),$$

or, equivlently,

$$E(X_i^v \mid F_{i-1}) = 0 \quad \text{a.s.} \qquad (|v| = 1,\ i \in \mathbb{N}) . \tag{3.2}$$

In the following, we say that a sequence $(X_i)_{i\in\mathbb{N}}$ satisfies a generalized Lindeberg condition of order $r \in \mathbb{N}$, iff for each $\delta > 0$, $n \to \infty$,

$$\sum_{i=1}^{n} \int_{\|x\|_B \geq \delta/\varphi(n)} \|x\|_B^r \, P_{X_i}(dx) = o_\delta\left(\sum_{i=1}^{n} E[\|X_i\|_B^r]\right) . \tag{3.3}$$

THEOREM 3.1. <u>Let</u> $(X_i, F_i)_{i\in\mathbb{P}}$ <u>and</u> $(Z_i, G_i)_{i\in\mathbb{P}}$ <u>be two</u> MDS, $r \in \mathbb{N}$, <u>and</u>

$$E[\|X_i\|_B^r] < \infty, \qquad E[\|Z_i\|_B^r] < \infty \tag{3.4}$$

<u>for each</u> $i \in \mathbb{N}$, <u>as well as for</u> $1 \leq |v| \leq r$, $i \in \mathbb{N}$ <u>let</u>

$$E(X_i^v \mid F_{i-1}) = E(Z_i^v \mid G_{i-1}) = C_{i,v} \quad \text{a.s.} \tag{3.5}$$

<u>Assume that the</u> r.vs. X_i <u>and</u> Z_i <u>satisfy the generalized Lindeberg condition</u> (3.3) <u>of order</u> r. <u>Then</u> $f \in C_B^r$ <u>implies for</u> $n \to \infty$

$$\begin{aligned} &|E[f(\varphi(n)S_n)] - E[f(\varphi(n)T_n)]| \\ &= o_f\left(\varphi(n)^r \sum_{i=1}^{n} (E[\|X_i\|_B^r] + E[\|Z_i\|_B^r])\right) . \end{aligned} \tag{3.6}$$

PROOF. For the two B-valued MDS $(X_i, F_i)_{i\in\mathbb{P}}$ and $(Z_i, G_i)_{i\in\mathbb{P}}$ defined on the common probability space (Ω, A, P) there exist a further probability space $(\tilde{\Omega}, \tilde{A}, \tilde{P})$ and two sequences of independent B-valued r.vs. $(\tilde{X}_i)_{i\in\mathbb{P}}$ and $(\tilde{Z}_i)_{i\in\mathbb{P}}$ (the $\tilde{X}_i$ also being independent of the G_i, and the $\tilde{Z}_i$ independent of the F_i, respectively) such that $P_{\tilde{X}_i} = P_{X_i}$ as well as $P_{\tilde{Z}_i} = P_{Z_i}$, $i \in \mathbb{P}$. Defining $\tilde{S}_n := \sum_{i=1}^n \tilde{X}_i$ and $\tilde{T}_n := \sum_{i=1}^n \tilde{Z}_i$ first note that $f(\varphi(n)S_n)$ and $f(\varphi(n)T_n)$, as well as $f(\varphi(n)\tilde{S}_n)$ and $f(\varphi(n)\tilde{T}_n)$ are real integrable r.vs. for each $f \in C_B$. By the triangle inequality one easily sees that

$$|E[f(\varphi(n)S_n)] - E[f(\varphi(n)T_n)]|$$

$$\leq |E[f(\varphi(n)\tilde{S}_n)] - E[f(\varphi(n)\tilde{T}_n)]| + |E[f(\varphi(n)T_n)] - E[f(\varphi(n)\tilde{S}_n)]|$$

$$+ |E[f(\varphi(n)\tilde{S}_n)] - E[f(\varphi(n)\tilde{T}_n)]|$$

$$= I_1 + I_2 + I_3, \quad \text{say.}$$

Setting

$$R_{n,i} := \sum_{k=1}^{i-1} X_k + \sum_{k=i+1}^{n} \tilde{Z}_k \qquad (1 \leq i \leq n,\ n \in \mathbb{N}),$$

a double application of Taylors formula (2.2) yields for $f \in C_B^r$

$$f(\varphi(n)S_n) - f(\varphi(n)\tilde{T}_n)$$

$$= \sum_{i=1}^{n} \{f(\varphi(n)R_{n,i} + \varphi(n)X_i) - f(\varphi(n)R_{n,i} + \varphi(n)\tilde{Z}_i)\}$$

$$= \sum_{i=1}^{n} \sum_{j=1}^{r} \frac{1}{j!} \{f^{(j)}(\varphi(n)R_{n,i})[\varphi(n)X_i]^j - f^{(j)}(\varphi(n)R_{n,i})[\varphi(n)\tilde{Z}_i]^j\}$$

$$+ \sum_{i=1}^{n} \frac{1}{(r-1)!} \int_0^1 (1-t)^{r-1} \Big\{ f^{(r)}(\varphi(n)R_{n,i} + t\varphi(n)X_i)[\varphi(n)X_i]^r$$

$$- f^{(r)}(\varphi(n)R_{n,i})]\ [\varphi(n)X_i]^r \Big\}\, dt \ -$$

$$- \sum_{i=1}^{n} \frac{1}{(r-1)!} \int_0^1 (1-t)^{r-1} \Big\{ f^{(r)}(\varphi(n)R_{n,i} + t\varphi(n)\tilde{Z}_i)[\varphi(n)\tilde{Z}_i]^r$$

$$- f^{(r)}(\varphi(n)R_{n,i})[\varphi(n)\tilde{Z}_i]^r \Big\}\, dt \ .$$

Since $f \in C_B^r$, $f^{(r)}$ is uniformly continuous, so that for each $\varepsilon > 0$ there exists a $\delta = \delta(\varepsilon) > 0$ with $\| f^{(r)}(\varphi(n)R_{n,i} + t\varphi(n)X_i) - f^{(r)}(\varphi(n)R_{n,i})\|_{L_r} < \varepsilon$ for all $\| t\varphi(n)X_i\|_B < \delta$, or all $\|X_i\|_B < \delta/\varphi(n)$ because $t \in [0,1]$. In the same way

$$\| f^{(r)}(\varphi(n)R_{n,i} + t\varphi(n)\tilde{Z}_i) - f^{(r)}(\varphi(n)R_{n,i})\|_{L_r} < \varepsilon$$

for all $\|Z_i\|_B < \delta/\varphi(n)$, $i \in \mathbb{N}$.

Noting that for an arbitrary $g \in L_r$ there holds the inequality

$$|g[x]^r| \leqslant \|g\|_{L_r} \|x\|_B^r ,$$

one has, together with (3.4),

$$E\Big[|\{f^{(r)}(\varphi(n)R_{n,i} + t\varphi(n)X_i) - f^{(r)}(\varphi(n)R_{n,i})\}[\varphi(n)X_i]^r| \Big]$$

$$\leqslant E\Big[\| f^{(r)}(\varphi(n)R_{n,i} + t\varphi(n)X_i) - f^{(r)}(\varphi(n)R_{n,i})\|_{L_r} \varphi(n)^r \|X_i\|_B^r \Big]$$

$$= E\Big[\| f^{(r)}(\varphi(n)R_{n,i} + t\varphi(n)X_i) - f^{(r)}(\varphi(n)R_{n,i})\|_{L_r}$$

$$\varphi(n)^r \|X_i\|_B^r \{ \mathbb{1}_{\|X_i\|_B < \delta/\varphi(n)} + \mathbb{1}_{\|X_i\|_B \geqslant \delta/\varphi(n)} \} \Big]$$

$$\leqslant \varphi(n)^r \{\varepsilon\, E[\|X_i\|_B^r] + 2|f|_{C_B^r} \int_{\|x\|_B \geqslant \delta/\varphi(n)} \|x\|_B^r P_{X_i}(dx)\} ,$$

where $\mathbb{1}$ denotes the indicator function. Likewise one has with $P_{\tilde{Z}_i} = P_{Z_i}$,

$$E\Big[|(f^{(r)}(\varphi(n)R_{n,i} + t\varphi(n)\tilde{Z}_i - f^{(r)}(\varphi(n)R_{n,i}))[\varphi(n)\tilde{Z}_i]^r| \Big]$$

$$\leqslant \varphi(n)^r \{\varepsilon\, E[\|Z_i\|_B^r] + 2|f|_{C_B^r} \int_{\|x\|_B \geqslant \delta/\varphi(n)} \|x\|_B^r P_{Z_i}(dx)\} .$$

Since the X_i and Z_i satisfy the Lindeberg condition (3.3), this yields that

$$I_1 \equiv |E[f\varphi(n)S_n) - E[f(\varphi(n)\tilde{T}_n)]|$$

$$\leqslant \sum_{i=1}^{n} \sum_{j=1}^{r} \frac{\varphi(n)^j}{j!} |E[f^{(j)}(\varphi(n)R_{n,i})([X_i]^j - [\tilde{Z}_i]^j)]|$$

$$+ o(\varphi(n)^r \sum_{i=1}^{n} (E[\|X_i\|_B^r + E[\|Z_i\|_B^r]))$$

$$+ o(\varphi(n)^r\, 2|f|_{C_B^r} \sum_{i=1}^{n} (E[\|X_i\|_B^r] + E[\|Z_i\|_B^r])) \qquad (n \to \infty).$$

In view of the fact that for each $1 \leqslant j \leqslant r$ one has

$$E[f^{(j)}(\varphi(n)R_{n,i})\,([X_i]^j - [\tilde{Z}_i]^j)\,]$$

$$= \sum_{|v|=j} (E[X_i^v\, f^{[v]}(\varphi(n)R_{n,i})] - E[\tilde{Z}_i^v\, f^{[v]}(\varphi(n)R_{n,i})]),$$

compare [5], it remains to show that for $1 \leqslant i \leqslant n$, $n \in \mathbb{N}$, $1 \leqslant j \leqslant r$, and $|v| = j$

$$E[X_i^v f^{[v]}(\varphi(n)R_{n,i})] = E[\tilde{Z}_i^v f^{[v]}(\varphi(n)R_{n,i})] \tag{3.7}$$

For a set $\mathcal{C} \subset \mathcal{P}(\Omega)$ or a r.v. X let $\mathcal{A}(\mathcal{C})$ and $\mathcal{A}(X)$ be the σ-algebra generated by $\mathcal{C}$ and X, respectively. Setting

$$\mathcal{A}_{i,n} := \mathcal{A}(\mathcal{F}_{i-1} \cup \mathcal{A}(\tilde{Z}_{i+1},\dots,\tilde{Z}_n)),$$

one deduces by standard arguments for the conditional expectation of real r.vs.

$$E[X_i^v f^{[v]}(\varphi(n)R_{n,i})] - E[\tilde{Z}_i^v f^{[v]}(\varphi(n)R_{n,i})]$$

$$= E\{E[X_i^v f^{[v]}(\varphi(n)R_{n,i})\,|\,\mathcal{A}_{i,n}] - E[\tilde{Z}_i^v f^{[v]}(\varphi(n)R_{n,i})\,|\,\mathcal{A}_{i,n}]\} \tag{3.8}$$

$$= E\{f^{[v]}(\varphi(n)R_{n,i})[E(X_i^v \mid \mathcal{A}_{i,n}) - E(\tilde{Z}_i^v \mid \mathcal{A}_{i,n})]\}$$

since $f^{[v]}(\varphi(n)R_{n,i})$ is $(\mathcal{A}_{i,n}, \mathcal{B}_{\mathbb{R}})$ - measurable. Moreover $\mathcal{A}(\tilde{Z}_{i+1},\dots,\tilde{Z}_n)$ is independent of $\mathcal{A}(\mathcal{F}_{i-1} \cup \mathcal{A}(X_i))$. Therefore $E(X_i^v|\mathcal{A}_{i,n}) = E(X_i^v|\mathcal{F}_{i-1})$ a.s. and since $\mathcal{A}(\tilde{Z}_i)$ is independent of $\mathcal{A}_{i,n}$, one has that $E(\tilde{Z}_i^v|\mathcal{A}_{i,n}) = E(\tilde{Z}_i^v) = E(Z_i^v)$a.s..

Together with asumption (3.5) this implies that $E(\widetilde{X}_i^v | A_{i,n}) = E(\widetilde{Z}_i^v | A_{i,n})$ a.s., since $E[\widetilde{X}_i^v] = E[X_i^v] = E(X_i^v | F_{i-1}) = E(Z_i^v | G_{i-1}) = E[Z_i^v] = E[\widetilde{Z}_i^v]$ a.s.. Therefore (3.7) holds. Analoguously we can show that

$$I_2 = |E[f(\varphi(n)T_n)] - E[f(\varphi(n)\widetilde{S}_n]|$$

$$= o(\varphi(n)^r \sum_{i=1}^{n} (E[\|Z_i\|_B^r] + E[\|X_i\|_B^r])) \qquad (n \to \infty),$$

by choosing $R_{n,i} := \sum_{k=1}^{i-1} Z_k + \sum_{k=i+1}^{n} \widetilde{X}_k$ and $A_{i,n} := A(G_{i-1} \cup A(\widetilde{X}_{i+1}, \ldots, \widetilde{X}_n))$. In this case (3.5) implies $E(Z_i^v | A_{i,n}) = E(\widetilde{X}_i^v | A_{i,n})$ a.s..

Since $(\widetilde{Z}_i)_{i \in \mathbb{N}}$ and $(\widetilde{X}_i)_{i \in \mathbb{N}}$ are two sequences of independent mean-zero r.vs. one can tread the term $I_3 \equiv |E[f(\varphi(n)\widetilde{S}_n)] - E[f(\varphi(n)\widetilde{T}_n)]|$ in the same way as in [3.4]. This completes the proof of Thm. 3.1.

4. A type of CLT with rates for MDS

In this section, our general theorem will be applied to a concrete sequence of r.vs. Z_i and a concrete normalizing function $\varphi(n)$. If X is a Banach space valued r.v. with finite second moment $E(\|X\|_B^2) < \infty$, and $E(X) = 0$, the covariance functional is the symmetric continuous bilinear function

$$R_X(f^*, g^*) := E[f^*(X) g^*(X)] \qquad (f^*, g^* \in B^*).$$

Denote by X_R a Gaussian (distributed) r.v. with mean zero and covariance functional $R \equiv R_{X_R}$. (A mean zero Gaussian r.v. is uniquely determined by its covariance functional R.) It is wellknown that for a separable B-space the absolute moments of order s of any mean-zero Gaussian r.v. X_R are finite for all $s \geq 0$, i.e. $E[\|X_R\|_B^s] < \infty$. In the following theorem we have to assume that B is a Banach space of type 2. This space is characterized by the fact that for each r.v. $X : \Omega \to B$ with covariance functional R_X there exists a mean-zero Gaussian r.v. X_R with the same covariance functional $R = R_X$. For this material see especially [11, p. 36,46] and the literature cited there.

THEOREM 4.1. Let B be of type 2, $r \geq 2$, $(X_i, F_i)_{i \in \mathbb{P}}$ be a MDS with $E[\|X_i\|_B^r] < \infty$ and covariance functional R_i, $i \in \mathbb{N}$. Assume that

$$E(X_i^v \mid F_{i-1}) = E[X_{R_i}^v] \text{ a.s.} \qquad (1 \leq |v| \leq r,\ i \in \mathbb{N}) \tag{4.1}$$

as well as that the sequences $(X_i)_{i \in \mathbb{N}}$ and $(X_{R_i})_{i \in \mathbb{N}}$ satisfy the generalized Lindeberg condition (3.3) of order r. Then $f \in C_B^r$ implies $(n \to \infty)$

$$\begin{aligned} &|E[f(\varphi(n)S_n)] - E[f(X_{\varphi(n)^2 \sum_{i=1}^n R_i})]| \\ &= o_f(\varphi(n)^r \sum_{i=1}^n (E[\|X_i\|_B^r] + E[\|X_{R_i}\|_B^r])) . \end{aligned} \tag{4.2}$$

In particular, if the r.vs. X_i, $i \in \mathbb{N}$ are identically distributed with common covariance functional R, then (4.1) yields for each $f \in C_B^r$ and $\varphi(n) := n^{-1/2}$

$$|E[f(S_n/\sqrt{n})] - E[f(X_R)]| = o_f(n^{-(r-2)/2}) \qquad (n \to \infty) .$$

The Proof of Thm. 4.1 follows from Thm 3.1 noting that there exist n independent mean-zero Gaussian r.vs. X_{R_i} such that $P_{X_{\varphi(n)^2 \sum_{i=1}^n R_i}} = P_{\varphi(n) \sum_{i=1}^n X_{R_i}}$. If the r.v. X_i are identically distributed (3.3) is fullfilled perse.

Finally note that it would also be possible to formulate a version of the weak law of large numbers for martingales in the frame of Banach spaces as an application of theorem 3.1 provided that instead of condition (3.5) a weaker asumption upon the rate of growth of the difference $E(X_i^v \mid F_{i-1}) - E(Z_i^v \mid G_{i-1})$ will be posed. For details see [11].

This work as well as the paper [6] was supported by the DFG research grant Bu 166/33, which is gratefully acknowledged.

REFERENCES

[1] Basu, A.K., On the rate of approximation in the central limit theorem for dependent random variables and random vectors. preprint (1979).

[2] Bergstroem, H., A comparison method for distribution functions of independent and dependent random variables. Teor. Verojatnost. i Primenen. 15 442 - 468; Transl.: Theor. Probability Appl. 15 (1970), 430 - 457.

[3] Butzer, P.L. - Hahn, L. - Roeckerath, M.Th., General theorems on "little-o" rates of closeness of two weighted sums of independent Hilbert space valued random variables with application. J. Multivariate Anal. 9 (1979), 487 - 510.

[4] Butzer, P.L. - Hahn, L. - Roeckerath, M.Th., The stable limit laws and weak law of large numbers for Hilbert space with "large-O" rates. Festschrift volume on the occasion of the 75th birthday of E. Lukacs. Academic Press, New York (in print) (1981).

[5] Butzer, P.L. - Hahn, L. - Roeckerath, M.Th., Central limit theorem and weak law of large numbers with rates for martingales in Banach spaces, (to appear) (1981).

[6] Butzer, P.L. - Roeckerath, M.Th., Central limit theorem with large-O rates for martingales in Banach spaces. In: Proceedings of the Conference "Analytische Methoden der Wahrscheinlichkeitsrechnung", Lecture Notes Math. (in print) (1981).

[7] Dvoretzky, A., Asymptotic normality for sums of dependent random variables. Proc. Sixth Berkeley Symp. on Math. Stat. and Prob., Voll. II, (1970), 513 - 535.

[8] Erickson, R.V. - Quine, M.P. - Weber, N.C., Explicit bounds for the departure from normality of sums of dependent random variables. Acta Math. Acad. Sci. Hungar. 34 (1979), 27 - 32.

[9] Gaenssler, P. - Strobel, J. - Stute, W., On central limit theorems for martingale triangular arrays. Acta Math. Acad. Sci. Hungar. 31 (1978), 205 - 216.

[10] Paulauskas, V.I., On the closeness of the distribution of two weighted sums of independent random variables with values in Hilbert space. Litovsk. Mat. Sb. 15 (1975), 177 - 200.

[11] Roeckerath, M.Th., Der Zentrale Grenzwertsatz und das Schwache Gesetz der Großen Zahlen mit Konvergenzraten für Martingale in Banachräumen. Doctoral Dissertation, RWTH Aachen 1980.

[12] Trotter, H.F., An elementary proof of the central limit theorem. Arch. Math. 10 (1959), 226 - 234.

[13] Zolotarev, V.M., On the closeness of the distributions of two sums of independent random variables. Teor. Verojatnost. i Primenen. 10 (1965), 519 - 526.

IX Splines and Numerical Integration

APPROXIMATION UND TRANSFORMATIONSMETHODEN III

Walter Schempp
Lehrstuhl für Mathematik I
Universität Siegen
Siegen

The present paper is concerned with the inter-relation of the theory of univariate spline functions and the harmonic analysis. Specifically it deals with (I) the periodic spline interpolants with equidistant knots and uniformly spaced data, (II_1) the cardinal exponential splines, and (II_2) the cardinal logarithmic spline interpolants. The underlying groups and their associated transforms are (I) the Heisenberg group mod N and the finite Fourier cotransform, (II_1) the additive group $\mathbb{R}$ and the inverse Laplace transform, and (II_2) the multiplicative group $\mathbb{R}_+^\times$ and the inverse Mellin transform, respectively. The principle aim is to show how these transforms may be used to represent the splines of the type referred to above. Finally, the paper presents an "Erlanger Programm" for splines with "regular" knot sequences on $\mathbf{T}$, $\mathbb{R}$, and $\mathbb{R}_+^\times$, respectively.

1. Einleitung

Zu den derzeit wohl am besten verstandenen (univariaten) Splines gehören

(I) die periodischen Spline-Funktionen

und

(II) die kardinalen Spline-Funktionen.

Ein wichtiger Grund, weshalb für diese beiden Klassen von Splines eine abgerundete und wirksame Theorie besteht, ist in ihrem engen Zusammenhang zur harmonischen Analyse zu sehen. Im Fall (I) wird dieser Zusammenhang durch die endliche Fourier-Kotransformation geliefert, also durch die Fourier-Kotransformation auf der zyklischen Gruppe $\mathbb{Z}/N\mathbb{Z}$ der ganzen Zahlen mod N. Dabei zeigt sich jedoch, daß nicht die zyklische Gruppe $\mathbb{Z}/N\mathbb{Z}$ selbst, sondern die der geometrischen Anschauung nicht unmittelbar zugängliche, endliche nilpotente Heisenberg-Gruppe mod N die für die periodischen Splines "zuständige" Gruppe ist (Abschnitt 4). Im Falle (II) ist der Zusammenhang zur (kommutativen) harmonischen Analyse durch "diskontinuierliche Faktoren" gegeben, die mit Hilfe einer zur unterliegenden Gruppe gehörenden linearen Integraltransformation dargestellt werden. Wir behandeln (II_1) kardinale exponentielle Splines auf der additiven Gruppe $\mathbb{R}$ der reellen Zahlen mit Hilfe der inversen Laplace-Transformation (Abschnitt 6) und (II_2) kardinale logarithmische Interpolationssplines auf der multiplikativen Gruppe $\mathbb{R}_+^\times$ der strikt positiven reellen Zahlen mit Hilfe der inversen Mellin-Transformation (Abschnitt 7). Der letzte Abschnitt schließlich faßt die bei "regulären" Knotenfolgen zugrunde liegenden Strukturen tabellarisch zusammen.

Der Autor ist Herrn Professor Zvi Ziegler (Haifa/Israel) für seine Gastfreundschaft im Technion und Herrn Professor Paul R. Halmos (Bloomington/Indiana) für sein Interesse an dieser Arbeit und seine Ermutigungen sehr zu Dank verpflichtet.

2. Periodische Splines

Es sei $m \geq 1$ eine natürliche Zahl und $\mathfrak{S}_m(\mathbb{R};\mathfrak{k})$ der komplexe Vektorraum aller polynomialen Spline-Funktionen auf $\mathbb{R}$ vom Grad m zur unendlichen Knotenfolge $\mathfrak{k}$. Ist $\mathfrak{k}$ äquidistant und besitzt die "Gitterkonstante" $k = \frac{1}{N}$ $(N \in \mathbb{N},\ N \geq 1)$, gilt also

$$\mathfrak{k} = (nk)_{n \in \mathbb{Z}}, \tag{1}$$

so läßt sich jede Spline-Funktion $s \in \mathfrak{S}_m(\mathbb{R};\mathfrak{k})$ mit Hilfe des Schoenbergschen Basis-Splines $b_m \in \mathfrak{S}_m(\mathbb{R};\mathfrak{k})$ und einer durch s eindeutig bestimmten Folge $(\alpha_n)_{n \in \mathbb{Z}}$ komplexer Zahlen in der Form

$$s = \sum_{n \in \mathbb{Z}} \alpha_n b_m(.-nk) \tag{2}$$

als Linearkombination von Translationen von b_m (punktweise) darstellen. Der Basis-Spline b_m hat das kompakte Intervall $[0,(m+1)k]$ zum Träger, ist in seinem Inneren $]0,(m+1)k[$ strikt positiv und erfüllt die Standardisierungsbedingung $\int_{\mathbb{R}} b_m(t)dt = 1$.

Projiziert man die additive Gruppe $\mathbb{R}$ auf die Quotientengruppe $\mathbb{R}/\mathbb{Z}$, geht also zu der (als multiplikative Kreisgruppe aufgefaßten) eindimensionalen Torusgruppe $\mathbb{T}$ über, so kann $s \in \mathfrak{S}_m(\mathbb{R};\mathfrak{k})$ genau dann als Element des Spline-Raumes $\mathfrak{S}_m(\mathbb{T};\mathfrak{k}^{per})$ zur äquidistanten Knotenfolge $\mathfrak{k}^{per}=(e^{2\pi ink})_{0\leq n\leq N-1}$ aufgefaßt werden, falls die Koeffizientenfolge $(\alpha_n)_{n\in\mathbb{Z}}$ in (2) die Periodizitätsbedingung

$$\text{(3)} \qquad \alpha_n = \alpha_{n'} \quad \text{für } n \equiv n' \bmod N$$

erfüllt.

Ist eine Funktion $f\colon \mathbb{T} \to \mathbb{C}$ vorgegeben und verlangt man, daß f vom Spline $s \in \mathfrak{S}_m(\mathbb{T};\mathfrak{k}^{per})$ in den äquidistanten Punkten $(e^{2\pi i(a+pk)})_{0\leq p\leq N-1}$ ($a \in \mathbb{R}$) von $\mathbb{T}$ interpoliert wird, so erhält man für die komplexen Koeffizienten $(\alpha_n)_{n\in\mathbb{Z}}$ das lineare Gleichungssystem

$$\text{(4)} \qquad \sum_{n\in\mathbb{Z}} b_m(e^{2\pi i(a+(p-n)k)})\alpha_n = f(e^{2\pi i(a+pk)}) \qquad (p \in \mathbb{Z}).$$

Die wie folgt definierten komplexen Zahlen

$$\text{(5)} \qquad \beta_{pn}^{(m)} := \sum_{r\in\mathbb{Z}} b_m(e^{2\pi i(a+(p-n-rN)k)}) \qquad ((p,n) \in \mathbb{Z}\times\mathbb{Z})$$

erfüllen offenbar die Identität

$$\text{(6)} \qquad \beta_{pn}^{(m)} = \beta_{p'n'}^{(m)} \qquad \text{für } p-n \equiv p'-n' \bmod N$$

und gestatten wegen (3) das lineare Gleichungssystem (4) in der Form

$$\text{(7)} \qquad \sum_{0\leq n\leq N-1} \beta_{pn}^{(m)}\alpha_n = f(e^{2\pi i(a+pk)}) \qquad (0\leq p\leq N-1)$$

zu schreiben. Die zugehörende (von m abhängige) komplexe Koeffizientenmatrix

$$\text{(8)} \qquad \mathfrak{B}_m = (\beta_{pn}^{(m)})_{\substack{0\leq p\leq N-1\\ 0\leq n\leq N-1}}$$

ist wegen (6) zirkulant (Ahlberg-Nilson-Walsh [1]), d.h. sie ist dadurch bereits von den Elementen ihre 0-ten Zeile

$$\text{(9)} \qquad \beta_n^{(m)} := \beta_{0n}^{(m)} \quad (0\leq n\leq N-1)$$

eindeutig bestimmt, daß die p-te Zeile ($1 \leq p \leq N-1$) durch Übertragen des letzten Elementes der (p-1)-ten Zeile auf die erste Stelle und Verschieben aller übrigen Elemente der (p-1)-ten Zeile um jeweils eine Stelle nach rechts entsteht ("cyclic shift").

Es sei $\mathbb{C}[\mathbb{Z}/N\mathbb{Z}]$ die Gruppenalgebra der zyklischen Gruppe $\mathbb{Z}/N\mathbb{Z}$ über dem Körper $\mathbb{C}$. Bezüglich der kanonischen, durch die Restklassen mod N indizierten Basis von $\mathbb{C}[\mathbb{Z}/N\mathbb{Z}]$ ruft die Matrix $\mathfrak{B}_m$ einen Vektorraum-Endomorphismus B_m von $\mathbb{C}[\mathbb{Z}/N\mathbb{Z}]$ hervor. Offensichtlich existiert zu jeder Funktion f: $\mathbb{T} \to \mathbb{C}$ und zu jedem $a \in \mathbb{R}$ genau ein f in den äquidistanten Punkten $(e^{2\pi i(a+pk)})_{0 \leq p \leq N-1}$ von $\mathbb{T}$ interpolierender Spline $s \in \mathfrak{S}_m(\mathbb{T};\mathfrak{k}^{per})$, falls der Grad m und die Anzahl N der Knoten so gewählt sind, daß der Endomorphismus B_m invertierbar ist. Unser erstes Ziel besteht darin, die Struktur von B_m und damit auch die der periodischen Interpolationssplines $s \in \mathfrak{S}_m(\mathbb{T};\mathfrak{k}^{per})$ selbst von der Warte der harmonischen Analyse aus zu verstehen. Dazu hat man im Sinne des "Erlanger Programms" von Felix Klein zunächst die "zuständige" Gruppe ausfindig zu machen.

3. Die Heisenberg Gruppe A(G)

Es bezeichne G eine (additiv geschriebene) abelsche lokalkompakte topologische Gruppe, $\hat{G}$ die zu G duale Gruppe (ebenfalls additiv geschrieben) und, wie üblich, $(x,\hat{x}) \rightsquigarrow \langle x,\hat{x}\rangle$ die zur Dualität $(G,\hat{G})$ gehörende kanonische Abbildung von $G \times \hat{G}$ in $\mathbb{T}$. Nach Weil [10] wird die über G modellierte Heisenberg-Gruppe A(G) folgendermaßen konstruiert: Man wählt als den A(G) unterliegenden lokalkompakten topologischen Raum das cartesische Produkt $G \times \hat{G} \times \mathbb{T}$ und definiert mit Hilfe des Bicharakters $(G \times \hat{G}) \times (G \times \hat{G}) \ni ((x_1,\hat{x}_1), (x_2,\hat{x}_2)) \rightsquigarrow \langle x_1,\hat{x}_2\rangle \in \mathbb{T}$ durch die Vorschrift

$$(10) \qquad (x_1,\hat{x}_1,\zeta_1).(x_2,\hat{x}_2,\zeta_2) = (x_1+x_2,\hat{x}_1+\hat{x}_2, \langle x_1,\hat{x}_2\rangle\zeta_1\zeta_2)$$

eine multiplikative Verknüpfung auf A(G). Dann repräsentiert A(G) eine nicht-abelsche unimodulare lokalkompakte topologische Gruppe.

Die irreduziblen stetigen unitären Darstellungen von A(G) lassen sich klassifizieren. Konstruiert man mit Hilfe eines Haar-Maßes von G den zugehörenden komplexen Hilbert-Raum $L^2(G)$, so wird durch die Zuordnung

$$(11) \qquad W_o(x,\hat{x},\zeta): L^2(G) \ni g \rightsquigarrow (s \rightsquigarrow \zeta\langle s,\hat{x}\rangle g(x+s)) \in L^2(G)$$

eine irreduzible stetige unitäre Darstellung $(x,\hat{x},\zeta) \rightsquigarrow W_o(x,\hat{x},\zeta)$ von A(G) in $L^2(G)$ definiert. Man nennt W_o die Schrödinger-Darstellung von A(G) in $L^2(G)$. Nach dem Unitätssatz von Stone-von Neumann-Mackey ist jede irreduzible stetige unitäre Darstellung W von A(G) in einem komplexen Hilbert-Raum E mit der Eigenschaft $W(0,0,\zeta) = \zeta^1 \mathrm{id}_E$ für alle $\zeta \in \mathbb{T}$ (W subduziert den zentralen Charakter $\zeta \rightsquigarrow \zeta^1$) zur Schrödinger-Darstellung W_o unitär äquivalent.

Man sieht: Die Heisenberg-Gruppe A(G) ist "beinahe" abelsch, d.h. eine verhältnismäßig elementare nicht-abelsche lokalkompakte Gruppe. Sie spielt im Falle $G=\mathbb{R}$ vor allem in der Quantenmechanik eine entscheidende Rolle. Die drei-dimensionale reelle nilpotente Lie-Gruppe $A(\mathbb{R})$ wird auch Heisenberg-Weyl-Gruppe eines (einzigen) nicht-relativistischen Teilchens ohne Spin mit einem Freiheitsgrad genannt.

4. Die Heisenberg-Gruppe mod N

Wie in Abschnitt 2 sei $N \geq 1$ eine natürliche Zahl. Konstruiert man zur zyklischen Gruppe $G = \mathbb{Z}/N\mathbb{Z}$ der Ordnung N unter der diskreten Topologie die Gruppe $A(\mathbb{Z}/N\mathbb{Z})$ und bezeichnet mit T_N die zu $\mathbb{Z}/N\mathbb{Z}$ (und zu $\widehat{\mathbb{Z}/N\mathbb{Z}}$) isomorphe abgeschlossene Untergruppe von $\mathbb{T}$ der N-ten Einheitswurzeln, so heißt die Untergruppe

$$(12) \qquad \mathcal{N}(\mathbb{Z}/N\mathbb{Z}) = \{(x,\hat{x},\zeta) \in A(\mathbb{Z}/N\mathbb{Z}) | \zeta \in T_N\}$$

von $A(\mathbb{Z}/N\mathbb{Z})$ die Heisenberg-Gruppe mod N (Auslander [2]). Man überzeugt sich, daß $\mathcal{N}(\mathbb{Z}/N\mathbb{Z})$ durch die aus allen oberen Dreiecksmatrizen der Form

$$(13) \qquad \begin{pmatrix} 1 & y & z \\ 0 & 1 & x \\ 0 & 0 & 1 \end{pmatrix} \qquad (x,y,z \in \mathbb{Z}/N\mathbb{Z})$$

gebildete Untergruppe von $SL(3,\mathbb{Z}/N\mathbb{Z})$ realisiert werden kann. Versieht man den komplexen Vektorraum $\mathbb{C}[\mathbb{Z}/N\mathbb{Z}]$ mit dem kanonischen Skalarprodukt, so induziert der (abelsche) Charakter

$$(14) \qquad \chi_1 : \widehat{\mathbb{Z}/N\mathbb{Z}} \times T_N \ni (\hat{x},\zeta) \rightsquigarrow \zeta \in \mathbb{T}$$

mit Hilfe des Monomorphismus $\widehat{\mathbb{Z}/N\mathbb{Z}} \times T_n \ni (\hat{x},\zeta) \rightsquigarrow (0,\hat{x},\zeta) \in \mathcal{N}(\mathbb{Z}/N\mathbb{Z})$ und geeigneter Identifizierung des Darstellungsraumes eine irreduzible unitäre Darstellung

(15) $$W_1 = \mathrm{Ind}(\chi_1)$$

von $\mathcal{N}(\mathbb{Z}/N\mathbb{Z})$ in $\mathbb{C}[\mathbb{Z}/N\mathbb{Z}]$, welche zur Schrödinger-Darstellung von $\mathcal{N}(\mathbb{Z}/N\mathbb{Z})$ (unitär) äquivalent ist. Das Entsprechende gilt für die vom Charakter

(16) $$\chi_2 : \mathbb{Z}/N\mathbb{Z} \times T_N \ni (x,\zeta) \rightsquigarrow \zeta \in \mathbb{T}$$

mit Hilfe des Monomorphismus $\mathbb{Z}/N\mathbb{Z} \times T_N \ni (x,\zeta) \rightsquigarrow (x,0,\zeta) \in \mathcal{N}(\mathbb{Z}/N\mathbb{Z})$ und geeigneter Identifizierung des Darstellungsraumes induzierte irreduzible unitäre Darstellung

(17) $$W_2 = \mathrm{Ind}(\chi_2)$$

von $\mathcal{N}(\mathbb{Z}/N\mathbb{Z})$ in $\mathbb{C}[\mathbb{Z}/N\mathbb{Z}]$.

Nach dem Unitätssatz sind W_1, W_2 äquivalente Darstellungen von $\mathcal{N}(\mathbb{Z}/N\mathbb{Z})$ im komplexen Hilbert-Raum $\mathbb{C}[\mathbb{Z}/N\mathbb{Z}]$. Der unitäre Verflechtungsoperator $\bar{\mathfrak{F}}_N$ von W_1 und W_2 besitzt bezüglich der kanonischen Basis von $\mathbb{C}[\mathbb{Z}/N\mathbb{Z}]$ die zur Knotenfolge $\mathfrak{k}^{per}$ gehörende Vandermondesche Matrix $\frac{1}{\sqrt{N}}(e^{2\pi inpk})_{\substack{0\leq n\leq N-1\\ 0\leq p\leq N-1}}$, stimmt also mit der "endlichen" Fourier-Kotransformation der zyklischen Gruppe $\mathbb{Z}/N\mathbb{Z}$ überein.

Identifiziert man die isomorphen additiven Gruppen $\mathbb{Z}/N\mathbb{Z}$ und $\widehat{\mathbb{Z}/N\mathbb{Z}}$, so repräsentiert die Abbildung

(18) $$\sigma : (x,y,\zeta) \rightsquigarrow (y,-x,\langle x,y\rangle^{-1}\zeta)$$

einen Automorphismus von $\mathcal{N}(\mathbb{Z}/N\mathbb{Z})$, welcher eine fundamentale Symmetrie-Eigenschaft von $\mathcal{N}(\mathbb{Z}/N\mathbb{Z})$ wiedergibt. Man erhält insbesondere $\chi_1 = \sigma(\chi_2) = \chi_2 \circ \sigma$ und $\chi_2 = \sigma(\chi_1)$, d.h. σ vertauscht die W_1 und W_2 induzierenden Charaktere. Ferner erhält man mit den gemäß (9) und (5) definierten komplexen Zahlen $(\beta_n^{(m)})_{0\leq n\leq N-1}$ in der 0-ten Zeile von $\mathfrak{B}_m$ das folgende Resultat:

SATZ 1. <u>Für den dem periodischen Interpolationsspline</u> $s \in \mathfrak{S}_m(\mathbb{T};\mathfrak{k}^{per})$ <u>zugeordneten Endomorphismus</u> B_m <u>von</u> $\mathbb{C}[\mathbb{Z}/N\mathbb{Z}]$ <u>gelten die Beziehungen</u>

$$(19)\qquad \begin{aligned} B_m &= \sum_{0\leq n\leq N-1} \beta_n^{(m)} W_1(n,0,1) \\ &= \sum_{0\leq n\leq N-1} \beta_n^{(m)} W_2(0,n,1) \end{aligned}$$

mit den irreduziblen unitären Darstellungen (17) und (15) der Heisenberg-Gruppe $\mathcal{H}(\mathbb{Z}/N\mathbb{Z})$ mod N.

Wendet man auf die erste der Gleichungen (19) den Verflechtungsoperator $\overline{\mathfrak{F}}_N$ von W_1 und W_2 an, so erhält man das folgende bekannte Resultat:

SATZ 2. Die endliche Fourier-Kotransformation $\overline{\mathfrak{F}}_N$ diagonalisiert den Endomorphismus B_m von $\mathbb{C}[\mathbb{Z}/N\mathbb{Z}]$; die Eigenwerte von B_m werden durch die Summen $\left(\sum_{0\leq n\leq N-1} \beta_n^{(m)} e^{2\pi i npk}\right)_{0\leq p\leq N-1}$ gegeben.

Aus den vorstehenden Sätzen wird der Zusammenhang zwischen den zu äquidistanten Knotenfolgen $\mathfrak{k}^{per}$ und äquidistanten Datensätzen auf $\mathbf{T}$ konstruierten periodischen Interpolationssplines auf der einen Seite und der Heisenberg-Gruppe $\mathcal{H}(\mathbb{Z}/N\mathbb{Z})$ mod N und ihrer (im Raum $\mathbb{C}[\mathbb{Z}/N\mathbb{Z}]$ operierenden) Schrödinger-Darstellung auf der anderen Seite deutlich. Als Anwendung kann man aus ihnen hinreichende Bedingungen für den Grad m und die Anzahl N der Knoten auf $\mathbf{T}$ herleiten, um die Existenz und Eindeutigkeit des periodischen Interpolationssplines $s \in \mathfrak{S}_m(\mathbf{T};\mathfrak{k}^{per})$ zu sichern. Details sollen an anderer Stelle wiedergegeben werden.

5. Diskontinuierliche Faktoren

Wie bereits in Abschnitt 1 erwähnt, wird der Zusammenhang zwischen den kardinalen Spline-Funktionen und der (kommutativen) harmonischen Analyse durch sog. diskontinuierliche Faktoren gegeben, die man etwa von der Behandlung von Einschaltproblemen in der Elektrotechnik kennt.

Für jede natürliche Zahl $m\geq 1$ erhält man durch eine Variablen-Transformation die bekannte Identität

$$(20)\qquad \int_{\mathbb{R}_+} e^{-zx} x^m dx = \frac{m!}{z^{m+1}} \qquad (\mathrm{Re}\ z > 0).$$

Die Gleichung (20) besagt, daß die einseitige Laplace-Transformierte des Monoms $x \rightsquigarrow x^m$ im Punkt z der offenen rechten komplexen Halbebene mit $m!/z^{m+1}$ übereinstimmt. Zur additiven lokalkompakten Gruppe $\mathbb{R}$ gehören, weil die Charaktere von $\mathbb{R}$ durch $x \rightsquigarrow e^{ixy}$ $(y \in \hat{\mathbb{R}} = \mathbb{R})$ gegeben sind, die (klassische) Fourier-Transformation und, durch Fortsetzen der Charaktere in der komplexen y-Ebene zur Erfassung nicht-unitärer Darstellungen, die zweiseitige Laplace-Transformation. Definiert man für jede reell-wertige Funktion f ihren Positivteil gemäß $f_+ = \sup(f,0) = \frac{1}{2}(f+|f|)$, so folgt aus (20) durch Anwenden der inversen Laplace-Transformation die Identität

$$(21)\qquad x_+^m = \frac{m!}{2\pi i} \int_{L_1} \frac{e^{xz}}{z^{m+1}} dz \qquad (x \in \mathbb{R}),$$

wobei $L_1 = \{z \in \mathbb{C} \mid \mathrm{Re}\ z = c\}$, $c > 0$, eine beliebige vertikale Gerade in der offenen rechten komplexen Halbebene bezeichnet. Entsprechend gilt die komplexe Linienintegral-Darstellung

$$(22)\qquad (-x)_+^m = (-1)^{m+1} \frac{m!}{2\pi i} \int_{L_2} \frac{e^{xz}}{z^{m+1}} dz \qquad (x \in \mathbb{R})$$

mit einer vertikalen Geraden L_2 in der offenen linken komplexen Halbebene.

Auf ähnliche Weise kann man vorgehen, wenn man die multiplikative Gruppe $\mathbb{R}_+^\times$ der streng positiven reellen Zahlen zugrunde legt. Bezeichnet $(\Gamma_m)_{m \geq 1}$ die Folge der Partialprodukte in der klassischen Gaußschen Produktdarstellung der Γ-Funktion, also

$$(23)\qquad \Gamma_m:\ z \rightsquigarrow \frac{m!\,m^z}{\prod_{0 \leq k \leq m}(z+k)} \qquad (m \geq 1),$$

so erhält man durch schrittweise partielle Integration der Funktion $t \rightsquigarrow (1-\frac{t}{m})^m t^{z-1}$ die Identität

$$(24)\qquad \int_0^m (1-\frac{t}{m})^m t^z \frac{dt}{t} = \Gamma_m(z) \qquad (\mathrm{Re}\ z > 0).$$

Die Rolle der Fourier-Transformation auf der additiven lokalkompakten Gruppe $\mathbb{R}$ wird von der Mellin-Transformation auf der multiplikativen lokalkompakten Gruppe $\mathbb{R}_+^\times$ übernommen, weil log: $\mathbb{R}_+^\times \to \mathbb{R}$ einen topologischen Isomorphismus definiert. Aus (24) erhält man durch Anwenden der inversen Mellin-Transformation die Identität

$$(1-\frac{t}{m})_+^m = \frac{1}{2\pi i}\int_L \Gamma_m(z)t^{-z}dz \qquad (t \in \mathbb{R}), \tag{25}$$

wobei L eine vertikale Gerade in der offenen rechten komplexen Halbebene bezeichnet. Die komplexe Linienintegral-Darstellung (25) des diskontinuierlichen Faktors $t \rightsquigarrow (1-\frac{t}{m})_+^m$ ($m \geq 1$) scheint, obschon sehr naheliegend, dennoch neu zu sein (vgl. [6]).

6. Kardinale Exponentielle Splines

Es bezeichne $h \neq 0$ eine feste komplexe Zahl. Jede Lösung $s_m \in \mathcal{G}_m(\mathbb{R};\mathbb{Z})$ der homogenen linearen Differenzen-Gleichung

$$f(x+1) - hf(x) = 0 \qquad (x \in \mathbb{R}) \tag{26}$$

heißt nach Schoenberg [9] ein kardinaler exponentieller Spline vom Grad $m \geq 1$ und Gewicht h. Mit Hilfe der diskontinuierlichen Faktoren (21) und (22) und der Differenzen-Gleichung (26) gewinnt man den folgenden Darstellungssatz (vgl. [4]):

SATZ 3. <u>Die kardinalen exponentiellen Splines</u> $s_m \in \mathcal{G}_m(\mathbb{R};\mathbb{Z})$ <u>vom Grad</u> $m \geq 1$ <u>und Gewicht</u> $h \in \mathbb{C}^\times - \mathbb{T}$ <u>besitzen die komplexe Kurvenintegral-Darstellung</u>

$$s_m(x) = C_{m,h}(1-\frac{1}{h})^{m+1}\frac{1}{2\pi i}\int_P \frac{e^{(x+1)z}}{(e^z-h)z^{m+1}}dz \qquad (x \in \mathbb{R}). \tag{27}$$

<u>Dabei bezeichnet</u> $C_{m,h} \in \mathbb{C}$ <u>eine beliebige Konstante und</u> P <u>den aus zwei antiparallelen Graden bestehenden positiv orientierten Rand eines abgeschlossenen vertikalen Streifens in der offenen rechten bzw. linken komplexen Halbebene, je nach dem, ob</u> $|h|>1$ <u>oder</u> $0<|h|<1$ <u>gilt, der die "kritische" Gerade</u> $\{w \in \mathbb{C} \mid \mathrm{Re}\, w = \log|h|\}$ <u>in seinem Innern enthält.</u>

Falls $s_m(0) \neq 0$ zutrifft, kann in (27) die Konstante $C_{m,h}$ so gewählt werden, daß $s_m \in \mathfrak{S}_m(\mathbb{R};\mathbb{Z})$ die Folge $(h^n)_{n \in \mathbb{Z}}$ in den Punkten $n \in \mathbb{Z}$ interpoliert. Mit Hilfe von Satz 3 und den Eigenschaften der Euler-Frobenius-Polynome (vgl. [5]) läßt sich durch Anwenden des Residuen-Satzes beweisen, daß die kardinalen exponentiellen Interpolationssplines punktweise die Konvergenz-Eigenschaft $\lim_{m\to\infty} s_m(x) = h^x (x \in \mathbb{R})$ erfüllen, falls das Gewicht $h \in \mathbb{C}-\mathbb{T}$ nicht zur abgeschlossenen reellen Halbgeraden $\mathbb{R}_-$ gehört. Eine Erweiterung dieses Resultats wird mit Hilfe von (27) in der Arbeit [7] bewiesen.

7. Kardinale Logarithmische Interpolationssplines

Es sei $h_0 > 1$ eine feste Schrittweite und $\mathfrak{k}_0$ die Knotenfolge $(h_0^n)_{n \in \mathbb{Z}}$ in der offenen reellen Halbgeraden $\mathbb{R}_+^\times$. Jede Lösung $S_m \in \mathfrak{S}_m(\mathbb{R}_+^\times;\mathfrak{k}_0)$ der inhomogenen linearen Differenzengleichung

$$f(h_0 x) - f(x) = 1 \qquad (x \in \mathbb{R}_+^\times), \tag{28}$$

welche die Interpolationsbedingung $S_m(h_0^n) = n (n \in \mathbb{Z})$ erfüllt, heißt nach Newman-Schoenberg [3] ein kardinaler logarithmischer Interpolationsspline vom Grad $m \geq 1$ und der Schrittweite h_0. Führt man die leicht modifizierte Logarithmusfunktion

$$f_0 : \mathbb{R}_+^\times \ni x \rightsquigarrow \frac{\log x}{\log h_0} \in \mathbb{R} \tag{29}$$

ein, die offenbar ebenfalls die Differenzengleichung (28) und die Interpolationsbedingung $f_0(h_0^n) = n \ (n \in \mathbb{Z})$ erfüllt, so liefern der diskontinuierliche Faktor (25) und die Differenzengleichung (28) den folgenden Darstellungssatz:

SATZ 4. <u>Die kardinalen logarithmischen Interpolationssplines</u> $S_m \in \mathfrak{S}_m(\mathbb{R}_+^\times;\mathfrak{k}_0)$ <u>vom Grad</u> $m \geq 1$ <u>und der Schrittweite</u> $h_0 > 1$ <u>besitzen die komplexe Kurvenintegral-Darstellung</u>

$$S_m(x) = \frac{1}{2\pi i} \int_Q \Gamma_m(z) h_0^{-z f_0(m)} \frac{1-x^{-z}}{1-h_0^{-z}} \, dz \qquad (x \in \mathbb{R}_+^\times). \tag{30}$$

Dabei bezeichnet Q den positiv orientierten Rand eines abgeschlossenen vertikalen Streifens, der von den beiden antiparallelen Geraden $Q_1 = \{w \in \mathbb{C} \mid \operatorname{Re} w = c\}$, $c > 0$, und $Q_2 = \{w \in \mathbb{C} \mid \operatorname{Re} w = d\}$, $-1<d<0$, begrenzt wird.

Die Gleichung (30) zeigt, daß im Falle der kardinalen logarithmischen Interpolationssplines die imaginäre Achse $\{w \in \mathbb{C} \mid \operatorname{Re} w = 0\}$ die "kritische" Gerade ist. Mit Hilfe von Satz 4, des Residuen-Satzes und eines mit der Gleichverteilung auf $\mathbb{T}$ zusammenhängenden Dichtheitsarguments läßt sich das "Newman-Schoenberg-Phänomen" beweisen: Es gilt $\lim_{m\to\infty} S_m(x) = f_0(x)$ für $x \in \mathbb{R}_+^\times$ genau dann, wenn der Punkt x mit einem Knoten zusammenfällt, also $x \in \mathfrak{k}_0$ zutrifft.

8. Zusammenfassung

Eine Übersicht über die behandelten Splines, die zugehörenden Gruppen und Transformationen sowie Literaturhinweise werden in der nachstehenden Tabelle gegeben, die man gewissermaßen als Ansatz zu einem "Erlanger Programm" für Splines zu "regulären" Knotenfolgen betrachten kann.

	Spline	Gruppe	Transformation	Lit.
I	Periodischer Spline zu äquidistanten Knoten	Heisenberg-Gruppe $\mathcal{N}(\mathbb{Z}/N\mathbb{Z})$ mod N	Endliche Fourier-Kotransformation	[7]
II_1	Kardinaler exponentieller Spline	Additive Gruppe $\mathbb{R}$ der reellen Zahlen	Inverse Laplace-Transformation	[4] [5]
II_2	Kardinaler logarithmischer Spline	Multiplikative Gruppe $\mathbb{R}_+^\times$ der positiven reellen Zahlen	Inverse Mellin-Transformation	[6]

Für eine Ankündigung der angesprochenen Ergebnisse sei auf den Artikel [8] verwiesen.

LITERATUR

[1] Ahlberg, J.H. - Nilson, E.N. - Walsh, J.L., The Theory of Splines and Their Applications. (Mathematics in Science and Engineering, Vol. 38) Academic Press, New York/London 1967.

[2] Auslander, L., Lecture Notes on Nil-Theta Functions. (CBMS Regional Conference Series in Math., No. 34) American Mathematical Society, Providence, R.I. 1977.

[3] Newman, D.J. - Schoenberg, I.J., Splines and the logarithmic function. Pacific J. Math. 61 (1975), 241-258.

[4] Schempp, W., Cardinal exponential splines and Laplace transform. J. Approx. Theory (to appear).

[5] Schempp, W., A contour integral representation of Euler - Frobenius polynomials. J. Approx. Theory (to appear).

[6] Schempp, W., Cardinal logarithmic splines and Mellin transform. J. Approx. Theory (to appear).

[7] Schempp, W., On cardinal exponential splines of higher order. J. Approx. Theory (to appear).

[8] Schempp, W., Contour integral representation of cardinal spline functions. C.R. Math. Pep. Acad. Sci. Canada 2 (1980), 165-170.

[9] Schoenberg, K.J., Cardinal interpolation and spline functions IV. The exponential Euler splines. In: Linear Operators and Approximation I, Butzer, P.L. - Kahane, J.P. - Sz.-Nagy, B. eds. (ISNM, Vol. 20), 382-404. Birkhäuser Verlag, Basel/Stuttgart 1972.

[10] Weil, A., Sur certains groupes d'opérateurs unitaires. Acta Math. 111 (1964), 143-211. Auch in: OEuvres Scientifiques - Collected Papers, Vol. III, pp. 1-69. Springer-Verlag, New York/Heidelberg/Berlin 1980.

Die Teile I und II der vorliegenden Arbeit sind in den Bänden ISNM, Vol. 42 (1978) 299-305 und Vol. 52 (1980), 277-281 des Birkhäuser Verlags, Basel/Stuttgart, erschienen.

UNIQUENESS OF OPTIMAL PIECEWISE POLYNOMIAL L_1 APPROXIMATIONS FOR GENERALIZED CONVEX FUNCTIONS

John B. Kioustelidis
Department of Applied Mathematics
National Technical University
Athens

It is shown that the optimal piecewise m^{th} degree polynomial L_1-approximation of a generalized convex function f ($f^{(m+1)}$ positive) is unique, if log $f^{(m+1)}$ is concave.

1. Introduction

The uniqueness of optimal piecewise polynomial approximations with free knots has already been investigated by Meinardus [1, p. 188] for the case of the uniform norm. Uniqueness is in this case guaranteed, if $f^{(m+1)}$ is positive and polynomials of degree not exceeding m are used.

More recently, Barrow, Chui, Smith and Ward [2] have answered the question of uniqueness of piecewise linear approximations for convex functions in the case of the L_2- and L_1-norms. Their main instrument is the concept of the topological degree of a mapping, which allows under certain conditions the determination of the exact number of zeros of the mapping in some region. Using the same technique Chow [3] has shown that optimal piecewise m^{th} degree polynomial L_2-approximations of generalized convex functions ($f^{(m+1)} > 0$) with concave log $f^{(m+1)}$ are unique. Here we prove the same result for the L_1-case. The notation used is not the spline notation but a simpler one.

2. The Problem, Characterization of the Solutions

Let $I = [a,b]$ be a given real interval and U_n denote the set of all possible partitions of I into maximally n subintervals

(1) $U_n := \left\{ \{u_i, i = 0(1)n\} : u_o = a,\ u_n = b,\ u_{i-1} \leqslant u_i,\ i = 1(1)n \right\}$.

Also let P_m be the linear space of polynomials of degree at most m.

By P(I) denote the set of piecewise continuous functions over I, and by $P_{m,u}(I)$ the set of functions $q \in P(I)$, which are described in each half-open subinterval $I_i := [u_{i-1}, u_i)$ of the partition U_n by some polynomials $p_i \in P_m$

(2) $P_{m,u}(I) := \{q \in P(I) : q(x) = p_i(x)\ \forall\ x \in [u_{i-1}, u_i),\ p_i \in P_m, i = 1(1)n\}$.

The problem under consideration is the following:

For given $f \in C(I)$ and $n \in \mathbb{N}$ find $\hat{u} \in U_n$ and $\hat{q} \in P_{m,\hat{u}}(I)$, which minimize the L_1-norm

(3) $$\| f - q \|_{1,I} = \int_I |f(x) - q(x)| dx$$

of $f - q$ over all $u \in U_n$ and $q \in P_{m,u}(I)$. Any solution of this problem is called optimal segmented (or piecewise) m^{th} degree polynomial L_1-approximation of f with n segments. The existence of solutions to this problem has been established in [4] and [6]. It has also been shown that any solution $\hat{q}$ has distinct knots [5], and that it is characterized by continuity of the pointwise error modulus ($|f - \hat{q}| \in C(I)$) [4], [6].

In the case of generalized convex functions ($f^{(m+1)}(x) > 0$ for $x \in I$) we have the following more specific results [6] (the second one being due to S.N. Bernstein).

PROPOSITION 1. <u>Let</u> $f \in C^{m+1}(I)$ <u>with</u> $f^{(m+1)}$ <u>positive in</u> I, <u>and suppose that</u> q <u>is an optimal piecewise</u> m^{th} <u>degree polynomial</u> L_1<u>-approximation of</u> f <u>with</u> n <u>segments: Then</u> q <u>has the following properties:</u>

(4) a) $$f(u_i) - p_i(u_i) = (-1)^{m+1}(f(u_i) - p_{i+1}(u_i)), \qquad i = 1(1)n-1$$

b) (S.N. Bernstein) <u>The polynomial</u> $p_i(x)$, $i = 1(1)n$, <u>which describes the optimal approximation in the interval</u> $[u_{i-1}, u_i)$ <u>is the interpolation polynomial of</u> f <u>at the</u> m + 1 <u>points</u>

(5) $$x_{ik} = \frac{u_i + u_{i-1}}{2} + \frac{u_i - u_{i-1}}{2} \cos\left(\frac{k\pi}{m+2}\right) \qquad k = 1(1)m+1 .$$

3. Uniqueness of Optimal Segmented Approximations

In order to establish the uniqueness of the optimal segmented L_1-approximation of generalized convex functions, under certain conditions we are going to derive first an expression for the approximation error and need the following lemmata:

LEMMA 1. *Let*

$$x_k = c + d \cos t_k, \qquad k = 0(1)m+2 \tag{6}$$

with

$$t_k = (k\pi)/(m+2). \tag{7}$$

Then for any polynomial of degree at most m *there holds*

$$H(p) := \sum_{k=0}^{m+1} (-1)^k \int_{x_{k+1}}^{x_k} p(x)dx = 0 . \tag{8}$$

PROOF. It is sufficient to prove the validity of this identity for the powers of $x-c$. Let

$$p_s(x) = (x-c)^{s-1}, \qquad s = 1(1)m+1 . \tag{9}$$

Then it is easy to show that

$$H(p_s) = -\frac{d^s}{s}\left\{1+(-1)^{m+s} + 2\sum_{k=1}^{m+1} (-1)^k \cos^s t_k\right\}. \tag{10}$$

Expressing the terms $\cos^s t_k$ by sums of cosines and interchanging the summations we obtain

$$H(p_s) \equiv 0 . \tag{11}$$

LEMMA 2. *For any function* g *there holds*

$$g(x_0) + (-1)^m g(x_{m+2}) + 2\sum_{k=1}^{m+1} (-1)^k g(x_k) \equiv \tag{12}$$

$$\equiv \sum_{k=0}^{m+1} (-1)^k (g(x_k) - g(x_{k+1})) \equiv 2(m+2)d^{m+2} g[x_o, x_1, \ldots, x_{m+2}] ,$$

where x_k are the points defined by (6), (7).

PROOF. The identity follows from the fact that

$$g[x_o, x_1, \ldots, x_{m+2}] \equiv \sum_{k=0}^{m+2} \frac{g(x_k)}{\omega'(x_k)} \tag{13}$$

with

$$\omega(x) := \prod_{i=0}^{m+2} (x - x_i) = d^{m+3} [(\tfrac{x-c}{d})^2 - 1] U_{m+1}(\tfrac{x-c}{d}) , \tag{14}$$

where $U_{m+1}(t)$ is the second kind Chebyshev polynomial of degree $m+1$:

$$U_{m+1}(t) := \frac{\sin((m+2)\arccos t)}{\sin(\arccos t)} . \tag{15}$$

For later use we note also

LEMMA 3. For an arbitrary function $h(x)$ and arbitrary points $z_1, z_2, \ldots, z_\ell$, it follows from

$$g(x) = (x - z_1)h(x) \tag{16}$$

that

$$g[z_1, z_2, \ldots, z_\ell] \equiv h[z_2, z_3, \ldots, z_\ell] . \tag{17}$$

PROOF. This is an immediate consequence of the relation

$$g[z_1, z_2, \ldots, z_\ell] \equiv \sum_{k=1}^{\ell} \frac{g(z_k)}{\prod_{i \neq k} (z_k - z_i)} . \tag{18}$$

THEOREM 1. Let $f \in C^{m+1}(J)$ with $f^{(m+1)}$ positive in the interval $J = [v, u]$. Also let

$$F(x) := \int_v^x f(t)dt, \qquad x \in J . \tag{19}$$

If p is the best m^{th} degree polynomial L_1-approximation to f in J

then the approximation error is

$$E_J = \|f-p\|_{1,J} = \sum_{k=0}^{m+1} (-1)^k (F(x_k) - F(x_{k+1})) \tag{20}$$

$$= F(x_o) + (-1)^m F(x_{m+2}) + 2 \sum_{k=1}^{m+1} (-1)^k F(x_k)$$

$$= 2(m+2) d^{m+2} F[x_o, x_1, \ldots x_{m+2}] \tag{21}$$

where the points x_k *are given by* (6), (7) *with*

$$c = (u+v)/2, \quad d = (u-v)/2. \tag{22}$$

PROOF. According to the above mentioned result of S.N. Bernstein, p interpolates f at the points x_k, $k = 1(1)m+1$, defined by (6), (7) and (22). Using the interpolation error formula and the fact that $f^{(m+1)}$ is positive we see that

$$\operatorname{sign}\{f(x) - p(x)\} = \operatorname{sign}\left\{\prod_{j=1}^{m+1} (x-x_j)\right\} \doteq (-1)^k \text{ for } x \in (x_{k+1}, x_k) . \tag{23}$$

Therefore,

$$E_J \equiv \sum_{k=0}^{m+1} \int_{x_{k+1}}^{x_k} |f(x) - p(x)|dx = \sum_{k=0}^{m+1} (-1)^k \int_{x_{k+1}}^{x_k} \{f(x) - p(x)\}dx . \tag{24}$$

Formula (20) now follows immediately with the help of Lemma 1, while formula (21) is a direct application of Lemma 2 to formula (20).

COROLLARY 1. *Let* $f \in C^{m+1}(I)$ *with* $f^{(m+1)}$ *positive in* I. *Then:*

a) *The error of any optimal segmented* m^{th}*-degree polynomial* L_1*-approximation (with* n *segments)* q *of* f *is*

$$E = \sum_{i=1}^{n} \left(F(x_{io}) + (-1)^m F(x_{i,m+2}) + 2 \sum_{k=1}^{m+1} (-1)^k F(x_{ik}) \right) \tag{25}$$

with

$$F(x) := \int_a^x f(t)dt , \tag{26}$$

(27) $$x_{ik} = c_i + d_i \cos t_k = u_{i-1} \sin^2 \frac{t_k}{2} + u_i \cos^2 \frac{t_k}{2}, \qquad k = 0(1)m+2$$

and

(28) $$c_i = \frac{u_i + u_{i-1}}{2}, \quad d_i = \frac{u_i - u_{i-1}}{2}, \qquad i = 1(1)n .$$

b) The partition knots u_i, $i = 1(1)n-1$, fulfill the equations

(29) $$0 = g_i(u) := f(x_{io}) + (-1)^m f(x_{i,m+2}) + 2 \sum_{k=1}^{m+1} (-1)^k \left(\cos^2 \frac{t_k}{2} f(x_{ik}) + \sin^2 \frac{t_k}{2} f(x_{i+1,k}) \right), \qquad i = 1(1)n-1 .$$

PROOF. The relation (25) follows by applying formula (20) in each partition interval and summing up the errors. The equations (29) follow either by equating the derivatives of the error E with respect to the variables u_i to zero, or by using the continuity conditions (4).

Under the assumptions of Corollary 1, the question, whether the optimal segmented approximation is unique or not, reduces to the question, whether the mapping g occuring in (29) has only one zero point in the region

(30) $$G := \{u \in \mathbb{R}^{n-1} : a \leq u_1 \leq u_2 \leq \ldots \leq u_{n-1} \leq b\}$$

or not (note that $u_o = a$, $u_n = b$). This question can be answered in some cases by considering the topological degree of the mapping g with respect to the region G and the value vector θ (θ is here the zero vector):

(31) $$\deg(g,G,\theta) = \sum_{k=1}^{N} \operatorname{sign} \det g'(u^{(k)}),$$

where $u^{(k)}$, $k = 1(1)N$, are the solutions of the equation:

(32) $$g(u) = \theta$$

in the region G, and g' the jacobian matrix of g. (For an introduction to the concept of the topological degree and its properties see [5, chapter 6].) If the determinant of g'(u) is always positive at the zero points of g(u), then obviously $\deg(g,G,\theta)$ is equal to N, i.e., to the number of solutions of (32) in G. Our first step is thus to determine conditions for the positivity of

$\det g'(u^{(k)})$. In order to determine $\deg(g,G,\theta)$ we then use the property of homotopy invariance of the topological degree [5, p. 156]. This property can be expressed in the following way:

If the mappings $g : G \to \mathbb{R}^{n-1}$ and $\tilde{g} : G \to \mathbb{R}^{n-1}$ are continuous and the mapping

$$A_t(u) := tg(u) + (1-t)\tilde{g}(u) \tag{33}$$

has no zero points on the boundary ∂G of G for all $t \in [0,1]$, then $\deg(A_t,G,\theta)$ is constant for $t \in [0,1]$ and more specifically,

$$\deg(g,G,\theta) = \deg(A_1,G,\theta) = \deg(A_o,G,\theta) = \deg(\tilde{g},G,\theta). \tag{34}$$

The mapping $\tilde{g}$ is chosen so that the number of its solutions in G can be easily determined. It is the mapping which corresponds to the continuity conditions (29) for the approximation of the function $\tilde{f}(x) = x^{m+1}$.

Following the above procedure, we first determine the jacobian $g'(u)$. This is a tridiagonal matrix, whose nonzero elements in each row are

$$\frac{\partial g_i}{\partial u_{i-1}}, \quad \frac{\partial g_i}{\partial u_i} \quad \text{and} \quad \frac{\partial g_i}{\partial u_{i+1}}.$$

From (29) it follows that

$$\frac{\partial g_i}{\partial u_{i-1}} = 2\sum_{k=1}^{m+1} (-1)^k \cos^2\frac{t_k}{2}\sin^2\frac{t_k}{2} f'(x_{ik}) \tag{35}$$

$$= \frac{2}{4d_i^2}\sum_{k=1}^{m+1} (-1)^k h(x_{ik}),$$

where

$$h(x) := (x_{io}-x)(x-x_{i,m+2})f'(x) \tag{36}$$

(note that $x_{io} = u_i$ and $x_{i,m+2} = u_{i-1}$). Applying successively Lemma 2 and Lemma 3 on this expression we have

$$\frac{\partial g_i}{\partial u_{i-1}} = \frac{2(m+2)}{4d_i^2} d_i^{m+2} h[x_{io},\dots,x_{i,m+2}] = -(m+2)d_i^m f'[x_{i1},\dots,x_{i,m+1}]. \tag{37}$$

In the same way it follows that

$$(38)\qquad \frac{\partial g_i}{\partial u_{i+1}} = 2\sum_{k=1}^{m+1} (-1)^k \sin^2\frac{t_k}{2}\cos^2\frac{t_k}{2} f'(x_{i+1,k})$$

$$= -(m+2)d_{i+1}^m f'[x_{i+1,1},\ldots,x_{i+1,m+1}]$$

while

$$(39)\qquad \frac{\partial g_i}{\partial u_i} = f'(x_{io}) + 2\sum_{k=1}^{m+1}(-1)^k\cos^4\frac{t_k}{2} f'(x_{ik}) + (-1)^m f'(x_{i+1,m+2})$$

$$+\, 2\sum_{k=1}^{m+1}(-1)^k \sin^4\frac{t_k}{2} f'(x_{i+1,k}) .$$

Now we prove

LEMMA 4. _Let_ $f\in C^{m+2}(I)$ _with_ $f^{(m+1)}$ _positive and_ $\log f^{(m+1)}$ _concave_ ($f^{(m+2)}/f^{(m+1)}$ _nonincreasing_) _in_ I. _If_ $g(u)=\theta$, _then the determinant of_ $g'(u)$ _is positive_.

PROOF. It is sufficient to show that $g'(u)$ has positive diagonal elements and is diagonally dominant with strict inequality in the first and last row. By a standard modification of Gerschgorin's theorem it follows then that the eigenvalues of $g'(u)$ lie in the open right half of the complex plane, i.e., have positive real parts. Since all elements of $g'(u)$ are real, the complex eigenvalues come in conjugate pairs, and the product of the eigenvalues, i.e., $\det g'(u)$, is therefore positive.

Using the generalized theorem of Rolle [8, p. 58] we see that the divided differences in (37) and (38) are equal to some intermediate values of $f^{(m+1)}$ in the interval I_i, resp. I_{i+1} and therefore positive. Thus, the expressions (37) and (38) are negative. Positivity of the diagonal elements and (strict in the first and last row) diagonal dominance are therefore established at the same time, if we show that

$$(40)\qquad s_i := \frac{\partial g_i}{\partial u_{i-1}} + \frac{\partial g_i}{\partial u_i} + \frac{\partial g_i}{\partial u_{i+1}} \geqslant 0 , \qquad i=2(1)n-2 ,$$

and

$$(41)\qquad \frac{\partial g_1}{\partial u_1} + \frac{\partial g_1}{\partial u_2} > 0 , \quad \frac{\partial g_{n-1}}{\partial u_{n-2}} + \frac{\partial g_{n-1}}{\partial u_{n-1}} > 0 .$$

By means of (35), (38) and (39) it follows that

$$(42)\qquad s_i = f'(x_{io}) + 2\sum_{k=1}^{m+1} (-1)^k \cos^2 \frac{t_k}{2} f'(x_{ik}) + (-1)^m f'(x_{i+1,m+2})$$

$$+ 2\sum_{k=1}^{m+1} (-1)^k \sin^2 \frac{t_k}{2} f'(x_{i+1,k})$$

$$= \frac{1}{2d_i}\left\{ h_1(x_{io}) + 2\sum_{k=1}^{m+1} (-1)^k h_1(x_{ik})\right\}$$

$$(43)\qquad + \frac{1}{2d_{i+1}}\left\{(-1)^m h_2(x_{i+1,m+2}) + 2\sum_{k=1}^{m+1} (-1)^k h_2(x_{i+1,k})\right\},$$

where

$$(44)\qquad h_1(x) := (x-x_{i,m+2})f'(x),\quad h_2(x) := (x_{i+1,0}-x)f'(x) .$$

Then, by means of Lemma 2 and Lemma 3, it follows from (43) that

$$(45)\qquad s_i = (m+2)d_i^{m+1}\, f'[x_{io},\dots,x_{i,m+1}]$$

$$- (m+2)d_{i+1}^{m+1}\, f'[x_{i+1,1},\dots,x_{i+1,m+2}] .$$

Using the same technique as above we can show that the system (29) is equivalent to

$$(46)\qquad d_i^{m+1}\, f[x_{io},\dots,x_{i,m+1}] = d_{i+1}^{m+1}\, f[x_{i+1,1},\dots,x_{i+1,m+2}],\quad i = 1(1)n-1.$$

Therefore,

$$(47)\qquad s_i = P_i(A_i - B_{i+1}),$$

where

$$(48)\qquad P_i = \frac{(m+2)d_i^{m+1}}{f[x_{io},\dots,x_{i,m+1}]} > 0 ,$$

and

$$(49)\qquad A_i = \frac{f'[x_{io},\dots,x_{i,m+1}]}{f[x_{io},\dots,x_{i,m+1}]},\quad B_{i+1} = \frac{f'[x_{i+1,1},\dots,x_{i+1,m+2}]}{f[x_{i+1,1},\dots,x_{i+1,m+2}]} .$$

Using integral representations for the divided differences in A_i and B_{i+1} (see [8, p. 56]), and writing $f^{(m+2)}$ in the form $f^{(m+1)}(f^{(m+2)}/f^{(m+1)})$ we see that A_i is a weighted average of $f^{(m+2)}/f^{(m+1)}$ in I_i and B_{i+1} is a weighted average of the same function in the interval I_{i+1}. Therefore, since $f^{(m+2)}/f^{(m+1)}$ is nonincreasing, s_i is nonnegative.

In order to establish the validity of the inequalities (41) we note that inequality (40) is also valid for $i=1$ and $i=n-1$. Since $-\partial g_1/\partial u_o$ and $-\partial g_{n-1}/\partial u_n$ are positive the inequalities (41) follow immediately.

Now we consider the mapping (33). Since $\tilde{f}(x)=x^{m+1}$ it follows that

$$A_{t,i}(u) := H(x_{io}) + (-1)^m H(x_{i,m+2}) + 2\sum_{k=1}^{m+1}(-1)^k\left(\cos^2\frac{t_k}{2}H(x_{ik}) + \sin^2\frac{t_k}{2}H(x_{i+1,k})\right), \qquad i=1(1)m-1, \tag{50}$$

where

$$H(x) := tf(x) + (1-t)\tilde{f}(x) = tf(x) + (1-t)x^{m+1}. \tag{51}$$

We now show:

LEMMA 5. <u>For any</u> $t\in[0,1]$ <u>the mapping</u> $A_t(u)$ <u>has no zero points on the boundary</u> ∂G <u>of</u> G.

PROOF. Using Lemmata 2 and 3 we can show in the same way as previously that

$$A_{t,i}(u) = (m+2)d_i^{m+1}H[x_{io},\dots,x_{i,m+1}] - (m+2)d_{i+1}^{m+1}H[x_{i+1,1},\dots,x_{i+1,m+2}]. \tag{52}$$

By means of the generalized theorem of Rolle it follows that

$$A_{t,i}(u) = (m+2)(d_i^{m+1}H^{(m+1)}(\xi_i) - d_{i+1}^{m+1}H^{(m+1)}(\eta_i)), \qquad i=1(1)n-1, \tag{53}$$

for some $\xi_i\in I_i$ and $\eta_i\in I_{i+1}$. Noting that $f^{(m+1)}$ is positive and $\tilde{f}^{(m+1)}$ is equal to $(m+1)!$ we see that

$$H^{(m+1)}(x) = tf^{(m+1)}(x) + (1-t)(m+1)! > 0 \qquad \text{for all } t\in[0,1]. \tag{54}$$

The boundary ∂G of G contains exactly the points $u \in G$ with

$$d_i = 0 \quad (u_{i-1} = u_i) \text{ for some } i,\ i = 1(1)n \text{ (with } u_o = a,\ u_n = b).$$

From

(55) $$A_{t,i}(u) = 0, \qquad i = 1(1)n-1$$

and $d_j = 0$ for any j it follows because of (53), and (54) that

(56) $$d_i = 0 \qquad \text{for } i = 1(1)n.$$

However, these equations are equivalent to

(57) $$a = u_o = u_1 = \dots = u_n = b,$$

which is obviously impossible. Therefore, A_t cannot vanish on G for any $t \in [0,1]$.

Invariance of the topological degree under the homotopy (33) has thus been established, and therefore

(58) $$\deg(g,G,\theta) = \deg(\tilde{g},G,\theta).$$

For the mapping $\tilde{g}$ there holds

(59) $$\tilde{g}_i(u) := (m+1)!\ (d_i^{m+1} - d_{i+1}^{m+1}), \qquad i = 1(1)n-1.$$

Its only zero point is given by

(60) $$d_i = \frac{b-a}{n}, \qquad i = 1(1)n,$$

i.e.,

(61) $$u_k = a + k\,\frac{b-a}{n}, \qquad k = 1(1)n-1.$$

The determinant of its jacobian at this point is positive. Therefore, according to (31) we have

(62) $$\det(\tilde{g},G,\theta) = 1,$$

and because of (58)

(63) $$\deg(g,G,\theta) = 1.$$

Thus, we have established the following result.

THEOREM 2. Let $f \in C^{m+2}(I)$ with $f^{(m+1)}$ positive and $\log f^{(m+1)}$ concave in I. The optimal segmented m-th degree polynomial L_1-approximation of f with n segments is then unique, and can be determined by solving the system (29).

As said before, for m = 1 this result is already given in [2, pp. 1141-1142].

REFERENCES

[1] Meinardus, G., Approximation of Functions: Theory and Numerical Methods. Springer, New York 1967.

[2] Barrow, D.L. - Chui, C.K. - Smith, P.W. - Ward, J.D., Unicity of best mean approximation by second order splines with variable knots. Math. Comp. 32 (1978), 1131-1143.

[3] Chow, J., Uniqueness of best $L_2[0,1]$ approximation by piecewise polynomials with variable breakpoints. Dissertation, Texas A & M University 1978.

[4] Kioustelidis, J.B., Optimal segmented approximations. Computing 24 (1980), 1-8.

[5] Kioustelidis, J.B., The distinctness of the knots of optimal segmented approximations. Submitted for publication in Computing.

[6] Kioustelidis, J.B., Optimal segmented polynomial L_s-approximations. To be published in Computing.

[7] Ortega, J.M. - Rheinhold, W.C.: Iterative Solution of Nonlinear Equations in Several Variables. Academic Press, New York / London 1970.

[8] Stummel, F. - Hainer, K., Praktische Mathematik. Teubner, Stuttgart 1971.

THE DOMINATED INTEGRAL OF FUNCTIONS OF TWO VARIABLES

Narayan S. Murthy
Dept. of Mathematics
Univ. of Rhode Island
Kingston, R.I. 02881
U.S.A.

Charles F. Osgood
Naval Research Lab.
Washington, D. C. 20375
U.S.A.

Oved Shisha
Dept. of Mathematics
Univ. of Rhode Island
Kingston, R.I. 02881
U.S.A.

The dominated integral of a function of two real variables is introduced along the lines of [2].

The concept of the dominated integral of a function of a real variable was recently introduced and studied in [2], [3] and [1]; in particular, its relationship with numerical quadrature of improper integrals has been investigated. The purpose of the present paper is to generalize this concept, along the lines of [2], to functions of two real variables.

DEFINITION 1. *Let f be a complex function on* $I = (0,1] \times (0,1]$. *A dominated integral of f on I is a number* $\Delta(f)$ *having the property: For each* $\varepsilon > 0$ *there exist* δ *and* χ, $0 < \delta < 1$, $0 < \chi < 1$, *such that*

$$(1) \qquad \left|\Delta(f) - \sum_{j=1}^{m} \sum_{k=1}^{n} f(P_{j,k})(x_j - x_{j-1})(y_k - y_{k-1})\right| < \varepsilon$$

whenever $0 < x_0 < x_1 < \ldots < x_m = 1$, $0 < y_0 < y_1 < \ldots < y_n = 1$, $x_0 < \chi$, $y_0 < \chi$; *and* $x_{j-1}/x_j > 1-\delta$, $y_{k-1}/y_k > 1-\delta$, $P_{j,k} \in [x_{j-1}, x_j] \times [y_{k-1}, y_k]$ *for* $j = 1,2,\ldots,m$, $k = 1,2,\ldots,n$. *Dominant integrability of f on I means existence of such a* $\Delta(f)$. *If it exists, it is clearly unique.*

THEOREM 1. *Let f be a complex function on I, dominantly integrable there. Then* (i) *f is Riemann integrable on each* $[a,1] \times [b,1]$, $0 < a < 1$, $0 < b < 1$, (ii) *f is summable on I, and* (iii) $\Delta(f) = \iint_I f(x,y)\,dx\,dy$.

If f is a complex function, defined and bounded on a nonempty set S, we denote by $w(f,S)$ the oscillation of f on S, i.e., $\sup_{P,Q\in S}|f(P)-f(Q)|$.

Given a complex function f, defined and bounded on each closed subset of I, and given sequences $0<x_0<x_1\ldots<x_m=1$, $0<y_0<y_1<\ldots<y_n=1$ $(m\geq 1,\ n\geq 1)$, we set

$$OS(f;\ x_0,x_1,\ldots,x_m;\ y_0,y_1,\ldots,y_n)$$
$$=\sum_{j=1}^{m}\sum_{k=1}^{n} w(f,[x_{j-1},x_j]\times[y_{k-1},y_k])\cdot(x_j-x_{j-1})(y_k-y_{k-1}).$$

(OS stands for "oscillation sum".)

DEFINITION 2. _A complex function_ f _satisfies on_ I _the Riemann condition for the dominated integral_ (RCDI) _iff the following two conditions hold:_

(i) f _is defined on_ I _and bounded on each of its closed subsets; and_

(ii) _for each_ $\varepsilon>0$ _there exists_ δ, $0<\delta<1$, _such that if_ $0<x_0<\ldots<x_m=1$, $0<y_0<y_1\ldots<y_n=1$ $(m\geq 1,\ n\geq 1)$; $x_{j-1}/x_j>1-\delta$ _for_ $j=1,2,\ldots,m$; _and_ $y_{k-1}/y_k>1-\delta$ _for_ $k=1,2,\ldots,n$, _then_

$$OS(f;\ x_0,x_1,\ldots,x_m;\ y_0,y_1,\ldots,y_n)<\varepsilon.$$

THEOREM 2. _A complex function on_ I _is dominantly integrable there iff it satisfies there_ RCDI.

COROLLARY 1. _If a complex function_ f _is dominantly integrable on_ I, _so is_ $|f|$.

PROOF OF COROLLARY 1. Theorem 2 and the inequality

$$\big||f(P)|-|f(Q)|\big|\leq|f(P)-f(Q)|.$$

LEMMA 1. _Let_ f _be dominantly integrable on_ I. _For every_ $(x,y)\in I$ _set_

$$\hat{f}(x,y)=\sup\{|f(u,v)|:\ x\leq u\leq 1,\ y\leq v\leq 1\}$$

(_see Theorem_ 2 _and Definition_ 2, (i)). _Then_ $\hat{f}$ _is dominantly integrable on_ I.

DEFINITION 3. _A complex function_ f _is absolutely dominantly integrable on_ I _iff it is Riemann integrable on each_ $[a,1]\times[b,1]$, $0<a<1$, $0<b<1$, _and_ $|f|$ _is dominantly integrable on_ I.

DEFINITION 4. *A complex function f has property D (for "dominated") on I iff it is Riemann integrable on each $[a,1]\times[b,1]$, $0<a<1$, $0<b<1$, and there exists a real function $g(x,y)$ which, on I, is monotone nonincreasing in x and in y, summable, and satisfies $g \geqslant |f|$.*

THEOREM 3. *The following are equivalent: (i) dominant integrability on I; (ii) absolute dominant integrability on I; (iii) property D on I; and (iv) Riemann integrability on each $[a,1]\times[b,1]$, $0<a<1$, $0<b<1$, along with domination of absolute value on I by some function, dominantly integrable there.*

(That (i) implies (ii) follows from Theorem 1 and Corollary 1. That (ii) implies (iii) follows from Lemma 1 and Theorem 1, (ii). By Lemma 1 applied to the dominating function, we see that (iv) implies (iii). That (i) implies (iv) is seen by Theorem 1, (i) and Corollary 1, letting the absolute value of the function dominate itself. Thus it merely remains to prove that (iii) implies (i).)

PROOF OF THEOREM 1, (i). Let $0<a<1$, $0<b<1$. As we shall see, it suffices to prove the following statement: For each $\varepsilon>0$ there exists $\delta_1(\varepsilon)>0$ such that if $a=x_0<x_1<\ldots<x_m=1$, $b=y_0<y_1<\ldots<y_m=1$; $x_{j-1}/x_j>1-\delta_1(\varepsilon)$, and $y_{j-1}/y_j>1-\delta_1(\varepsilon)$ for $j=1,2,\ldots,m$, and $P_{j,k}$, $Q_{j,k}$ are points of $[x_{j-1},x_j]\times[y_{k-1},y_k]$ for $j,k=1,2,\ldots,m$, then

$$(2)\qquad \left|\sum_{j=1}^{m}\sum_{k=1}^{m}[f(P_{j,k})-f(Q_{j,k})](x_j-x_{j-1})(y_k-y_{k-1})\right|<\varepsilon.$$

Indeed, assume its truth. Let $\varepsilon>0$, and set $\delta_2=\delta_1(\varepsilon)\min(a,b)$. Choose $a=x_0<x_1<\ldots<x_m=1$, $b=y_0<y_1<\ldots<y_m=1$ with $x_j-x_{j-1}\leqslant\delta_2$ and $y_j-y_{j-1}\leqslant\delta_2$ for $j=1,2,\ldots,m$. Then $x_{j-1}/x_j>1-\delta_1(\varepsilon)$, $y_{j-1}/y_j>1-\delta_1(\varepsilon)$ for $j=1,2,\ldots,m$, and hence, if $P_{j,k}$, $Q_{j,k}$ are as above, (2) holds. This clearly implies that f is bounded on $[a,1]\times[b,1]$ and
$OS(Re(f);\, x_0,x_1,\ldots,x_m;\, y_0,y_1,\ldots,y_m)\leqslant\varepsilon$,
$OS(Im(f);\, x_0,x_1,\ldots,x_m;\, y_0,y_1,\ldots,y_m)\leqslant\varepsilon$. Hence f is Riemann integrable on $[a,1]\times[b,1]$. What remains is to prove the above statement.

Given $\varepsilon>0$, choose δ and χ, both in $(0,1)$, so that (1) holds under the conditions following it, with ε replaced by $\varepsilon/2$. Set $\delta_1(\varepsilon)=\delta$, and choose a positive integer N such that $[1-(\delta/2)]^N \max(a,b) < \chi$. If x_j, y_j, $P_{j,k}$ and $Q_{j,k}$ are as in the statement, set $x_j^* = [1-(\delta/2)]^{N-j}a$, $y_j^* = [1-(\delta/2)]^{N-j}b$ for $j = 0,1,\dots,N-1$; $x_j^* = x_{j-N}$, $y_j^* = y_{j-N}$ for $j = N,N+1,\dots,N+m$; $P_{j,k}^* = Q_{j,k}^* = (x_j^*, y_k^*)$ for $j = 1,2,\dots,N$, $k = 1,2,\dots,N+m$ and for $j = 1,2,\dots,N+m$, $k = 1,2,\dots,N$; $P_{j,k}^* = P_{j-N,k-N}$, $Q_{j,k}^* = Q_{j-N,k-N}$ for $j,k = N+1,N+2,\dots,N+m$. Then

$$\left|\sum_{j=1}^{m}\sum_{k=1}^{m}[f(P_{j,k}) - f(Q_{j,k})](x_j - x_{j-1})(y_k - y_{k-1})\right|$$

$$= \left|\sum_{j=1}^{N+m}\sum_{k=1}^{N+m}[f(P_{j,k}^*) - f(Q_{j,k}^*)](x_j^* - x_{j-1}^*)(y_k^* - y_{k-1}^*)\right| < \varepsilon.$$

LEMMA 2. _Let_ f _be_ _dominantly_ _integrable_ _on_ I. _Then_ _it_ _satisfies_ _there_ RCDI.

PROOF. By Theorem 1, (i), f is bounded on each closed subset of I. Given $\varepsilon > 0$, choose δ,χ as in the last proof. Let $0 < x_0 < \dots < x_m = 1$, $0 < y_0 < \dots < y_n = 1$ $(m \geqslant 1,\ n \geqslant 1)$; $x_{j-1}/x_j > 1-\delta$ for $j = 1,2,\dots,m$; and $y_{k-1}/y_k > 1-\delta$ for $k = 1,2,\dots,n$. Choose a positive integer N such that $[1-(\delta/2)]^N \max(x_0,y_0) < \chi$, and set $x_j = [1-(\delta/2)]^{-j}x_0$, $y_j = [1-(\delta/2)]^{-j}y_0$, $j = -1,-2,\dots,-N$. Let $P_{j,k}$, $Q_{j,k}$ be arbitrary points in $[x_{j-1},x_j] \times [y_{k-1},y_k]$, $j = -N+1,-N+2,\dots,m$, $k = -N+1,-N+2,\dots,n$. Then

$$\sum_{j=-N+1}^{m}\sum_{k=-N+1}^{n}[\mathrm{Re}\ f(P_{j,k}) - \mathrm{Re}\ f(Q_{j,k})](x_j - x_{j-1})(y_k - y_{k-1}) < \varepsilon,$$

which clearly implies that also

$$\sum_{j=-N+1}^{m}\sum_{k=-N+1}^{n}|\mathrm{Re}\ f(P_{j,k}) - \mathrm{Re}\ f(Q_{j,k})|(x_j - x_{j-1})(y_k - y_{k-1}) < \varepsilon.$$

Similarly,

$$\sum_{j=-N+1}^{m}\sum_{k=-N+1}^{n}|\mathrm{Im}\ f(P_{j,k}) - \mathrm{Im}\ f(Q_{j,k})|(x_j - x_{j-1})(y_k - y_{k-1}) < \varepsilon.$$

Hence

$$\sum_{j=-N+1}^{m}\sum_{k=-N+1}^{n}|f(P_{j,k}) - f(Q_{j,k})|(x_j - x_{j-1})(y_k - y_{k-1}) < 2\varepsilon,$$

and therefore

$$OS(f;\ x_o,x_1,\dots,x_m;\ y_o,y_1,\dots,y_n)$$

$$\leq \sum_{j=-N+1}^{m} \sum_{k=-N+1}^{n} w(f,[x_{j-1},x_j] \times [y_{k-1},y_k])(x_j - x_{j-1})(y_k - y_{k-1}) \leq 2\varepsilon.$$

We shall use the following simple

LEMMA 2a. _Let_ $r \leq s$, $\rho \leq \sigma \leq \tau$, _and let_ f _be a complex function, defined and bounded on_ $[r,s] \times [\rho,\tau]$. _Then_

$$\sup\{|f(u,v)|:\ r \leq u \leq s,\ \rho \leq v \leq \tau\} - \sup\{|f(u,v)|:\ r \leq u \leq s,\ \sigma \leq v \leq \tau\}$$

$$\leq w(f,[r,s] \times [\rho,\sigma]).$$

PROOF. Denote by s_1 the first sup, by s_2 the second. We may assume $s_1 > s_2$. Then

$$s_1 = \sup\{|f(u,v)|:\ r \leq u \leq s,\ \rho \leq v \leq \sigma\}, \qquad s_2 \geq \inf\{|f(u,v)|:\ r \leq u \leq s,\ \rho \leq v \leq \sigma\}.$$

So $\quad s_1 - s_2 \leq$ last sup $-$ last inf

$$= w(|f|,[r,s] \times [\rho,\sigma]) \leq w(f,[r,s] \times [\rho,\sigma]).$$

Another simple result we use is

LEMMA 2b. _Let_ $r < s$, $\rho < \sigma$, _and let_ $f(x,y)$ _be a real function on_ $J = [r,s] \times [\rho,\sigma]$, _monotone nonincreasing in_ x _and in_ y _there. Then_ f _is Riemann integrable on_ J.

PROOF. Let $\varepsilon > 0$. Choose an integer $n \geq 1$ with

$$(2n-1)(s-r)(\sigma-\rho)[f(r,\rho) - f(s,\sigma)]/n^2 < \varepsilon,$$

and observe that, by the monotonicity, the left hand side is $\geq$

$$\sum_{j=1}^{n} \sum_{k=1}^{n} w(f,[x_{j-1},x_j] \times [y_{k-1},y_k])(x_j - x_{j-1})(y_k - y_{k-1})$$

where, for $j = 0,1,\dots,n$, $x_j = r+j(s-r)n^{-1}$, $y_j = \rho+j(\sigma-\rho)n^{-1}$.

For convenience of the reader we state here the definition of dominant integrability on $(0,1]$.

DEFINITION 5. Let f be a complex function on (0,1]. Dominant integrability of f on (0,1] means existence of a number Δ having the property: For each $\varepsilon > 0$ there exist δ and χ, $0 < \delta < 1$, $0 < \chi < 1$, such that

$$|\Delta - \sum_{j=1}^{n} f(\xi_j)(x_j - x_{j-1})| < \varepsilon$$

whenever $0 < x_o < x_1 < \ldots < x_n = 1$, $x_o < \chi$, $x_{j-1} \leqslant \xi_j \leqslant x_j$, and $x_{j-1}/x_j > 1-\delta$, $j = 1,2,\ldots,n$.

By Theorem 3 of [2], a complex function f on (0,1] is dominantly integrable there iff it is Riemann integrable on each [a,1], $0 < a < 1$, and there exists a function g, monotone nonincreasing, summable and satisfying $g \geqslant |f|$ on (0,1].

LEMMA 3. Let f satisfy, on I, RCDI. Then $\hat{f}$ is summable there.

PROOF. We first show that

$$f_1(y) \equiv \sup\{|f(1,v)|: y \leqslant v \leqslant 1\}$$

is dominantly integrable on (0,1]. Using (ii) of Definition 2, take, for our f, δ corresponding to $\varepsilon = 1$, and set $\gamma = \delta/2$. Let $0 < \eta \leqslant 1$, and let N be an integer $\geqslant 1$ with $(1-\gamma)^N < \eta$.

Set $\eta_k = (1-\gamma)^{N-k}$, $k = 0,1,\ldots,N$. Then

$$OS(f;\ 1-\gamma,1;\ \eta_o,\eta_1,\ldots,\eta_N) < 1.$$

For every $y \in (0,1]$ let

$$g(y) = \sup\{|f(u,v)|: 1-\gamma \leqslant u \leqslant 1,\ y \leqslant v \leqslant 1\}.$$

Then, by Lemma 2a,

$$\int_{\eta}^{1} g(y)\,dy \leqslant \int_{\eta_o}^{1} g(y)\,dy \leqslant g(1) + \sum_{k=1}^{N} [g(\eta_{k-1}) - g(\eta_k)]\,\eta_k$$

$$= g(1) + \gamma^{-1} \sum_{k=1}^{N} [g(\eta_{k-1}) - g(\eta_k)](\eta_k - \eta_{k-1})$$

$$\leqslant g(1) + \gamma^{-2} \sum_{k=1}^{N} w(f,[1-\gamma,1] \times [\eta_{k-1},\eta_k])\gamma(\eta_k - \eta_{k-1}) < g(1) + \gamma^{-2},$$

and so, g is summable on (0,1]. Since $g \geq |f_1|$ throughout (0,1], f_1 is dominantly integrable there.

By Definition 5, there are δ_1 and χ in (0,1), and an A such that $\sum_{k=1}^{n} f_1(y_{k-1})(y_k - y_{k-1}) < A$ whenever $0 < y_0 < y_1 < \ldots < y_n = 1$, $y_0 < \chi$, and $y_{k-1}/y_k > 1-\delta_1, k = 1,2,\ldots,n$. Let

$$1 - \min(\delta, \delta_1) < \beta < 1,$$

and let N_1 be an integer ≥ 1 with $\beta^{N_1} < \chi$. For $k = 1,2,\ldots$, and every $x \in (0,1]$ let

$$g_k(x) = \sup\{|f(u,v)| : x \leq u \leq 1,\ \beta^k \leq v \leq \beta^{k-1}\}.$$

Let $m \geq 1$, $n \geq N_1$ be integers. Then

$$\sum_{k=1}^{n} [g_k(1) + \sum_{j=1}^{m} \{g_k(\beta^j) - g_k(\beta^{j-1})\}\beta^{j-1}]\beta^{k-1}$$

$$= (1-\beta)^{-1} \sum_{k=1}^{n} [g_{n+1-k}(1) + (1-\beta)^{-1} \sum_{j=1}^{m} \{g_{n+1-k}(\beta^{m+1-j}) - g_{n+1-k}(\beta^{m-j})\}$$

$$\times(\beta^{m-j} - \beta^{m+1-j})](\beta^{n-k} - \beta^{n+1-k})$$

$$\leq (1-\beta)^{-1} \sum_{k=1}^{n} f_1(\beta^{n+1-k})(\beta^{n-k} - \beta^{n+1-k})$$

$$+ (1-\beta)^{-2} \sum_{j=1}^{m} \sum_{k=1}^{n} w(f;[\beta^{m+1-j}, \beta^{m-j}] \times [\beta^{n+1-k}, \beta^{n-k}])$$

$$\cdot (\beta^{m-j} - \beta^{m+1-j})(\beta^{n-k} - \beta^{n+1-k})$$

$$< (1-\beta)^{-1}A + (1-\beta)^{-2}.$$

Letting $m \to \infty$, we obtain

$$\sum_{k=1}^{n} (\int_{0}^{1} g_k(x)dx)\beta^{k-1} \leq (1-\beta)^{-1}A + (1-\beta)^{-2}.$$

Since, for every $x \in (0,1]$ and every integer $k \geq 1$, $\hat{f}(x,\beta^k) \leq \sum_{j=1}^{k} g_j(x)$, we have

$$\int_{\beta^n}^{1}\left(\int_0^1 \hat{f}(x,y)dx\right)dy \leqslant \sum_{k=1}^{n}\int_0^1 \hat{f}(x,\beta^k)dx(\beta^{k-1}-\beta^k) \leqslant \sum_{k=1}^{n}\sum_{j=1}^{k}\left(\int_0^1 g_j(x)dx\right)(\beta^{k-1}-\beta^k)$$

$$\leqslant \sum_{k=1}^{n}\left(\int_0^1 g_k(x)dx\right)\beta^{k-1} \leqslant (1-\beta)^{-1}A + (1-\beta)^{-2}.$$

Therefore $\int_0^1 (\int_0^1 \hat{f}(x,y)dx)dy < \infty$. By Lemma 2b, $\hat{f}(x,y)$ is measurable on $I = \bigcup_{j=2}^{\infty} [1/j,1] \times [1/j,1]$. Hence $\hat{f}$ is summable on I.

LEMMA 4. Let f satisfy, on I, RCDI, and let $0 < a < 1$, $0 < b < 1$. Then f is Riemann integrable on $[a,1] \times [b,1]$.

PROOF. Given $\varepsilon > 0$, take, using Definition 2,(ii), a corresponding δ, and choose $a = x_0 < x_1 < \ldots < x_m = 1$, $b = y_0 < y_1 < \ldots < y_m = 1$ so that $x_j - x_{j-1}$ and $y_j - y_{j-1}$ are $\leqslant \delta \min(a,b)$ for $j = 1,2,\ldots,m$. Then $x_{j-1}/x_j > 1-\delta$, $y_{j-1}/y_j > 1-\delta$, $j = 1,2,\ldots,m$, and hence $OS(f;\ x_0,x_1,\ldots,x_m;\ y_0,y_1,\ldots,y_m) < \varepsilon$, which proves the Lemma.

We shall use the following technical

LEMMA 5. Suppose f is a complex function, Riemann integrable on each $[a,1] \times [b,1]$, $0 < a < 1$, $0 < b < 1$. For every $\varepsilon, \varepsilon_1$ $(\varepsilon > 0,\ 0 < \varepsilon_1 < 1)$ there exists δ in $(0,1/2]$ such that if

$$(3)\quad \begin{cases} 0 < x_0 < x_1 < \ldots < x_m = 1,\ 0 < y_0 < y_1 < \ldots < y_n = 1\ (m \geqslant 1,\ n \geqslant 1);\ x_{j-1}/x_j > 1-\delta, \\ y_{k-1}/y_k > 1-\delta,\ P_{j,k} \in [x_{j-1},x_j] \times [y_{k-1},y_k] \text{ for } j = 1,2,\ldots,m,\ k = 1,2,\ldots,n; \\ \text{and } x_{m_1-1} \leqslant \varepsilon_1 < x_{m_1},\ y_{n_1-1} \leqslant \varepsilon_1 < y_{n_1} \text{ for some } m_1,\ n_1, \end{cases}$$

then

$$(4)\qquad \left|\iint_R f(x,y)dx\,dy - \sum_{j=m_1}^{m}\sum_{k=n_1}^{n} f(P_{j,k})(x_j - x_{j-1})(y_k - y_{k-1})\right| < \varepsilon$$

where $R = [x_{m_1-1},1] \times [y_{n_1-1},1]$.

PROOF. Let $\delta \in (0,1/2]$. If (3) holds, then $x_j - x_{j-1} < \delta$ for $j = 1,2,\ldots,m$, $y_k - y_{k-1} < \delta$ for $k = 1,2,\ldots,n$, and the left hand side of (4) is $\leqslant$

$$\Big|\iint_Q f\,dx\,dy - (x_{m_1}-\varepsilon_1)(y_{n_1}-\varepsilon_1)f(\varepsilon_1,\varepsilon_1) - \sum_{j=m_1+1}^{m} (x_j - x_{j-1})(y_{n_1}-\varepsilon_1)f(x_{j-1},\varepsilon_1)$$

$$- \sum_{k=n_1+1}^{n} (x_{m_1}-\varepsilon_1)(y_k - y_{k-1})f(\varepsilon_1,y_{k-1})$$

$$- \sum_{j=m_1+1}^{m} \sum_{k=n_1+1}^{n} f(P_{j,k})(x_j - x_{j-1})(y_k - y_{k-1})\Big| + 6\delta\hat{f}(\varepsilon_1/2,\varepsilon_1/2)$$

where $Q = [\varepsilon_1,1]\times[\varepsilon_1,1]$ (an "empty" $\sum$ is 0). Using the Riemann integrability of f on Q, the desired conclusion follows.

LEMMA 6. Let a complex function f be Riemann integrable on each $[a,1]\times[b,1]$, $0<a<1$, $0<b<1$, and let $\hat{f}$ be summable on I. Then f is dominantly integrable and summable on I and $\Delta(f)$ of Definition 1 is $\int_I\int f$.

Theorems 1 and 2 are fully established by Lemmas 2-4 and 6.

PROOF OF LEMMA 6. Since $|f|\leqslant\hat{f}$ throughout I, f is summable there. Given $\varepsilon>0$, choose $\varepsilon_1\in(0,1)$ such that

$$\iint_L \hat{f} < \varepsilon \qquad \text{where } L=[(0,\varepsilon_1)\times(0,1)]\cup[(0,1)\times(0,\varepsilon_1)].$$

Take δ guaranteed by Lemma 5, and suppose $0<x_0<x_1<x_2<\dots<x_m=1$, $0<y_0<y_1<\dots<y_n=1$, $x_0<\varepsilon_1$, $y_0<\varepsilon_1$; $x_{j-1}/x_j>1-\delta$, $y_{k-1}/y_k>1-\delta$, $P_{j,k}\in[x_{j-1},x_j]\times[y_{k-1},y_k]$ for $j=1,2,\dots,m$, $k=1,2,\dots,n$. Let $x_{m_1-1}\leqslant\varepsilon_1<x_{m_1}$, $y_{n_1-1}\leqslant\varepsilon_1<y_{n_1}$. Denote

$$\Lambda = \{(j,k):\ 1\leqslant j<m_1 \text{ and } 1\leqslant k\leqslant n, \text{ or } 1\leqslant j\leqslant m \text{ and } 1\leqslant k<n_1\}.$$

Then

$$(5)\quad \begin{aligned} &\Big|\iint_I f(x,y)\,dx\,dy - \sum_{j=1}^{m}\sum_{k=1}^{n} f(P_{j,k})(x_j - x_{j-1})(y_k - y_{k-1})\Big| \\ &< \iint_L \hat{f}(x,y)\,dx\,dy + \Big|\sum_{(j,k)\in\Lambda} f(P_{j,k})(x_j - x_{j-1})(y_k - y_{k-1})\Big| + \varepsilon. \end{aligned}$$

The last absolute value is $\leqslant$

$$\sum_{(j,k)\in\Lambda} \hat{f}(P_{j,k})(x_j - x_{j-1})(y_k - y_{k-1})$$

$$\leqslant \sum_{(j,k)\in\Lambda} \hat{f}((1-\delta)x_j,(1-\delta)y_k)(x_j - x_{j-1})(y_k - y_{k-1}) \leqslant (1-\delta)^{-2} \iint_L \hat{f}(x,y)\,dx\,dy < 4\varepsilon.$$

Hence the left side of (5) is $< 6\varepsilon$.

LEMMA 7. Property D on I implies dominant integrability there.

Lemma 7 establishes completely Theorem 3.

Suppose f is dominantly integrable on I. By Definition 4 (with $g = \hat{f}$), Lemma 2b, Theorem 2 and Lemma 3, $\hat{f}$ has property D on I. By Lemma 7, $\hat{f}$ is dominantly integrable on I, which proves Lemma 1.

PROOF OF LEMMA 7. Let f have property D on I, and let g be as in Definition 4. Then $\hat{f} \leqslant g$ throughout I, and hence $\hat{f}$ is summable there. By Lemma 6, f is dominantly integrable on I.

REFERENCES

[1] Lewis, J.T. - Osgood, C.F. - Shisha, O., Infinite Riemann sums, the simple integral, and the dominated integral. In General Inequalities 1, E.F. Beckenbach, ed. (ISNM, vol. 41). Birkhäuser Verlag, Basel/Stuttgart, 1978, 233-242.

[2] Osgood, C.F. - Shisha, O., The dominated integral. J. Approximation Theory 17 (1976), 150-165.

[3] Osgood, C.F. - Shisha, O., Numerical quadrature of improper integrals and the dominated integral. J. Approximation Theory 20 (1977), 139-152.

BERNSTEIN - POLYNOME, 1912 - 1955

Eberhard L. Stark
Lehrstuhl A für Mathematik
RWTH
Aachen

This paper may be considered as a first attempt in writing down the story of the Bernstein polynomials. It is based more on the bibliographical background than on the trace of the mathematical development. The latter, due to the lack of space, is postponed to another occasion. It leads from the original paper (1912) of S.N. BERNSTEIN to the natural caesura as given by the book (1953) of G.G. LORENTZ. Finally, the suprisingly large quantity of contributions to the subject is indicated by the additional pages of the bibliography ending in 1955.

In Dutzenden von Büchern über Approximationstheorie, Numerik, Analysis, etc. und mehreren hundert Zeitschriftenartikeln wird die mit Recht so gerühmte Arbeit [1] von S.N. BERNSTEIN (1880 - 1968) mit dem elementaren wahrscheinlichkeitstheoretischen Beweis des Approximationssatzes von K. WEIERSTRASS zitiert: entscheidendes Hilfsmittel sind die auf der Binomial - (Bernoulli -) Verteilung beruhenden "Bernstein - Polynome"

$$B_n(f;x) := \sum_{k=o}^{n} f(\tfrac{k}{n})\binom{n}{k} x^k (1-x)^{n-k} \qquad (n \in \mathbb{N};\ f \in C[0,1]).$$

Erstaunt wird man jedoch feststellen müssen, daß die bibliographischen Angaben sowohl bezüglich der Zeitschrift selbst als auch bezüglich des Erscheinungsjahres stark variieren. Die Erklärung ist darin zu finden, daß die meisten der Autoren diese nur zwei Seiten umfassende Originalarbeit nie in den Händen gehabt haben! Dies wird klar, wenn man die ungeahnten Schwierigkeiten bei der Beschaffung der Zeitschrift oder einer Kopie der Arbeit in Betracht zieht. Exemplare zumindest dieses unter zweisprachigem Titel erschienenen Bandes der

≫ Communications de la Societé mathématique de Kharkow ≪

bzw. [1])

СООБЩЕНIЯ
ХАРЬКОВСКАГО
МАТЕМАТИЧЕСКАГО ОБЩЕСТВА
ВТОРАЯ СЕРIЯ, ТОМЬ XIII, 1913

dürften - außerhalb der UdSSR und selbst dort - nur noch in wenigen und dazu unvermuteten Bibliotheken vorhanden sein.

Wesentlich einfacher zu beschaffen sind BERNSTEINs "Gesammelte Werke" [A2]: in ihnen sind jedoch die ursprünglich fremdsprachlichen Veröffentlichungen BERNSTEINs einheitlich nur in russischer Übersetzung - ohne gewisse Wiederholungen - zu finden. (Die provisorische englische Teilausgabe [A3] der Gesammelten Werke enthält lediglich die Übersetzungen der in russischer Sprache abgefaßten Publikationen - bzgl. der nicht erfaßten Arbeiten wird - welch' Ironie! - auf die Originalveröffentlichungen verwiesen!) Die bibliographischen Datierungen werden in [A2] wie auch in den BERNSTEIN gewidmeten Jubiläumsartikeln etc. (mit entsprechendem Werksverzeichnis, vgl. [A6] - [A13]) rein chronologisch vorgenommen: so wird [1] unter 1912, dem tatsächlichen Erscheinungsjahr des ersten Heftes, No 1, aufgeführt, obwohl der Band 13 dieser Zeitschrift redaktionell unter der Jahreszahl 1913 registriert ist (s. obiges Zitat).

Die Verwirrung wird noch dadurch vergrößert, daß wieder zahlreiche Autoren bzgl. des erstmaligen Auftretens der Bernstein-Polynome auf die im wesentlichen inhaltsgleichen Arbeiten [A1] bzw. [2], auch aus dem Jahre 1912, verweisen (letztere wird in [A2, I, p. 567] unter Nr. 43 zeitlich vor [A1] unter Nr. 46 eingeordnet). Seiner Dissertation [2] fügt BERNSTEIN selbst die Anmerkung bei: "... In addition, I deem it necessary to note that with the exception of the two Appendices to the fourth and fifth chapters the present work is a minimally edited translation of my monograph of the same title, which won the prize of the Belgian Academy, to which it was sent

[1])Wörtliche Wiedergabe der Angaben des Titelblattes dieses Bandes (einschließlich also der Abweichungen von der heutigen Rechtschreibung).

in June of 1911." [A2,I, p. 12, footnote 2;A3, p. 3/4] [2)] Festzustellen ist jedoch, daß die Bernstein - Polynome in [A1] überhaupt nicht auftreten, sondern gerade im Appendix zu Kapitel V unter der Überschrift "Expansion of arbitrary functions in normal series" [A2, I, p. 79 - 84; A3, p. 68 - 73]; diesem Problemkreis ist dann auch die ausführlichere Darstellung in [3] gewidmet. Abschließend sei darauf hingewiesen, daß die in [1] bzw. [2] mittels der Bernstein - Polynome geführten Beweise des Weierstraß - Satzes erheblich voneinander abweichen - was aus vielen Zitaten auch nicht hervorgeht!

Nach diesem Versuch, die "Entstehungsgeschichte" der Bernstein-Polynome zu rekonstruieren, sollen daran anknüpfend die Anfänge ihrer äußerst interessanten "Ausbreitungsgeschichte" verfolgt werden. Vor allem beabsichtigt ist auch, eine, cum grano salis, möglichst lückenlose Bibliographie der Bernstein - Polynome bis zum Jahre 1955 einschließlich zusammenzustellen. (Daß zahlreiche, insbesondere russische Publikationen trotz aller Bemühungen nicht beschaffbar waren, sondern nur durch Referate, Zitate, etc. als existent nachgewiesen werden konnten, dürfte in der Natur der Sache liegen. Allerdings erhebt sich dabei die Frage, worin der Wert von z.B. nach 1945 erschienenen Arbeiten zu sehen ist, die nicht allgemein zugänglich sind ?)

Die Bibliographie nach 1955 (vorerst) abzubrechen, wird durch mehrere Gründe gerechtfertigt. Das Erscheinen der Monographie "Bernstein Polynomials" [90] von G.G. LORENTZ (siehe dazu auch den Jubiläumsband [A14]) im Jahre 1953 - und deren Auswirkungen ab 1955 - stellt eine natürliche Zäsur dar.[3)] -

---

2) Auf dem Titelblatt zu [A1]: "Mémoire couronné par la Classe des sciences, dans sa séance du 15 décembre 1911." bzw. aus dem Inhaltsverzeichnis"..., médaille d'or en 1911." - Die öffentliche Verteidigung der Doktor - Dissertation erfolgte in Har'kov am 19. Mai 1913; siehe den aus diesem Anlaß von BERNSTEIN gehaltenen Vortrag [A2, I, p. 209 - 214; A3, p. 109-114].

3) Der erste Satz der Einleitung zu [90, p. vii] "This contribution attempts to give an exhaustive exposition of main facts about the Bernstein polynomials and to discuss some of their applications in Analysis.", mag angesichts der Fülle des heute zu verzeichnenden Materials als zu weitgesteckt erscheinen (so werden z.B. [36],[54],[58],[80] vermißt); jedoch waren offenkundig die Intentionen dieses so verdienstvollen Buches andere - auf neuere Entwicklungen, auch abstrakter Art, hinführende - als Vollständigkeit der Literatur.

Publikationen über den Bernstein - Polynomen nachempfundenen Verallgemeinerungen von Approximationsoperatoren - bzgl. der ersten Ansätze siehe [52], [73],[82],[88],[92],[93] - mehren sich. Dies wird insbesondere durch das Testfunktionenkriterium für positive lineare Operatoren von H. BOHMAN - P.P. KOROVKIN (1952/53) gefördert, das gerade auch die Bernstein - Polynome als deren einfachstem Prototyp unter einem völlig neuen Aspekt erscheinen läßt; vgl. insbesondere auch [80]. - Schließlich ermöglicht die große Zahl der bis in das Jahr 1955 zu datierenden Arbeiten eine sich durchaus bestätigende Extrapolation auf die Flut der Arbeiten über Bernstein - Polynome und deren Verallgemeinerungen in den nachfolgenden Jahren.

Zu Beginn dieses Abschnitts stehe kommentarlos das "erstaunliche Kurz" - Referat von Prof. D. Sintzov (Charkow)[4] in "Fortschritte der Mathematik" 43/1912 (1915) 301 (also auch hier die Einordnung von [1] in das Jahr 1912): " Ist F(x) eine kontinuierliche Funktion, dann genügen die Polynome

$$E_n = \sum_{o}^{n} F(\tfrac{m}{n})\, C_n^m x^m (1-x)^{n-m} ,$$

welche in der Wahrscheinlichkeitsrechnung auftreten, der Ungleichung

$$|F(x) - E_n| < \varepsilon ."$$

(Diese Besprechung ist unter dem Abschnitt "Kombinationslehre und Wahrscheinlichkeitsrechnung" nachzulesen, nicht - wie vom Thema her üblich und zu erwarten - im Abschnitt "Reihen"!)

In einem Lehrbuch erscheinen die Bernstein - Polynome zum ersten Mal erstaunlicherweise bereits im Jahre 1913, und zwar im zweiten Band von R. d'ADHEMARs "Leçons sur les Principes d' Analyse" [4], einem von mehreren seinerzeitigen Lehr- und Übungsbüchern desselben Verfassers (der zu [4] gehörende erste Band enthält zu dieser Zeit schon Beweisskizzen zum Weierstraß-Approximationssatz mittels der singulären Integrale von Weierstraß,

---

4) D.M. Sincov (1867 - 1946); s. Bernštein, S.N. - L.Ja. Giršval'd: Obituary: D.M. Sincov. Uspehi Mat. Nauk 2, no. 4 (20)(1947) 191 - 206 (Russ.). MR 10, 420.

Landau - Stieltjes und de La Vallée Poussin !); zu bemerken ist allerdings, daß dieses besagte Zitat in einer eigenen Abhandlung [3] mit BERNSTEIN als Autor auftritt, die als Anhang dem Band [4] beigegeben wurde: "... Serge Bernstein ouvre des voies nouvelles, en substituant certaines *Séries de polynomes* au développement taylorien. ... je fais pressentir la valeur des idées de M.S. Bernstein, en résumant quelques pages de sa thèse" [4,p. vi]. Festzustellen ist auch, daß [4] den Bernstein - Polynomen zu keiner durchgreifenden Verbreitung verhalf: Hinweise auf [4] sind lediglich im Lehrbuch (1925) von W. SIERPIŃSKI [10,p. 228] sowie bei I. CHLODOVSKY [13] (hier wiederum der einzige Verweis auf [10]) und im FdM - Referat über die Arbeit [26] von A. WUNDHEILER durch BERNSTEIN selbst (die Ergebnisse in [26] seien in [3] enthalten !) zu finden.

Es ist dann eine lange (kriegsbedingte ?) Pause zu verzeichnen. Erst 1921 wird in der berühmten Arbeit von F. HAUSDORFF [7, p. 104] beiläufig, unter Hinzufügung eines knappen Beweises, auf [1] verwiesen. Es folgt [8]. Die nächste, äußerst exponierte Erwähnung des Bernsteinschen Satzes findet sich in dem 1924 herausgegebenen Heft [9] über - in moderner Terminologie - "Approximationstheorie" der "Encyklopädie der mathematischen Wissenschaft". Der Bearbeiter der deutschen Ausgabe, A. ROSENTHAL, verweist - nach einer Aufzählung von rund zwei Dutzend Beweisen des Weierstraß - Satzes in einer Fußnote [9, p. 1148] auf die BERNSTEINsche Arbeit: "937)$_*$ Es sei noch erwähnt, daß *S. Bernstein*, Communications Soc. math. de Kharkow (2) 13 (1912/13), p. 1/2, einen Beweis von Satz 1. mit Hilfe der Wahrscheinlichkeitsrechnung erbracht hat. - ...*"; [2] wird nicht erwähnt. Der Zusammenhang des BERNSTEINschen Verfahrens mit der Interpolationsformel von E. BOREL (1905) - für $f \in C[0,1]$ lassen sich stets Polynome $P_{\mu,\nu}(x)$ konstruieren, so daß $f(x) = \lim_{\nu=\infty} \sum_{\mu=0}^{\mu=\nu} f(\frac{\mu}{\nu})\, P_{\mu,\nu}(x)$ gilt bei gleichmäßiger Konvergenz - wird wieder mit einer zusätzlichen Fußnote [9, p. 1155] gewürdigt: "962 a)$_*$ *S. Bernstein* $^{937)}$. Hier ein ganz besonders einfacher Ausdruck für $P_{\mu,\nu}(x)$, nämlich

$$P_{\mu,\nu}(x) = \binom{\nu}{\mu} x^{\mu}(1-x)^{\nu-\mu}. \quad ^{*}"$$

Doch auch dieses Werk blieb bzgl. der Bernstein - Polynome (jedenfalls, was rückverweisende Zitate in anderen Quellen anbelangt) ohne (die gebührende) Wirkung.

Umso durchschlagenderen Erfolg zeitigte dann 1925 das Erscheinen der ersten Auflage von G. PÓLYA - G. SZEGÖs "Aufgaben und Lehrsätze aus der Analysis, I" [11, p. 66]: drei isolierte Aufgaben (144 - 146) am Ende des 3. Kapitels / 2. Abschnitt sind dem Satz von BERNSTEIN gewidmet; auf die Originalquelle [1] (1912 !) wird in den Lösungen [11, p. 230] verwiesen. Wegen der schon damals offensichtlichen Unzugänglichkeit der BERNSTEINschen Arbeit verweisen in der nun folgenden Zeit zahlreiche nicht - russische Autoren (ehrlicherweise !) zumindest auch auf diese Quelle; siehe z.B. [12],[18],[23].

Hinzuzufügen bleibt im Rahmen dieser Urgeschichte der Bernstein - Polynome, daß der Begriff "S. Bernsteinsche Polynome" wohl von F. HAUSDORFF [8, p. 243, Fußnote](1923) geprägt wurde; bei I. CHLODOVSKY [13] finden sich "polynômes de M.S. Bernstein"([11] wird nicht erwähnt); in [23] treten erstmals "Bernstein polynomials" auf.

Die mathematische Entwicklung der Bernstein - Polynome (einschließlich der auch in diesem Zeitraum schon zahlreichen Parallelentwicklungen, Wiederentdeckungen, etc.) chronologisch zu verfolgen, ist an dieser Stelle aus Platzgründen - noch - nicht möglich. Zusätzlich wird das Vorhaben dadurch erschwert, daß viele Arbeiten - noch - nicht zugänglich sind; und selbst deren teilweise verfügbaren Referate enthalten widersprüchliche Wertungen.

Von Interesse wäre auch zu ergründen, warum zahlreiche Bücher und Übersichtsartikel (bis 1955), die den Weierstraß-Approximationssatz in aller Ausführlichkeit behandeln, die Bernstein - Polynome übergehen ? (!)

Der Verfasser dankt allen, die mitgeholfen haben, diese Bibliographie wenigstens auf den hier vorgelegten Stand zu bringen. Stellvertretend gilt ein besonderes Wort des Dankes Prof. J. Musielak, Poznań, für die entsprechenden Kopien aus dem Buch [10]: für lange Zeit erschien es hoffnungslos, auch nur ein Exemplar dieses Bandes ausfindig zu machen. Das Literaturverzeichnis wird Lücken und Fehler enthalten. Der Verfasser bittet alle am Thema Interessierten, ihn darauf hinzuweisen und ihm Quellen zu und Belege von fehlenden Arbeiten (insbesondere auch der mit * gekennzeichneten) zugänglich zu machen.

Abschließend sei vermerkt, daß diese Arbeit aus Anlaß von BERNSTEINs 100. Geburtstag, der im Jahre 1980 ebenso wie der von L. FEJÉR (1880 - 1959) und F. RIESZ (1880 - 1956) gefeiert werden konnte, niedergeschrieben wurde: so sei sie - neben J.L.B. Cooper, dem dieser Tagungsband als ganzes gewidmet ist - auch dem Entdecker der Bernstein - Polynome mitzugedacht.

Démonstration du théorème de Weierstrass fondée sur le calcul des probabilités.

Je me propose d'indiquer une démonstration fort simple du théorème suivant de Weierstrass:

Si $F(x)$ est une fonction continue quelconque dans l'intervalle 01, il est toujours possible, quel que petit que soit ε, de déterminer un polynome $E_n(x) = a_0x^n + a_1x^{n-1} + \ldots + a_n$ de degré n assez élevé, tel qu' on ait

$$|F(x) - E_n(x)| < \varepsilon$$

en tout point de l'intervalle considéré.

A cet effet, je considère un évenement A, dont la probabilité est égale à x. Supposons qu'on effectue n expériences et que l'on convienne de payer à un joueur la somme $F\left(\frac{m}{n}\right)$, si l'évenement A se produit m fois. Dans ces conditions, l'espérance mathématique E_n du joueur aura pour valeur

$$E_n = \sum_{m=0}^{m=n} F\left(\frac{m}{n}\right) . C_n^m x^m (1-x)^{n-m}. \tag{1}$$

Or, il résulte de la continuité de la fonction $F(x)$ qu'il est possible de fixer un nombre δ, tel que l'inégalité

$$|x - x_0| \leqq \delta$$

entraine

$$\left| F(x) - F(x_0) \right| < \frac{\varepsilon}{2}$$

de sorte que, si $\bar{F}(x)$ désigne le maximum et $\underline{F}(x)$ le minimum de $F(x)$ dans l'intervalle $(x - \delta, x + \delta)$, on a

$$\bar{F}(x) - F(x) < \frac{\varepsilon}{2}, \; F(x) - \underline{F}(x) < \frac{\varepsilon}{2}. \tag{2}$$

Soit de plus η la probabilité de l'inégalité $\left| x - \frac{m}{n} \right| > \delta$, et L le maximum de $|F(x)|$ dans l'intervalle 01.

On aura alors,

$$\underline{F}(x) . (1 - \eta) - L . \eta < E_n < \bar{F}(x) . (1 - \eta) + L . \eta. \tag{3}$$

Mais, en vertu du théorème de Bernoulli, on pourra prendre n assez grand pour avoir

$$\eta < \frac{\varepsilon}{4L}. \tag{4}$$

L'inégalité (3) se mettra donc sucessivement sous la forme

$$F(x) + (\underline{F}(x) - F(x)) - \eta(L + \underline{F}(x)) < E_n < F(x) + (\bar{F}(x) - F(x)) + \eta(L - \bar{F}(x))$$

et ensuite

$$F(x) - \frac{\varepsilon}{2} - \frac{2L}{4L}\varepsilon < E_n < F(x) + \frac{\varepsilon}{2} + \frac{2L}{4L}\varepsilon;$$

donc

$$|F(x) - E_n| < \varepsilon \tag{5}$$

Or E_n est manifestement un polynome de degré n.

Le théorème est donc démontré.

J'ajouterai seulement deux remarques.

Les polynomes approchés $E_n(x)$ sont surtout commodes, il me semble, lorsqu'on connait exactement ou approximativement les valeurs de $F(x)$ pour $x = \frac{m}{n}$ $(m = 0, 1, \ldots n)$.

La formule (1) et l'inégalité (5) montrent que, quelle que soit la fonction continue $F(x)$, on a

$$F(x) = \lim_{n=\infty} \sum_{m=0}^{m=n} F\left(\frac{m}{n}\right) . C_n^m x^m (1-x)^{n-m}.$$

S. Bernstein.

LITERATUR

Den bibliographischen Angaben wurden, soweit dies möglich war, die entsprechenden Referate hinzugefügt; dabei bedeuten (wie üblich und in zeitlicher Reihenfolge):

FdM	Jahrbuch über die Fortschritte der Mathematik;
Zbl	Zentralblatt für Mathematik und ihre Grenzgebiete;
MR	Mathematical Reviews;
RZM	Referativnyi Žurnal. Matematika.

Die Abkürzungen der Zeitschriften sind die in den MR gebräuchlichen bzw. - falls die entsprechenden Zeitschriften ihr Erscheinen eingestellt haben - jenen angeglichen.
Ein Stern soll andeuten, daß die so gekennzeichnete Publikation dem Verfasser in keiner Form (Original, Kopie, "Gesammelte Werke", etc.) zur Verfügung stand: ihre Existenz ist (also) lediglich durch Referate oder Zitate nachgewiesen.

BIBLIOGRAPHIE: Bernstein-Polynome, 1912-1955

[1] BERNSTEIN, S., Démonstration du théorème de Weierstrass fondée sur le calcul des probabilités.
Communications de la Societé Mathématique de Kharkow (2) 13, no. 1 (1913) 1-2 (1912) = [A 2, I,(4), 105-106]. FdM 43, 301.

[2] BERNSTEIN, S.N., On the best approximation of continuous functions by polynomials of a given degree (Russ.).
Comm. Soc. Math. Har'kov (2) 13, no. 4-5 (1913) 49-194 (1912) = [A 2, I,(3), 11-104]. FdM 43, 493.

[3] BERNSTEIN, S., Sur les series normales.
In: [4], pp. 259-283. FdM 44, 323, 456.

[4] D'ADHEMAR, R., Leçons sur les Principes d'Analyse, II.
Gauthier-Villars, Paris 1913, vii+297 pp.. FdM 44, 322, 456.

[5] BERNSTEIN, S., Sur la représentation des polynomes positifs.
Comm. Soc. Math. Har'kov (2) 14 (1915) 227-228 = [A 2,I,(19), 251-252]. FdM 48 (1926) 1371.

[6] BERNSTEIN, S., Quelques remarques sur l'interpolation.
Comm. Soc. Math. Har'kov (2) 15 (1917) 49-61, 208 = [A 2,I,(20), 253-263]. FdM 48, 311, 1372.

[7] HAUSDORFF, F., Summationsmethoden und Momentfolgen. I. Math. Z. 9 (1921) 74-109.

[8] HAUSDORFF, F., Momentprobleme für ein endliches Intervall. Math. Z. 16 (1923) 220-248. FdM 49, 193.

[9] FRÉCHET, M. - A. ROSENTHAL *), II. C. 9c. Funktionenfolgen (abgeschlossen im Juli 1923; als Heft 7 ausgegeben am 1.IV. 1924). In: Encyklopädie der Mathematischen Wissenschaften mit Einschluß ihrer Anwendungen; 2. Band in 3 Teilen: Analysis. Teubner Verl., Leipzig 1923-1927, xiv, pp. 675-1648; pp. 1136-1187 (1924). FdM 50, 176.

*) Nach dem französischen Artikel von M. Fréchet in Poitiers (jetzt in Straßburg) bearbeitet von A. Rosenthal in Heidelberg (p. 854 und Fußnote, p. 851).

[10] SIERPIŃSKI, W., Analiza, Vol. I, Part II (Polish). Warsaw 1925^2, v+278 pp. FdM 51, 176.

[11] PÓLYA, G. - G. SZEGÖ, Aufgaben und Lehrsätze aus der Analysis, I. (Grundlehren Math. Wiss. 19) Springer Verl., Berlin-Göttingen-Heidelberg 1925, xvi+338 pp.. FdM 51, 172.

[12] WIGERT, S., Réflexions sur le polynome d'approximation $\sum_{\nu=0}^{n}\binom{n}{\nu}\varphi(\frac{\nu}{n})x^{\nu}(1-x)^{n-\nu}$. Ark. Mat. Astronom. Fys. 20A, no. 5 (1927) 1-15. FdM 53, 237.

[13] CHLODOVSKY, I., Sur la représentation des fonctions discontinues par les polynômes de M.S. Bernstein. Fund. Math. 13 (1929) 62-72. FdM 55, 169.

[14] BERNSTEIN, S., Quelques remarques sur les polynomes d'ecart minimum à coefficient entiers (Russ.). Dokl. Akad. Nauk SSSR (A) 1930, no. 16 (1930) 411-418 = [A 2,I,(43), 468-471, 562-563]. FdM 57, 1403.

[15] KANTOROVIČ, L.V. (L. Kantorowitsch), Sur certains développements suivant les polynomes de la forme de S.N. Bernstein, I (Russ.). Dokl. Akad. Nauk SSSR (A) 1930, no. 21 (1930) 563-568. FdM 57,1393.

[16] KANTOROVIČ, L.V. (L. Kantorowitsch), Sur certains développements suivant les polynomes de la forme de S.N. Bernstein, II (Russ.). Dokl. Akad. Nauk SSSR (A) 1930, no. 22 (1930) 595-600. FdM 57, 1393.

[17] VORONOVSKAJA, E.V. (E. Voronowsky), Transformation d'une série de fonctions au moyen des différences de ses termes (Russ.). Dokl. Akad. Nauk SSSR (A) 1930, no. 26 (1930) 693-700. FdM 57, 1404.

[18] WRIGHT, E.M., The Bernstein approximation polynomials in the complex plane.
J. London Math. Soc. 5 (1930) 265 - 269. FdM 56, 964.

[19] KANTOROVIČ, L.V., Sur la convergence de la suite des polynômes de S. Bernstein en dehors de l'intervalle fondamental (Russ.).
Izv. Akad. Nauk SSSR (7) Otd. Mat. Estest. Nauk 1931, no. 8 (1931) 1103 - 1115. FdM 57, 1412; Zbl 3, 304.

[20] KANTOROVIČ, L.V., Quelques observations sur l'approximation de fonctions au moyen de polynomes à coefficients entiers (Russ.).
Izv. Akad. Nauk SSSR (7) Otd. Mat. Estest. Nauk 1931, no. 9 (1931) 1163 - 1168. FdM 57, 1403; Zbl 3, 391.

[21] BERNSTEIN, S., Sur une modification de la formule d'interpolation de Lagrange.
Comm. Soc. Math. Har'kov. (4) 5 (1932) 49 - 57 = [A 2,II, (54), 130 - 140]. FdM 58, 262; Zbl 5, 12.

[22] BERNSTEIN, S., Complément à l'article de E. Voronovskaya ≪ Détermination de la forme asymptotique de l'approximation des fonctions par les polynômes de M. Bernstein ≫ (French; Russ.sum.).
Dokl. Akad. Nauk SSSR (A) 1932, no. 4 (1932) 86 - 92. = [A 2,II (57), 155 - 158 + Bem.]. FdM 58, 1062; Zbl 5, 13.

[23] HILDEBRANDT, T.H., On the moment problem for a finite interval.
Bull. Amer. Math. Soc. 38 (1932) 269 - 270. FdM 58, 432; Zbl 4, 207.

[24] VORONOVSKAJA, E.V., Détermination de la forme asymptotique de l'approximation des fonctions par les polynômes de M. Bernstein (Russ.).
Dokl. Akad. Nauk SSSR (A) 1932, no. 4 (1932) 79 - 85. FdM 58, 1062; Zbl 5, 12.

[25] WIGERT, S., Sur l'approximation par polynomes des fonctions continues.
Ark. Mat. Astronom. Fys. 22B, no. 9 (1932) 1 - 4. FdM 58, 262; Zbl 4, 106.

[26] WUNDHEILER, A., Une démonstration simple de la formule d'interpolation de S. Bernstein.
Enseignement. Math. (1) 31 (1932) 75 - 77. FdM 59, 283; Zbl 6, 159.

[27] HILDEBRANDT, T.H. - SCHOENBERG, I.J., On linear functional operations and the moment problem for a finite interval in one or several dimensions.
Ann. of Math. (2) 34 (1933) 317 - 328. FdM 59, 410; Zbl 6, 402.

[28] JACKSON, D., A proof of Weierstrass's theorem.
Amer. Math. Monthly 41 (1934) 309 - 312 = In: Selected Papers on Calculus (Ed. T.M. Apostol et al.) Math. Ass. Amer., Belmont, Cal. 1969, xv + 397 pp.; pp. 227 - 231. FdM 60, 211; Zbl 9, 158.

[29] KANTOROVITCH, L. (L.V. KANTOROVIČ), La représentation explicite d'une fonction measurable arbitraire dans la forme de la limite d'une suite de polynômes (French and Russ.).
Mat. Sb. 41 (1934) 503-506, 507-510. FdM 60, 978; Zbl 11, 15.

[30] GONČAROV, V.A., Theory of Interpolation and Approximation of Functions (Russ.).
Gos. Izdat. Tehn.-Teor. Lit., Moscow 1954, 327 pp. (1. Ed. 1934). Zbl 57, 298; MR 16, 803.

[31] LANDAU, E., Einführung in die Differentialrechnung und Integralrechnung.
P. Noordhoff N.V., Groningen-Batavia 1934, 368 pp. FdM 60, 167; Zbl 8, 303.

[32] POPOVICIU, T., Sur l'approximation des fonctions convexes d'ordre supérieur.
Mathematica (Cluj) 10 (1935) 49-54. FdM 61, 295; Zbl 10, 295.

[33](*) VITALI, G. - G. SANSONE, Moderna Teoria delle Funzioni di Variabile Reale, II. Sviluppi in Serie di Funzioni Ortogonali.
Nicola Zanichelli Ed., Bologna, 1. Ed.* 1935, vi+310 pp.; 2. Ed.* 1946, viii+511 pp.; 3. Ed. 1952, viii+614 pp. FdM 62, 189; Zbl 16, 157; MR 7, 434/13, 741.

[34] BERNSTEIN, S., Sur le domaine de convergence des polynomes
$B_n f(x) = \sum_0^m f(m/n)\, C_n^m\, x^m(1-x)^{n-m}$.
C.R. Acad. Sci. Paris 202 (1936) 1356-1358 = [A 2,II,(64), 184-186]. FdM 62, 334; Zbl 14, 13.

[35] BERNSTEIN, S., Sur la convergence de certaines suites de polynomes.
J. Math. Pures Appl. (9) 15 (1936) 345-358 = [A 2, II,(65), 187-197]. FdM 62, 335; Zbl 15, 100.

[36]* CHLODOVSKIĬ, I.N., On some properties of the polynomials of S.N. Bernšteĭn (Russ.).
In: Proceedings of the First All-Soviet Mathematics Conference (Har'kov, 24.-29. Juni 1930) (Russ.). ONTI, Moskau-Leningrad 1936. †)

†) Dieser häufig zitierte Beitrag wird durchgehend unter 1930 angeführt; vgl. hier u.a. BERNSTEINs Referat über [25], [30; 1954, p. 114] etc.; der zugeordnete Tagungsband ist aber offensichtlich erst 1936 erschienen, s. [A 2,I, p. 500; II, p. 187], [A 3, p. 145, 215]. Zur Tagung selbst vgl. "Le premier congrès des mathématiciens de l'U.R.S.S., Kharkoff, juin 1930" in Enseignement Math. 29 (1930) 338-340 (FdM 57, 48).

[37] HAVILAND, E.K., On the momentum problem for distribution functions in more than one dimension. II.
Amer. J. Math. 58 (1936) 164 - 168. FdM 62, 483; Zbl 15, 109.

[38] CHLODOVSKY, I., Sur le développement des fonctions définies dans un intervalle infini en séries de polynomes de M.S. Bernstein.
Compositio Math. 4 (1937) 380 - 393. FdM 63 , 237; Zbl 16, 354.

[39] KANTOROVIČ, L., On the moment problem for a finite interval (Russ.).
Dokl. Akad. Nauk SSSR 14 (1937) = Transl. C.R. (Doklady) Acad. Sci. URSS (N.S.) 14 (1937) 531 - 537; 16 (1937) 147. FdM 63, 387; Zbl 16, 353.

[40] LORENTZ, G. (G.R. Lorenc), Zur Theorie der Polynome von S. Bernstein*) (Russ.sum.).
Mat. Sb. (N.S.) 2 (44) (1937) 543 - 556. FdM 63, 236; Zbl 17, 395.

*) Fußnote: "Diese Arbeit bildete den Hauptinhalt meiner Kandidatdissertation (Verteidigt in Leningrad, den 28. April 1936)."

[41]* POPOVICIU, T., Despre cea mai bună aproximatie a funcţiilor continue prin polinoame (Roum.; = On the Best Approximation of Continuous Functions by Polynomials).
Inst. Arte Grafice, Cluj 1937, 66 pp. (Monogr. Mat. Publ. Sect. Mat. Univ. Cluj 3). FdM 63, 959.

[42] BERNSTEIN, S., Constructive theory of functions of a real variable (Russ.).
In: Mathematics and Natural Sciences in the USSR (Russ.). Izdat. Akad. Nauk SSSR, Moscow - Leningrad 1938; pp. 36 - 41 = [A 2, II, (78), 295 - 300].

[43] CHLODOVSKY, I., Le problème des moments et les polynômes de S. Bernstein (Russ.).
Dokl. Akad. Nauk SSSR 19 (1938) = Transl. C.R. (Doklady) Acad. Sci. URSS (N.S.) 19 (1938) 659 - 661. FdM 64, 407; Zbl 20, 41.

[44] KAC, M., Une remarque sur les polynomes de M.S. Bernstein.
Studia Math. 7 (1938) 49 - 51 = In: Marc Kac: Probability, Number Theory, and Statistical Physics. Selected Papers (Ed. K. Baclawski- M.D. Donsker) MIT Press, Cambridge (Mass.) - London 1979, xxxviii + 529 pp.; pp. 61 - 63.

[45] WEGMÜLLER, W., Ausgleichung durch Bernstein - Polynome.
Mitt. Verein. Schweiz. Versich.-Math. 36 (1938) 15 - 59. FdM 64, 1028; Zbl 19, 316.

[46] FAVARD, J., Sur l'interpolation.
Bull. Soc. Math. France 67 (1939) 102 - 113. FdM 65, 1196; Zbl 23, 24; MR 1, 54.

[47] FELDHEIM, E., Théorie de la Convergence des Procédés d'Interpolation et de Quadrature Mecanique.
Gauthier-Villars, Paris 1939, 91 pp. FdM 65, 245; Zbl 21, 397.

[48] KAC, M., Reconnaissance de priorité relative à ma Note "Une remarque sur les polynomes de M.S. Bernstein".
Studia Math. 8 (1939) 170. FdM 65, 248; Zbl 20, 212.

[49]* AHIEZER, N.I., Lectures on the Theory of Approximation (Russ.).
Har'kov 1940, 136 pp. Zbl 60, 169; MR 3, 234.

[50] CHLODOVSKY, I., Certaines propriétés interpolatoires des fonctions absolument monotones de deux variables (Russ.).
Dokl. Akad. Nauk SSSR 28 (1940) = Transl. C.R. (Doklady) Acad. Sci. URSS (N.S.) 28 (1940) 387-390. FdM 66, 243; Zbl 26, 304; MR 2, 361.

[51] FRÉCHET, M., Commentaire sur les formules d'interpolation.
In: Jubilé Scientifiqué de M. Émile Borel. Gauthier-Villars, Paris 1940, 418 pp.; pp. 197-198 = In: Oevres de Émile Borel, I. Ed. du Centre National de la Recherche Scientifique, Paris 1972, xvi + 634 pp. (1 plate); pp. 219-220. Zbl 60, 14; MR 1, 128.

[52] MIRAKJAN, G., Approximation des fonctions continues au moyen de polynômes de la forme
$e^{-nx}\sum_{k=0}^{m} C_{k,n} x^{k}$ (Russ.).
Dokl. Akad. Nauk SSSR 31 (1941) = Transl. C.R. (Doklady) Acad. Sci. URSS (N.S.) 31 (1941) 201-205. FdM 67, 216; MR 2, 363.

[53] WIDDER, D.V., The Laplace Transform.
Princeton Univ. Press, Princeton 1941, x + 406 pp. MR 3, 232.

[54] POPOVICIU, T., Sur l'approximation des fonctions continues d'une variable réelle par les polynomes.
Ann. Sci. Univ. Jassy 28 (1942) 208; Zbl 61, 133; MR 8, 266.

[55] BERNSTEIN, S.N., Sur les domaines de convergence des polynomes
$\sum_{0}^{n} C_{n}^{m} f(m/n) x^{m} (1-x)^{n-m}$ (Russ.; French sum.).
Izv. Akad. Nauk SSSR Ser. Mat. 7 (1943) 49-88 = [A 2, II, (81), 310-348]. Zbl 61, 144; MR 5, 180, 328.

[56] SHOHAT, J.A. - J.D. TAMARKIN, The Problem of Moments.
Amer. Math. Soc., New York 1943, xiv + 140 pp. Zbl 41, 433; MR 5, 5/13, 1138.

[57] FAVARD, J., Sur les multiplicateurs d'interpolation.
J. Math. Pures Appl. 23 (1944) 219-247. MR 7, 436.

[58] NATANSON, I.P., On some estimations connected with singular integral of C. de la Vallée-Poussin (Russ.).
Dokl. Akad. Nauk SSSR 45 (1944) 290 - 293 = Transl. C.R. (Doklady) Acad. Sci. URSS (R.S.) 45 (1944) 274 - 277. Zbl 60, 259; MR 6, 267.

[59] POPOVICIU, T., Les Fonctions Convexes.
(Actualités Scientifiques et Industrielles 992; Exposés sur la Théorie des Fonctions XVII). Hermann, Paris 1944, pp. 5 - 76.
Zbl 60, 149; MR 8, 319.

[60] HERZOG, F. - J.D. HILL, The Bernstein polynomials for discontinuous functions.
Amer. J. Math. 68 (1946) 109 - 124. Zbl 61, 133; MR 7, 440.

[61]* IPATOV, A.F., On the convergence of the S.N. Bernšteĭn polynomials for functions of two variables (Russ.).
Petrozavodsk. Učen. Zap. Gos. Univ. 2, no. 4 (1947) 53 - 57.
Zit. in [A 5, II, p. 282].

[62]* SOKOLOV, I.G., The approximation of certain classes of functions by Bernšteĭn polynomials (Russ.).
L'vov, Dokl. i Soobšč. Gos. Univ. 1 (1947)? - ?. Zit. in [A 5, II, p. 653]. (!)

[63]* SOKOLOV, I.G., On the approximation of functions satisfying a Lipschitz condition by Bernšteĭn polynomials (Ukrain.).
L'vov, Učen. Zap. Gos. Univ. Ser. Fiz.-Mat. 5, no. 1 (1947) 5 - 9.
Zit. in [A 4, p. 468],[A 5,II, p. 653].

[64] ACHIESER, N.I., Vorlesungen über Approximationstheorie.
Akademie Verl., Berlin 1953, ix + 309 pp. (Orig. Russ. Ed.: Moscow - Leningrad 1947, 323 pp.). Zbl 31, 157 / 52, 290; MR 10, 33/15, 867/ 20 # 1872.

[65] HIRSCHMAN, I.I.Jr. - D.V., WIDDER, Generalized Bernstein polynomials.
Duke Math. J. 16 (1949) 433 - 438. Zbl 33, 111; MR 11, 29.

[66] FAVARD, J., Sur l'approximation dans les espaces vectoriels.
Ann. Mat. Pura Appl. (4) 29 (1949) 259 - 291. Zbl 36, 204; MR 11,669.

[67] SOKOLOV, I.G., Approximation of functions with a given modulus of continuity by Bernšteĭn polynomials (Russ.).
L'vov, Učen. Zap. Gos. Univ. Ser. Fiz.-Mat. 12, no. 3 (1949) 45-52.
Zit. in [A 5,II, p. 653].

[68] NATANSON, I.P., Konstruktive Funktionentheorie.
Akademie Verl., Berlin 1955, xiv + 515 pp. (Orig. Russ. Ed.: Moscow 1949, 688 pp.; Hung. Transl. Budapest 1952, 517 pp.). Zbl 41, 186 - 189; MR 16, 1100/11, 591/ 15, 306.

[69] FAVARD, J., Remarques sur l'approximation des fonctions continues.
Acta Sci. Math. (Szeged) 12A (1950) 101 - 104. Zbl 37, 328; MR 12, 176.

[70] GELFOND, A.O., On the generalized polynomials of S.N. Bernšteĭn (Russ.).
Izv. Akad. Nauk SSSR Ser. Mat. 14 (1950) 413-420. Zbl 39, 68; MR 12, 332.

[71] LEVI, E., Sopra un'applicazione dei polinomi di Bernstein all'approssimazione in media delle funzioni sommabili.
Atti Accad. Naz. Lincei Rend. Cl. Sci. Fis. Mat. Natur. (8) 9 (1950) 242-246. Zbl 39, 292; MR 12, 701.

[72] NATANSON, I.P., Theorie der Funktionen einer reellen Veränderlichen. Akademie Verl., Berlin 1954 (1961) xi + 478 pp. (xii + 590 pp.) (Orig. Russ. Ed.: Gos. Izdat. Tehn.-Teor. Lit., Moscow - Leningrad 1950, 399 pp.; Engl. Transl. New York 1955, 277 pp.) Zbl 39, 282; MR 12, 598 / 16, 120, 804.

[73] SZÁSZ, O., Generalization of S. Bernstein's polynomials to the infinite interval.
J. Res. Nat. Bur. Standards Sect. B 45 (1950) 239 - 245 = Collected Mathematical Papers (Ed. H.D. Lipsich). Hafner, New York 1955, xiv + 1432 pp. (1 plate); pp. 1401 - 1407. MR 13, 648.

[74] BUTZER, P.L., On Bernstein Polynomials.
Ph.D. Thesis, University of Toronto, 1951, 76 pp.

[75] DINGHAS, A., Über einige Identitäten vom Bernsteinschen Typ.
Norske Vid. Selsk. Forh. 24, no. 21 (1951) 96 - 97. Zbl 46, 273; MR 14, 167.

[76] GÁL, I.S., Sur la convergence d'interpolations lineaires. III. Fonctions continues.
C.R. Acad. Sci. Paris 233 (1951) 1001 - 1003. Zbl 44, 68; MR 13, 549.

[77] KINGSLEY, E.H., Bernstein polynomials for functions of two variables of class $C^{(k)}$.
Proc. Amer. Math. Soc. 2 (1951) 64 - 71. Zbl 43, 290; MR 13, 128.

[78] LEVI, E., Ancora sopra un'applicazione dei polinomi di Bernstein all' approssimazione in media delle funzioni sommabili.
Atti Accad. Naz. Lincei Rend. Cl. Sci. Fis. Mat. Natur (8) 10 (1951) 360 - 364. Zbl 42, 218; MR 13, 342.

[79] LORENTZ, G.G., Deferred Bernstein polynomials.
Proc. Amer. Math. Soc. 2 (1951) 72 - 76. Zbl 43, 290; MR 13, 17.

[80] POPOVICIU, T., Asupra demonstraţiei teoremei lui Weierstrass cu ajutorul polinoamelor de interpolare (Roum.; Russ. and Engl. sum.).
Lucrările Sesiunii Generale Stiinţifice din 2. - 12. iunie 1950; pp. 1664 - 1667 (1951).

[81] RADON, J., Zur Polynomentwicklung analytischer Funktionen.
Math. Nachr. 4 (1951) 156 - 157. Zbl 42, 82; MR 12, 606.

[82] BOHMAN, H., On approximation of continuous and of analytic functions. Ark. Mat. 2 (1952) 43 - 56. Zbl 48, 299; MR 14, 254.

[83] BUTZER, P.L., Dominated convergence of Kantorovitch polynomials in the space L^p.
Trans. Roy. Soc. Canada (III, 3) 46 (1952) 23 - 27. Zbl 48, 46; MR 14, 641.

[84] GELFOND, A.O., Differenzenrechnung.
Dt. Verl. Wiss., Berlin 1958, viii + 336 pp. (Orig. Russ. Ed.: Moscow - Leningrad 1952; 479 pp.). Zbl 47, 332/80, 76; MR 14, 759/ 20 # 1121.

[85] GONTSCHAROW, W.I., Elementare Funktionen einer reellen Veränderlichen, Grenzwerte von Folgen und Funktionen. Der allgemeine Funktionsbegriff.
In: Enzyklopädie der Elementarmathematik, III, Analysis (Ed. P.S. Alexandroff et al.) Dt. Verl. Wiss., Berlin 1958, ix + 536 pp.; pp. 1 - 280. (Orig. Russ. Ed.: Moscow - Leningrad 1952, 559 pp.; pp. 9 - 296) MR 14, 1070.

[86] BUTZER, P.L., On two-dimensional Bernstein polynomials.
Canad. J. Math. 5 (1953) 107 - 113. Zbl 50, 70; MR 14, 641; RZM 1953 # 169.

[87] BUTZER, P.L., Linear combinations of Bernstein polynomials.
Canad. J. Math. 5 (1953) 559 - 567. Zbl 51, 50; MR 15, 309; RZM 1956 # 7969.

[88] IZUMI, SHIN-ICHI, On an approximation problem in the theory of probability.
Tôhoku Math. J. (2) 5 (1953) 22 - 28. Zbl 51, 48; MR 15, 217. RZM 1956 # 5313.

[89]* KIPRIJANOV, I.A., On polynomials like S.N. Bernšteĭn's for functions of two variables (Russ.).
Kazan. Učen. Zap. Gos. Univ. 113, no. 10 (1953) 193 - 207. MR 17, 728; RZM 1954 # 4787.

[90] LORENTZ, G.G., Bernstein Polynomials.
(Math. Expositions 8) Univ. of Toronto Press, Toronto 1953, x + 130 pp.. Zbl 51, 50; MR 15, 217; RZM 1955 # 166.

[91] McSHANE, E.J., Order - Preserving Maps and Integration Processes.
(Annals of Mathematics Studies 31) Princeton Univ. Press, Princeton, N.J. 1953, vi + 136 pp.. Zbl 51, 293; MR 15, 19; RZM 1955 # 5168.

[92] MIRAKJAN, G.M., On a convergent process of approximation of continuous functions (Russ.; Armen. sum.).
Akad. Nauk Armjan. SSR Dokl. 16 (1953) 33 - 37. Zbl 103, 288; MR 16, 575; RZM 1953 # 1147.

[93] BUTZER, P.L., On the extensions of Bernstein polynomials to the infinite interval.
Proc. Amer. Math. Soc. 5 (1954) 547 - 553. Zbl 56, 287; MR 16, 128; RZM 1955 # 166.

[94]* IPATOV, A.F., On the S.N. Bernšteĭn polynomials of bounded functions of two variables (Russ.).
Petrozavodsk. Učen. Zap. Gos. Univ. 3, no. 4 (1954) 16 - 51. RZM 1955 # 3698.

[95] MOLDOVAN, E., Observaţii asupra unor procedee de interpolare generalizate (Russ. and French sum.).
Bul. Şti. Secţ. Şti. Mat. Fiz. 6 (1954) 477 - 482. Zbl 59, 51; MR 16, 694; RZM 1956 # 327.

[96] TEMPLE, W.B., Stieltjes integral representation of convex functions.
Duke Math. J. 21 (1954) 527 - 531. Zbl 58, 50; MR 16, 22; RZM 1955 # 3810.

[97] BUTZER, P.L., Summability of generalized Bernstein polynomials, I.
Duke Math. J. 22 (1955) 617 - 623. Zbl 65, 297; MR 17, 476; RZM 1957 # 2199.

[98]* GUSEĬNOV, G.A., On the approximation of discontinuous functions by generalized polynomials of S.N. Bernšteĭn type (Russ.).
Trudy Azerbaĭdžan. Gos. Ped. Inst. 2 (1955) 133 - 145. MR 20 #5384; RZM 1957 # 318.

[99]* GUSEĬNOV, G.A., On the approximation of summable semicontinuous and measurable functions by generalized polynomials of S.N. Bernšteĭn type (Russ.).
Trudy Azerbaĭdžan. Gos. Ped. Inst. 2 (1955) 163 - 180. MR 20 #5384; RZM 1957 # 319.

[100]* IPATOV, A.F., The process of S.N. Bernšteĭn in points of polygonal proper discontinuities of a function f(x,y) to be approximated (Russ.).
Petrozavodsk. Učen. Zap. Gos. Univ. 4, no. 4 (1955) 13 - 30 (1957). RZM 1958 # 3653.

[101]* IPATOV, A.F., Estimation of the error and order of approximation of functions of two variables by the S.N. Bernšteĭn polynomials (Russ.).
Petrozavodsk. Učen. Zap. Gos. Univ. 4, no. 4 (1955) 31 - 48 (1957). MR 23 # A 2681; RZM 1958 # 3654.

[102]* IPATOV, A.F., Some theorems concerning the convergence of the polynomials $B_{n,m}(f;x,y)$ formed from $f(r,s) \in \overline{S}$ and of the derivatives of these polynomials (Russ.).
Petrozavodsk. Učen. Zap. Gos. Univ. 4, no. 4 (1955) 49 - 58 (1957). MR 23 # A 2680; RZM 1958 # 3655.

[103]* ZIDKOV, G.V., Remark on Bernšteĭn polynomials (Russ.). Grodnensk. Učen. Zap. Gos. Ped. Inst. 1955, no. 1 (1955) 31 - 33. MR 18, 574; RZM 1956 # 8717.

ANHANG

[A 1] BERNSTEIN, S.N., Sur l'ordre de meilleure approximation des fonctions continues par des polynomes de degré donné. Mem. Cl. Sci. Acad. Roy. Belgique (2) 4 (1912) 1 - 103. FdM 42, 435.

[A 2] BERNŠTEĬN, S.N., Collected Works, Vol. I,II; The Constructive Theory of Functions [1905 - 1930; 1931 - 1953] (Russ.). Izdat. Akad. Nauk SSSR, Moscow 1952, 581 pp. (1 plate); 1954, 627 pp.. Zbl 47, 73/56, 60; MR 14, 2/16, 433.

[A 3] BERNSTEIN, S.N., Collected Works: Volume I. Constructive Theory of Functions (1905 - 1930). U.S. Atomic Energy Commission, Technical Information Service Extension, Oak Ridge, Tennessee. Translation Series. Office of Technical Services, Dept. of Commerce, Washington, D.C./AEC - tr - 3460, vi + 221 pp., without year.

[A 4] NIKOL'SKIĬ, S.M., The approximation of functions of a real variable by polynomials (Russ.). In: Thirty Years of Mathematics in the USSR: 1917 - 1947 (Ed. A.G. Kuroš et al.) (Russ.). Gos. Izdat. Teh.-Teor. Lit., Moscow - Leningrad 1948, 1044 pp.; pp. 288 - 318.

[A 5] LOZINSKIĬ, S.M. - I.P. NATANSON, Metric and constructive theory of functions of a real variable (Russ.). In: Forty Years of Mathematics in the USSR: 1917 - 1957; Vol. I: Survey articles; Vol. II: Bibliography (Ed. A.G. Kuroš et al.) (Russ.). Gos. Izdat. Fiz.-Mat. Lit., Moscow 1959, 1002 pp./819 pp.; pp. 295 - 379 (I). Zbl 191, 275.

[A 6] GONČAROV, V.L. - A.N. KOLMOGOROV, The sixtieth birthday of Sergeĭ Natanovič Bernšteĭn (Russ.). Izv. Akad. Nauk Ser. Mat. 4 (1940) 249 - 260 (1 plate). FdM 66, 21; Zbl 24, 224; MR 2, 114.

[A 7] KUZMIN, R.O., The mathematical works of S.N. Bernšteĭn (Russ.). Uspehi Mat. Nauk 8 (1941) 3 - 7. FdM 67, 24; Zbl 60, 14; MR 3, 98.

[A 8] GONČAROV, V.L., Sergeĭ Natanovič Bernšteĭn (To the seventieth birthday) (Russ.). Uspehi Mat. Nauk 5, no. 3 (37)(1950) 172 - 183 (1 plate). Zbl 36, 146.

[A 9] The seventieth birthday of Sergeĭ Natanovič Bernšteĭn (Russ.). Izv. Akad. Nauk Ser. Mat. 14 (1950) 193 - 198 (1 plate). Zbl 36, 5; MR 11, 707.

[A 10] ACHIEZER, N.I., The work of academian S.N. Bernšteĭn on the constructive theory of functions (for his seventieth birthday) (Russ.).
Uspehi Mat. Nauk 6, no. 1 (41)(1951) 3 - 67. Zbl 45, 336; MR 12, 808.

[A 11]* AHIEZER, N.I., Academian S.N. Bernšteĭn and his work on the constructive theory of functions (Russ.).
Izdat. Har'kov. Gos. Univ., Har'kov 1955, 112 pp. (1 plate). MR 17, 697; RZM 1956 # 2126.

[A 12] ALEKSANDROV, P.S. - N.I. AHIEZER - B.V. GNEDENKO - A.N. KOLMOGOROV, Sergeĭ Natanovič Bernšteĭn: Obituary (Russ.).
Uspehi Mat. Nauk 24, no. 3 (147)(1969) 211 - 218 (1 plate) = Russian Math. Surveys 24 (1969) 169 - 176. Zbl 174, 4; MR 40 # 1248.

[A 13] KOLMOGOROV, A.N. - Ju. V. LINNIK - Ju.V. PROHOROV - O.V. SARMANOV, Sergeĭ Natanovič Bernšteĭn (Russ.).
Teor. Verojatnost. i Primenen. 14 (1969) 113 - 121 (1 plate). MR 39 # 5303.

[A 14] SHISHA, O. et al., J. Approximation Theory 13 (1975): Dedicated to Professor George G. Lorentz on the occasion of his sixty-fifth birthday.
J. Approximation Theory 13 (1975) pp. 1 - 16, etc. (1 plate).

NEW AND UNSOLVED PROBLEMS

1. C. BENNETT - R. SHARPLEY: An Approximation Problem in Interpolation Theory

The basic ingredient in the Peetre-Lions method of interpolation of operators is to solve the following problem:
Identify the K-functional

$$K(f,t;X_o,X_1) = \inf_{f=g+h} \{\|g\|_o + t\|h\|_1\}$$

as some "analytic measurement" of f; e.g. $K(f,t;L^1,L^\infty) = \int_o^t f^*(s)ds$.

PROBLEM. Identify $K(f,t;H^1, \mathrm{BMO})$.

2. W.R. BLOOM: Modulus of Continuity of Trigonometric Polynomials

Let T_n denote the set of trigonometric polynomials on the real line $\mathbb{R}$ of degree at most n. It is known that for each $p \in [1,\infty]$ there exists a constant C_p such that for all $t \in T_n$ and $a \in \mathbb{R}$,

$$\|{}_a t - t\|_p \leq C_p\, \omega_n(a)\, \|t\|_p ,$$

where ${}_a t : x \to t(x-a)$ and $\omega_n(a) = \max\{|\exp\{ika\} - 1| : |k| \leq n\}$, [see W.R. Bloom, J. Austral. Math. Soc. 17(1974), Remark 2.4, p.96], where it is observed that C_p can be taken not exceeding $3\sqrt{2}$.

PROBLEM. Find the minimum possible value of C_p for each p.

3. Z. CIESIELSKI: On BMO Space

Let $t_1 = 0$, $t_n = (2\nu-1)/2^{\mu+1}$ for $n = 2^\mu+\nu$, $1 \leq \nu \leq 2^\mu$, μ,ν - being integers.

Let B_1 be the periodic Bernoulli polynomial of degree 1 i.e.,

$$B_1(t)=\frac{1}{2}-t \text{ for } 0\leqslant t<1, \quad B_1(t)=B_1(t+1).$$

The BMO space is considered on $T=\langle 0,1)$.

PROBLEM. Show that there is an absolute $C<\infty$ such that

$$\left\|\sum_{k=1}^{n}\pm[B_1(\cdot)-B_1(\cdot-t_k)]\right\|_{BMO}\leqslant C$$

holds for all choices of signs and for all $n\geqslant 1$.

4. Z. CIESIELSKI: On Interpolation by Splines

Let $\{s_n;\ n=0,1,\ldots\}$ be a dense sequence in $\langle 0,1\rangle$ such that $s_0=0$, $s_1=1$, $\{s_0,\ldots,s_n\}\subsetneqq\{s_0,\ldots,s_{n+1}\}$ for $n\geqslant 0$. Let S_n^r be the set of all polynomial splines of order r corresponding to the simple nodes $\{s_2,\ldots,s_n\}$. Let P_nf, for given $f\in W_1^r\langle 0,1\rangle$, be the unique spline in S_n^{2r} such that:

$$D^jP_nf(0)=D^jP_nf(1)=0,\quad j=0,\ldots,r-1,$$

$$P_nf(s_k)=f(s_k),\quad k=2,3,\ldots,n.$$

PROBLEM: Show that

$$D^rP_nf(t)\to D^rf(t) \text{ a.e. in } \langle 0,1\rangle \text{ as } n\to\infty.$$

5. H.G. FEICHTINGER: On the Minimal Deviation of Convolution Squares

Given $m\in\mathbb{N}$ one may define

$$d_m := \inf\{\|g-g*g\|_1,\ g\in L^1(\mathbb{R}^m),\ g\geqslant 0,\ \int_{\mathbb{R}^m}g(x)dx=1\}.$$

One can show that one has

$$1/4\leqslant d_m\leqslant d_{m-1}\leqslant\ldots\leqslant d_1<1/3.$$

QUESTIONS. i) Can the above estimates be improved?

ii) What is the numerical value of d_m?

iii) Does d_m actually depend on m?

iv) Is the infimum attained?

v) If the answer to iv) is yes, can the corresponding functions be characterized, maybe even as the (normalized) dilations of a single function?

6. P.R. HALMOS: Invariant Subspaces Via Convexity (Dedicated to D.P. Milman)

This is not a problem but the hope of a research program. Can the Krein-Milman theorem be applied to the invariant subspace problem for operators on Hilbert space?

If A is an operator on H, a subspace is invariant under A if and only if the projection P whose range is that subspace satisfies the equation AP = PAP. Projections are sometimes (often?) the extreme points of algebraically characterized sets of operators. Proposed program: carry out an experimental study on many operators whose invariant subspaces are known, for each such operator form the (weakly closed?) set whose extreme points are exactly the projections onto those subspaces, and thus try to discover a natural way of associating a convex set with (some?, many?, all?) operators, whose extreme points are exactly the invariant projections. If successful, then, presumably the steps can be reversed: given A, manufacture a convex set, use Krein-Milman, and thus end up with non-trivial invariant subspaces (or an example where only trivial ones exist?).

7. P.R. HALMOS: Cyclic Vectors

If S is the unilateral shift on a Hilbert space H ($Se_n = e_{n+1}$, $n = 0,1,2,..$, where $\{e_0, e_1, e_2, ...\}$ is an orthonormal basis for H), does the direct sum $S \oplus S^*$ acting on $H \oplus H$ have a cyclic vector? (Note: S has cyclic vectors, and so does S^*. The direct sum $S \oplus S$ does n o t have a cyclic vector; the direct sum $S^* \oplus S^*$ does.)

The question is due to C. Foiaş and D. Voiculescu.

8. G.G. LORENTZ: About Interpolation

We consider the Lagrange interpolation polynomial as a function of knots $X : -1 \leqslant x_1 \leqslant x_2 < \ldots < x_n \leqslant +1$. For a function f, analytic on [-1,+1],

$$(*) \qquad f(z) - P_{m-1}(f;X,z) = \frac{1}{2\pi i} \int_C \frac{\omega(t) - \omega(z)}{\omega(t)(t-z)} f(t)dt,$$

where the contour C contains [-1,+1] inside and $\omega(z) = (z-x_1)\ldots(z-x_m)$. In this case, P_{m-1} is an analytic function of $x_1,\ldots,x_m$, even if some of them coinside. Also, $P_{m-1}(f;X,z) = \tilde{V}(f;X,z)/V(X)$, where V(X) is the Vandermonde determinant, and $\tilde{V}$ is some other determinant. In this representation, V(X) cancells out. Again, P_{m-1} is an analytic function of X. In particular, P_{m-1} is a continuous function of X for $-1 \leqslant x_1 \leqslant x_2 \leqslant \ldots \leqslant x_m \leqslant +1$.

For the polynomial of Birkhoff interpolation, $P_n(f;E,X,t) = \tilde{D}(f;E',X,t)/D(E,X)$. One can show that D(E,X) cancells out here exactly when E is an Hermitian matrix. Therefore, in general, P(f;E,X,t) is a meromorphic function of X. Nevertheless, S.D. Riemenschneider and myself have shown that, for conservative matrices E and $f \in C^n$, P(f) is a continuous function of X for $-1 \leqslant x_1 \leqslant \ldots \leqslant x_m \leqslant 1$ (knots are allowed to coincide!).

QUESTION. What corresponds to (*) for Birkhoff approximation?

9. P. MASANI: Banach-Space Valued Stationary Measures

Let

$$\mathscr{P} = \{(a,b] : a,b \in \mathbb{R} \ \& \ a \leqslant b\}$$

$$\hat{\mathscr{P}} = \{\textstyle\bigcup_1^\infty P_k : P_k \in \mathscr{P}\}$$

$$\hat{\mathscr{P}}_0 = \{S : S \in \hat{\mathscr{P}} \ \& \ \exists P \in \mathscr{P} \ni S \subseteq P \ \& \ P \setminus S \in \hat{\mathscr{P}}\}$$

$\mathscr{X}$ = an infinite dimensional Banach space.

PROBLEM. Show that $\exists$ a strongly continuous unitary representation $U(\cdot)$ of $\mathbb{R}$ over $\mathscr{X}$ and $\exists$ a set $S \in \hat{\mathscr{P}}_0$ such that

$$\text{Range} \int_S U(t)dt \not\subseteq \mathscr{D}_A,$$

where A is the infinitesimal generator of $U(\cdot)$.

REMARKS. For the definitions, and for the relevance of the result in the theory of stationary measures, see [Measure Theory, Ed. D. Kölzow, Springer Lecture Notes #794, 198o, pp. 295-3o9]. There the validity of the result for $\mathcal{X}$ = a Hilbert space and for $\mathcal{X}$ = any l_p space, $1 \leqslant p < \infty$, is stated.

1o. P. MASANI: Extreme Points in Banach-Graphs

Let $(\mathcal{X}, \; |.| \; , \; \text{corr})$ be a Banach graph,

$\bar{U}$ = the closed unit ball in $\mathcal{X}$,

$\partial\bar{U}$ = the boundary of $\bar{U}$,

$\partial_e\bar{U}$ = the set of extreme points of $\bar{U}$,

$\mathcal{X} = \{x : \; x \in \mathcal{X} \;\; \& \;\; x \operatorname{corr} x \}$.

PROBLEM. Characterize the Banach graphs $(\mathcal{X}, \; |.| \; , \text{corr})$ for which

$$\mathcal{X}_o \cap \partial\bar{U} \subseteq \partial_e\bar{U}. \tag{1}$$

REMARKS. For the definitions, see [Linear Spaces and Approximation. Ed. P.L. Butzer - B.Sz.-Nagy, Birkhäuser, 1978, pp. 71-89]. When $\mathcal{X}$ is the Marcinkiewicz Banach space and "corr" stands for Wiener correlatedness, the validity of (1) has been established by K.S. Lau [paper to appear]. When $\mathcal{X}$ is the Banach space of bounded countably additive measures on a σ algebra with values in a Hilbert space and "corr" stands for "biorthogonal", the validity of (1) has been shown by P. Ressel [unpublished report]. For $\mathcal{X} = \text{Cl}(\mathcal{H},\mathcal{H})$, $\mathcal{H}$ = a Hilbert space, and "A corr B" meaning "A commutes with B*", (1) is false.

11. F. MÒRICZ: On the Convergence of Double Orthogonal Series

QUESTION 1. Does there exist a double orthonormal system $\{\varphi_{ik}(x,y)\}_{i,k=1}^{\infty}$ on the unit cube $I^2 = [0,1] \times [0,1]$, which is

i) uniformly bounded,

ii) complete in $L^2(I^2)$, and

iii) the double series $\sum_{i=1}^{\infty}\sum_{k=1}^{\infty} a_{ik}\varphi_{ik}(x,y)$ converges a.e. (regularly or only in Pringsheim's sense) for every double sequence $\{a_{ik}\}_{i,k=1}^{\infty}$ of coefficients such that $\sum_{i=1}^{\infty}\sum_{k=1}^{\infty} a_{ik}^2 < \infty$?

REMARK. The double Haar system violates (i), the double Rademacher system violates (ii), and the double trigonometric system violates (iii).

Let $0 < \lambda_{ik} \uparrow \infty$ as $\min(i,k) \to \infty$ and let $\{\lambda_{ik}\}$ behave "fairly well" (e.g. let $\Delta_{1,1}\lambda_{i,k} := \lambda_{i+1,k+1} - \lambda_{i+1,k} - \lambda_{i,k+1} + \lambda_{ik} \geqslant 0$ for every i and k).

QUESTION 2. Is it true or not that if a double orthonormal system $\{\varphi_{ik}(x,y)\}_{i,k=1}^{\infty}$ is such that for every double sequence $\{a_{ik}\}_{i,k=1}^{\infty}$ of coefficients, $\sum_{i=1}^{\infty}\sum_{k=1}^{\infty} a_{ik}^2 < \infty$, we have the estimate

$$\sum_{i=1}^{m}\sum_{k=1}^{n} a_{ik}\varphi_{ik}(x,y) = O_x(\sqrt{\lambda_{mn}}) \quad \text{a.e.,}$$

then the double series $\sum_{i=1}^{\infty}\sum_{k=1}^{\infty} a_{ik}\varphi_{ik}(x,y)$ converges a.e. (regularly or only in Pringsheim's sense) for every sequence $\{a_{ik}\}_{i,k=1}^{\infty}$ with $\sum_{i=1}^{\infty}\sum_{k=1}^{\infty} a_{ik}^2\lambda_{ik} < \infty$?

REMARK. In case $\lambda_{ik} \equiv 1$ the conclusion holds true.

12. J. MUSIELAK: On Hardy Spaces

Let H^p, $0 < p < \infty$, be the Hardy space of analytic functions in the unit disc, then $\rho(r,f) = \int_0^{2\pi} |f(re^{it})|^p\, dt$ is a non-decreasing function of $r \in [0,1)$ and so $\lim_{r\to 1-} \rho(r,f) = \sup_{0\leqslant r<1} \rho(r,f)$ for every $f \in H^p$.

PROBLEM. Do there exist non-constant, positive continuous functions p(t) on $[0,2\pi]$ such that $\rho(r,f) = \int_0^{2\pi} |f(re^{it})|^{p(t)}\, dt$ has both the above properties for all analytic f for which $\sup_{0\leqslant r<1} \rho(r,f) < \infty$, or that at least $\lim_{r\to 1-}\rho(r,f)$ exists? If yes, then give sufficient conditions on p(t) in order that this be true.

13. R.S. PHILLIPS: On Dual Subspaces

Let H be a Hilbert space and E an orthogonal projection. Set

$$I(f,g) = (Ef,g) - ((1-E)f,g) .$$

Then I is an indefinite form and we can define the notion of positive and negative subspaces relative to I: P is a positive subspace if $I(f,f) \geq 0$ for all $f \in P$ and N is an negative subspace if $I(f,f) \leq 0$ for all $f \in N$. P and N are called dual if $I(P,N) = 0$. A pair P,N is called a maximal dual pair if P is positive, N is negative and $N = P^{\perp}$, $P = N^{\perp}$ (here $\perp$ means orthogonal complement with respect to I).

Next we introduce a commutative algebra A of operators on N which is closed relative to I - adjoints: $T \to T^{o}$ when $I(f,Tg) = I(T^{o}f,g)$. We say N is invariant under A if $TN \subset N$ for all T in A.

PROBLEM. Given a dual pair of subspaces N,P both invariant under A. Does there exist a maximal dual pair N', P' invariant under A such that $N' \supset N$ and $P' \supset P$?

14. W. SCHEMPP: On Patil Type Approximations

Let $B_n = \{z \in \mathbb{C}^n; |z| < 1\}$ denote the open unit ball in the space $\mathbb{C}^n (n \geq 1)$ and $\partial B_n = S_{2n-1} = U(n)/U(n-1)$ its boundary sphere. Denote by $\mathcal{H}^p(B_n)$ the Hardy space of exponent $p \in [1, +\infty]$ modelled on B_n and let $f \rightsquigarrow \tilde{f}$ be the isometric embedding of $\mathcal{H}^p(B_n)$ into the complex Lebesgue space $L^p(S_{2n-1})$. If Ω denotes a subset of S_{2n-1} of Lebesgue surface measure > 0 then $\tilde{f} \mid \Omega = 0$ implies $f = 0$. In the case $n = 1$, $p \in]1, \infty[$, D.J. Patil [Bull. Amer. Math. Soc. 78 (1972), 617-62o] has pointed out a constructive algorithm to recapture the function $f \in \mathcal{H}^p(B_1)$ form its boundary values $\tilde{f} \mid \Omega$ on the set Ω. For related work in the case $n = 1$ see [S.E. Zarantonello, Pacific J. Math. 79. (1978) 271-282] and [Quantitative Approximation, Ed. R.A. DeVore - K. Scherer, Academic Press, 198o, pp. 291-3oo], where BMO techniques are used to study certain Patil type approximations in the case $p = 1$.

QUESTION. Does there exist an extension of the Patil procedure to the case $n > 1$?

REMARK. For the polydisc case, see [D.J. Patil, Trans. Amer. Math.Soc. 188

(1974), 97-1o3].

15. W. SPLETTSTÖSSER: On the Quantization Error

With this problem we wish to draw the reader's attention to a certain type of (approximation) error which does not seem to be familiar in approximation theory but is so in several of the applied fields. For instance, for the digital processing of signal functions, samples $f(t_k)$ of the signal $f(t)$ are taken which are quantized afterwards, i.e., they are replaced by $Q_\varepsilon[f(t_k)]$ which is the multiple nearest to $f(t_k)$ of a given $\varepsilon > 0$ (ε being the quantization step size). The question then is whether it is possible to represent the signal f in terms of these quantized sampled values, thereby keeping the reconstruction error small. The mathematical problem, which is also the question concerning the influence of small deviations of the sampled values, now reads as follows.

PROBLEM. Given any (interpolating) operator of the form $I_n f(t) = \sum f(t_k) s_{k,n}(t)$ with f belonging to an appropriate function space, is is true that

$$\| I_n f - I_n(Q_\varepsilon(f)) \| = \mathcal{O}(\varepsilon) \qquad (\varepsilon \to 0),$$

at least for large n; what is the $\mathcal{O}$-constant?

REMARKS. For $I_n = B_n$ being the Bernstein polynomial operator (in terms of which the question has been raised at the conference) the answer is trivially "yes", because the B_n are positive operators. In case $I_n(f)$ are the Shannon sampling series the problem has been dealt with in [P.L. Butzer - W. Splettstößer, Signal Processing 2 (198o), 1o1-112]. In this respect the socalled "jitter" error, considerd in the latter paper, would also be of interest.

16. B.SZ.-NAGY: About the Corona Theorem for Matrices

In his paper: "A corona theorem for countably many functions" in "Integral Equations and Operator Theory", vol.3, no 1(198o), Marvin Rosenblum extends Carleson's theorem as follows.

Consider the operator $a = [a_1, a_2 \ldots]^T$ from E^1 to E^∞ (E^n: Hilbert coordinate space of dim. n), with $a_k = a_k(z) \in H^\infty(D)$. In order that there exist an operator $c = [c_1, c_2, \ldots]$ from E^∞ to E^1 with $c\,a = 1$ with norm $\|c\|$ bounded on the unit disc D it is necessary and sufficient that a should have a positive lower bound on D, i.e.

$$\inf_{z \in D} \sum_{k=1}^{\infty} |a_k(z)|^2 > 0.$$

Some estimates are also given.

QUESTION. Does this result generalize to matrices over $H^\infty(D)$, finite or even infinite. More precisely, suppose $A = A(z)$ is a bounded operator from E^n to E^m (n,m possibly ∞), with bound, say equal to 1, i.e., with $\|A(z)x\| \leqslant \|x\|$ for any $X \in E^n$ and $z \in D$, and with

$$\inf_{z \in D} \inf_{\substack{x \in E^n \\ \|x\| = 1}} \|A(z)x\| = \delta > 0.$$

Does there follow the existence of an analoguous matrix over $H^\infty(D)$, $C = C(z)$, so that

1) $\|C(z)y\| \leqslant \gamma \|y\|$ for all $y \in E^m$ and $z \in D$, and with some γ independent of z and y.
2) $C(z)\,A(z) = I_n$.

If this is the case, give estimates for γ (in terms of n, m, δ, if possible).

17. M. WOLFF: Spektrum und Störungstheorie für stark stetige Halbgruppen positiver linearer Operatoren

Sei E ein Banachverband und $\mathcal{T} = (T_t)_{t \geqslant 0}$ eine stark-stetige Halbgruppe von positiven Operatoren auf E. Sei A ihr Generator, $\sigma(A)$ das Spektrum von A, $s(A) = \sup\{\mathrm{Re}\,z : z \in \sigma(A)\}$ die Spektralschranke und $\omega_o = \inf\{(1/t) \ln \|T_t\| : T > 0\}$ der Typ von $\mathcal{T}$.

FRAGE 1.1. Gibt es auf $E = L^p([0,1])$ ein solches $\mathcal{T}$ mit $s(A) < \omega_o$ (für $p = 1, \infty$ gilt stets $s(A) = \omega_o$ für alle solche Halbgruppen $\mathcal{T}$; so muß also $1 < p < \infty$ gelten, das Problem ist selbst für $p = 2$ offen)?

FRAGE 1.2. Sei jetzt E beliebig. Für alle $0 < x \in E$, $0 < x' \in E'$ (Dualraum) sei $\sup\{\langle T_t x, x'\rangle : t > 0\} \gneq 0$. Gilt dann $s(A) = \omega_o$? Auch dies Problem ist selbst für $E = L^2([0,1])$ offen.

FRAGE 1.3. Ist $\mathcal{T} = (T_t)_{t \in \mathbb{R}}$ eine stark stetige Gruppe positiver Operatoren, so ist $s(A) \in \mathbb{R} \cap \sigma(A)$. Gilt hierfür stets $s(A) = \omega_o$? Auch dies Problem ist selbst für $E = L^2([0,1])$ offen.

Sei $B = A + U$, wo A der Generator einer stark stetigen Halbgruppe $\mathcal{T}$ von positiven Operatoren auf dem Banachverband E und U ein (beschränkter) positiver linearer Operator auf E ist. Sei $s(A) = -\infty$.

FRAGE 2. Welche Bedingungen muß U erfüllen, damit $s(B) \gneq -\infty$ ist ?

Für $E = L^p(\Omega)$, wo Ω ein beschränktes Gebiet von $\mathbb{R}^n$ ist $(1 \leq p < \infty)$, ist dies Problem von Interesse in der Theorie über die lineare Boltzmann-Gleichung. Das Problem ist selbst für den Fall eines kompakten Operators U nur in Spezialfällen gelöst.

M.Z. NASHED: Regularizability of Ill-Posed Operator Equations

The concepts and methods used in the analysis and regularization of ill-posed problems have stimulated in recent years advances in some areas of operator and approximation theory. For some perspectives, see [A.N. Tikhonov - V.Y. Arsenin, Winston & Sons, Washington, DC, 1977], [Generalized Inverses and Applications, Ed. M.Z. Nashed, Academic Press, New York, 1976], [M.Z. Nashed, In: Constructive and Computational Methods for Differential and Integral Equations, Springer Lecture Notes # 43o, 1974] The following problem arising from the theory of ill-posed operator equations seems to be still open.

Let A be a one-to-one mapping from a Banach space X into a Banach space Y. The operator equation $Af = g$ is said to be r e g u l a r i z a b l e if there exists a one-parameter family of mappings T_α: $Y \to X$, $0 < \alpha \leq 1$ for which

$$\lim_{\alpha \to 0} (\sup\{\|f - T_\alpha g\| : g \in Y, \|g - Af\| \leq \alpha\}) = 0$$

for all $f \in X$.

There are examples of nonregularizable operator equations where A is a bounded linear operator (even compact) from a Banach space X into a Banach space Y. However, in all known examples the space X is nonseparable (see [V̌.A. Vinokurov, Soviet Math. Dokl, 11 (1970), 1495-1496]). So the question is:

QUESTION. Let A be a one-to-one bounded linear operator on a s e p a r a b l e Banach space X into a Banach space Y, with the range of A nonclosed. Is the operator equation $Af = g$ regularizable?

ERRATA

Some corrections to papers which appeared in earlier volumes of Oberwolfach conference proceedings (see the list in the preface) are given here.

K. ISHIGURO - W. MEYER-KÖNIG: Über das Verträglichkeitsproblem bei den Kreisverfahren der Limitierungstheorie. ISNM 25 (1974), 547-558 .

In der letzten Zeile auf Seite 550 muß $|z+1/6| > 1/9$ ersetzt werden durch $|z+2/9| > 1/9$. In der ersten Zeile auf Seite 554 muß S_{-3}^{R} ersetzt werden durch S_{3}^{R}.

W. SPLETTSTÖSSER: Some extensions of the sampling theorem. ISNM 40 (1978), 615-628.

An additive term is missing in formula (4.7); it must read correctly

$$f''(t) = \lim_{W\to\infty} \{2 \sum_{\substack{k=-\infty \\ k\neq 0}}^{\infty} f(t+\frac{k}{W}) \frac{(-1)^{k+1}}{(k/W)^2} - \pi W^2 \frac{f(t)}{3}\} \qquad (t \in \mathbb{R}). \tag{4.7}$$

On page 625 there are two formulae numbered (5.7), the second of which has to be changed into

$$f(t) = \lim_{W\to\infty} \frac{1}{4} \sum_{k=-\infty}^{\infty} f(\frac{k}{W}) \frac{\sin \frac{3\pi}{4}(Wt-k) \sin \frac{\pi}{4}(Wt-k)}{[\frac{\pi}{4}(Wt-k)]^2} \qquad (t \in \mathbb{R}). \tag{5.8}$$

Alphabetical list of papers

Mathematics subject classification numbers*

*according to the 1980 Math. Subject classification of Mathematical Reviews and Zentralblatt für Mathematik. Classification numbers were given by the authors. Numbers following subjects indicate the first page of the respective paper.

Key words and phrases*

*Given by the authors; numbers indicate first page of respective paper.